Lecture Notes in Physics

New Series m: Monographs

Editorial Board

H. Araki, Kyoto, Japan
E. Brézin, Paris, France
J. Ehlers, Potsdam, Germany
U. Frisch, Nice, France
K. Hepp, Zürich, Switzerland
R. L. Jaffe, Cambridge, MA, USA
R. Kippenhahn, Göttingen, Germany
H. A. Weidenmüller, Heidelberg, Germany
J. Wess, München, Germany
J. Zittartz, Köln, Germany

Managing Editor

W. Beiglböck
Assisted by Mrs. Sabine Lehr
c/o Springer-Verlag, Physics Editorial Department II
Tiergartenstrasse 17, D-69121 Heidelberg, Germany

Springer
Berlin
Heidelberg
New York
Barcelona
Budapest
Hong Kong
London
Milan
Paris
Santa Clara
Singapore
Tokyo

The Editorial Policy for Monographs

The series Lecture Notes in Physics reports new developments in physical research and teaching - quickly, informally, and at a high level. The type of material considered for publication in the New Series m includes monographs presenting original research or new angles in a classical field. The timeliness of a manuscript is more important than its form, which may be preliminary or tentative. Manuscripts should be reasonably self-contained. They will often present not only results of the author(s) but also related work by other people and will provide sufficient motivation, examples, and applications.
The manuscripts or a detailed description thereof should be submitted either to one of the series editors or to the managing editor. The proposal is then carefully refereed. A final decision concerning publication can often only be made on the basis of the complete manuscript, but otherwise the editors will try to make a preliminary decision as definite as they can on the basis of the available information.
Manuscripts should be no less than 100 and preferably no more than 400 pages in length. Final manuscripts should preferably be in English, or possibly in French or German. They should include a table of contents and an informative introduction accessible also to readers not particularly familiar with the topic treated. Authors are free to use the material in other publications. However, if extensive use is made elsewhere, the publisher should be informed. Authors receive jointly 50 complimentary copies of their book. They are entitled to purchase further copies of their book at a reduced rate. As a rule no reprints of individual contributions can be supplied. No royalty is paid on Lecture Notes in Physics volumes. Commitment to publish is made by letter of interest rather than by signing a formal contract. Springer-Verlag secures the copyright for each volume.

The Production Process

The books are hardbound, and quality paper appropriate to the needs of the author(s) is used. Publication time is about ten weeks. More than twenty years of experience guarantee authors the best possible service. To reach the goal of rapid publication at a low price the technique of photographic reproduction from a camera-ready manuscript was chosen. This process shifts the main responsibility for the technical quality considerably from the publisher to the author. We therefore urge all authors to observe very carefully our guidelines for the preparation of camera-ready manuscripts, which we will supply on request. This applies especially to the quality of figures and halftones submitted for publication. Figures should be submitted as originals or glossy prints, as very often Xerox copies are not suitable for reproduction. For the same reason, any writing within figures should not be smaller than 2.5 mm. It might be useful to look at some of the volumes already published or, especially if some atypical text is planned, to write to the Physics Editorial Department of Springer-Verlag direct. This avoids mistakes and time-consuming correspondence during the production period.
As a special service, we offer free of charge LATEX and TEX macro packages to format the text according to Springer-Verlag's quality requirements. We strongly recommend authors to make use of this offer, as the result will be a book of considerably improved technical quality.
Manuscripts not meeting the technical standard of the series will have to be returned for improvement.
For further information please contact Springer-Verlag, Physics Editorial Department II, Tiergartenstrasse 17, D-69121 Heidelberg, Germany.

Michail Zak Joseph P. Zbilut
Ronald E. Meyers

From Instability to Intelligence

Complexity and Predictability in Nonlinear Dynamics

Springer

Authors

Michail Zak
Jet Propulsion Laboratory
California Institute of Technology
4800 Oak Grove Drive, MS 525-3660
Pasadena, CA 91109, USA

Joseph P. Zbilut
Department of Molecular Biophysics and Physiology
Rush University, 1653 W. Congress
Chicago, IL 60612, USA

Ronald E. Meyers
US Army Research Laboratory
2800 Powder Mill Rd.
Adelphi, MD 20783, USA

Cataloging-in Publication Data applied for.

Die Deutsche Bibliothek - CIP-Einheitsaufnahme

Zak, Michail:
From instability to intelligence : complexity and predictability in nonlinear dynamics / Michail Zak ; Joseph P. Zbilut ; Ronald E. Meyers. - Berlin ; Heidelberg ; New York ; Barcelona ; Budapest ; Hong Kong ; London ; Milan ; Paris ; Santa Clara ; Singapore ; Tokyo : Springer, 1997
(Lecture notes in physics : N.s. M, Monographs ; 49)
ISBN 3-540-63055-4

ISSN 0940-7677 (Lecture Notes in Physics. New Series m: Monographs)
ISBN 3-540-63055-4 Edition Springer-Verlag Berlin Heidelberg New York

Typesetting: Camera-ready by authors
Cover design: *design & production* GmbH, Heidelberg
SPIN: 10550887 55/3144-543210 - Printed on acid-free paper

Dedication

To the Memory of my Father
Fate, Time, Occasion, Chance and Change? –To this All things are subject... –Prometheus Unbound
M. Zak

To Barbara, Anna, and Joe
Ein Blick von dir, ein Wort mehr unterhält Als alle Weisheit dieser Welt.
–Faust
J.P. Zbilut

To my wife Genevieve; children, Marc, Chris, Carole; and brothers Richard, and Sheridan
In the realm of Nature there is nothing purposeless, trivial, or unnecessary. -Maimonides
R.E. Meyers

1

Preface

So far as the laws of mathematics refer to reality, they are not certain. And so far as they are certain, they do not refer to reality. *–A. Einstein*

The word "instability" in day-to-day language is associated with something going wrong or being abnormal: exponential growth of cancer cells, irrational behavior of a patient, collapse of a structure, etc. This book, however, is about "good" instabilities, which lead to change, evolution, progress, creativity, and intelligence; they explain the paradox of irreversibility in thermodynamics, the phenomena of chaos and turbulence in classical mechanics, and non-deterministic (multi-choice) behavior in biological and social systems.

The concept of instability is an attribute of dynamical models that describe change in time of physical parameters, biological or social events, etc. Each dynamical model has a certain sensitivity to small changes or "errors" in initial values of its variables. These errors may grow in time, and if such growth is of an exponential rate, the behavior of the variable is defined as unstable. However, the overall effect of an unstable variable upon the dynamical system is not necessarily destructive. Indeed, there always exists such a group of variables that do not contribute to the energy of the system. In mechanics such variables are called ignorable or cyclic. Usually, an ignorable variable characterizes orientations of a vector or a tensor with respect to a certain frame of reference. An exponential growth of such a variable does not violate the boundedness of the energy, so even if the instability persists, the system still continues to function. However, its behavior can be significantly different from the pre-instability state in the same way in which a turbulent flow is different from a laminar one. If the original system is conservative (for instance, as a set of molecules in a potential field), its post-instability behavior may attain some dissipative features: the mean, or regular component characterizing macroscopic properties will loose some portion of its initial energy to irregular fluctuations, and this will lead to irreversibility of the motion, despite the fact that the original system was fully reversible. Hence, the instability of ignorable variables can "convert" a deterministic process into a stochastic one whose mean behavior is significantly different from the original one. Based upon this paradigm one can introduce a chain of irreversible processes of increasing complexity which can be interpreted as an evolutionary process.

When dynamical models simulate biological, or social behavior, they should include the concept of "discrete events", i.e., special critical states which give rise to branching solutions, or to bifurcations. To attain this property, such systems must contain a "clock" – a dynamical device that generates a global rhythm. During the first half of the "clock's" period, a critical point is stable, and therefore, it attracts the solution; during the second half of this period, the "clock" destabilizes the critical point, and the solution escapes it in one of several possible directions. Obviously, the condition of uniqueness of the solution at the critical points must be relaxed. Thus, driven by alternating stability and instability effects, such systems perform a random walk-like behavior whose complexity can match the complexity of biological and social worlds.

Finally, based upon these paradigms, one can develop a phenomenological approach to cognition to include "quantum-like" features.

Contents

2
Introduction

Certainly no subject or field is making more progress on so many fronts at the present moment, than biology... –*R. Feynman*

The beginnings of the 20th century witnessed a curious development in the history of science. The applications of statistics to the diverse phenomena of the biological and social sciences were about to explode as a result of the work of such people as Pearson and Fisher. On the other hand, the world of physical sciences was still avoiding the use of stochastic models, although Boltzmann and Gibbs had supplied sufficient reason not to do so. Now, near the end of the 20th century, the state of affairs has changed considerably. The physical sciences have come to appreciate the significant insights into low dimensional systems which appear random, while at the same time, the biological and social sciences are increasingly interested in deterministic descriptions of what appear to be very complex phenomena. The intersection of these two approaches would appear to be the often neglected, sometimes unwanted phenomenon of "noise." Often relegated to status as nuisance, noise has become more appreciated really as "that which we cannot explain." And in this explanation noise has become recognized as a possible deterministic system itself, perhaps involved with quantum effects, with a complicated description that interacts with observables on a variety of length scales. At the juncture between the physical and biological this noise creates myriad effects which ultimately redound to the very basic ideas regarding the constitution of what we know as living matter.

That this should be so is not surprising. Although it was not that long ago that scientists felt that a understanding of classical Newtonian laws of motion could provide the key to understanding all of existence, the current climate appreciates that with some modifications, this might still be true: the movements of ions through cellular channels are being investigated by biologists with a seriousness that would be the envy of an experimental physicist. Indeed some of the very time-honored models of the physical sciences such as spin lattices are being used to explore this area. And why not? At this level the very fundamental laws of physics control discrete molecular events which have profound importance for living tissues. Ultimately, the dynamics at this level govern the way neurons, and other humoral agents orchestrate the myriad events to maintain the human organism. "Neural nets" are once again being studied as true models of the nervous system, not only by biologists, but by physicists as well.

The flurry of activity in this broad area is not unremarkable given that biological systems are often poorly defined. Until the present, most of our understanding of biological systems has been defined by phenomenological descriptions guided by statistical results. Linear models with little consideration of underlying processes have tended to inform such processes. What is more frustrating has been the failure of such models to explain transitional, and apparently aperiodic changes of observed records. The resurgence of nonlinear dynamics has provided an opportunity to explain these processes more systematically, and with a formal explanation of transitional phenomena.

Certainly, nonlinear dynamics is not a panacea. Linear descriptions do, in fact, account for many biological and social processes. Additionally, there is the danger to assume that chaotic correspondence with experimental data "explains" the system. Scientists are all too familiar with the pitfalls of model-making. Mathematics is the language of science, but the language is not the science. Physics itself is replete with examples of this tension between mathematics and reality. Consider for example the debates regarding delta functions, and "infinitesimals." It was Einstein himself who cautioned about the interface between mathematics and the physical sciences (Einstein 1983).

At the same time there is the ever present concern that by learning about the intricacies of the processes, we neglect the global kinetics of a system. Continuing evidence suggests that there is a constant interplay between microscopic and macroscopic length scales, as well as randomness to create enormous variety and patterns in biology. And perhaps this is the important point that has emerged in this last decade of the century: we have traditionally maintained a perspective of looking for order, and disregarding randomness and instability as a nuisance; whereas the correct perspective may be to see this nuisance as an active process which informs order and vice versa.

The perspective we take here is to attempt to understand biological systems in a unique way, and this unique way involves the admittance of singularities both mathematically and biologically. In this endeavor we refer to the comments made by James Clerk Maxwell over a century ago when he pointed out (Campbell and Garnett 1884): "Every existence above a certain rank has its singular points: the higher the rank the more of them. At these points, influences whose physical magnitude is too small to be taken account of by a finite being, may produce results of the greatest importance. All great results produced by human endeavor depend on taking advantage of these singular states when they occur."

Certainly, biological organisms are of a high rank, and indeed, many of these singularities have already been uncovered. From a topological perspective Winfree (1987) has demonstrated time and again that biological oscillators admit singularities. Other work has argued from first principles and experimentation that physiological singularities must exist in order for the organisms to maintain adaptability (Zbilut et al. 1996, Zbilut et al. 1995). What has not been adequately appreciated is the reconciliation between classical Newtonian dynamics and these biological phenomena. This monograph represents a modest attempt in this direction. In order to proceed, certain problems in classical dynamics need to be highlighted.

Classical dynamics describes processes in which the future can be derived from the past, and past can be traced from future by time inversion. Because of such determinism, classical dynamics becomes fully predictable, and therefore it cannot explain the emergence of new dynamical patterns in nature, in biological, and in social systems. This major flaw in classical dynamics has attracted attention of many outstanding scientists (Gibbs, Planck, Prigogine, etc.). Recent progress in understanding the phenomenology of nonlinear dynamical systems was stressed by the discovery and intensive studies of chaos which, in a addition to a fundamental theoretical impact, has become a useful tool for several applied methodologies. However, the, actual theory of chaos has raised more questions than answers. Indeed, how fully deterministic dynamical equations with small uncertainties in initial conditions can produce random solutions with a stable probabilistic structure? And how this structure can be predicted? What role does chaos play in information processing performed by biological systems? Does it contribute into elements of creativity, or irrationality (or both!) in the activity of a human brain? All these questions, and many others which are related to them, will be discussed in this monograph.

The monograph treats unpredictability in nonlinear dynamics, and its applications to information processing. The main emphasis is on intrinsic stochasticity caused by the instability of governing dynamical equations. This approach is based upon a revision of the mathematical formalism of Newtonian dynamics, and, in particular, upon elimination of requirements concerning differentiability, which is some cases lead to unrealistic solutions.

This new mathematical formalism allows us to reevaluate our view on the origin of chaos and turbulence, on prediction of their probabilistic structures, and on their role in information processing in biological systems.

For our treatment of the material, we emphasize physical insight rather than mathematical rigor, with few exceptions, when a mathematical "way of thinking" is really important.

The monograph is intended to be a self-contained text for physicists, biologists, and engineers who wish to apply methods of nonlinear dynamics to the problems of artificial intelligence, neural networks, and information processing in general. It may also be useful to scientists who are interested in the theory of chaos and turbulence, and their applications in physics, biology and social sciences (Gregson 1993).

The core ideas of the monograph are based upon the authors' publications in mathematical, physical, biological, and engineering journals.

The book consists of two parts: Part I, "Predictability in Classical Dynamics" (Chap. 3–5); and Part II, "Terminal (Non-Lipschitz) Dynamics in Biology and Physics" (Chap. 6–15).

Chapter 3 presents some basic concepts and related open questions.

Chapter 4 is devoted to the concept of instability in dynamical systems with the main emphasis on orbital and Hadamard's instabilities. It demonstrates that the requirement of differentiability in some cases is inconsistent with the physical nature of motions, and it may lead to unrealistic solutions. Special attention is paid to the fact that instability is not an invariant of motion: it depends upon the class of functions in which the motion is described. This prepares the reader to the possibility of elimination of certain types of instabilities (and in particular, those which lead to chaos and turbulence) by enlarging the class of functions using the Reynolds-type transformation in combination with a "stabilization principle": the additional terms (the so-called Reynolds stresses) are found from the condition that they suppress the original instability. Based upon these ideas, a new approach to chaos and turbulence, as well as the new mathematical formalism for nonlinear dynamics, termed the "stabilization principle," are discussed in Chap. 5.

Chapter 6 situates these problems in the broader biological context. It introduces biological evidence for the existence of dynamical data which are best understood in the context of nondeterministic dynamics when uniqueness criteria for differential equations are relaxed.[1] Examples of experimental data are presented, as well as strategies to detect such dynamics. The

[1] The requirement for uniqueness is traditionally met by requiring that the solutions of differential equations adehere to the Lipschitz conditions (i.e., boundedness of derivatives). For reasons explained in the text, we feel that by imposing such a restriction, certain real phenomena are not adequately modeled. In the past, to describe the dynamics whereby the Lipschitz condition is relaxed, we have used terms such as "non-Lipschitz," "nonderterministic," and "terminal." We have never been completely satisfied with these terms, since they each fail to convey the complete meaning intended. For this monograph we have decided to use both terminal and non-Lipschitz, depending upon what properties of a given phenomenon should be emphasized. Although some may find these terms awkward, other terms, such as "nondeterministic" have established

discrepancy between these data and traditional mathematical views are emphasized, with a view toward Chap. 7.

Chapter 7 formally introduces and elucidates another discrepancy between mathematical formalisms and physical reality: the Lipschitz condition which provides the uniqueness of solutions, and which is actually responsible for determinism and reversibility of Newtonian dynamics. In addition to this, it leads to such idealizations as infinite time of approaching equilibria, the impossibility of passing over an attractor, etc. As shown in this chapter, violation of the Lipschitz condition at equilibria eliminates these idealizations, and allows one to explain such "trivial" physical phenomena as snap of a whip, etc. But as a "side effect," it leads to irreversibility and unpredictability of a "terminal" version of Newtonian dynamics.

Chapter 8 is devoted to a new architecture of neural nets based upon the terminal version of Newtonian dynamics introduced in Chap 7. Due to violations of the Lipschitz conditions at certain critical points, the neural network forgets its past as soon as it approaches these points: the solution at these points branches, and the behavior of the dynamical system becomes unpredictable. Since any vanishingly small input applied at critical points causes a finite response, such an unpredictable system can be controlled by a microdynamical device which operates by sign strings. These sign strings, as a "genetic code," uniquely define the system behavior by specifying the direction of the motions in the critical points. By changing the combinations of signs in the code strings, the system can reproduce any prescribed behavior of a prescribed accuracy. That is why unpredictable systems driven by sign strings are extremely flexible and are highly adaptable to environmental changes. The supersensitivity of critical points to external inputs appears to be an important tool for creating chains of coupled subsystems of different scales whose range is theoretically unlimited.

Due to existence of the critical points, the neural network becomes a weakly coupled dynamical system: its neurons (or groups of neurons) are uncoupled (and therefore, can perform parallel tasks) within the periods between the critical points, while the coordination between the independent units (i.e., the collective part of the performance) is carried out at the critical points where the neural network is fully coupled. As a part of the new architecture, weakly coupled neural networks acquire the ability to be activated not only by external inputs, but also by internal periodic rhythms. Such a spontaneous performance resembles the brain activities, as is suggested in Chap. 9.

One of the most fundamental features of the new architecture is "terminal chaos," or the existence of a "stochastic" attractor, which incorporates an element of "irrationality" into the dynamical behavior, and can be as-

meanings, and, moreover, do not characterize the consequences derived from a relaxation of the Lipschitz conditions.

sociated with creativity of neural network performance. Terminal chaos is generated at critical points by the sign strings of the genetic code, and its behavior qualitatively resembles "classical" chaos. However, terminal chaos is characterized by the following additional property: it has a fully predictable probabilistic structure, and therefore, it can be learned, controlled and exploited for information processing. It appears that terminal chaos significantly increases the capability of dynamical systems via a more compact and effective storage of information. It provides an ability to create a hierarchical parallelism in the neurodynamical architecture which is responsible for the tremendous degree of coordination among the individual parts of biological systems. It also implements a performance of high level cognitive processes such as formation of classes of patterns, i.e., formation of new logical forms based upon a "generalization" procedure.

Chapter 10 introduces a discrete event form of these dynamics.

Chapter 11 uses the discrete events of a heartbeat as a paradigm for terminal dynamics (attractor and repeller), based upon experimental observations.

Chapter 12 extends the results to thermodynamics and Chap. 13 to viscous flows.

Chapter 14 considers an observation linking non-Lipschitz dynamics, and quantum computing.

Finally, Chapter 15 demonstrates some computational advantages of the stabilization principle and terminal dynamics, while the Epilogue situates these perspectives within the realm of biology..

As prerequisites for reading this book, a working knowledge of classical mechanics is necessary. The mathematical background is no more elaborate than the one required for general engineering.

Acknowledgments. At various times, we have profited from discussions with our colleagues and friends, as well as their individual responses to our professional requests. We would like to take this opportunity to acknowledge their importance in the preparation of this monograph.

M. Zak would like to cite Jacob Barhen and E. Oblow for their continual support; B. Cetin, H.K. Liu, K. Goser, S. Gulati, N. Toomarian and E. Hupkens for their implementations of non-Lipschitz dynamics; A. Zak for discussion of basic concepts; and Jack Miller for computational support.

J.P. Zbilut would like to thank Chuck Webber, Jr. for his innumerable questions, comments, and long friendship. K. Andreoli, R.S. Eisenberg, J. Keithley, J. Llewellyn, L. Lawson, S. Shott, and M. Norušis gave their support in many different ways. Dave Dixon has provided on-going interest and stimulating insights. D. Corcos and M. Shapiro were kind enough to extend their expertise with arm motion recordings. C. Grebogi, and D. Ruelle were willing to listen in an unbiased way, and have contributed unknowingly to this effort.

R.E. Meyers thanks Keith Deacon for his skillful navigation around the myriad of computers and programs we have used. Appreciation is also expressed to Marc Meyers for his typing and research.

Support for some of our research came from agencies of the U.S. Department of Defense, including the Innovative Science and Technology Office of the Ballistic Missile Defense Organization, the U.S. Army Research Laboratory, and the U.S. Army/ASAS Project Office, through agreements with the National Aeronautics and Space Administration. The NASA/JPL Center for Space Microelectronics Technology's (CSMT) long-term support and commitment to the research presented in this book are greatly acknowledged. Support for the recent theoretical work on Quantum Intelligence and Quantum-Inspired Computing was provided in part by the JPL Director's Discretionary Fund (DDF). Helpful collaboration was also received from Carl Chen of Brookhaven National Laboratory. J.P. Zbilut and R.E. Meyers also acknowledge support from the National Science Foundation, Grant SGER # 9708044 (CISE/IRIS).

Any faults with this monograph are ours, and any successes are due to the kindness of our colleagues, and friends.

Part I

Predictability in Classical Dynamics

3

Open Problems in Dynamical Systems

You believe in a God who plays dice, and I in complete law and order. –*A. Einstein*

3.1 Models of Dynamical Systems

3.1.1 Mathematical Formulation and Dynamical Invariants

Dynamics describes the motion of systems, i.e. the time evolution of its parameters. The time variable t can be discrete or continuous. In discrete-time dynamical systems, the rate of change of their parameters x is defined only for discrete values of t. These systems can be presented as the iteration of a function:

$$x_{t+1} = v(x_t, t), \quad t = 0, 1, 2, ..., \text{etc.} \tag{3.1}$$

i.e. as difference equations.

In continuous-time dynamical systems the rate of change of x is defined for all values of t; such systems can be modelled by ordinary differential equations.

$$\frac{dx}{dt} = \dot{x} = v(x, t) \tag{3.2}$$

or by partial differential equations:

$$\dot{x} = v(x, x', x'', ..., t), \; x' = \frac{\partial x}{\partial s}, \; x'' = \frac{\partial^2 x}{\partial s^2}, ... \tag{3.3}$$

if the rate of change, in addition, depends upon distributions of x over space coordinates s. In (3.1–3.3), x represents the state of the dynamical system.

Continuous-time dynamical system theory has adopted the basic mathematical assumptions of the theory of differential equations such as differentiability of the parameters (with respect to time and space) "as many times as necessary," the boundedness of the velocity gradients $\partial\dot{x}/\partial x$ (the Lipschitz conditions)(Coddington and Levinson 1955) etc. Under these assumptions, the existence, uniqueness and stability of solutions describing the behavior of dynamical systems has been studied. However, the dynamical systems cannot be identified with the mathematical models, i.e., with the differential equations. Indeed, dynamical systems are characterized by scalars, vectors, or tensors which are invariant with respect to coordinate transformations. Hence, (3.2) or (3.3) model a dynamical system only if they preserve these invariants after any (smooth) coordinate transformation. For instance, any model of a mechanical system must be derivable from variational principles which are expressed via the mechanical invariants (kinetic and potential energy, dissipation functions, etc.). In other words, the difference between dynamical systems and the corresponding differential equations is similar to the difference between a matrix as an object of algebra, and a second rank tensor as an object of geometry: the same tensor can be modelled by different matrices depending on choices of coordinates; however, all of these matrices must have the same eigenvalues. Continuing this analogy, it can be expected that the parameters x in (3.2), or (3.3) can be decomposed (at least, in principle) into "invariant" and "non-invariant" components, in the same way in which a matrix A can be decomposed into invariant (diagonal $\tilde{A}$) and coordinate-dependent $(\theta, \theta)^1$ components:

$$A = \theta \tilde{A} \theta^{-1} \tag{3.4}$$

3.1.2 Ignorable Coordinates and Orbital Instability

In mechanical systems, "non-invariant" components of x can be associated with ignorable (or cyclic) coordinates which do not enter the Lagrangian function explicitly, and therefore, do not affect the energy of the system. For non-conservative systems, in addition to this, the generalized forces corresponding to these coordinates are zero. In terms of Lagrange equations, this property is expressed as the conservation of generalized ignorable impulses ρ (Gantmacher 1970):

$$\frac{\partial L}{\partial q_\alpha} = 0, \; Q_\alpha = 0, \text{ i.e., } \frac{\partial L}{\partial \dot{q}_\alpha} = \rho_\alpha = \text{Const}, \; \alpha = 1, 2, ..., m \tag{3.5}$$

unlike the equations for the position coordinates which, in general, do not preserve the position impulses ρ:

$$\frac{d}{dt}\frac{\partial L}{\partial \dot{q}_k} - \frac{\partial L}{\partial q_k} = Q_k, \quad k = 1, 2, ..., n \tag{3.6}$$

Here L is the Lagrangian, q_α and q_k are ignorable and position coordinates, respectively, Q_k are non-potential components of generalized forces.

3.1.3 Open Problems

As will be shown in the next chapter, the evolution of ignorable coordinates may be fundamentally different from the evolution of non-ignorable (or position) coordinates. For instance, the growth of position coordinates is limited by the boundedness of the system energy, and consequently, their instability cannot persist: the system must find an alternative stable state. In contradistinction to this, the instability of ignorable coordinates (which is called an orbital instability) can persist all the time without having an alternative stable state. In particular, the indifference of the energy to changes of ignorable parameters is responsible for such phenomena as turbulence, chaos, failure of differentiability and uniqueness of solutions. In turn, the occurrence of these phenomena questions the basic mathematical assumptions about the class of functions in which the dynamical systems are described. The existence of two different types of parameters in dynamical systems raises some other questions: can instability of ignorable coordinates develop independently of the behavior of the position coordinates? Is instability of ignorable coordinates an invariant of the frame of reference, or of the class of functions in which motions are studied? Can the instability of ignorable coordinates be eliminated by change of motion representation? The answers to these questions will be discussed in the following chapters.

3.2 Self-Organization in Autonomous Dynamical Systems

3.2.1 Static Attractors

Let[1] us start with a particular case of (3.2):

$$\dot{x} = \sin x \tag{3.7}$$

and find equilibrium points, i.e., points with zero velocity:

[1] The term "self-organized" has acquired several unique definitions in the literature. We make no connection to these unique specifications.

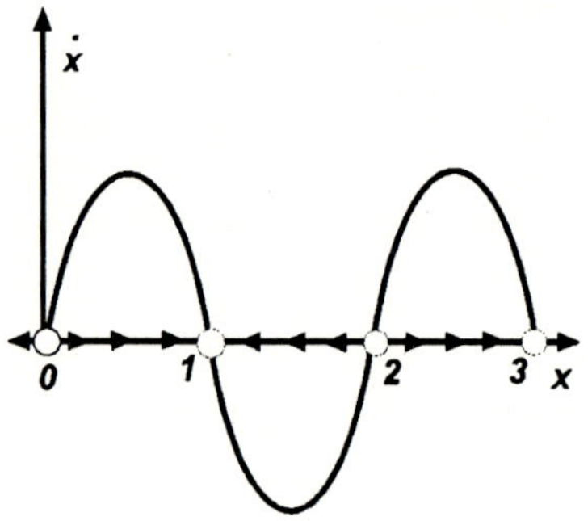

Fig. 3.1. Static attractors and repellers

$$\dot{x} = \sin x_k = 0,\ x_k = \frac{\pi k}{2},\ k = \cdots - 2, -1, 0, 1, \cdots \tag{3.8}$$

One can verify that for

$$k = ... - 3, -1, 1, 3, ... = \frac{\pi(2n+1)}{2},\quad n = \cdots - 2, -1, 0, 1, 2, \cdots \tag{3.9}$$

the equilibrium points are stable. They are called static attractors (Fig. 3.1).

If initially the system (3.1) was located at an attractor, say the attractor 1, it will remain there forever. If the system started from a position between the points 0 and 2 (the basin of attraction), it will eventually approach the attractor 1 (during an infinitely large period of time). Hence, a dynamical system in an open interval between two static attractors cannot pass either of them, and it cannot escape the basin of attraction. The basins of attractors are separated by static repellers – the unstable equilibrium points $\cdots$-2, 0, 2, 4,$\cdots$.

Static attractors and repellers, in general, can be found in autonomous dynamical systems, i.e. systems which are not subjected to any external influences that depend on the time, so that in (3.2):

$$\frac{\partial v}{\partial t} \equiv 0 \tag{3.10}$$

The local structure of v within a small neighborhood of a static attractor is characterized by eigenvalues with negative real parts of the matrix $\|\partial v/\partial x\|$

$$\operatorname{Re} \lambda_i < 0 \tag{3.11}$$

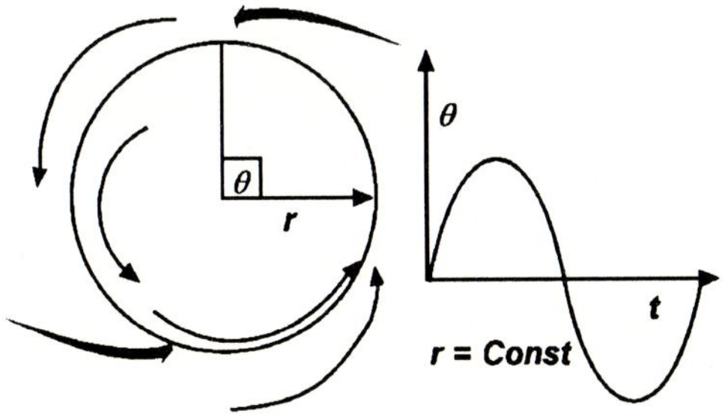

Fig. 3.2. Periodic attractor in phase (left) and actual spaces (right)

[Strictly speaking, some of $\operatorname{Re}\lambda_i$ may be zero if the local stability is defined by the higher order terms in Taylor decomposition of the function $v(x)$.]

A sufficient (but not necessary) condition for (3.10) is that v represents a gradient of some scalar u:

$$v = -\nabla u(x) \tag{3.12}$$

3.2.2 Periodic Attractors

We proceed with our overview with the following dynamical system:

$$\dot{r} = -\sin r, \quad \dot{\theta} = \omega = \text{Const} \tag{3.13}$$

where r and θ are polar coordinates.

Obviously, r has the same attractors as x in the previous example [see (3.1)]. However, in this case these attractors will not be static, since the total velocity of the system is not zero:

$$v = r\omega \neq 0 \text{ if } r \neq 0 \tag{3.14}$$

Hence, the dynamical system approaches states which are characterized by periodic motions:

$$\theta = \theta_0 + \omega t \tag{3.15}$$

with the period $2\pi/\omega$. Such states are called periodic attractors (Fig. 3.2).

Returning to (3.13), one can easily identify r and θ as position and ignorable coordinates, respectively. Indeed, the Lagrangian and generalized forces for this dynamical system are:

$$L = \frac{1}{2}(\dot{r}^2 + \theta^2), Q_r = \dot{r} \cos r, \; Q_\theta = 0 \tag{3.16}$$

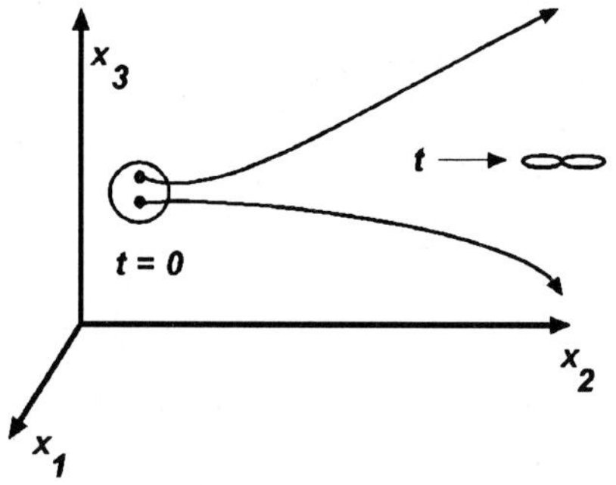

Fig. 3.3. Exponential divergence of trajectories in phase space

and therefore,

$$\frac{d}{dt}\frac{\partial L}{\partial \dot{q}_r} = Q_r \neq 0, \quad \frac{d}{dt}\frac{\partial L}{\partial \dot{q}_\theta} = Q_\theta = 0 \tag{3.17}$$

It is important to emphasize that the position coordinate r is stable at the attractors, while the ignorable coordinate θ is at the boundary of stability: any small error in θ will increase linearly (but not exponentially) in time.

3.2.3 Chaotic Attractors

Indifference of energy of a dynamical system to an unlimited growth of ignorable coordinates raises the following question: do there exist such states where all the position coordinates are stable, but some of the ignorable coordinates are unstable? Numerical experiments give positive answers to this question. These states are associated with chaotic behavior. Unlike periodic attractors, here any small error in initial values of ignorable coordinates increases exponentially (but not linearly) with time, so that two motion trajectories which initially were indistinguishable (because of finite scale of observation), diverge exponentially, and therefore, a behavior of the dynamical system becomes unpredictable (Fig. 3.3). But is such a "multivaluedness" of trajectories consistent with the basic mathematical assumptions about motions of dynamical systems? This problem will be discussed in the next sections in connection with predictability in classical dynamics.

In actual space, chaotic behavior is characterized by aperiodic oscillations with a continuous power spectrum (Fig. 3.4).

A sufficient but not necessary criterion for chaos is very simple: some of the local eigenvalues corresponding to ignorable coordinates must have

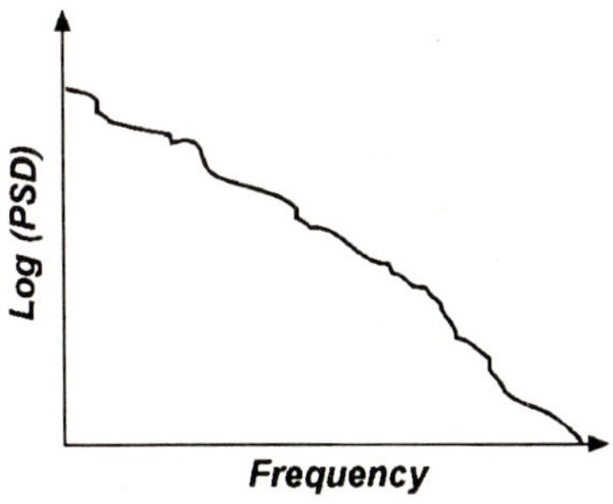

Fig. 3.4. Continuous power spectrum for a chaotic motion

positive real parts in those areas of phase space where the motion can occur. In other words, the orbital instability must persist. As mentioned above, such a situation is physically possible since ignorable coordinates do not effect the energy of the system. However, actually this criterion is too strong: chaos may emerge even under weaker conditions. As shown in dynamical systems theory (Lichtenberg and Lieberman 1983), the sufficient and necessary condition for chaos is associated with global behavior of trajectories, i.e., with their exponential divergence averaged over an orbit. This behavior is characterized by Liapunov exponents σ :

$$\sigma = \lim_{\substack{t\to\infty \\ d(0)\to 0}} \left(\frac{1}{t}\right) \ln\left(\frac{d(t)}{d(0)}\right) \tag{3.18}$$

where $d(0)$ and $d(t)$ represent initial and current distances, respectively, between two neighboring trajectories in certain directions.

Unfortunately, the main problem with this criterion is that the Liapunov exponents, in general, cannot be analytically expressed via the parameters of the dynamical system, and this makes prediction and analysis of chaos a hard task.

There are many different types of chaotic behavior, and one of them is a chaotic attraction. In order to describe this phenomenon, we have to start with some general remarks.

Any attractor is a property of a dissipative dynamical system where the total energy is not conserved, while within the basin of attraction it decreases. The change of energy in dynamical systems can be measured by the change of volume occupied by motion trajectories (corresponding to different initial conditions) in phase space. The instantaneous rate of change of this volume for a dynamical system (3.2) is given by an invariant:

$$\text{div } v = \sum_i \lambda_i \tag{3.19}$$

where λ_i are the eigenvalues of the matrix $\|\partial v/\partial x\|$, while the global rate averaged over an orbit is expressed via the Liapunov exponents:

$$\overline{\text{div } v} = \lim_{t\to\infty} \left(\frac{1}{t}\right) \ln \left|\frac{\Delta V(t)}{\Delta V(0)}\right| = \sum_i \sigma_i \tag{3.20}$$

Here V is the change of volume in phase space.

It can be shown that for an n-dimensional dynamical system, there are n real Liapunov exponents σ_i that can be ordered as:

$$\sigma_1 \geq \sigma_2 \geq ...\sigma_n \tag{3.21}$$

and one of the exponents, representing the direction along the trajectories, is zero.

A dynamical system will be locally, or globally, dissipative, if

$$\sum \lambda_i < 0 \text{ or } \sum \sigma_i < 0 \tag{3.22}$$

respectively. As follows from (3.22), the dissipativity of a dynamical system does not exclude the possibility that some of the Liapunov exponents are positive (such a possibility is excluded only for $n < 3$). Hence, chaotic attraction emerges from a coexistence of two conflicting phenomena: dissipation which contracts the volume occupied by the motion trajectories, and instability which diverges these trajectories in certain directions. As a result of this, the trajectories are mixing, and a limit set, i.e. a chaotic attractor, attains a monstrous configuration. An appropriate quantitative description of this configuration requires new mathematical tools based upon fractal geometry.

Obviously, a replacement of all the positive Liapunov exponents by zeros leads to periodic or multi-periodic attractors which represent the boundary between predictable and unpredictable motions.

Finally, if $(n - 1)$ Liapunov exponents of an n-dimensional dynamical system are negative (the nth exponents must be zero), then the attractor is static.

Since the classification and qualitative description of attractors in dissipative dynamical systems can be based only upon the invariants – the Liapunov exponents, all the attractors with their basins represent invariant sets of states which are defined by the property that if any system is in such a set at some time, then it remains in that set for all times. As will be demonstrated below, such a temporal self-organization when a dissipative dynamical system acquires a temporal structure without specific interference from the outside is a powerful tool for information processing.

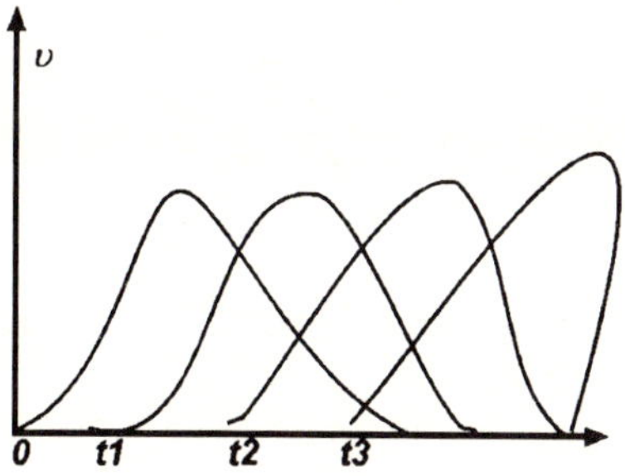

Fig. 3.5. Formation of shock waves

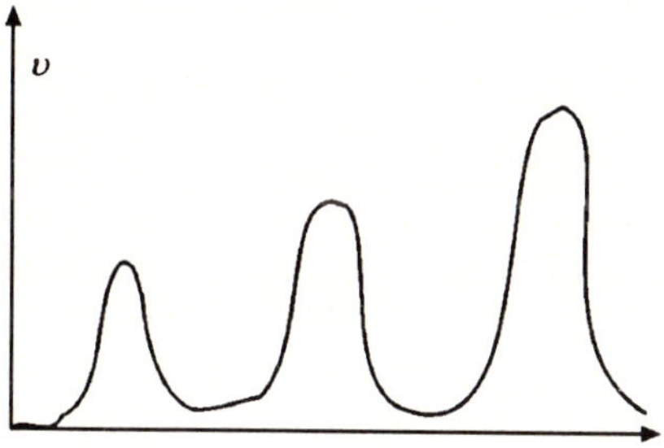

Fig. 3.6. Formation of solitons

3.2.4 Distributed Systems

There are two types of distributed systems – hyperbolic and parabolic – which can model dynamical behaviors. (Elliptic equations are ill-posed for time evolution processes.) Distributed dynamical systems can have all three types of attractors discussed above. But in addition to that, they may exhibit more sophisticated behaviors such as turbulence (whose relation to chaos is still disputed), and Hadamard's instability (Zak 1994a), which is associated with failure of hyperbolicity and transition to ellipticity (Zak 1994b), formation of cumulative effects (Zak 1970), shock waves, solitons, etc. (Zak 1983). However, the basic property distinguishing distributed dynamical systems as a possible tool for information processing is spatial coherence, and spatial self-organization.

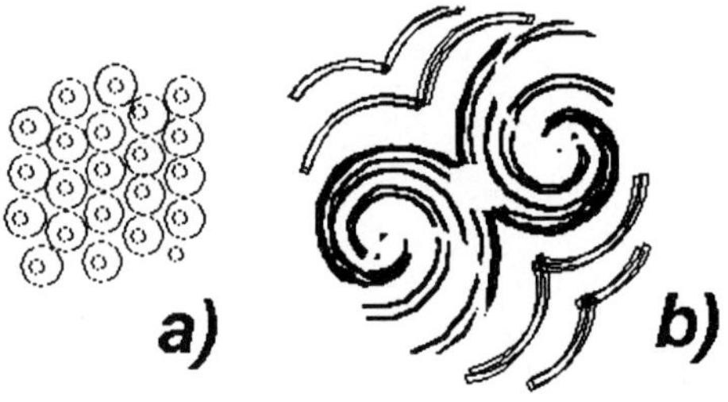

Fig. 3.7. Typical patterns in fluid dynamics (a), and spiral waves in the Belousov-Zhabotinsky reaction (b)

Hyperbolic equations describe wave propagation, and the simplest example of spatial self-organization can be associated with formation of shock waves in a one-dimensional flow of free particles (Fig. 3.5)

$$\dot{v} + vv' = 0 \tag{3.23}$$

A combination of shock wave effect and the wave dispersion leads to another type of spatial self-organization – a soliton (Fig. 3.6) described by the Korteweg-de Vries equation (which is of the parabolic type)

$$\dot{v} + vv' + v''' = 0 \tag{3.24}$$

It is interesting to note that both (3.23) and (3.24) are non-dissipative.

More sophisticated spatial and temporal-spatial patterns can appear in dissipative distributed systems of parabolic type (Taylor vortices, spiral waves, etc., Fig. 3.7).

Actually all the spatial effects in distributed dynamical systems result from additional mathematical restrictions requiring differentiability of dynamical parameters with respect to spatial coordinates. But are these restrictions always consistent with physical nature of motions? The following example shows that such restrictions may lead to unrealistic solutions.

Consider an ideal filament stretched in a vertical direction, as shown in Fig. 3.8. Let us crosscut it in a middle point and observe the behavior of upper and lower parts. The lower part will be folding up in a "thick point," losing differentiability of its configuration. The upper part will preserve differentiability of its configuration in an open interval (which does not include the free end), but at the end small initial disturbances will accumulate and become infinitely large (snap of a whip). Both of these effects are lost in the dynamical model based upon differentiability of the dynamical parameters

Fig. 3.8. Loss of smoothness of filament configuration

(for the lower part of the filament) and upon the Lipschitz condition at the free end (for the upper part of the filament) (Zak 1970).

Does this inconsistency between the mathematical formalism of classical dynamics link to unpredictability? We will discuss this problem in the next chapters.

3.2.5 Dynamical Systems and Information Processing

Self-organization in dynamical systems can be exploited for information processing in the following way.

Consider a gradient dynamical system (3.12):

$$\dot{x} = -\nabla u(x, T) \tag{3.25}$$

where x is an n-dimensional state vector, T is an m-dimensional matrix of control parameters. This system has static attractors at the minimum of the scalar function $u(x)$ (Fig. 3.9).

Each attractor can be envisaged as an n-dimensional vector representing a pattern, i.e. an ordered set of real numbers encoding some information about a certain object. Suppose, for instance, that the attractors 1 and 2 in Fig. 3.9 represent the letters "A" and "B," respectively. It is reasonable to assume that all the vectors within the basins of attraction represent the same letters, respectively, but with possible distortions. Then, any distorted letter "A" introduced to the dynamical system in the form of an initial vector x_0, located within the basin of attraction 1 (Fig. 3.9), will be attracted to the point 1; and in the same way, any distorted letter "B" will be attracted to the point 2.

Three important elements of the information processing should be emphasized in this dynamical procedure. Firstly, each attractor plays the role

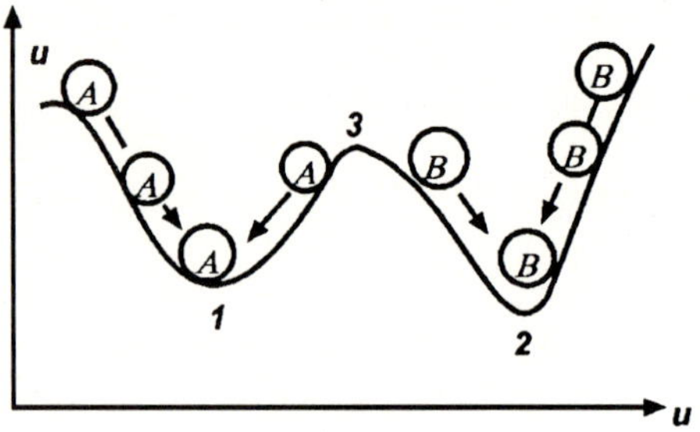

Fig. 3.9. Landscape of gradient system

of a memory storage for a prescribed pattern. Secondly, two different attractors perform a pattern classification: each pattern is put at the corresponding place. Thirdly, a process of attraction can be envisaged as a generalization procedure: all insignificant features of an initially distorted pattern are suppressed, and therefore, the attractor becomes a "typical" representative of a class, i.e., a collection of all possible versions of a certain concept (say, of all the possible writing of the letter "A"). Obviously the landscape of this dynamical system (see Fig. 3.9) can be prescribed or changed via appropriate adjustment of the control matrix T.

Formation of classes is associated with higher level cognitive processes such as generalization and abstraction. More sophisticated versions of these processes can be implemented by periodic and chaotic attractors. In order to illustrate this, we will introduce a dissipative nonlinear dynamical system with a constant control matrix:

$$\dot{x}_i = v_i(x_j,\ T_{ij}),\ \ i,j = 1,2,...,n \tag{3.26}$$

We will assume that the control parameters T_{ij} are chosen such that the system (3.26) has a set of m static attractors at the points

$$x_i = \tilde{x}_i^k \tag{3.27}$$

It is easily verifiable that such a property imposes certain constraints upon the control parameters T_{ij}. Indeed, for a set of m equilibrium points $\tilde{x}_i^k (k = 1,2,...,m)$ one obtains $m \times n$ constraints following from (3.26):

$$0 = v_i(\tilde{x}_j^k,\ T_{ij}),\ \ i,j = 1,2,...,n;\ \ k = 1,2,...,m \tag{3.28}$$

In order to provide stability of the equilibrium points $\tilde{x}_j^k$ the control parameters must also satisfy the following $m \times n$ inequalities:

$$\mathrm{Re}\,\lambda_i^k < 0, \; i = 1, 2, ...n; \; k = 1, 2, ..., m \tag{3.29}$$

in which λ_i^k are the eigenvalues of the matrices $\|\partial v_i / \partial x_j\|$ at the fixed points $\tilde{x}_i^k$.

How can the number of parameters T_{ij} be minimized without a significant loss of the quality of a prescribed performance?

Let us assume that the vectors $\tilde{x}_j^k$ have some characteristics in common, for instance, their ends are located on the same circle of a radius r_0 (Fig. 3.2a) i.e., (after proper choice of coordinates):

$$\sum_{i=1}^{2} (\tilde{x}_j^k)^2 = r_0^k, \; k = 1, 2, ..., m \tag{3.30}$$

If for the patterns represented by the vectors $\tilde{x}_i^k$ the property (3.30) is much more important than their angular coordinates $\theta^k (\theta^{k_1} \neq \theta^{k_2}$ if $k_1 \neq k_2)$, then it is more economical to store the circle $r = r_0$ instead of storing m static attractors with at least $2 \times m$ control parameters T_{ij}. Indeed, in this case one can "afford" to eliminate unnecessary coefficients by reducing its structure to the simplest form:

$$\dot{r} = r(r - r_0)(r - 2r_0), \quad \dot{\theta} = \omega = \mathrm{Const} \tag{3.31}$$

Equations (3.31) have a periodic attractor

$$r = r_0, \; \theta = \omega t \tag{3.32}$$

which generates harmonic oscillations with frequency ω. But what is the role of these oscillations in the logical structure of the dynamical system performance? The transition to the form (3.31) can be interpreted as a generalization procedure in the course of which a collection of unrelated vectors $\tilde{x}_i^k$, is united into a class of vectors whose lengths are equal to r_0. Hence, in terms of symbolic logic, the circle $r = r_0$ is a logical form for the class of vectors to which the concept (3.30) applies. In other words, the oscillations (3.32) represent a higher level cognitive process associated with generalization and abstraction. During these processes, the point describing the motion of (3.31) in the phase space will visit all those and only those vectors whose lengths are equal to r_0; thereby the dynamical system "keeps in mind" all the members of the class.

Suppose that a bounded set of isolated static attractors which can be united in a class occupies a more complex subspace of the phase space, i.e., instead of a circle (3.32) the concept defining the class is:

$$\varphi(\tilde{x}_1^k, \tilde{x}_2^k, ...\tilde{x}_n^k) = r, \; k = 1, 2, ..., n \tag{3.33}$$

Then the formation of the class will be implemented by storing a surface:

$$\varphi(x_1, x_2, ..., x_n) = r \tag{3.34}$$

as a limit set of the corresponding dynamical system while all the control parameters which impose constraints on the velocities along the surface (3.34) will be eliminated.

The character of the motion on the limit set depends upon the properties of the surface (3.34). If (by proper choice of coordinates) this surface can be approximated by a topological product of $(n - 1)$ circles (i.e., by an $n - 1$-dimensional torus) then the motion is quasi-periodic; it generates oscillations with frequencies which are dense in the reals. If the surface (3.34) is more complex and is characterized by a fractal dimension, the motion on such a limit set must be chaotic: it generates oscillations with a continuous spectrum. In both cases the motion is ergodic: the point describing the motion in the phase space sooner or later will visit all the points of the limit set, i.e., the dynamical system will "keep in mind" all the members of the class.

Returning to the periodic attractor (3.31), it is interesting to note that the position coordinate r [see (3.16–3.17)] represents an invariant characterizing the class of patterns, while the ignorable coordinate θ is associated with numbering of members of this class. This means that the energy and information in the dynamical system (3.31) have the same invariants. Such a property is not a coincidence: any dynamical system performing a generalization procedure via attractions should operate with position coordinates as invariants representing classes, and with ignorable coordinates representing the "lists of members" of these classes.

Phenomenological resemblance between information processing by nonlinear dynamics and the human brain activity gave motivation for introducing a special type of dynamical systems – neural nets which are supposed to link a nonlinear dynamics paradigm and biological neural systems.

3.3 Neural Nets with Temporal Self-Organization

3.3.1 Introduction

One of the oldest and the most challenging problems in cognitive science is to understand the relationships between cognitive phenomena and brain functioning. Since ancient times the "mystery of mind" has attracted philosophers, neuroscientists, psychologists, and later, mathematicians, and physicists. On the line with attempts to understand and to simulate brain activity itself there have been many successes in developments of brain-style information processing devices which focus on brain-inspired modeling rather than modeling of brain as a part of a human body. The most powerful information processing device of this kind is a digital computer which has revolutionized the science and technology of our century and even changed the life-style of the whole society. The digital computer became the first candidate for the human brain modeling. Artificial intelligence researchers predicted that "thinking machines" will take over our mental work. Futurologists have proclaimed the birth of a new species, *machina sapiens* that will share our place as the intelligent sovereigns of our earthly domain.

Notwithstanding some achievements in "thinking machines"development, it seems very unlikely that digital computers, with their foundations on rigid, cold logic, and full predictability, can model even the simplest biological systems which are flexible, creative, and to a certain degree, irrational. In addition to this, the main brain characteristics which contribute to information processing are different from those of digital computers. Indeed, neurons, as the basic hardware of the brain, are a million times slower than the information processing elements of serial computers. This slow speed is compensated by units which are highly interconnected. Hence, the brain succeeds through massive parallelism of a large number of slow neurons, and therefore, the mechanisms of mind are most likely best understood as resulting from the cooperative activity of very many relatively simple processing units working in parallel rather than by fast, but sequential processing units of digital computers. There are many indirect evidences that the structure of computational procedures in digital computers and brains are also different: instead of calculating a solution using sequences of rigid rules, the primary mode of computation in the brain is rather associated with a relaxation procedure, i.e., with setting into a solution in the same way in which a dynamical system converges to an attractor. Another difference between the digital computer and the brain is in the mechanisms of learning and memory storing. There are a number of facts suggesting that knowledge is in the connections between the neurons rather than in the neurons themselves, while these connections have a clear geometric and topological structure. Such a distributed memory storage is responsible for the graceful degradation phenomenon when the system performance gradually deteriorates as more and more neural units are destroyed, but there is no single critical point where performance breaks down. Based upon this kind of representation of the distributed memory, the learning procedure can be understood as the gradual modification of the connections strengths during a relaxation-type dynamical process.

Along with the abstract model of a computer as a formal machine that could be programmed to carry out any effective procedure, introduced by Alan Turing (1937) another potential candidate for brain simulation was developed: in 1943 McCulloch and Pitts published the paper (1943), "A Logical Calculus of the Ideas Immanent in Nervous Activity." In this paper the authors offered their formal model of the neuron as a threshold logic unit. They demonstrated that each Turing matching program could be implemented using a finite network of their formal neurons. In support to the idea of neural networks, Hebb published "The Organization of Behavior" (1949) which provided the inspiration for many computational models of learning. However, it took about thirty years until neural networks became a potential competitor to digital computers in regard to simulation of brain performance.

Two main factors significantly contributed to the "second birth" of neural networks. The first factor is associated with the pioneering work of Carver Mead on the design of neural networks and their implementation in analog VLSI systems. In his work, he has shown how the powerful organizing principles of nervous systems can be realized in silicon integrated circuits. The second factor based upon the progress in dynamical system theory. In the past, most theoretical studies of dynamical systems have been concerned with modeling of energy transformations. However, in recent years several attempts have been made to exploit the phenomenology of nonlinear dynamical systems for information processing as an alternative to the traditional paradigm of finite-state machines. There are many evidences coming from the analysis of electroencephalogram data that human brain activity resembles a dissipative nonlinear adaptive system. In contradistinction to finite-state machines which operate by simple bits of information, the nonlinear dynamics paradigm operates in terms of complex "blocks" of information which resemble patterns of practical interest.

3.3.2 Neural Net as a Dynamical System

The current artificial neural nets can be considered as massively parallel adaptive dynamical systems modelled on the general features of biological neural networks, that are intended to interact with the objects of real world in the same way the biological systems do. As a dynamical system, a neural net is characterized by nonlinearity and dissipativity which provide the existence of at least several attractors. There are many different modifications of neural nets. In this monograph we will be interested only in those neural net architectures which do not contain any man-made devices (such as digital devices), and therefore are suitable for circuit implementations. Such a neural net (which in literature are called continuously updated recurrent neural nets) can be represented by the following dynamical system (Cohen and Grossberg 1983; Hopfield 1984):

$$\tau_i u_i = -u_i + \sigma(\sum_i T_{ij} u_j), \ \tau_i > 0 \tag{3.35}$$

where u_i are state variables, or mean soma potentials, characterizing the neuron activities; T_{ij} are constant control parameters representing the weights of synaptic interconnections; τ_i are suitable time constants; and $\sigma(\cdot)$ is a sigmoid functions having a saturated nonlinearity (usually $\tau(x) = \tanh \beta x$, where $\beta = \text{Const} > 0$ is an additional control parameter).

The invariant (3.19) characterizing local dissipativity of the system (3.35) is expressed explicitly via its parameters:

$$\text{div } \dot{u} = \sum_i \frac{1}{\tau_i}(-1 + \frac{\beta T_{ii}}{\cos h^2 \sum_j T_{ij} u_j}) \tag{3.36}$$

A necessary (but not sufficient) condition that the system (3.35) has attractors is that there are some domains in phase space where the invariant (3.36) is negative.

If the matrix T is symmetric

$$T_{ij} = T_{ji} \tag{3.37}$$

then (3.35) can be represented in the form of a gradient system (3.12), and therefore, it can have only static attractors. In the basin of a static attractor the invariant (3.36) must be negative.

Since the system (3.35) is nonlinear, it can have more than one attractor; consequently, in some domains of phase space, the invariant (3.36) may be positive or zero.

Equations (3.1) present the neural net in its natural form in the sense that u_i and T_{ij} correspond to physical parameters: neuron potentials and synaptic interconnections, respectively. However, it is important to emphasize that the relationship between the invariants of the vector u_i – and the tensor T_{ij} are not preserved by the coordinate transformation, i.e., (3.35) do not possess an invariant tensor structure. Consequently, the column u_i and the matrix T_{ij} cannot be treated as a vector and a tensor, respectively.

In most applications, the neural nets performance is associated with convergence to attractors (pattern recognition, optimization, decision making, control, associative memory, generalization, etc.). The locations of attractors and their basins in phase space can be prescribed by an appropriate choice of the synaptic weights T_{ij}, i.e. by solving inverse dynamical problems. However, since the dimensionality of neural nets is usually very high (in biological systems it is of order of 10^{11} with the number of synaptic interconnections of order of 10^{15}), the straight-forward analytical approach can be very expensive and time consuming. An alternative way to select synaptic weights in order to do a specific task was borrowed from biological systems. It is based upon iterative adjustments of T_{ij} as a result of comparison of the net output with known correct answers (supervised learning) or as a result of creating of new categories from the correlations of the input data when correct answers are not known (unsupervised learning). Actually, the procedure of learning is implemented by another dynamical system with the state variables T_{ij} which converges to certain attractors representing the desired synaptic weights T_{ij}^0.

3.3.3 Supervised Learning

Most of the supervised learning methods are based upon the famous hypothesis of Hebb (1949): the synaptic weights T_{ij} change is proportional to the product of input and output signals. Analytically this rule follows from the least-square error measure between the actual and required outputs $E(T_{ij})$ and the minimization procedure leads to a gradient dynamical system of the type (3.12):

$$T_{ij} = -\xi \frac{\partial E}{\partial T_{ij}}, \ \xi > 0 \tag{3.38}$$

The system converges to a static attractor corresponding to a minimum of the total error measure E which is supposed to give the best choice of synaptic weights T_{ij} for a required task.

There are several problems associated with this approach. Firstly, for neural nets of the form (3.35) the function $E(T_{ij})$ is not explicit, and computations of the gradient E are laborious and expensive. Secondly, since the function $E(T_{ij})$ is not quadratic, it can have many local minima, and consequently, the solution to the system (3.38) does not necessarily correspond to global minima, i.e., it may not be the best.

A considerable effort has recently been devoted to the development of efficient computational methodologies for learning recurrent back-propagation (Pineda 1987), a joint operator's approach, etc. (Toomarian and Barhen 1992). And despite the fact that all of them hardly represent the way the biological system learn, they significantly contributed in the progress of artificial neural nets theory.

It should be mentioned that as a "side-effect," learning theory developed a powerful computational tool for identification and control of dynamical systems whose analytical structure is not fully understood.

3.3.4 Unsupervised Learning

In unsupervised learning there is no feedback from outputs to inputs: the network must discover for itself new patterns, new categories, new generalizations. In this section we will illustrate unsupervised learning based upon a velocity field approach (Zak and Toomarian 1990). The neural network will be presented in the following form

$$\dot{u}_i + ku = \sum_{j=1}^{n} T_{ij}\sigma(u_j) + I_i \tag{3.39}$$

in which u is an n-dimensional vector function of time, representing the neuron activity; T_{ij} is a constant matrix whose elements represent synaptic interconnections between the neurons; σ is a sigmoid function; I_i is the constant exterior input to each neuron; and k is a positive constant.

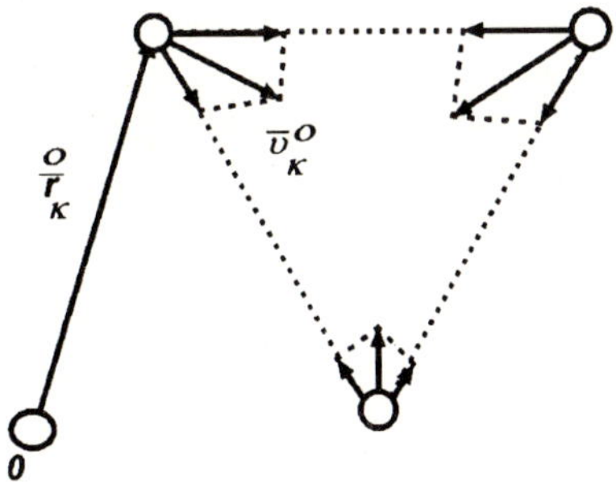

Fig. 3.10. Interpolation nodes

Let us consider a pattern vector $\widetilde{u}$ represented by its end point in an n-dimensional phase space, and suppose that this pattern is introduced to the neural net in the form of a set of vectors – examples $u^{(k)}$, $k = 1, 2, ..., k$ (Fig. 3.10). The difference between these examples which represent the same pattern can be caused not only by noisy measurements, but also by the invariance of the pattern to some changes in the vector coordinates (for instance, to translations, rotations etc.). If the set of the points $u^{(l)}$ is sufficiently dense, it can be considered as a finite-dimensional approximation of some subspace $\theta^{(l)}$.

Now the goal of this study is formulated as following: find the synaptic interconnections T_{ij} and the input to the network I_i such that any trajectory which is originated inside of $\theta^{(l)}$ will be entrapped there (Fig. 3.11). In such a performance the subspace $\theta^{(l)}$ practically plays the role of the basin of attraction to the original pattern $\widetilde{u}$. However, the position of the attractor itself is not known in advance: the neural net has to create it based upon the introduced representative examples. Moreover, in general the attractor is not necessarily static it can be periodic, or even chaotic.

The achievement of the goal formulated above would allow one to incorporate into a neural net a set of attractors representing the corresponding clusters of patterns, where each cluster is embedded into the basin of its attractor. Any new pattern introduced to such a neural net will be attracted to the closest attractor. Hence, the neural net would learn examples to perform content-addressable memory and pattern recognition.

The approach is based upon the utilization of the original clusters of the example points u^k as interpolation nodes of the velocity field in phase space. The assignment of a certain velocity to an example point imposes a corresponding constraint upon the unknown synaptic interconnections T_{ij} and input I_i via (3.39). After these unknowns are found. the velocity field in phase space is determined by (3.39). Hence, the main problem is

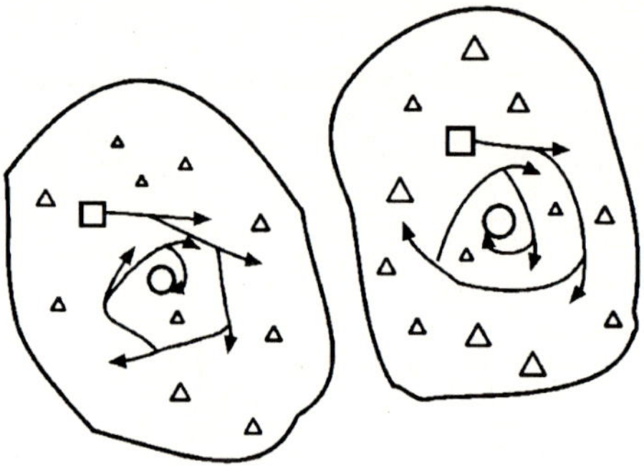

Fig. 3.11. Two dimensional vectors as examples u^k and formation of clusters θ

to assign velocities at the point examples such that the required dynamical behavior of the trajectories formulated above is provided.

One possibility for the velocity selection is based upon a gravitational attraction between the examples. Suppose that each example-point $u^{(k)}$ is attracted to all the other points $u^{(k')}(k' \neq k)$ such that its velocity is found by the same rule as a gravitational force (Fig. 3.11):

$$v_i^{(k)} = v_0 \sum_{\substack{k'=1 \\ k' \neq k}}^{k} \frac{u_i^{(k')} - u_i^{(k)}}{\left[\sum_{j=1}^{n} (u_j^{(k')} - u_j^{(k)})^2 \right]^{3/2}} \tag{3.40}$$

in which v_0 is a constant scale coefficient.

Actual velocities at the same points are defined by (3.39) rearranged as:

$$\dot{u}_i^k = \sum_{j=1}^{n} T_{ij} \sigma(u_j^{(k)} - u_{0i}) - k(u_i^{(k)} - u_{0i}), \quad \begin{array}{rl} i = & 1, 2, ..., n \\ k = & 1, 2, ..., k \end{array} \tag{3.41}$$

The objective is to find synaptic interconnections T_{ij} and center of gravity u_{0i} such that they minimize the distance between the assigned velocity (3.40) and actual calculated velocities (3.41)

Introducing the error measure:

$$E = \frac{1}{2} \sum_{k=1}^{k} \sum_{i=1}^{n} (v_i^{(k)} - u_i^{(k)})^2 \tag{3.42}$$

one can find T_{ij} and u_{0i} from the condition:

$$E \to \min$$

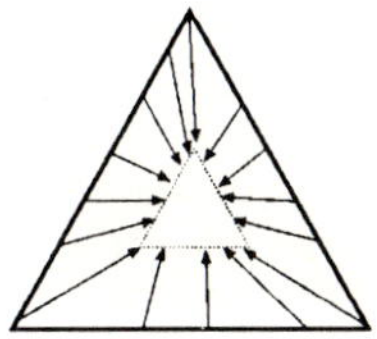

Fig. 3.12. Velocities at boundaries are directed toward the interior of the cluster

i.e., as the static attractor of the dynamical system:

$$\dot{u}_{0i} = -\alpha^2 \frac{\partial E}{\partial u_{0i}} \tag{3.43}$$

$$\dot{T}_{ij} = -\alpha^2 \frac{\partial E}{\partial T_{ij}} \tag{3.44}$$

in which α is a time scale parameter for learning. By appropriate selection of this parameter the convergence of the dynamical system can be considerably improved (Barhen et al. 1989).

The static attractor of (3.43, 3.44) is unique, since the error measure is a quadratic function of T_{ij} and u_{0i}. As follows from (3.41)

$$\frac{\partial \dot{u}_i^{(k)}}{\partial u_j^{(k)}} = T_{ij} \frac{d\sigma_j^{(k)}}{du_j^{(k)}}, \quad i \neq j \tag{3.45}$$

Since $\sigma(u)$ is a monotonic function, sgn $\frac{d\sigma_j^{(k)}}{du_j^{(k)}}$ is constant which in turn implies that

$$\text{sgn}\ \frac{d\sigma_j^{(k)}}{du_j^{(k)}} = \text{Const}, \quad i \neq j \tag{3.46}$$

Applying this result to the boundary of the cluster one concludes that the velocity at the boundary is directed inside of the cluster (Fig. 3.12).

For numerical illustration of the learning concept described above, 6 points were selected in the two dimensional space, (i.e., two neurons) which constructs two separated clusters [(Fig. 3.13), points 1–3 and 16–18 (three points are the minimum to form a cluster in two dimensional space)]. The assigned velocity v_i^k calculated based on (3.40) and $v_0 = 0.04$ are shown in dotted line.

For a random initialization of T_{ij} and u_{0i} the error measure E decreases sharply from an initial value of 10.608 to less than 0.04 in about 400 iterations and at about 2000 iterations the final value of 0.0328 has been achieved (Fig. 3.14).

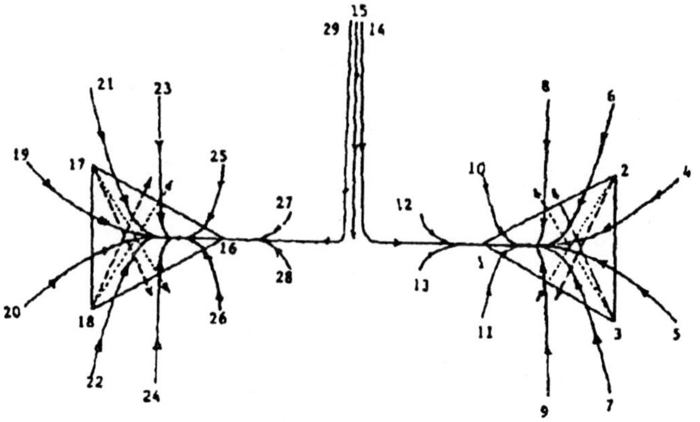

Fig. 3.13. Cluster 1 (1-3) and Cluster 2 (16-19). Assigned velocity (··). Calculated velocity (—). Activation dynamics initiated at different points

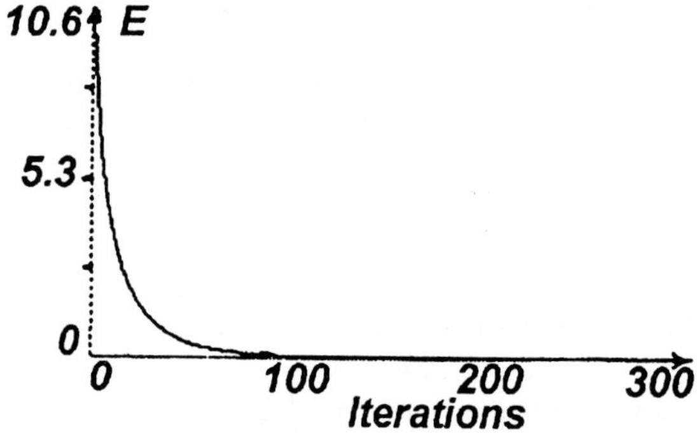

Fig. 3.14.

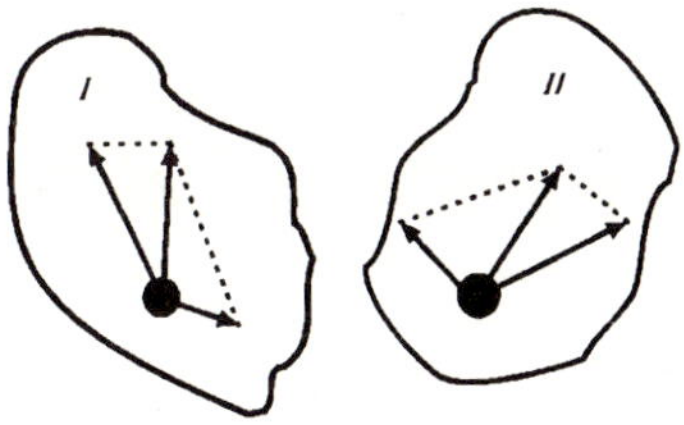

Fig. 3.15. Self-categoraization effect

To carry out numerical integration of the differential equations, a first order Euler numerical scheme with time step of 0.01 was used. In this simulation the scale parameter a^2 was kept constant and set to one. By substituting the calculated T_{ij} and u_{0i} into (3.41) for point u^k, ($k = 1, 2, 3, 26, 17, 18$), one will obtain the calculated velocities at these points (shown as dashed lines in Fig. 3.13). As one may notice, the assigned and calculated velocities are not exactly the same. However, this small difference between the velocities are of no importance as long as the calculated velocities are directed toward the interior of the cluster. This directional difference of the velocities is one of the reasons that the error measure E did not vanish. The other reason is the difference in the value of these velocities, which is of no importance either, based on the concept developed.

In order to show that for different initial conditions, (3.46) will converge to an attractor which is inside one of the two clusters, this equation was started from different points (14–15, 19–29). In any case, the equation converges to either (0.709, 0.0) or (–0.709, 0.0). However, the line $x = 0$ in this case is the dividing line, and all the points on this line will converge to u_0.

The decay coefficient k and the gain of the hyperbolic tangent were chosen to be 1 in this simulation. However, during the course of this simulation it was observed that the system was very sensitive to these parameters as well as v_0 which calls for further study in this area.

The "gravitational" power 3/2 taken in (3.40) can be changed. It is easy to conclude that increase of this power will lead to more subtle clustering of patterns since the attraction "force" will vanish sharply with the distance between "examples" (Fig. 3.15). On the contrary, decrease of this power will be less discriminative, and eventually it will lead to a personalization process when the original set of examples are represented by one typical pattern (Fig. 3.16).

Thus from a dynamical viewpoint, both supervised and unsupervised learnings are implemented by convergence to attractors, i.e., by systems with temporal organizations.

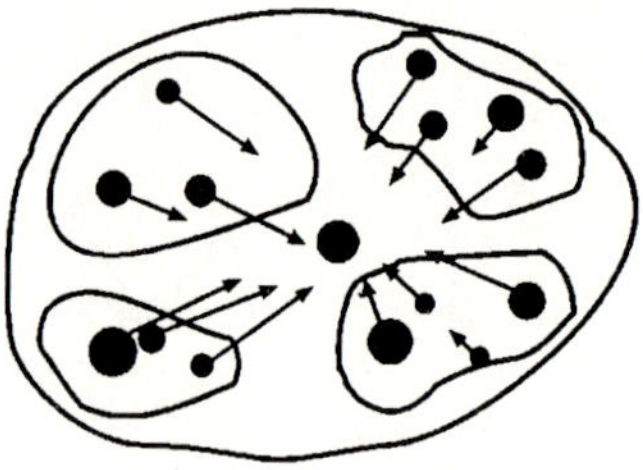

Fig. 3.16. Generalization effect

3.4 Neural Nets with Spatial Organization

3.4.1 Introduction

The traditional approach to neurodynamics is based upon dynamical system theory which operates by ordinary differential equations. We will start with the additive model as a typical representative of artificial neural networks.

$$u_i = -u_i + \sum_{j=1}^{n} T_{ij}\sigma(u_j) + I_i,\ i = 1, 2, ..., n \tag{3.47}$$

in which is the mean soma potential $u_i(t)$ of the ith neuron, T_{ij}, are constant synaptic interconnections, $\sigma(u)$ is a sigmoid function, I_i is an external input.

We will emphasize one important property of the system (3.47): the neuron locations are not associated with any metric, and no distance is defined between them: the neurons are labelled in an arbitrary fashion, and therefore, u_i and T_{ij} are not necessarily smooth functions of i and i,j respectively. This means that the information is processed only through the values of the mean soma potentials, u_i but not through the locations of the neurons. Obviously (3.47) does not have a mechanism for smoothing out the distributions of u_i and T_{ij} over the indexes i and j. In order to express this concept into a more rigorous way, let us turn to the continuum analog of (3.47) which reads:

$$\frac{\partial u(s,t)}{\partial t} = -u(s,t) + \int_{s_1} T(s, s_1, t)\sigma[u(s_1, t)]ds_1 + I \tag{3.48}$$

where the integration is performed over a space coordinate s. Equation (3.48) was introduced and discussed by Amari (1983), and analyzed by Sakaguchi (1990) as a possible model for a nerve field. Clearly (3.48) is different from the system (3.47) in the same way in which the eigenvalues of the kernel $T(s, s_1)$ are different from the eigenvalues of the matrix

T_{ij}. At the same time, (3.48) is less general than (3.47) because its derivation requires, at least, integrability of the functions $u(s,t)$ and $T(s,s_1,t)$ over s and s_1, although these functions can be nondifferentiable. Let us assume that initially the functions $u(s,0)$ and $T(s,s_1,0)$ are nondifferentiable. Then, in general they will remain nondifferentiable during all the time since, as in (3.47), here there are no mechanisms to smooth them out. The means that the dynamics (3.48) can run smoothly in phase space and approach, for instance, and attractor

$$u(s) = \int_{x_1} T(s,s_1)\sigma[u(s_1,t)]ds_1 + I \tag{3.49}$$

as a point in phase space but in actual space this attractor $u = u(s)$ will have a monstrous shape and therefore, will not process any useful information.

Our main objective now is to incorporate into (3.47) and (3.48) such mechanisms that would smooth out the shapes of static attractors in actual space and allow one to specify these shapes in advance in order to process information (Zak 1991c).

3.4.2 Neurodynamics with Diffusion

We will start with the continuous version of a neural network (3.48) and modify it as follows:

$$\frac{\partial u}{\partial t} = \alpha^2\frac{\partial^2 u}{\partial s^2} - u + \int_{s_1} T\sigma(u)ds_1 + I \tag{3.50}$$

In order to clarify the effect of the additional term $\partial^2 u/\partial s^2$ let us recall the properties of the diffusion (or thermoconductivity) equation.

$$\frac{\partial u}{\partial t} = a^2\frac{\partial^2 u}{\partial s^2} \tag{3.51}$$

This equation smooths out any discontinuity of u with respect to s providing gradual spreading of the variations of u along s. The distance of penetration of these variations into regions of uniform u after a time t is of order of $a\sqrt{t}$ while the rate of penetration decreases as t increases. The distribution of u over s tends asymptotically to a similarity form depending on $x/a\sqrt{t}$ alone, irrespective of the initial form of the transition (Fig. 3.17). The same smoothing-out effect will appear in (3.50) if s^2 is large enough in comparison to the synaptic interconnections T with the only difference that the final distribution $u(s)$ will depend upon the rest of the terms in this equation (Fig. 3.18).

The finite-dimensional version of (3.50), i.e., the modified version of (3.47) reads

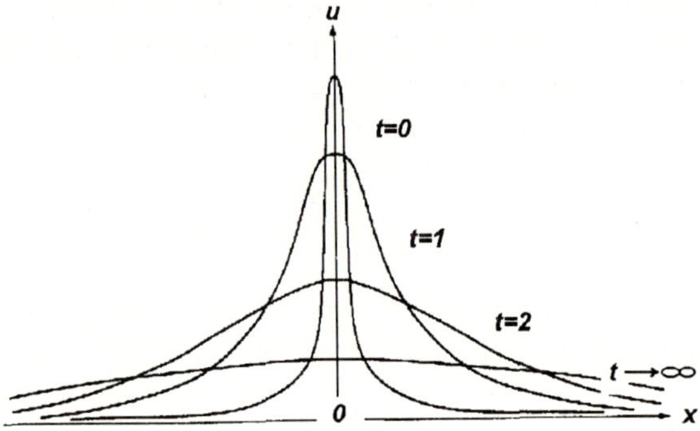

Fig. 3.17. Smoothing-out of δ function

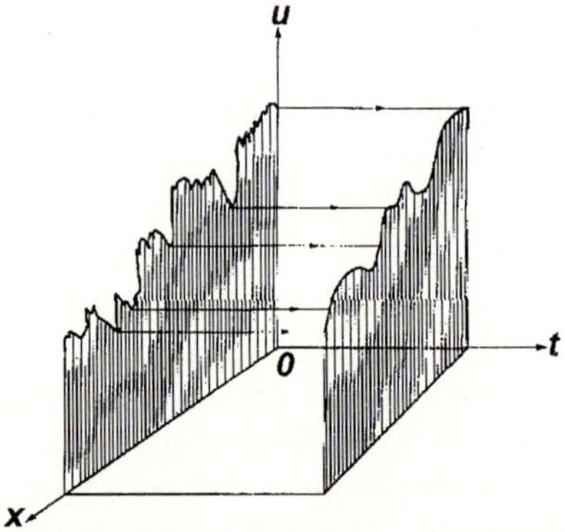

Fig. 3.18. Smoothing out of small changes of the image by diffusion

$$\dot{u}_i = -u_i - \alpha_i^2(u_{i-1} - 2u_i + u_{i-1}) + \sum_{j=1}^{n} T_{ij}\sigma(u_j) + I_i \tag{3.52}$$

where α_i play the role of adjustable coefficients as T_{ij} do.

The advantage of (3.52) over its conventional version (3.47) can be illustrated by the following example. Let us assume that (3.52) must converge to a static attractor.

$$u_i = \overset{\circ}{u}_i \text{ at } \dot{u}_i \to 0, \; i = 1, 2, ..., n \tag{3.53}$$

which corresponds to a "picture" in actual space given by a discrete function:

$$u = u(i), \; i = 1, 2, ..., n \tag{3.54}$$

Each equation in (3.54) represents a constraint imposed on T_{ij}: the more the number of n of these constraints the more non-zero coefficients T_{ij} will be required to satisfy them.

Consider the case when the function (3.54) is piecewise monotonic, i.e., it is monotonic between some critical points.

$$i_q < i_2, ... < i_m, \; m \ll n \tag{3.55}$$

Then instead of constraints (3.53) imposed upon T_{ij} and s_t^2, one can require weaker constraints:

$$u_i = \overset{\circ}{u}_i \text{ at } \dot{u}_i \to 0, \; i = 1,2,..., m \ll n \tag{3.56}$$

providing the convergence to (3.54) only in the critical points (3.55), while the values at the rest of the points will result from smoothing-out the differences between the corresponding critical points. Clearly the values of s_i at the critical points must be zero providing "thermoisolation" between different branches of the curve (3.54). Since the number of the constraints (3.56) is significantly smaller than the number of the constraints (3.53), the number of the required T_{ij} in the presence of local interconnections will be much smaller.

Thus, the incorporation of the diffusion into neurodynamical architecture, i.e., by introducing additional local interconnections between neurons, one can significantly decrease the number of synaptic interconnections required for processing "smooth" topographical images in comparison to the case when the local interconnections are not incorporated.

The information processing performed by diffusion neurodynamics has a strong connection with the first step of vision which is the edge detection, i.e., marking the borders that separate clusters of significantly different pixel values. In fact, our visual system is designed primarily to detect changes in, rather than absolute values of, light signals. There are two steps in edge detection. First, the detector blurs the entire image by adding to each pixel value the average of the pixel values in a cluster surrounding it. In other words, it smooths-out small (high frequency) changes of the image. In the model (3.50) this procedure is performed by the diffusion due to the local interconnections represented by the term $\partial^2 u/\partial s^2$. Second, it enhances large changes in the intensity signal. In the model (3.50) this effect is conducted through the local interconnection strengths α_i at the critical points which represents the borders separating clusters of different intensity.

Diffusion neurodynamics simulates another important characteristic of vision: the continuity constraint. This constraint expresses the fact that most surfaces are smooth, so that if one point in an image is assigned a certain distance from the eye, nearby points will be assigned similar distances. Indeed, the local interconnection term $\partial^2 u/\partial s^2$ provides the smoothness, or differentiability of the solution with respect to spatial variables, i.e., in actual space. This means that the information about the neuron potential of a point in actual space allows one to receive information about the neuron potentials of its close neighbors. Loosely speaking, this can be done by using Taylor expansion based upon the spatial derivatives at the original point. Such a procedure actually eliminates combinatorial explosions in the number of parameters characterizing neural performances. On the contrary, when the continuity constraint fails [see (3.47)], i.e., when the neurodynamics does not have local interconnections the solution may be not differentiable with respect to spatial coordinates. In this case neurons from the same neighborhood may have significantly different and totally unrelated potential values. Therefore, the information about the neuron potential of a point in actual space gives no information about the neuron potentials of its closes neighbors at all. This means that the number of parameters characterizing neural performances can grow exponentially with the increase of dimensionality of the problem.

Thus local interconnection in the diffusion neurodynamics implement a natural constraint which allows one to reduce the number of possible interpretation of an image by ruling out some of them as physically impossible

It is worth noting that the smoothing-out effects require $\alpha \gg T$, i.e., the "specially organized," or differential, local interconnections represented by the diffusive terms are significantly stronger than the "additive" synaptic interconnections represented by the integral term.

3.4.3 Neurodynamics with Dispersion

Another type of smoothing-out effect can be associated with the local interconnections represented by the third spatial derivative.

We will turn to (3.48) again and modify it in the following way:

$$\frac{\partial u}{\partial t} = \gamma \frac{\partial^3 u}{\partial s^3} - u + \int_{s_1} T\sigma(u) ds_1 + I \tag{3.57}$$

The effect of the term $\partial^3 u / \partial s^3$ can be analyzed using the simplest version of the dispersion equation:

$$\frac{\partial u}{\partial t} = \gamma \frac{\partial^3 u}{\partial s^3} \tag{3.58}$$

This equation has a solution in the form of superposition of dispersive waves:

$$u = u_0 e^{ks - \omega t}, \quad \omega = -\gamma k^3 \tag{3.59}$$

each travelling with its own phase speed:

$$c(k) = \frac{\omega(k)}{k} = -\gamma k^2 \tag{3.60}$$

As time evolves, these different component modes "disperse" with the result that a single concentrated hump, for example, disperses into a whole oscillatory train. However, in contrast to the diffusion, here the total energy is not dissipated, but is rather dispersed over a larger area. Qualitatively the same smoothing-out effect will appear in (3.57), although the solution, in general, will not have the form (3.59).

In this section we will concentrate our attention on the high-order linear dispersion neurodynamics and demonstrate its application in a dynamical way of a Fourier expansion of an arbitrary function.

Consider the dispersion equation:

$$\frac{\partial u}{\partial t} = \gamma \frac{\partial^3 u}{\partial s^3} + b \frac{\partial^5 u}{\partial s^5} \tag{3.61}$$

which has the following phase speed [compared with (3.60)]:

$$c(k) = -ak^2 + bk^4 \tag{3.62}$$

If $a = b$ then $c(k = 1) = 0$ which means that the mode with the wave number $k = 1$:

$$u = u_0^{(1)} e^{is} \tag{3.63}$$

remains motionless while all the other modes move away from the origin $s = 0$.

Analogously, if

$$a = bk^{*2} \tag{3.64}$$

then the only mode with the wave number $k = k^*$.

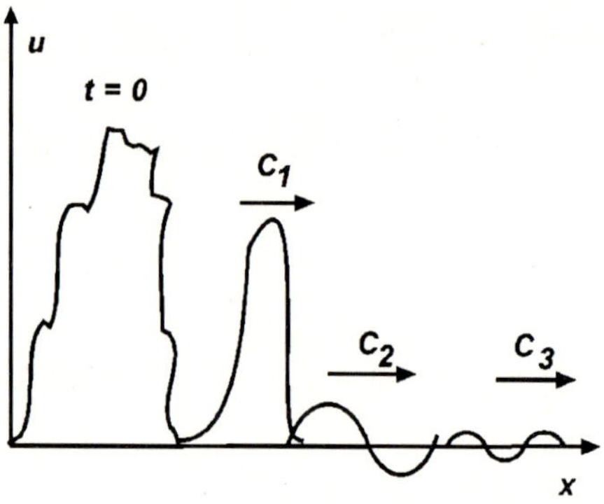

Fig. 3.19. Image dispersion

$$u = u_0^{k^*} e^{ik^* s} \tag{3.65}$$

remains motionless while all the others move away, etc.

Hence the system of equations:

$$\frac{\partial u_j}{\partial t} = \gamma \frac{\partial^3 u_j}{\partial s^3} + b \frac{\partial^5 u_j}{\partial s^5} \tag{3.66}$$

with $a_j = j^2 b$ will decompose an arbitrary function $u(s)$ into the Fourier series if $u(s)$ is introduced into (3.66) as the initial condition $u(t = 0) = u(s)$ (Fig. 3.19).

3.4.4 Convective Neurodynamics

In the previous sections we have analyzed the effects of local synaptic interconnections represented by he second and the third spatial derivatives and it has been demonstrated the smoothing-out property of these effects. It appears that the effect of local synaptic interconnections represented by the first derivative $\partial u / \partial s$ is more complex: it can smooth out spatial discontinuities, but it also can amplify them via the mechanism of shock waves. But the most important property of the first order local interconnections is that they can represent a convection velocity which drives the image in actual space and therefore, the neurodynamical architecture with convection can process moving images (Zak 1991c).

Let us start with the simplest convective model in its continuous version:

$$\frac{\partial u}{\partial t} + v \frac{\partial u}{\partial s} = 0, \;\; v = v(s, t) \tag{3.67}$$

whose discrete version is $u_i = -v_i(u_i = u_{i-1})$.

It is easily verifiable that the solutions to this equation can be written in the following form:

$$u = \overset{\circ}{u}\,(s - \int_0^t v\;dt), \quad ds = v\;dt \tag{3.68}$$

in which

$$u = \overset{\circ}{u}\,(s) \tag{3.69}$$

is the initial condition at $t = 0$.

As follows from these equations, the initial distribution (3.69) of the neuron potential u over the s coordinate moves with the convective velocity v along the s axis. Depending on the sign of the derivative $\partial v/\partial s$ the image $u(s)$ is contracted ($\partial v/\partial s < 0$), elongated ($\partial v/\partial s > 0$), or translated ($\partial v/\partial s = 0$). Obviously, in the first case all the discontinuities are amplified, in the second case they are smoothed-out, and in the last case they remain unchanged.

The convective velocity $v(s,t)$ which actually defines the type of image time- and space-transformations can be prescribed in advance, or it can be learned in the same way as the synaptic interconnections are. In the last case the convective velocity can be represented as a variable in learning neurodynamics with the architecture (3.50):

$$\frac{\partial v}{\partial t} = \alpha\frac{\partial^2 v}{\partial s^2} - v + \int_{s_1} \widetilde{T}\sigma(v)ds_1 + \widetilde{I} \tag{3.70}$$

while the values of the corresponding static attractors of (3.70):

$$v = \overset{\circ}{v}\,(s) \tag{3.71}$$

should be substituted in the activation neurodynamics (3.67). The attractor (3.71) is supposed to capture the basic features of the image motion to be stored and recognized.

It is interesting to note that those points s^* where the velocity is zero $v(s^*)$ possess some critical properties: as follows from (3.67), the derivative $\partial u/\partial s$ in these points is not defined. A geometrical meaning of this can be clarified by the following example.

Let us take the initial image in the form of a unit square on the plane us:

$$u(s) = \begin{cases} 1 & \text{at } 0 < s < 1 \\ 0 & \text{at } s < 0 \text{ and } s > 1 \end{cases} \tag{3.72}$$

and

$$v = 1 - s \tag{3.73}$$

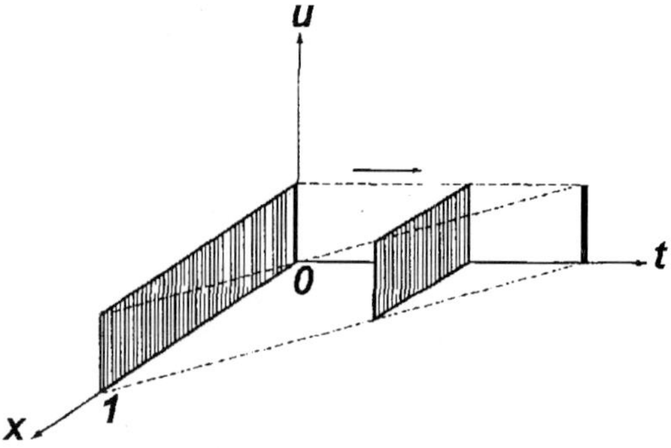

Fig. 3.20. Image convection

Since $v > 0$ for $s < 1$, all the points of the square (3.68) will move to the right and stop at the critical point $x = 1$ where $v = 0$. In other words, the original square will be absorbed by its right side $u = 1$ at $x = 1$ and therefore, it will degenerate into a line (Fig. 3.20).

Hence the critical points of the velocity field play the role similar to attractors in phase space: the image "converges" to them, although the mechanism of this convergence is different from those in phase space. This property of the critical point can be exploited for feature extraction: since at the final state the original image will have non-zero values of u only at the critical points of the velocity field, these values can represent essential properties of the image if the location of the critical points in actual space is appropriately selected.

So far it was implied that the initial state of the image (3.69) was prescribed. However, one can assume that this image is performed by a neurodynamics (3.50) or (3.57) as a static attractor. Then the combined set of (3.50), (3.67), and (3.70) performs the following information processing: (3.50) stores the image as a static attractor, and (3.70) stores the velocity field which drives the motion of this image. Finally, the convective neurodynamics (3.67) reproduces the stored motion $v(s)$ of the stored image $u(s)$.

Let us now discuss some biological aspect of the convective neurodynamics. Spatial changes in intensity are not the only kind of changes for which the visual system is on the lookout: temporal changes in the light signal are also very important. The retinal cells continuously compare intensity values, and send out transient responses that encode changes in intensity values rather than the intensity values themselves. If the intensity values stay the same, the cell's response falls to zero and the image fades. In convectional neurodynamics model (3.67) all the temporal changes of the image are conducted through the convective velocity v. Moreover, a special selection of the velocity structure [see (3.73)] provides formation of new image edges by degeneration of a line into a point.

3.4.5 Coupled Phase-Space Neurodynamics

The most powerful tools for information processing performed by dynamical systems are based upon the concept of an attractor as a limit set which does not depend upon variations of initial conditions. In the previous section, in addition to the attraction in phase space we introduced a new type of dynamical attraction performed by convection in actual space. However, in the neurodynamical architecture, represented by (3.50), (3.67) and (3.70), these two types of attractors were utilized sequentially: firstly, the attractor in phase space $u(s)$ processed the initial shape of the image (3.69), and the attractor in phase space $v(s)$ processed the velocity field of the image motion; and only after that (3.67) performs the attraction in actual space. In other words, the attractions in phase and actual spaces remain uncoupled. In this section we will introduce a more complex phenomenon of coupled attractions in phase and actual spaces inspired by the theory of shock waves and solitons (Zak 1991c).

We will start with the nonlinear convection described by the equation:

$$\frac{\partial u}{\partial t} + u\frac{\partial u}{\partial s} = 0 \tag{3.74}$$

Introducing the Lagrangian coordinate s_0:

$$s = s(s_0, t), \quad \frac{ds}{dt}|_{s_0} = \text{Const} = u \tag{3.75}$$

one can rewrite (3.74) as:

$$\left.\frac{du}{dt}\right|_{s_0} = \text{Const} = 0 \tag{3.76}$$

Hence, (3.74) describes the flow of free (non-interacting) particles moving with constant velocities. The properties of (3.74) are well-known (Whitham 1972): the nonlinearity in the convection term leads to the concentration of a disturbance and to formation of shock waves. In contrast to this, diffusion and dispersion lead to smoothing-out of this disturbance, and, as a result, some steady waves of permanent form are possible. It can be proved (Whitham 1972) that the following equation:

$$\frac{\partial u}{\partial t} + a_1 u\frac{\partial u}{\partial s} - a_2^2\frac{\partial^2 u}{\partial s^2} + a_3^2\frac{\partial^3 u}{\partial s^3} = 0 \tag{3.77}$$

has solutions in the form of steady wave of permanent form moving with a constant velocity. If $a_1 = 1$, $a_3 = 0$ (the Burger's equation), an original step function diffuses into a steady profile (Fig. 3.21).

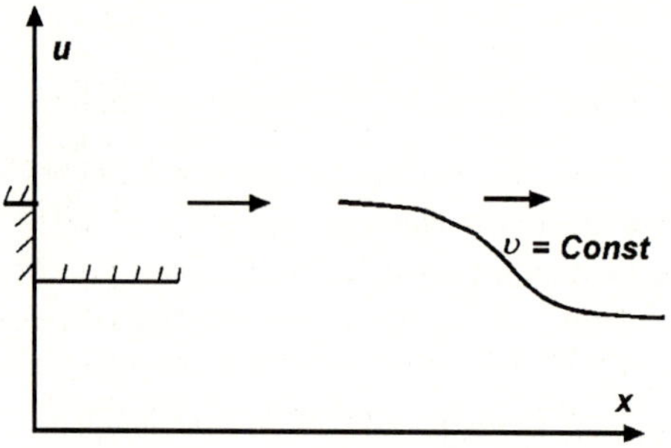

Fig. 3.21. Weak shock wave formation

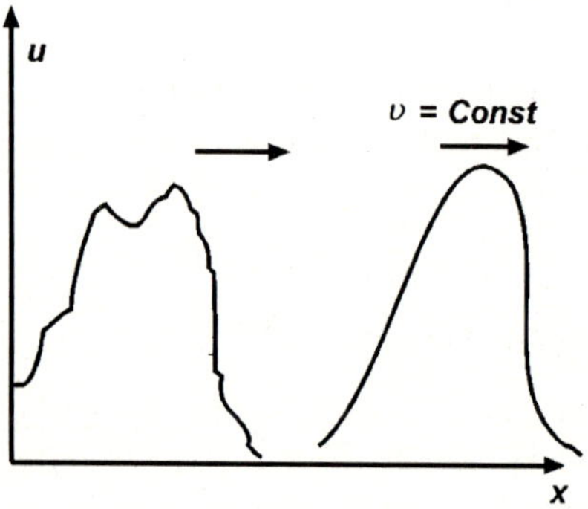

Fig. 3.22. Soliton formation

$$u = u_1 + \frac{u_2 - u_1}{1 + \exp \frac{u_2 - u_1}{2a_2^2}\left(s - \frac{u_1 + u_2}{2}t\right)} \tag{3.78}$$

in which the step function at $t = 0$ is:

$$u(s) = \begin{cases} u_1, & s > 0 \\ u_2 > u_1, & s < 0 \end{cases} \tag{3.79}$$

If $a_2 = 0$ (the Korteweg-de Vries equation), an original distribution $u(s)$ breaks up into a system of solitons – steady waves of permanent form (Fig. 3.22).

For instance, if

$$u, \frac{\partial u}{\partial s}, \frac{\partial^2 u}{\partial s^2} \to 0 \text{ as } s - ct \to \pm\infty \tag{3.80}$$

the solution to (3.77) with $a_1 = -6$, $a_2 = 0$, $a_3 = 1$ is a soliton (Fig. 3.22):

$$u(s,t) = -\frac{1}{2}c \operatorname{sech}^2\left[\frac{1}{2}\sqrt{c(s-ct-\text{Const})}\right] \tag{3.81}$$

in which

$$c = \text{Const} \geqslant 0 \tag{3.82}$$

Thus, in both cases initial distributions of $u(s)$ at $t = 0$ are attracted to the steady solution (3.78) and (3.81) and therefore, these solutions can be associated with the concept of an attractor

The remarkable property of this attraction is that both solution (3.78) and (3.81) are localized in actual space (see Figs. 3.21 and 3.22) which results from coupling between the attractions in phase space and in actual space. Due to the attraction properties the effect of weak shock waves (the Burger's equation) and of solitons (the Korteweg-de Vries equation) can be exploited for information processing in the same way in which a static attractor in phase space is exploited for associative memory and pattern recognition with the only difference that instead of a pattern vector, the attractors (3.78) and (3.81) operate with a dynamical behavior $u(s,t)$ which is localized in space.

However, in addition to this "operational" difference, there are some more significant differences between the phase-space attractors and nonlinear-wave attractors (3.78) and (3.81).

Firstly, the weak shock wave attractors do not completely "forget" the initial conditions (although the attraction is driven by the energy dissipation). Indeed, as follows from (3.81), the solution slightly depends upon the initial condition u_1 and u_2 unless $a_2 \to \infty$.

Secondly, the soliton solutions are sensitive to boundary conditions. For instance, any changes in the boundary conditions (3.80) may lead to a totally different system of solitons, while this dependence is much more complex than the dependence of the basin of a phase-space attractor on the initial conditions. Strictly speaking, solitons are not "true" attractors since the Korteweg-de Vries equation is not dissipative at all. The attraction to solitons is driven not by the energy dissipation, but by the nonlinear convection "balanced" by the wave dispersion.

Nevertheless, from the viewpoint of information processing both of these nonlinear-wave effects can play the role of attractors.

The concept of coupled dynamical attractions in phase and actual spaces can be associated with the interaction of spatial and temporal intensity changes in vision process: the retinal cells designed to detect changes in light intensity across time and space; as the eye flickers restlessly in its socket, edges in space move across photoreceptor and are converted into edges in time.

3.4.6 General Model

In the previous sections spatial coordinates have been associated with actual physical space, and therefore, spatial self-organization has been linked to vision and image processing. However, one can introduce a n-dimensional abstract space with coordinates $s_1, ..., s_n$ such that the neuron potential $u(t, s_1, ..., s_n)$ is a piecewise differentiable function of all its arguments. General mathematical formulation for such a model which still preserves the simplicity of a single neuron nonlinear performance can be presented as:

$$\frac{\partial u}{\partial t} = -u+\sigma\left(\sum_{i=1}^{n}\alpha_i\frac{\partial u}{\partial s_i}+\sum_{\substack{i=1\\j=1}}^{n}\alpha_{ij}\frac{\partial^2 u}{\partial s_1 \partial s_2}+\sum_{i=1,...,i_n}^{n}\alpha_{i_1,...,i_n}\frac{\partial^n u}{\partial s_{i_1}, ..., \partial s_{i_n}}\right) \tag{3.83}$$

where σ is a sigmoid function, and α_i, α_{ij},..., etc. are time-independent functions of $s_1, ..., s_n$.

As in the previous one- and two-dimensional models here the coefficients α_i, α_{ij},... implicitly contribute to the strengths of local interconnections. Indeed, as follows from finite dimensional representations:

$$\begin{aligned}(\alpha_1\tfrac{\partial u}{\partial s_1})_i &= \alpha_{1(i)}\tfrac{u_{i+1}-u_{i-1}}{2h}\\ (\alpha_{11}\tfrac{\partial^2 u}{\partial s_1^2})_i &= \alpha_{11(i)}\tfrac{u_{i+1}-2u_i+u_{i-1}}{h^2}\end{aligned} \tag{3.84}$$

etc.

The interconnection strength T_{ij} between a pair of neurons u_i and u_j depends on a linear combination of $\alpha_i, \alpha_{ij}, ...$, etc. But due to the differential character of these interconnections, their strengths monotonically decrease with the increase of the "distance" between the neurons in the same way in which the coefficients of the Taylor series for a differentiable function do. In other words, for each prescribed performance, the architecture (3.83) minimized the strength energy (Zak 1989e):

$$s = \frac{1}{2}\sum_{\substack{i=1\\j=1}}^{n} T_{ij}^2(i-j)^2 \rightarrow \min \tag{3.85}$$

as a measure of interconnections between neurons.

In this respect one can suggest an interesting analogy with the Gauss principle formulated for mechanical systems with geometrical constraints (Appel 1953). According to this principle, actual motion for such a system corresponds to the least constraints, i.e., to the least difference from the free (unconstrained) motion. If by a free motion in neurodynamics one understands the motion of uncoupled neurons ($T_{ij} = 0$ if $i \neq j$) then the strengths T_{ij} can be associated with constraints imposed upon the free motions, while the minimum of the strength energy (3.85) corresponds to the least constraints. In terms of neurodynamics it means that for each performance a neural net will "select" such interconnections T_{ij} which impose the least constraints upon the free motions of its neurons if these constraints are measure by the strength energy s. In other words, neurons "try" to act with the least "disagreement" with their neighbors (Zak 1991a).

Now we can return to the problem of selection of spatial coordinates $s_1, ..., s_n$ in order to construct the model (3.83).

Obviously we will exclude from the consideration the trivial case when s_1 and s_2 are the physical space coordinates since in this case their choice is natural. For the case of abstract "space" coordinates the choice can be based upon the "least constraints principle": for a certain class of performances the coordinates $s_1, ..., s_n$ must be selected such that the interconnections T_{ij} found from (3.83) minimize the strength energy (3.85).

In our opinion, in general, this is a hard problem: as we noticed in Sec. 4.1, the solution to such a problem (which is equivalent to finding continuity constraints) would eliminate combinatorial explosion in the number of parameters characterizing neural performances. However, in certain particular situations this problem can be solved, and one will arrive at the system (3.83). From the operational viewpoint this system is similar to (3.47): a certain type of performance can be achieved by an appropriate selection of the constants $\alpha_i, \alpha_{ij}, ...,$ etc. (for instance, as a result of a learning procedure). However, in contradistinction to (3.47), here all the attractors will have a piecewise smooth "spatial" image.

3.4.7 Limitations of the Classical Approach

The biggest promise of neural networks as computational tools lies in the hope that they will resemble the information processing in biological systems. Notwithstanding many successes in this direction, it is rapidly becoming evident that current models based upon classical dynamical system theory are characterized by some limitations.

We will analyze these limitations using the additive model as a typical representative of artificial neural networks:

$$\dot{u}_i + u_i = \sum_{j-1}^{n} T_{ij}\sigma(u_j) + I_i, \quad i = 1, 2 \ldots n \tag{3.86}$$

in which $u_i(t)$ is the mean soma potential of the ith neuron, T_{ij} are constant synaptic interconnections, $v(u)$ is a sigmoid function, I_i is an external input.

Firstly, the neurons performance in this model is collective, but not parallel: any small change in the activity of an ith neuron instantaneously effects all other neurons:

$$\frac{\partial \dot{u}_i}{\partial u_i} = \frac{dv}{du_j} T_{ij} \neq 0 \tag{3.87}$$

In contrast to this, all biological systems exhibit both collective and parallel performances. For instance, the right and the left hands are mechanically independent (i.e., their performance is parallel), but at the same time their activity is coordinated by the brain; that makes their performance collective.

Secondly, the performance of the model (3.86) is fully prescribed by initial conditions. The system never "forgets" these conditions: it carries their "burden" up to $t \to \infty$. In order to change the system performance, the external input must overpower the "inertia of the past." In contrast to this, biological systems are much more flexible: they can forget (if necessary) the past adapting their behavior to environmental changes.

Thirdly, the features characterizing the system (3.86) are of the same scale: they are insulated from the microworld by a large range of scales. At the same time, biological systems involve mechanisms that span the entire range from the molecular to the macroscopic.

Can these limitations be removed within the framework of classical dynamics? The answer is no. Indeed, all the systems considered in classical dynamics satisfy the Lipschitz condition which guarantees the uniqueness of the solutions subject to prescribed sets of initial conditions. For the system (3.86) this condition requires that all the derivatives $\partial \dot{u}_i / \partial u_j$ exist and are bounded:

$$\left| \frac{\partial \dot{u}_i}{\partial u_j} \right| < \infty \tag{3.88}$$

The uniqueness of the solution

$$u_i = u_i(t, u_1, \ldots, u_n), \quad i = 1, 2, \ldots, n \tag{3.89}$$

subject to the initial conditions $u_i(i = 1, 2, \ldots, n)$ can be considered as a mathematical interpretation of rigid, predictable behavior of the corresponding dynamical system.

Therefore, all the limitations of the current neural net models mentioned above are inevitable consequences of the Lipschitz condition (3.88), and therefore of determinism of classical dynamics.

Classical dynamics describes processes in which the direction of time does not matter: its governing equations are invariant with respect to time inversion, in the sense that the time backward motion can be obtained from the governing equations by time inversion $t \to -t$ As stressed by Prigogine (1980), in this view future and past play the same role: nothing can appear in future which could not already exist in past since the trajectories followed by particles can never cross. This means that classical dynamics cannot explain the emergence of new dynamical patterns in nature in the same way in which nonequilibrium thermodynamics does. This is why the discovery of chaotic motions (which could lead to unpredictability in classical dynamics) has "shaken-up" the scientific community, and the number of publications in the area of chaos is still growing. However, notwithstanding the many successes of applications of chaos to modeling complex phenomena, many open problems remain. Indeed, so far relationships between the parameters of chaotic dynamical systems and probabilistic structures of the solutions are not available. Moreover, based upon Gödel's incompleteness theorem (1931), da Costa and Doria (1991) recently presented a rigorous proof of the algorithmic impossibility of deciding whether a given equation has chaotic domains or not in the class of "elementary" functions. A similar view was expressed by Zak (1970, 1982a,b,c) who attempted to enlarge the class of functions to study chaos and turbulence. Hence, even if a chaotic dynamical system describing a given nondeterministic process is found, its usefulness for prediction of the motion depends upon the values of positive Liapunov exponents. To complete this line of argumentation, one should recall that the Navier-Stokes equations are known for more than a century, but it does not help much in predictions of turbulent motions for even reasonably short time intervals.

Several fundamental questions concerning the origin of chaos and turbulence are still unanswered, and one of them is the following: how a fully deterministic dynamical equation whose solution subject to prescribed initial conditions is unique, can generate randomness without external random inputs. In order to elucidate the situation, let us consider a steady laminar flow whose instability is characterized by an exponential multiplier:

$$\widetilde{v} = v_0 e^{\mu t}, \ 0 < \mu < \infty \tag{3.90}$$

Obviously, the solution with infinitely close initial condition

$$\widetilde{v} = v_0 + \varepsilon, \ \varepsilon \to 0 \tag{3.91}$$

will remain infinitely close to the original one:

$$|\widetilde{v} - v| = \varepsilon e^{\mu t} \to 0 \text{ if } \varepsilon \to 0, \ t \leq N < \infty \tag{3.92}$$

during all the bounded time intervals. This means that random solutions can result only from random initial conditions when ε in (3.91) is small, but finite rather than infinitesimal.

The same arguments can be applied to discrete chaotic systems if divergence of actual trajectories in (3.90) is replaced by divergence of trajectories in configuration space.

Thus, as in stochastic differential equations, the changes in initial conditions for chaotic equations must be finite, although they may be humanly indistinguishable. However, unlike stochastic equations, the phenomenon of unpredictability in chaotic systems has a different origin: it is caused by exponential amplifications of the initial changes due to the mechanism of instability. Indeed, if two trajectories initially are "very close," and then they diverge exponentially, the same initial conditions can be applied to either of them, and therefore, the motion cannot be traced. From a mathematical viewpoint, such a multivaluedness can be interpreted as a failure of differentiability of the solution which means that the instability represents a boundary of applicability of the original model with the framework of differentiable dynamics.

But then two arguments can be brought up. Firstly, from the mechanical viewpoint, stability is not an invariant of motion: it depends upon the frame of reference. For instance, the same inviscid flow can be stable in Eulerian representation and unstable in the Lagrangian one (Arnold 1988), or in a frame of reference moving with the streamlines (Zak 1990c). This leads to the following question: is it possible to find such a (non-inertial) frame of reference in which the inertia forces would stabilize the motion, i.e., they would eliminate all the positive Liapunov exponents? The answer to this question was given by Zak (1985c). He introduced a specially selected rapidly oscillating frame of reference in which the originally chaotic motion was stabilized by inertia forces coupled with the motion itself. In other words, he found a frame of reference which provides the best view of the motion. However, there was a certain price paid for this representation: the component of the solution corresponding to the transport motion with the frame of reference contained the function $\sin \omega t \to \infty$ which is actually multivalued. Indeed for any arbitrarily small interval $\Delta\ t$ there always exists such a large frequency $\omega > \Delta\ / 2\pi$ that within this interval the function runs through all its values. This means that in order to eliminate chaos one has to enlarge the class of smooth functions by introducing non-differentiable functions, and this leads us to the second question: is chaos an invariant of motion or is it an attribute of a mathematical model? From the mathematical viewpoint the concept of stability is related to a certain class of function, or a type of space, and therefore, the same solution can be stable in one space and unstable in another depending upon the distance between two solutions. Hence, the occurrence of chaos in description of mechanical motions means only that these motions cannot be properly described by smooth functions if the scale of observations is limited. These arguments can be linked to Gödel's incompleteness theorem, and the Richardson's (1968) proof that the theory of elementary functions in classical analysis is undecidable. Indeed, classical dynamics, in addition to Newton's laws, is based upon certain assumptions of a pure mathematical nature. They restrict the class of functions that describes the motions, to functions of sufficient smoothness. Such artificial limitations which do not follow from axioms of mechanics may become inconsistent with the physical nature of motions. As shown by Zak (1974; 1982a,b,c; 1985a,b), these inconsistencies lead to instabilities (in the class of smooth functions) of equations which govern turbulent and chaotic motions.

The first step toward the enlarging the class of functions for modeling turbulence was made by Reynolds (1895) who decomposed the velocity field into the mean and pulsating components, and actually introduced a multivalued velocity field. However, this decomposition brought new unknowns without additional governing equations, and this created a "closure" problem. In 1986 (a,b), Zak showed that the Reynolds equations can be obtained by referring the Navier-Stokes equations to a rapidly oscillating frame of reference, while the Reynolds stresses represent the contribution of inertia forces. From this viewpoint the "closure" has the same status as proof of Euclid's parallel postulate, since the motion of the frame of reference can be chosen arbitrarily. In other words, the closure of Reynolds equations represents a case of undecidability in classical mechanics. However, based upon the interpretation of the Reynolds stresses as inertia forces, it is reasonable to choose the motion of the frame of reference such that the inertia forces eliminate the original instability. In other words, the enlarged class of functions should be selected such that, the solution to the original problem in this class of functions will not possess an exponential sensitivity to changes in initial conditions. This stabilization principle has been formulated and applied to chaotic and turbulent motions by Zak (1984; 1985b; 1986a,b; 1989a; Zak and Meyers 1996). As shown there, the motions which are chaotic (or turbulent) in the original frame of reference can be represented as a sum of the mean motion and rapid fluctuations, while both components are uniquely defined. It is worth emphasizing that the amplitude of velocity fluctuation is proportional to the degree of the original instability, and therefore, the rapid fluctuations can be associated with the measure of the uncertainty in the description of the motion. It should be noted that both "mean" and "fluctuation" components representing the originally chaotic motion are stable, i.e., they are not sensitive to changes of initial conditions, and are fully reproducible.

Thus, chaos as a supersensitivity to initial conditions can be eliminated by describing the originally chaotic motion in enlarged class of functions, for instance, by performing a Reynolds-type transformation and applying the stabilization principle. Nevertheless, the new deterministic representation will still contain an uncertainty coming from the "lack of knowledge" about initial conditions. However,this uncertainty has a subjective, rather than objective nature: as stressed by Ford (1988) randomness in chaotic motions is not an attribute of dynamics itself, but rather a result of its mathematical treatment, i.e., chaos only makes predictions difficult, but not impossible.

Turning back to our original problem of unpredictability and irreversibility in Newtonian dynamics one might ask now: are there some additional mathematical restrictions in Newtonian dynamics which do not have enough physical "ground"? As shown by Zak (1989b,c,d; 1990a,b; 1991a; 1992b; 1993a), there are such restrictions. And one of them is the Lipschitz condition which requires that for a dynamical system:

$$\dot{x}_i = v_i(x_i, ..., x_n), \ i = 1, 2, ..., n \tag{3.93}$$

all the derivatives

$$\left|\frac{\partial v_i}{\partial x_j}\right| < \infty, \; i,j = 1,2,...,n \tag{3.94}$$

must be bounded.

This condition allows one to describe the Newtonian dynamics within the framework of classical theory of differential equations which guarantees its reversibility and predictability. This, in turn, leads to such effects as infinite time of approaching an attractor, infinite time for escape of a repeller if changes in initial conditions are infinitesimal [see (3.90–3.92)], intractability of two trajectories which originally are "very close," but diverge exponentially, etc.

Hence, there are variety of phenomena whose explanations cannot be based directly upon the classical dynamics: in addition they require some "words" about a scale of observation, "very close" trajectories, etc.

Turning to governing equations of classical dynamics:

$$\frac{d}{dt}\frac{\partial L}{\partial \dot{q}_i} = \frac{\partial L}{\partial q_i} - \frac{\partial R}{\partial \dot{q}_i}, \; i = 1,2,...,n \tag{3.95}$$

where L is the Lagrangian, q_i, q are the generalized coordinates and velocity, and R is the dissipation function, one should recall that the structure of $R(q_i, ..., q_n)$ is not prescribed by Newtons laws: some additional assumptions are to be made in order to define it. The "natural" assumption (which has never been challenged) is that these functions can be expanded in Taylor series with respect to equilibrium states:

$$\dot{q}_i = 0 \tag{3.96}$$

Obviously this requires the existence of the derivative:

$$\left|\frac{\partial^2 R}{\partial \dot{q}_i \, \partial \dot{q}_j}\right| < \infty \text{ at } \dot{q}_i \to 0 \tag{3.97}$$

The departure from this condition was proposed by Zak (1992b), who introduced the following function:

$$R = \frac{1}{k+1}\sum_i \alpha_i \left|\frac{\partial r_i}{\partial q_j}\dot{q}_j\right|^{k+1} \tag{3.98}$$

in which

$$k = \frac{p}{p+2} < 1, \; p \gg 1 \tag{3.99}$$

while p is a large odd number.

By selecting large p, one can make k close to 1 so that (3.98) is almost identical to classical one (when $k = 1$) everywhere excluding a small neighborhood of the equilibrium point $q_j = 0$, while at this point:

$$\left| \frac{\partial^2 R}{\partial \dot{q}_i \, \partial \dot{q}_j} \right| \to \infty \text{ at } \dot{q}_j \to 0 \tag{3.100}$$

Hence, the condition (3.93) is violated, the friction force

$$F_i = \frac{\partial R}{\partial \dot{q}_i} \tag{3.101}$$

grows sharply at the equilibrium point, and then it gradually approaches its "classical" value. This effect can be interpreted as a mathematical representation of a jump from static to kinetic friction.

It appears that this "small" difference between the friction forces at $k = 1$ and $k < 1$ leads to *fundamental changes in Newtonian dynamics.*

Firstly, the time of approaching attractors as well as the time of escaping repellers becomes theoretically finite. Secondly, at repellers the solution becomes totally unpredictable within the deterministic mathematical framework, but it remains fully predictable in probabilistic sense. In contrast to "classical" chaos, here the randomness is generated by the differential operator itself as a result of the failure of uniqueness conditions at the equilibrium points.

In the next chapters the problem of unpredictability in dynamics and in neural systems will be discussed in the context of this new mathematical formalism.

4
Instability in Dynamics

A very slight cause, which escapes us, determines a considerable effect which we cannot help seeing, and then we say this effect is due to chance. *–Henri Poincaré*

4.1 Basic Concepts

Most dynamical processes are so complex that their universal theory, capturing all the details during all time periods, is unimaginable. This is why the purpose of mathematical modeling is to extract only fundamental aspects of the process, and to neglect insignificant features, without losing core information. But "insignificant features" is not a simple concept. In many cases even vanishingly small forces can cause large changes in dynamical system parameters, and such situations are intuitively associated with the concept of instability. Obviously destabilizing forces cannot be considered as "insignificant features," and therefore, they cannot be ignored. But since they may be humanly indistinguishable, in the very beginning, there is no way to incorporate them into a model. This simply means that a model is not adequate for quantitative description of the corresponding dynamical process: it must be changed or modified. However, instability delivers important qualitative information: it manifests the boundaries of applicability of an original model.

We will distinguish short and long-term instabilities. Short-term instability occurs when a system has alternative stable states. For dissipative systems such states can be represented by static or periodic attractors. For instance, a gradient system (3.12) has an unstable equilibrium point 3 (Fig. 3.9), and eventually, it will approach one of two attractors: 1 or 2, depending upon initial uncertainties of its state 3. In the very beginning of the post-instability transition period, the unstable motion cannot be traced quantitatively, but it becomes more and more deterministic as it approaches the attractor. Hence, a short-term instability does not necessarily require a model modification. Usually this type of instability is associated with bounded deviation of position coordinates (see 3.6) whose changes affect the energy of the system. Indeed, if the growth of a position coordinate persists, the energy of the system would become unbounded.

Long term instability occurs when a system does not have an alternative stable state. Such instabilities can be associated only with ignorable coordinates (see 3.5) since these coordinates do not effect the energy of the system. Long term instability will be one of the main subjects of this monograph.

4.2 Orbital Instability

4.2.1 Ignorable Coordinates

As mentioned in the previous chapter (see 3.5), the coordinate q_α is called ignorable if it does not enter the Lagrangian function L as well as non-conservative generalized forces Q:

$$\frac{\partial L}{\partial q_\alpha} = 0, \quad Q_\alpha = 0 \tag{4.1}$$

therefore,

$$\frac{\partial L}{\partial \dot{q}_\alpha} = P_\alpha = \text{Const} \tag{4.2}$$

i.e., the generalized ignorable impulse P_α is constant.

As follows from (4.2), there exist such states of dynamical systems (called stationary motions) that all the position coordinates retain constant value while the ignorable coordinates vary in accordance with a linear law. For example, a regular precession of a heavy symmetric gyroscope is a stationary motion characterized by the equation:

$$\theta = \text{Const}, \quad \dot{\psi} = \text{Const}, \quad \dot{\phi} = \text{Const} \tag{4.3}$$

where the angle of precession ψ and the angle of pure rotation ϕ are ignorable coordinates, while the angle of rotation θ – an angle formed by the axis of the gyroscope and the vertical – is a position coordinate.

Obviously, stationary motions are not stable with respect to ignorable velocities: a small change in $\dot{q}_\alpha$ at $t = 0$ yields, as time progresses, an

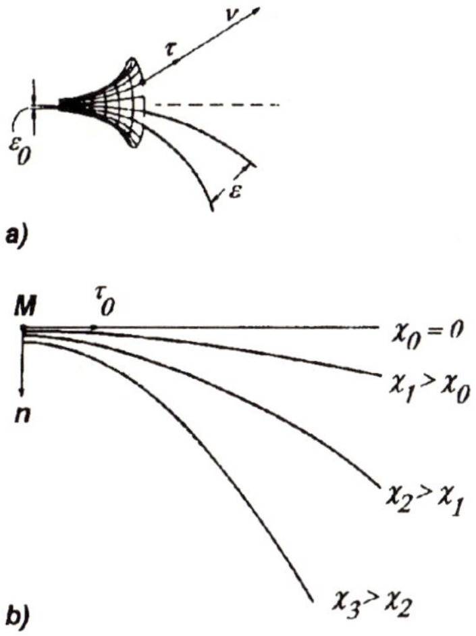

Fig. 4.1.

arbitrarily large change in the ignorable coordinates themselves. However, since this change increases linearly (but not exponentially), the motion is still considered predictable. In particular, the Liapunov exponents for stationary motions are zero [see (3.18)]:

$$\sigma = \lim_{d(0)\to 0,\ t\to\infty} \lim\left(\frac{1}{t}\right) \ln\frac{d(0)t}{d(0)} = 0 \tag{4.4}$$

However, in case of nonstationary motions, the ignorable coordinate can exhibit more sophisticated behaviors. In order to demonstrate this, let us consider an inertial motion of a particle M of unit mass on a smooth pseudosphere S having a constant negative curvature (Fig. 4.1a) :

$$G_0 = \text{Const} < 0 \tag{4.5}$$

Remembering that trajectories of inertial motions must be geodesics of S, we will compare two different trajectories assuming that initially they are parallel and that the distance between them, ε_0, is very small.

As shown in differential geometry, the distance between such geodesics will exponentially increase:

$$\varepsilon = \varepsilon_0 e^{\sqrt{-G_0}t}, \quad G_0 < 0 \tag{4.6}$$

Hence, no matter how small the initial distance ε_0, the current distance ε tends to infinity.

Let us assume now that the accuracy to which the initial conditions are known is characterized by L. It means that any two trajectories cannot be distinguished if the distance between them is less than L, i.e. if:

$$\varepsilon < L \tag{4.7}$$

The period during which the inequality (4.7), holds has the order:

$$\Delta t \sim \frac{1}{|\sqrt{-G_0}|} \, ln\frac{L}{\varepsilon_0} \tag{4.8}$$

However, for

$$t >> \Delta t \tag{4.9}$$

these two trajectories diverge such that they can be distinguished and must be considered as two different trajectories. Moreover, the distance between them tends to infinity even if ε_0 is small (but not infinitesimal). This is why the motion, once recorded, cannot be reproduced again (unless the initial conditions are known exactly), and consequently, it attains stochastic features. The Liapunov exponent for this motion is positive and constant:

$$\sigma = \lim_{t\to\infty,\ d(0)\to 0} \left(\frac{1}{t}\right) ln\frac{\varepsilon_0 e^{\sqrt{-G_0}t}}{\varepsilon_0} = \sqrt{-G_0} = \text{Const} > 0 \tag{4.10}$$

Let us introduce a system of coordinates at the surface S: the coordinate q_1 along the geodesic meridians, and the coordinate q_2 along the parallels. In differential geometry such a system is called semi-geodesical. The square of the distance between adjacent points on the pseudosphere is:

$$ds^2 = g_{11}\, dq_1^2 + 2g_{12}dq_1dq_2 + g_{22}dq_2^2 \tag{4.11}$$

where

$$g_{11} = 1,\ g_{12} = 0,\ g_{22} = -\frac{1}{G_0}e^{-2\sqrt{-G}q_1} \tag{4.12}$$

The Lagrangian for the inertial motion of the particle M on the pseudosphere is expressed via the coordinates and their temporal derivatives as:

$$L = g_{ij}\dot{q}_i\dot{q}_j = \dot{q}_1^2 - \frac{1}{G_0}e^{-2\sqrt{-G}q_1}\dot{q}_2^2 \tag{4.13}$$

and, consequently,

$$\frac{\partial L}{\partial q_2} = 0 \tag{4.14}$$

while

$$\frac{\partial L}{\partial q_1} \neq 0,\ \text{if}\ \ \dot{q}_2 \neq 0 \tag{4.15}$$

Hence, q_1 and q_2 play roles of position and ignorable coordinates, respectively. Therefore, an inertial motion of a particle on a pseudosphere is stable with respect to the position coordinate q_1, but it is unstable with respect to the ignorable coordinate. However, in contradistinction to the stationary motions considered above, here the instability is characterized by exponential growth of the ignorable coordinate, and this is why the motion becomes unpredictable. It can be shown that such a motion becomes stochastic (Arnold 1988).

Instability with respect to ignorable coordinates can be associated with orbital instability. Indeed, turning to the last example, one can represent the particle velocity $\boldsymbol{v}$ as the product:

$$\boldsymbol{v} = |v|\tau \tag{4.16}$$

In the course of the instability, the velocity magnitude $|v|$, and consequently, the total energy, remain unchanged, while all the changes affect only τ, i.e. the direction of motion. In other words, orbital instability leads to redistribution of the total energy between the coordinates, and it is characterized by positive Liapunov exponents.

4.2.2 Orbital Instability of Inertial Motions

The results described above were related to inertial motions of a particle on a smooth surface. However, they can be easily generalized to motions of any finite-degree-of-freedom mechanical system by using the concept of configuration space. Indeed, if the mechanical system has N generalized coordinates $q^1 (i = 1, 2, ..., N)$ and is characterized by the kinetic energy:

$$W = \alpha_{ij} \dot{q}^i \dot{q}^j \tag{4.17}$$

then the configuration space can be introduced as an N-dimensional space with the following metric tensor:

$$g_{ij} = a_{ij} \tag{4.18}$$

while the motion of the system is represented by the motion of the unit-mass particle in this configuration space.

In order to continue the analogy to the motion of the particle on a surface in actual space we will consider only two-dimensional subspaces of the N-dimensional configuration space, without loss of generality. Indeed, a motion which is instable in any such subspace, has to be qualified as an unstable in the entire configuration space.

Now the Gaussian curvature of a two-dimensional configuration subspace (q^1, q^2) follows from the Gauss formula:

$$G = \frac{1}{a_{11}a_{22} - a_{12}^2} \left[\left(\frac{\partial^2 a_{12}}{\partial q^1 \partial q^2} - \frac{1}{2} \frac{\partial^2 a_{11}}{\partial q^2 \partial q^2} - \frac{1}{2} \frac{\partial^2 a_{22}}{\partial q^1 \partial q^1} \right) \right]$$

$$-\Gamma^{\gamma}_{12}\Gamma^{\delta}_{12}\, a_{\gamma\delta} - \Gamma^{\alpha}_{11}\Gamma^{\beta}_{22} a_{\alpha\beta} \tag{4.19}$$

where the connection coefficients Γ^l_{sk} are expressed via the Christoffel symbols:

$$\Gamma^l_{sk} = \frac{1}{2} a^{lp} \left(\frac{\partial a_{sp}}{\partial q^k} + \frac{\partial a_{kp}}{\partial q^s} - \frac{\partial a_{sk}}{\partial q^p} \right) \tag{4.20}$$

while

$$a^{\alpha\beta}a_{\beta\gamma} = a^{\alpha}_{\gamma} = \begin{matrix} 0 \text{ if } & \alpha \neq \gamma \\ 1 \text{ if } & \alpha = \gamma \end{matrix} \tag{4.21}$$

Thus, the Gaussian curvature of these subspaces depends only on the coefficients a_{ij}, i.e. it is fully determined by the kinematical structure of the system [see (4.17)]. In case of inertial motions, the trajectories of the representative particle must be geodesics of the configuration space. Indeed, as follows from (4.15):

$$\frac{d\tau}{dt} = \frac{d\tau}{ds}\dot{s} = 0 \text{ if } \dot{\boldsymbol{v}} = 0, \text{ and } |\boldsymbol{v}| = |\dot{s}| = \text{Const} \neq 0 \tag{4.22}$$

where s is the arc coordinate along the particle trajectory:

$$ds = a_{ij}dq^i dq^j \tag{4.23}$$

But then:

$$\frac{d\tau}{ds} = 0 \tag{4.24}$$

which is the condition that the trajectory is geodesic.

If the Gaussian curvature which is uniquely defined by the parameters of the dynamical system a_{ij}, is negative:

$$G < 0 \tag{4.25}$$

then the trajectories of inertial motions of the system originated at close, but different points of the configuration space diverge exponentially from each other, and the motion becomes unpredictable and stochastic. Some examples of orbital instability in inertial motions are discussed by Zak (1985b).

As an example let us consider a two bar linkage, i.e., a system of two rigid rods AB and CD connected by an ideal hinge B and rotating about a vertical axis x normal to the plane ABC (Fig. 4.2). Setting

$$q^1 = \varphi_1, \; q^2 = \varphi_2 \tag{4.26}$$

one obtains their kinetic energy:

$$2W = a_{11}\dot{\varphi}_1^2 + a_{12}\dot{\varphi}_1\dot{\varphi}_2 + a_{22}\dot{\varphi}_2^2 \tag{4.27}$$

where

$$a_{11} = (I_1 + mr^2), \; a_{12} = mrl\cos(\varphi_2 - \varphi_1), \; a_{22} = I_2$$

while I_1 and I_2 are the moments of inertia of the rods AB and BC; with respect to the vertical axes passing through the points A and B, respectively, m is the mass of the rod BC, r is the length of the rod AB, and l is the distance between point B and the center of inertia of the rod BC.

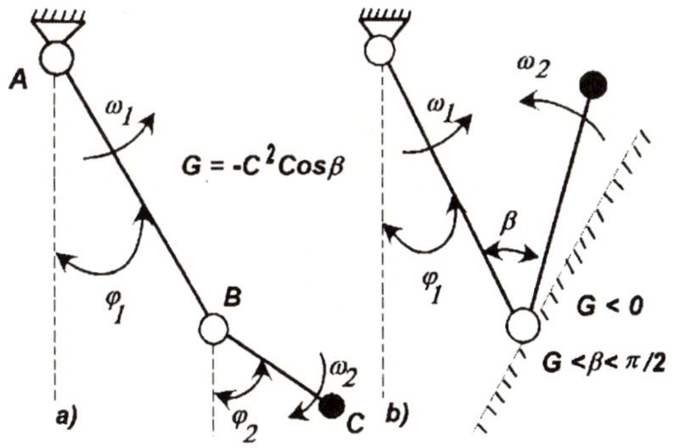

Fig. 4.2.

Taking into account that

$$\Gamma^1_{11} = \frac{-mrl\sin(\varphi_2 - \varphi_1)}{a^2} a_{12}, \ \Gamma^1_{11} = \frac{-mrl\sin(\varphi_2 - \varphi_1)}{a^2} a_{22} \tag{4.28}$$

$$\Gamma^2_{11} = \frac{-mrl\sin(\varphi_2 - \varphi_1)}{a^2} a_{11}, \ \Gamma^2_{22} = \frac{-mrl\sin(\varphi_2 - \varphi_1)}{a^2} a_{12} \tag{4.29}$$

$$\Gamma^1_{12} = \Gamma^2_{12} = 0, \ a^2 = a_{11}a_{22} - a^2_{12} \tag{4.30}$$

one arrives at the following expression for the Gaussian curvature of the configuration space [see (4.19)]:

$$G = \frac{mrl}{a^2} \left\{ 1 + \frac{[mrl\sin(\varphi_2 - \varphi_1)]^2}{a^2} \right\}^2 \cos(\varphi_2 - \varphi_1) \tag{4.31}$$

where

$$G < 0 \text{ if } \pi > |\varphi_2 - \varphi_1| > \frac{\pi}{2} \tag{4.32}$$

and

$$G > 0 \text{ if } |\varphi_2 - \varphi_1| < \frac{\pi}{2} \tag{4.33}$$

Let us eliminate the condition (4.33) by introducing a bar 0–0 normal to AB and rigidly attached to it such that the bar BC reflects from 0–0 (without loss of energy). In this case the condition (4.24) is satisfied, and therefore the inertial motions of two-bar linkage (with reflections from 0–0) are stochastic (Fig. 4.2,b).

Next we will consider a symmetric rigid body rotating about its center of gravity (Fig. 4.3). Determining its position by Euler's angles

$$\theta = q^1, \ \psi = q^2, \ \phi = q^3 \tag{4.34}$$

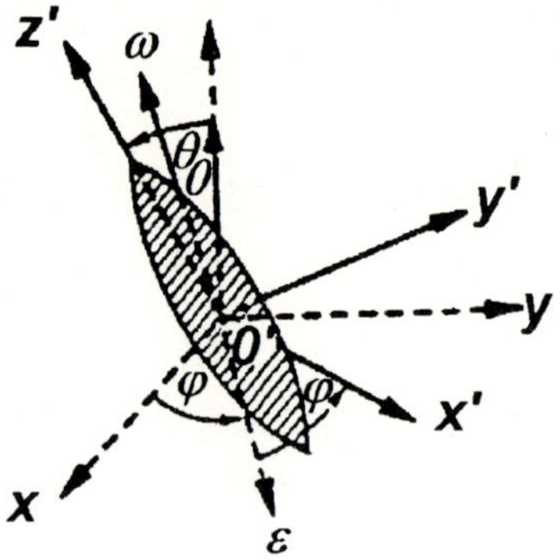

Fig. 4.3.

one obtains the following expression for the kinetic energy:

$$W = \frac{1}{2}\left[A(\dot{\theta}^2 + \dot{\psi}^2 \sin^2\theta) + C(\dot{\phi} + \dot{\psi}\cos\theta)^2\right] \tag{4.35}$$

in which A and C are the axial moments of inertia. Then the metric coefficients, the Christoffel symbols and the Gaussian curvature of two dimensional subspaces are, respectively:

$$a_{11} = A,\ a_{12} = 0,\ a_{22} = A\sin^2\theta + C\cos^2\theta$$

$$a_{13} = 0,\ a_{23} = C\cos\theta,\ a_{33} = 0 \tag{4.36}$$

$$\Gamma^1_{22} = \frac{C-A}{A}\sin\theta\cos\theta,\ \Gamma^1_{23} = \frac{C}{2A}\sin\theta$$

$$\Gamma^2_{21} = \frac{2A-C}{2A}\cot\theta,\ \Gamma^2_{31} = -\frac{C}{2A\sin\theta}$$

$$\Gamma^2_{21} = -\frac{1}{2\sin\theta}\left(\frac{A-C}{A}\cos^2\theta + 1\right),\ \Gamma^3_{31} = \frac{C}{2A}\cot\theta \tag{4.37}$$

$$G_{(12)} = \frac{1}{A(A\sin^2\theta + C\cos^2\theta}[2(A-C)\cos 2\theta$$

$$+\left(\frac{2A-C}{2A}\cot\theta\right)^2 (A\sin^2\theta + C\cos^2\theta)] \tag{4.38}$$

$$G_{(13)} = 0,\ G_{(23)} = 0 \tag{4.39}$$

Since the motion is inertial ($Q_1 = 0$) the condition of orbital instability can be presented as:

$$G_{(12)} < 0 \tag{4.40}$$

where $G_{(12)}$ is the Gaussian curvature in the subspace θ, ψ.

Assuming, for simplicity, that $2A = C$ one reduces (4.40) to the following:

$$\cos 2\theta > 0 \tag{4.41}$$

Thus, any motion in the subspace will be stochastic if θ is bounded:

$$0 < \theta < \frac{\pi}{4} \tag{4.42}$$

4.2.3 Orbital Instability of Potential Motions

Turning back to the motion of the particle M on a smooth pseudosphere (Fig. 4.1), let us depart from inertial motions and introduce a force $\mathbf{F}$ acting on this particle. For noninertial motions ($\mathbf{F} \neq 0$) the trajectories of the particle will not be geodesics, while the rate of their deviation from geodesics is characterized by the geodesic curvature χ. It is obvious that this curvature must depend on the forces $\mathbf{F}$:

$$\chi = \chi(\mathbf{F}) \tag{4.43}$$

Synge (1926) has shown that if the force $\mathbf{F}$ is potential:

$$\mathbf{F} = -\nabla\Pi \tag{4.44}$$

where Π is the potential energy, then the condition (4.26) is replaced by the following:

$$G_0 + 3\chi^2 + \frac{1}{W}\left(\frac{\partial^2\Pi}{\partial q^i \partial q^j} - \Gamma^k_{ij}\frac{\partial\Pi}{\partial q^k}\right)n^i n^j < 0;\; i,j = 1,2 \tag{4.45}$$

Here Γ^k_{ij} are defined by (4.21), and n^i are the contravariant components of the unit normal $\mathbf{n}$ to the trajectory.

The geodesical curvature χ in (4.45) can be expressed via the potential force $\mathbf{F}$:

$$\chi = \frac{\mathbf{F}\cdot\mathbf{n}}{2W} = -\frac{\nabla\Pi\cdot\mathbf{n}}{2W} \tag{4.46}$$

As follows from (4.45) and (4.46), the condition (4.45) reduces to (4.26) if $\mathbf{F} = 0$.

Suppose for example, that the following elastic force:

$$F = -\alpha^2\varepsilon, \alpha^2 = \text{Const} \tag{4.47}$$

proportional to the normal deviation ε from the geodesic trajectory is applied to the particle M moving on the smooth pseudosphere.

If the initial velocity is directed along one of the meridians (which are all geodesics), the unperturbed motion will be inertial, and its trajectory will coincide with this meridian since there $\varepsilon = 0$, and therefore, $\mathbf{F} = 0$.

In order to verify the orbital stability of this motion, let us turn to the criterion (4.45). Since:

$$\chi = 0, \text{ and } \frac{\partial \Pi}{\partial q^k} = F^k = 0 \tag{4.48}$$

for the unperturbed motion, one obtains the condition for orbital stability:

$$G_0 + \frac{\alpha^2}{2W} > 0, \text{ i.e. } \alpha^2 < -2WG, \quad G < 0 \tag{4.49}$$

where

$$W = \frac{1}{2} m v_0^2 \tag{4.50}$$

As in the case of inertial motions, the inequality:

$$\alpha^2 < -2WG_0 \tag{4.51}$$

leads to unpredictable (stochastic) motions which are characterized by the positive Liapunov exponent:

$$\sigma = \sqrt{G_0 - \frac{\alpha^2}{2W}} = \text{Const} > 0 \tag{4.52}$$

For pure inertial motions ($\alpha = 0$), (4.52) reduces to (4.5).

After the discovery of chaotic attractors, the stochastic motions which are generated by the instability and are characterized by positive Liapunov exponents, are called chaotic. Hence, the inequalities (4.26) and (4.51) can be associated with criteria of chaos: if the left hand part in (4.51) is bounded away from zero by a negative number in all the configuration space where the motion can occur, then the motion will be chaotic, and its positive Liapunov exponent will be:

$$\sigma \geq B^2 \tag{4.53}$$

Unfortunately, this criterion is too "strong" to be of practical significance: it is sufficient, but not necessary. Indeed, this criterion assumes that not only global, but also the local Liapunov exponents are positive in any point of the configuration space. At the same time, for many chaotic motions, local Liapunov exponents in certain domains of the configuration space are all negative, or zero, although some of the global exponents are still positive.

4.2.4 General Case

Following Synge, the results for the orbital instability of inertial and potential motions for a system of material points can be generalized to arbitrary motions. Based upon the concept of configuration space introduced in (4.18) and (4.19), the velocity vector of a representative point in its contravariant form is:

$$v^r = \dot{q}^r \tag{4.54}$$

its magnitude being:

$$v = (\alpha_{mn}\dot{q}^m\dot{q}^n)^{1/2} = \dot{s} = (2W)^{1/2} \tag{4.55}$$

The acceleration vector is defined as the contravariant time-flux of the velocity vector:

$$\tilde{v}^r = \dot{v}^r + \Gamma^r_{mn}v^m v^n = \ddot{q}^r + \Gamma^r_{mn}\dot{q}^m\dot{q}^n \tag{4.56}$$

where the connection coefficients Γ^r_{mn} are expressed via the Christoffel symbols by (4.21).

It should be noted that the time-differentiation in (4.56) takes into account the variable metric of the configuration space ($g_{ij} = \alpha_{ij}$) captured by the connection coefficients.

Since the motion of a system of material points in the configuration space with the metric (4.19) is represented by a unit-mass point, the momentum equation follows from the second Newton's law:

$$\ddot{q}^r + \Gamma^r_{mn}\dot{q}^m\dot{q}^n = Q^r \tag{4.57}$$

where Q^r is the force applied to the point. This force is expressed via the generalized forces Q_β as:

$$Q^\alpha = a^{\alpha\beta}Q_\beta \tag{4.58}$$

Let q^r be the coordinates of the representative point M moving along an undisturbed natural trajectory C, and $(q^r + \eta^r)$ the coordinates of the corresponding (simultaneous) point M^* of the disturbed natural trajectory C, while η^r is an infinitesimal disturbance vector. The condition for stability of the motion is that the magnitude of the disturbance vector should remain permanently small.

Let us substitute $(q^r + \eta^r)$ in (4.57) and obtain:

$$\ddot{q}^r + \ddot{\eta}^r + \overset{*}{\Gamma}{}^r_{mn}\,(\dot{q}^m + \dot{\eta}^m)(\dot{q}^n + \dot{\eta}^n) - (Q^r)^* = 0 \tag{4.59}$$

where the asterisk indicates quantities to be calculated at M^*. Expanding these quantities, and assuming that the force Q^r depend only upon the coordinates, but not the velocities, one has, after retaining only the first powers of small quantities:

$$\ddot{\eta}^r + 2\Gamma^r_{mn}\dot{\eta}^m\dot{q}^n + \frac{\partial\Gamma^r_{mn}}{\partial q^s}\eta^s\dot{q}^m\dot{q}^n - \frac{\partial Q^r}{\partial q^s}\eta^s = 0 \tag{4.60}$$

Let us turn now to the invariant quantities, i.e., the vectors of velocity and acceleration of the disturbance η^r:

$$\tilde{\eta}^r = \dot{\eta}^r + \Gamma^r_{mn}\eta^m\dot{q}^n \tag{4.61}$$

$$\tilde{\tilde{\eta}}^r = \ddot{\eta}^r + 2\Gamma^r_{mn}\dot{\eta}^m\dot{q}^n + \frac{\partial\Gamma^r_{mn}}{\partial q^s}\eta^m\dot{q}^n\dot{q}^s + \Gamma^r_{mn}\eta^m\ddot{q}^n + \Gamma^r_{s\ell}\Gamma^s_{mn}\eta^m\dot{q}^n\dot{q}^\ell \tag{4.62}$$

Substituting for $\ddot{q}_n$ from (4.57), we find:

$$\ddot{\eta}^r + 2\Gamma^r_{mn}\dot{\eta}^m\dot{q}^n = \tilde{\tilde{\eta}}^r - (\frac{\partial\Gamma^r_{ms}}{\partial q^n} - \Gamma^r_{s\ell}\Gamma^\ell_{mn} + \Gamma^r_{n\ell}\Gamma^\ell_{ms})\eta^s\dot{q}^m\dot{q}^n - \Gamma^r_{mn}\eta^m Q_n \tag{4.63}$$

Substitution in (4.60) gives:

$$\tilde{\tilde{\eta}}^r + G^r_{msn}\eta^s\dot{q}^m\dot{q}^n - Q^r_s\eta^s = 0 \tag{4.64}$$

where

$$G^r_{msn} = \frac{\partial\Gamma^r_{mn}}{\partial q^s} - \frac{\partial\Gamma^r_{ms}}{\partial q^n} + \Gamma^\ell_{mn}\Gamma^r_{s\ell} - \Gamma^\ell_{ms}\Gamma^r_{n\ell} \tag{4.65}$$

the mixed curvature tensor of the manifold of configurations with the metric (4.18–4.19). The particular form of this tensor for two-dimensional case in the Gaussian curvature being introduced by (4.20), and

$$Q^r_s = \frac{\partial Q^r}{\partial q^s} + \Gamma^r_{sn}Q^n \tag{4.66}$$

Equation (4.64) describes the time evolution of the disturbance vector η^r. It should be emphasized the invariant nature of this equation: it contains only geometrical (the curvature tensor G) and physical parameters which characterize the original system and its undisturbed motion.

Introducing a unit disturbance vector μ^r co-directional with η^r, so that:

$$\eta^r = \eta\mu^r \tag{4.67}$$

and

$$a_{mn}\mu^m\mu^n = 1 \tag{4.68}$$

Then

$$\tilde{\eta}^r = \dot{\eta}\mu^r + \eta\tilde{\mu}^r \tag{4.69}$$

and

$$\tilde{\tilde{\eta}}^r = \eta\mu^r + 2\eta\tilde{\mu}^r + \eta\tilde{\tilde{\mu}}^r \tag{4.70}$$

Therefore

$$a_{r\ell}\tilde{\tilde{\eta}}^r\mu^\ell = \ddot{\eta} + 2\dot{\eta}a_{r\ell}\tilde{\mu}^r\mu^\ell + \eta a_{r\ell}\tilde{\tilde{\mu}}^r\mu^\ell \tag{4.71}$$

But from (4.68) we obtain:

$$a_{mn}\,\tilde{\mu}^m\mu^n = 0 \tag{4.72}$$

and hence

$$a_{mn}\tilde{\tilde{\mu}}^m\mu^n + a_{mn}\tilde{\mu}^m\tilde{\mu}^n = 0 \tag{4.73}$$

Thus, (4.71) may be written:

$$a_{r\ell}\tilde{\tilde{\eta}}^r\mu^\ell = \ddot{\eta} - \eta a_{r\ell}\tilde{\mu}^r\tilde{\mu}^l \tag{4.74}$$

Now if we multiply (4.64) by μ^r and sum as indicated, we have by (4.74):

$$\ddot{\eta} - \eta a_{r\ell}\tilde{\mu}^r\tilde{\eta}^\ell + G_{rmsn}\mu^r\dot{q}^m\eta^s\dot{q}^n - Q_{rs}\mu^r\eta^s = 0 \tag{4.75}$$

which may be written:

$$\ddot{\eta} + \eta(G_{mns\ell}\mu^m\dot{q}^n\mu^s\dot{q}^\ell - \tilde{\mu}^2 - Q_{mn}\mu^m\mu^n) = 0 \tag{4.76}$$

where $G_{mns\ell}$ is the curvature tensor (4.65) expressed in a covariant form, and $\tilde{\mu}$ is the magnitude of the vector $\tilde{\mu}_r$.

Hence, we arrive at the invariant equation for the magnitude of the disturbance η. The invariant $G_{mns\ell}\mu^m\dot{q}^n\mu^s\dot{q}^\ell$ is equal to the Riemannian curvature of the manifold of configurations corresponding to the directions μ^r and $\dot{q}^r$, multiplied by a positive factor. Based upon (4.76), we can state the following result: if the Riemannian curvature of the manifold of configurations corresponding to every two-space element $x^m x^n$ containing the direction of the given trajectory is bounded away from zero by a constant negative value, and $Q_{mn}x^m x^n$ is bounded away from zero by a constant positive value in all the domains of the configuration space where the motion can occur, then the motion will be exponentially unstable; since this instability persists, the motion will attain stochastic features (as in the case of the inertial or potential motion of a particle on a smooth pseudosphere), and therefore, it will become chaotic. Actually the condition (4.45) which was formulated earlier without a proof, follows directly from (4.76) (Zak 1994a).

Obviously, the persistency of the instability in (4.76) can occur only due to a contribution of the exponential growth of the ignorable coordinates into the total magnitude of the disturbance vector η. For instance, in the case of inertial motion of the particle M on a smooth pseudosphere, the disturbance vector can be represented by the components ε and ν which are co-directional and normal to the unperturbed (geodesic) trajectory. The component ν corresponds to the ignorable coordinate, and its evolution is described by (4.76) which reduces to:

$$\ddot{\nu} + 2W_0 G_0 \nu = 0 \tag{4.77}$$

The exponential instability of ν when $G_0 = \text{Const} < 0$ leads to chaos. At the same time, the position coordinate ε is eliminated from (4.77) and it can be found from the energy conservation:

$$\varepsilon = \varepsilon_0 \frac{\dot{\varepsilon}}{\dot{S}_0} \tag{4.78}$$

where $\dot{S}_0$ and ε_0 are the initial conditions at $t = 0$ for the motion velocity along the trajectory, and the position coordinate of the disturbance vector, respectively. In spite of some limitations of the results described above (the conditions for chaos are sufficient, but not necessary, the forces Q^r depend only upon coordinates, but not upon velocities), they nevertheless elucidate physical origin of orbital instability, chaos, and consequently, of unpredictability of motions in classical dynamics.

4.3 Hadamard's Instability

4.3.1 General Remarks

The results presented in the previous section can be applied to distributed systems after a discretization technique which reduces them to a finite-dimensional systems. For instance, as noticed by Arnold (1988), an inviscid stationary flow with a smooth velocity field:

$$v_x = A \sin z + C \cos y, \; v_y = B \sin x + A \cos z, \; v_z = C \sin y + B \cos x \tag{4.79}$$

has chaotic trajectories $x(t)$, $y(t)$, $z(t)$ of fluid particles (Lagrangian turbulence) due to negative curvature of the configuration space which is obtained as a finite-dimensional approximation of a continuum. However, there are some special types of instability in distributed systems which can be lost in the course of the discretization, and they will be focused in this section.

As noticed in the previous section, the long-term instability which may lead to chaos, is associated with the orbital instability, i.e. with the instability of ignorable coordinates. However, in distributed systems described by partial differential equations, there is another possibility for long-term instability which is associated with the decrease of scale of motions, i.e. with the growth of spatial derivatives of the system parameters. In mathematical terms it means a failure of differentiability of the solutions to the corresponding governing equations. However, an unlimited growth of spatial derivatives must be consistent with the boundedness of energy. Indeed, the stresses in continuous media depend not upon displacements or velocities, but upon their gradients, i.e. upon their space derivatives. Hence, we have to find such situations when an unlimited growth of these derivative does not lead to unbounded stresses.

Turning to geometry of displacements and their gradients in continua, let us introduce the displacement vector:

$$\mathbf{u} = \mathbf{r} - \mathbf{r}_0 \tag{4.80}$$

where $\mathbf{r}_0$ and $\mathbf{r}$ are the radii-vectors of the same particle before and after deformation, respectively. In elastic bodies, the stress tensor depends upon the displacement gradient $\nabla\mathbf{u}$ via the strain tensor ε:

$$\varepsilon = \frac{1}{2}[\nabla\mathbf{u} + (\nabla\mathbf{u})^T + \mathbf{u}\cdot(\nabla\mathbf{u})^T] = \frac{1}{2}[\nabla\mathbf{r}\cdot(\nabla\mathbf{r})^T - \mathring{g}] \tag{4.81}$$

where $\mathring{g}$ is the unit (the initial state) tensor, while the current state metric tensor is defined as:

$$g = 2\varepsilon + \mathring{g} \tag{4.82}$$

The tensor-gradient $\nabla\mathbf{r}$ in (4.81) can be decomposed as:

$$\nabla\mathbf{r} = CB \tag{4.83}$$

where C is a symmetric tensor:

$$C = +[\nabla\mathbf{r}\cdot(\nabla\mathbf{r})]^{1/2} \tag{4.84}$$

and B is an orthogonal tensor:

$$B = +[\nabla\mathbf{r}\cdot(\nabla\mathbf{r})^T]^{1/2}\cdot(\nabla\mathbf{r})^T = (B^{-1}),\ \det B = 1 \tag{4.85}$$

As follows from (4.84), the strain tensor

$$\varepsilon = \frac{1}{2}(c^2 - \mathring{g}) \tag{4.86}$$

and consequently, the stress tensor depends only upon the symmetric part of the tensor-gradient $\nabla\mathbf{r}$, and does not depend upon its orthogonal component B which corresponds to rigid rotations of elementary volumes. However, indirectly an unlimited growth of these rotations can lead to unbounded stresses in three-dimensional elastic bodies. Indeed, as follows from the identity:

$$\nabla\times\nabla\mathbf{r} = 0 \tag{4.87}$$

the components of the tensor-gradient $\nabla\mathbf{r}$ must satisfy six additional constraints which are called the compatibility equations. Loosely speaking,

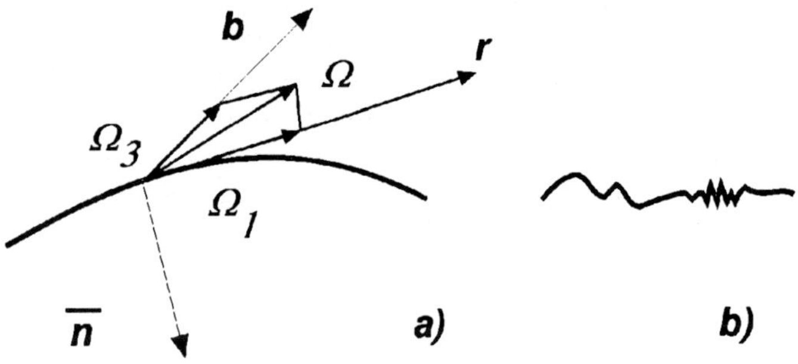

Fig. 4.4.

they follow from the requirement that after deformations the continuum should not have any "holes" or "cracks." In geometrical terms, (4.87) represent the fact that after deformations, the actual space remain Euclidean, i.e. the curvature tensor is zero:

$$R = 0 \tag{4.88}$$

However, six constraints imposed upon the tensor-gradient $\nabla \mathbf{r}$ by (4.87), or (4.88) are also not independent. Indeed, according to another identity:

$$\nabla \cdot \nabla \times \nabla \mathbf{r} \equiv 0 \tag{4.89}$$

which holds even if

$$\nabla \times \nabla \mathbf{r} \neq 0 \tag{4.90}$$

and which is equivalent to three scalar equations, only three of the six constraints (4.87) are true independent. In geometrical terms, (4.89) can be associated with the Bianchi identities (Fluge 1962).

Thus, nine components of the vector-gradient $\nabla \mathbf{r}$ must satisfy three independent compatibility equations, and therefore, if all six components of the stress tensor ε are given, then the remaining three components of $\nabla \mathbf{r}$ and consequently, all the rigid rotations, will be uniquely defined. This means that in isotropic three-dimensional elastic bodies, an unlimited decrease of scale of motions would lead to unbounded stresses which is physically impossible.

Let us turn to one-dimensional continua (filaments). In this case, rigid rotations define the external geometry of the model [the rotations about the binormal to the filament correspond to the first curvature, and the rotations about the tangent to the filament correspond to the second curvature, or twist, (Fig. 4.4a)] and they do not depend upon the elongations of the curve which define the stress. Indeed, let us introduce the filament equation in the form:

$$\mathbf{r} = \mathbf{r}(\psi, t), |\frac{\partial \mathbf{r}}{\partial \psi}| = 1 \tag{4.91}$$

where ψ plays the role of an Eulerian coordinate. Then the motions associated with changes of the internal geometry, and therefore, the stresses are described by the function:

$$\psi = \psi(\mathrm{s}, \mathrm{t}) \tag{4.92}$$

where s is a Lagrangian coordinate of individual particle. At the same time, the curvatures of the filament configurations can be expressed as:

$$\Omega_3 = \left|\frac{\partial^2 \mathbf{r}}{\partial \psi^2}\right|, \quad \Omega_1 = \frac{\left(\frac{\partial \mathbf{r}}{\partial \psi} \times \frac{\partial^2 \mathbf{r}}{\partial \psi^2}\right) \cdot \frac{\partial^3 \mathbf{r}}{\partial \psi^3}}{\left|\frac{\partial^2 \mathbf{r}}{\partial \psi^2}\right|} \tag{4.93}$$

Consequently, both curvatures are independent upon the internal geometry characterized by (4.92), and in particular, upon the stress defined by the derivative. This means that unlimited growth of the curvature may not cause stress at all, and therefore, the instability in the form of unlimited decrease of scale of motions is possible (Fig. 4.4b).

The situation becomes more complicated in two-dimensional continua (films, membranes). Here the internal geometry is defined by two dimensional versions of (4.80–4.87), while the external geometry is described by the coefficients of the second fundamental form:

$$b_{ij} = \frac{\partial^2 \mathbf{r}}{\partial \psi^i \partial \psi^j} \cdot \mathbf{n}, \quad i, j, = 1, 2 \tag{4.94}$$

where ψ^i are coordinates on the surface, and $\mathbf{n}$ is the unit normal to the surface.

However, these coefficients are not independent: they are coupled with the strains by the compatibility equations:

$$b_{11} b_{22} - b_{12}^2 = \Gamma_{12}^{v} \Gamma_{12}^{\delta} g_{\alpha\beta} - \frac{1}{2} \frac{\partial^2 g_{11}}{\partial \psi^{(2)^2}} + \frac{\partial^2 g_{12}}{\partial \psi^{(1)} \partial \psi^{(2)}}$$
$$-\frac{1}{2} \frac{\partial^2 g_{22}}{\partial \psi^{(1)^2}}, (v, \delta, \alpha, \beta = 1, 2) \tag{4.95}$$

$$\frac{\partial b_{il}}{\partial \psi^{(2)}} - \frac{\partial b_{i2}}{\partial \psi^{(1)}} = \Gamma_{i2}^1 b_{11}^i - \Gamma_{il}^2 b_{22} + (\Gamma_{i2}^2 - \Gamma_{il}^1) b_{12}, (i = 1, 2) \tag{4.96}$$

where

$$g_{11} = 1 + 2\varepsilon_{11}, \; g_{12} = 2\varepsilon_{12}, \; g_{22} = \overset{\circ}{g}_{22} + 2\varepsilon_{22} \tag{4.97}$$

The two dimensional Christoffel symbols are:

$$\Gamma^n_{ij} = \frac{1}{2} g^{n\ell} \left(\frac{\partial g_{\ell i}}{\partial \psi^{(j)}} + \frac{\partial g_{\ell j}}{\partial \psi^{(i)}} - \frac{\partial g_{ij}}{\partial \psi^{(\ell)}} \right), \; (n, i, j = 1, 2) \tag{4.98}$$

while

$$||g_{ij}|| = ||g^{ij}||^{-1} \tag{4.99}$$

Hence, in general, three coefficients b_{ij} are defined by the strains ε_{ij} from the equations (4.95–4.96), and consequently, change in b_{ij} affects the strains ε_{ij}.

Nevertheless, there are situations when the unlimited growth of curvature may not effect the stress at all. In order to describe this case, recall that on a surface with negative or zero Gaussian curvature

$$G = \frac{b_{11}b_{22} - b_{12}}{g_{11}g_{22} - g_{12}^2} \tag{4.100}$$

there exists a family of asymptotic lines where the second fundamental form is equal to zero:

$$b_{11} \tan^2 \phi + 2b_{12} \tan \phi + b_{22} = 0 \tag{4.101}$$

while the angle ϕ between an asymptotic line and the coordinate line ψ_1 is found as:

$$\tan \phi = -\frac{b_{12}}{b_{11}} \pm \frac{\sqrt{b_{11}^2 - b_{11}b_{22}}}{b_{11}}, \quad (b_{11} \neq 0) \tag{4.102}$$

Selecting the coordinate $\psi_1$as an asymptotic line, one obtains:

$$\tan \phi = 0 \tag{4.103}$$

and, as follows from (4.101) and (4.102):

$$b_{22} = 0 \tag{4.104}$$

Now it is obvious that along the asymptotic line the curvature b_{11} can be selected arbitrarily without affecting the parameters of the internal geometry g_{ij}, and consequently, the stress. Indeed, since $b_{22} = 0$, b_{11} is eliminated from (4.95). In addition to this, as follows from (4.96):

$$\frac{\partial b_{11}}{\partial \psi^{(2)}} - \frac{\partial b_{12}}{\partial \psi^{(1)}} = \Gamma^1_{12} b_{11} + (\Gamma^2_{11} - \Gamma^1_{11})\, b_{12} \tag{4.105}$$

$$\frac{\partial b_{12}}{\partial \psi^{(2)}} = \Gamma^1_{22} b_{11} + (\Gamma^2_{22} - \Gamma^1_{21})\, b_{12} \tag{4.106}$$

The derivative $\partial b_{11}/\partial \psi_1$ is not defined, i.e. that asymptotic line ϕ_1 coincides with the characteristic of the partial differential equations (4.105–4.106).

This means that the curvature b_{11} can be chosen arbitrarily along the asymptotic lines of the surface without effecting any parameters of the film including stresses. In other words, an unlimited growth of the curvature b_{11} may be consistent with the unboundedness of stresses and it can be associated with the formation of wrinkles along the asymptotic lines (Zak 1979a,b; 1994a).

So far we were concerned with elastic continua. Turning to fluids, one should recall that their stresses depend only upon the velocities, but not upon the displacements. This is why an unlimited growth of any component of the displacement vector (4.80), or of the tensor-gradient (4.83) is consistent with the unboundedness of stresses, and it can be associated with the Lagrangian turbulence.

In terms of velocities, the situation is different. In order to demonstrate this, recall that in viscous fluid the stress tensor depends upon the velocity gradient $\nabla \boldsymbol{v}$ via the time derivative of the strain tensor (4.80). The velocity gradient $\nabla \boldsymbol{v}$ has the same type of structure as the vector-gradient $\nabla \mathbf{r}$: it can be decomposed into a symmetric tensor of the rate of strain:

$$\dot{\varepsilon} = \frac{1}{2}[\nabla \boldsymbol{v} + (\nabla \boldsymbol{v})^T] \tag{4.107}$$

and an anti-symmetric tensor:

$$\omega = \frac{1}{2}[\nabla \boldsymbol{v} - (\nabla \boldsymbol{v})^T] \tag{4.108}$$

which is equivalent to the vector of vortex:

$$\omega = \frac{1}{2}\text{curl } \boldsymbol{v} = \frac{1}{2}\nabla \times \boldsymbol{v} \tag{4.109}$$

while

$$\nabla \boldsymbol{v} = \dot{\varepsilon} + \omega \tag{4.110}$$

Since

$$\nabla \times \nabla \boldsymbol{v} \equiv 0 \text{ and } \nabla \cdot \nabla \times \nabla \boldsymbol{v} \equiv 0 \text{ (even if) } \nabla \times \nabla \boldsymbol{v} \neq 0, \tag{4.111}$$

one comes to the same conclusion as in the case of the vector-gradient $\nabla \mathbf{r}$ [see (4.87) and (4.89)]: nine components of the tensors $\dot{\varepsilon}$ and ω are coupled by three compatibility equations. Hence, six components of the rate of strain tensor $\dot{\varepsilon}$ uniquely define the velocity gradient, and for this reason, an unlimited growth of the vortices in viscous fluids would lead to unlimited growth of stresses.

The situation becomes different in inviscid fluids where stress is defined only by a scalar – the divergency $\nabla \cdot \boldsymbol{v}$. But since any velocity field can be uniquely defined based upon two independent components of its gradient $\nabla \boldsymbol{v}$, which are the divergency $\nabla \cdot \boldsymbol{v}$ and the vorticity $\nabla \times \boldsymbol{v}$, one concludes that an unlimited growth of vorticity in inviscid fluid may not lead to unbounded stresses. This conclusion can be loosely applied to motions of viscous fluids characterized by high Reynolds number when viscous stress are ignorable in comparison to the inertia forces. In this case an "unlimited" growth of vortices can be associated with turbulence.

Thus, in this section we have analyzed a possibility "in principle" of an unlimited decrease of scale of motions in continua from the viewpoint of a consistency of this type of instability with the boundedness of stresses and energy. This means that if such an instability exists, it can be found only in one-or two-dimensional elastic models, or in fluid motions with high Reynolds number.

4.3.2 Failure of Hyperbolicity in Distributed Systems

Mathematical models of continua are based on the assumption that the functions describing their states can be differentiated "as many times as necessary" at any point exclusive of some special surfaces of discontinuities simulating shock waves or coinciding with the characteristics of the governing equations. In other words, these functions must be at least piece-wise differentiable. From the physical viewpoint it means that any point as a center of mass of an infinitesimal volume represents all the properties of this volume. Obviously, the assumption about smoothness of the functions allows us to use the mathematical technique of differential equations.

However, this artificial mathematical limitation follows neither from the principles of mechanics nor from the definition of a continuum. The price paid for such a mathematical convenience is instability (in the class of smooth functions) of the solutions to the corresponding governing equations in some regions of the parameters. This instability is characterized by unlimited decrease of the scale of the motions, in the course of which the derivatives of the corresponding functions tend to infinity although the functions themselves remain finite. In other words, the solution tends to "go out" from the class of differentiable functions.

What is the physical interpretation of such an instability? Does it mean that post-instability states do not have any physical meaning? Definitely not. First of all, it is important to emphasize that the concepts of stability or instability are associated with the corresponding class of functions. Hence, the instability mentioned above shows that the corresponding state of a continuum cannot be properly simulated by smooth functions.

Indeed, the class of possible motions of continua is broader than the class of differential functions: for instance, in a turbulent motion two particles located within the same infinitesimal volume can have significantly different velocities. The same phenomena can occur in thin films which are covered by wrinkles as a result of a local compression.

Thus, the instabilities mentioned above are caused by the contradiction between the artificially imposed mathematical limitation about smoothness of the functions describing media motions and physical nature of these motions. Consequently, giving up the assumption about differentiability, i.e., the enlarging class of functions in which solutions are sought is one of the possible ways of mathematical simulation of post-instability behavior of continua. Obviously, a new mathematical technique should be developed to describe the solutions which are not necessarily differentiable.

Most of the instability phenomena leading to unlimited decreasing of the scale of continua motions are associated with the failure of hyperbolicity of the corresponding governing equations, i.e., with the appearance of imaginary characteristic speeds (Zak 1982b,c).

In order to illustrate this, we will start with the governing equations of motion of elastic bodies in the following form:

$$\rho\frac{\partial^2 u_i}{\partial t^2} = \sum_{j=1}^{3}\frac{\partial}{\partial x_j}\left[\frac{\partial \Pi}{\partial(\frac{\partial u_i}{\partial x_j})}\right] + F_i, \; i = 1,2,3 \tag{4.112}$$

where u_i are the displacements, Π is the potential energy of strains, ρ is the density, F_i are the external forces, and x_i are the material coordinates, posing the initial value problem:

$$u_i^\circ = u_i\,|_{t=0} \begin{array}{ll} = \frac{1}{\lambda_0^2}\sin\lambda_0 x_1 & \text{if } |x_1| \leq x_0 \\ = 0 & \text{if } |x_i| > x_0, \; i = 1,2,3, \end{array} \tag{4.113}$$

$$\left(\frac{\partial u}{\partial t}\right)_{t=0} = 0, \quad i = 1,2,3 \tag{4.114}$$

with the parameter λ_0 can be made as large as necessary, i.e.,

$$\lambda_0 \to \infty \tag{4.115}$$

The region of the initial disturbance can be arbitrarily shrunk, i.e.,

$$|x_0| \to 0 \tag{4.116}$$

Consequently, the initial disturbances u_i and their first derivatives $\partial u_i/\partial x_i$ can be made as small as necessary. It means that for the corresponding infinitesimal period of time Δt_0 the equations (4.112) can be linearized and the solution subject to the initial conditions (4.113) can be sought in the form: $u_j = (x_i, t)$, i.e.,

$$\rho_0\frac{\partial^2 u_i}{\partial t^2} = \sum_{j=1}^{3} a_{ij}\frac{\partial^2 u_j}{\partial x_1^2}, \text{ while } \frac{\partial u_j}{\partial x_j}\Big|_{j\neq 1} \equiv 0 \tag{4.117}$$

where

$$a_{ij} = \frac{\partial^2 \Pi}{\partial(\frac{\partial u_i}{\partial x_1})\partial(\frac{\partial u_j}{\partial x_1})}\Bigg|_{\frac{\partial u_i}{\partial x_i},\,\frac{\partial u_j}{\partial x_1} = 0} \tag{4.118}$$

Let us assume that one of the eigenvalues of the matrix a_{ij} is negative:

$$\lambda_1 < 0 \tag{4.119}$$

Then the solution to the equation (4.117) will contain the term:

$$\frac{1}{\lambda_0^2} e^{\lambda_0 \sqrt{\frac{|\lambda_1|}{\rho_0}} \Delta t} \sin \lambda_0 x \tag{4.120}$$

which tends to infinity if $\lambda_0 \to \infty$ within an arbitrary short period of time Δt_0 and within an infinitesimal volume around the point x_i. Hence, one arrives at the following situation:

$$|u_i| \to \infty \tag{4.121}$$

in spite of the fact that

$$|u_i| \,|_{t=0} \to 0 \tag{4.122}$$

However, strictly speaking, because of utilization of the governing equation (4.112) in a linearized form, the condition (4.121) must be weakened:

$$|u_i| \neq 0 \text{ if } |u_i| \,|_{t=0} \to = 0 \tag{4.123}$$

The formula (4.123) shows that the appearance of negative eigenvalues of the matrix (4.118), and consequently, imaginary characteristic roots of the governing equation (4.112) (failure of its hyperbolicity) leads to the violation of a continuous dependence between the initial and transient disturbances during an arbitrary short period of time and within an arbitrarily selected volume. This type of instability was first observed by Hadamard in connection with the ill-posedness of the Cauchy problem for the Laplace equation. Further results with applications to the instability of a string, film and free surfaces of elastic bodies were reported by Zak (1982b,c).

The result formulated above was obtained under specially selected initial conditions (4.113), but it can be generalized to include any initial conditions. Indeed, for equations (4.117) let the initial conditions be arbitrarily defined by:

$$|u_i|_{t=0} = u_i^{00} \tag{4.124}$$

and the corresponding solution is:

$$u_i = f_i(x, t) \tag{4.125}$$

By altering the initial conditions to:

$$u(0, t) = u_i^0 + u_i^{00} \tag{4.126}$$

where u_i is defined in (4.113), we observe from the preceding argument by superposition that vanishingly small change in the initial conditions would lead to unboundedly large solutions.

To obtain a geometrical interpretation of the above described instability, let us turn to expression (4.120) of the solution and note that if the second derivatives $\partial^2 u_i/\partial t^2, \partial^2 u_i/\partial x_i^2$ are of order λ_0, then the first derivatives $\partial u_i/\partial t, \partial u_i/\partial x_i$ are of order 1, and u_i are of order $1/\lambda_0$. Hence, the period of time Δt_0 can be selected in such a way that the second derivatives will be as large as necessary, but the first derivatives and u_i are still sufficiently small. Taking into account that the original governing equation (4.114) is quasi-linear with respect to the second derivatives and, therefore, the linearization does not impose any restrictions on their values, one can conclude that the linearized equation (4.117) is valid for the solution during the above-mentioned period of time Δt_0. Turning to the formula (4.120) one can now interpret the solution by the function having an infinitesimal amplitude and changing its signs with an infinite frequency ($\nu = \lambda_0 \rightarrow \infty$). The first derivatives of this function $\partial_i/\partial t, \partial_i/x_i$ can be small and change their signs by finite jumps (with the same infinite frequency ν) so that the second derivatives $\partial^2 u_i/\partial t^2, \partial^2 u_i/\partial x_1^2$ at the points of such jumps are infinite. Thus, within an arbitrary small volume there is located an arbitrary large number of points at which the strains have jumps. From the mathematical point of view, the function describing such a field of displacements u_i is considered as a continuous but non-differentiable function. This function can be simulated, for instance by the function with a multivalued derivative.

4.3.3 The Criteria of Hadamard's Instability

Let us fix an arbitrary point M and an arbitrary direction x_i at this point in an elastic body. According to the above-formulated result, the instability at the point M in the x_1 direction results from the negative eigenvalues of the matrix:

$$a_{ij} = \{\frac{\partial^2 \Pi}{\partial(\frac{\partial u_i}{\partial x_1})\partial(\frac{\partial u_j}{\partial x_1})}\}\Big|_{\frac{\partial u_i}{\partial x_1}, \frac{\partial u_j}{\partial x_1} = 0} \qquad (4.127)$$

Assuming that unperturbed state at this point is characterized by the initial stresses:

$$T_{1j}^0 \neq 0,\ T_{2j}^0 = 0,\ T_{3j}^0 = 0,\ j = 1,2,3 \qquad (4.128)$$

but zero strains:

$$(\frac{\partial u_i}{\partial x_1})_0 = 0, \text{ i.e., } \varepsilon_{11}^0 = 0,\ \gamma_{12}^0 = 0,\ \gamma_{13}^0 \qquad (4.129)$$

let us utilize the following expression for a variation of the specific potential energy from the initial stresses defined by (4.128):

$$\delta\Pi = T_{11}\delta\varepsilon_{11} + T_{12}\delta\gamma_{12} + T_{13}\delta\gamma_{13} = \cdots \text{ etc.} = T_{ij}\delta\varepsilon_{ij} \qquad (4.130)$$

Taking into account that:

$$\varepsilon_{11} = \frac{\partial u_1}{\partial x_1} + \frac{1}{2}[(\frac{\partial u_1}{\partial x_1})^2 + (\frac{\partial u_2}{\partial x_1})^2 + (\frac{\partial u_3}{\partial x_1})^2]$$

$$\gamma_{1i} = \frac{\partial u_i}{\partial x_1} + \cdots \text{etc. } (i = 1, 2, 3), \text{ i.e.}$$

$$\delta\varepsilon_{11} = (1 + \frac{\partial u_1}{\partial x_1})\,\delta\frac{\partial u_1}{\partial x_1} + \frac{\partial u_2}{\partial x_1}\,\delta\frac{\partial u_2}{\partial x_2} + \frac{\partial u_3}{\partial x_1}\,\delta\frac{\partial u_3}{\partial x_1}$$

$$\delta\gamma_{1i} = \delta\frac{\partial u_i}{\partial x_1} \quad (i = 1, 2, 3)$$

$$T_{1i}\Big|_{\frac{\partial u_1}{\partial x_1} = 0} = 0 \tag{4.131}$$

one obtains for

$$\frac{\partial u_i}{\partial x_i} = 0$$

$$a_{11} = T_{11}^0 + \frac{\partial T_{11}}{\partial \varepsilon_{11}} \tag{4.132}$$

$$a_{22} = T_{11}^0 + \frac{\partial T_{12}}{\partial \gamma_{12}} = T_{11}^0 + \frac{1}{2}\frac{\partial T_{12}}{\partial \varepsilon_{12}} \tag{4.133}$$

$$a_{33} = T_{11}^0 + \frac{\partial T_{13}}{\partial \gamma_{13}} = T_{11}^0 + \frac{1}{2}\frac{\partial T_{13}}{\partial \varepsilon_{13}} \tag{4.134}$$

$$a_{12} = a_{21} = a_{13} = a_{31} = a_{23} = a_{32} = 0 \tag{4.135}$$

where the stresses T_{ij} are related to the local Cartesian coordinates x_i, x_2, x_3 at the point M.

Now the eigenvalues of the matrix (4.127) can be written in the form:

$$\lambda_i = a_{ii}\ , \text{ i.e.,}$$

$$\lambda_1 = T_{11}^0 + \frac{\partial T_{11}}{\partial \varepsilon_{11}} \tag{4.136}$$

$$\lambda_2 = T_{11}^0 + \frac{\partial T_{12}}{\partial \gamma_{12}} = T_{11}^0 + \frac{1}{2}\frac{\partial T_{12}}{\partial \varepsilon_{12}} \tag{4.137}$$

$$\lambda_3 = T_{11}^0 + \frac{\partial T_{13}}{\partial \gamma_{13}} = T_{11}^0 + \frac{1}{2}\frac{\partial T_{13}}{\partial \varepsilon_{13}} \tag{4.138}$$

Hence, the criteria of instability are:

$$T_{11}^0 < -\frac{\partial T_{11}}{\partial \varepsilon_{11}} \tag{4.139}$$

$$T_{11}^0 < -\frac{\partial T_{12}}{\partial \gamma_{12}} \tag{4.140}$$

$$T_{11}^0 < -\frac{\partial T_{13}}{\partial \gamma_{13}} \tag{4.141}$$

Each inequality leads to the failure of differentiability of the corresponding component of strains: $\varepsilon_{11}, \varepsilon_{12}$, or ε_{13}, while the potential energy $\Pi(\varepsilon_{1i})$ has a local maximum.

Recall that all the above-formulated results are related to an arbitrary point M_0 and arbitrary selected direction x_1, with the unidirectional initial stress T^0_{11}.

In the general case when all the components of the initial stresses are non-zero:

$$T^0_{ij} \neq 0 \tag{4.142}$$

one can decompose them into spherical and deviatoric parts:

$$T^0_{ij} = \frac{1}{3}T_0E + T^{0*}_{ij}, \; T^{0*}_{ij} = \text{dev } T^0_{ij} \tag{4.143}$$

where E is the unit tensor, and

$$T_0 = \frac{1}{3}\sum_{i=1}^{3} T_{ij}, \quad \sum_{i=1}^{3} T^*_{ij} = 0 \tag{4.144}$$

Now (4.130) can be rewritten in the following form:

$$\delta\Pi = T_0\delta\varepsilon_0 + T^*_{11}\delta\varepsilon_{11} + T^*_{12}\delta\varepsilon_{12} +, \cdots, \text{ etc.} = T_0\delta\varepsilon_0 + T^*_{ij}\delta\varepsilon_{ij} \tag{4.145}$$

where ε_0 is the spherical part of the strain tensor:

$$\varepsilon_0 = \frac{1}{3}\sum_{i=1}^{3} \varepsilon_{ij} \tag{4.146}$$

and instead of (4.132–4.134), one obtains:

$$\begin{aligned} a_{11} &= T^0_{11} + \frac{\partial T^*_{11}}{\partial \varepsilon_{11}} \\ a_{22} &= T^0_{11} + \frac{1}{2}\frac{\partial T^*_{12}}{\partial \varepsilon_{12}} \\ a_{33} &= T^0_{11} + \frac{1}{2}\frac{\partial T^*_{13}}{\partial \varepsilon_{13}} \end{aligned} \tag{4.147}$$

Consequently, the sufficient conditions of the instability in some directions at the fixed point for an isotropic elastic material for which the derivatives $\partial T_{ij}/\partial \varepsilon_{ij}$ do not depend on a selected direction x_1:

$$\hat{T}^0_{11} < -\frac{\partial T_{11}}{\partial \varepsilon_{11}} \tag{4.148}$$

$$\hat{T}^0_{11} < -\frac{1}{2}\frac{\partial T_{12}}{\partial \varepsilon_{12}} \tag{4.149}$$

$$\hat{T}^0_{11} < -\frac{1}{2}\frac{\partial T_{13}}{\partial \varepsilon_{13}} \tag{4.150}$$

where $\hat{T}^0_{11}$ is one of the principle deviatoric stresses (Zak 1994a).

The instability emerges in any direction if these inequalities are valid for all the principle deviatoric stresses $\hat{T}^0_{ii}$:

$$\hat{T}^0_{ii} < -\frac{1}{2}\frac{\partial T_{ij}}{\partial \varepsilon_{ij}} \quad (i \neq j), \text{ because usually}$$

$$\frac{\partial T_{ii}}{\partial \varepsilon_{ii}} > \frac{\partial T_{ij}}{\partial \varepsilon_{ij}} \ (i \neq j) \tag{4.151}$$

For Hook's material, the criteria of the instability are expressed in terms of Young's modulus E and Poisson's ratio ν since

$$\frac{\partial T_{ii}}{\partial \varepsilon_{ii}} = \frac{E(1-\nu)}{(\nu+1)(1-2\nu)}, \ \frac{\partial T_{ij}}{\partial \varepsilon_{ij}} = \frac{E}{\nu+1} = 2G \tag{4.152}$$

if the initial stress tensor is spherical ($T^0_{ij} = p$), where E, G, and μ are the Young's and shear modulus, and ν is the Poisson's ratio, respectively.

As a further generalization of the results discussed above, let us consider a porous elastic material filled with inviscid incompressible liquids having the densities $\rho_1, \rho_2, ..., \rho_m$ and relative velocities $\nu_1, \nu_2, ..., \nu_m$.

Their mean relative velocity is written in the following form:

$$\bar{v}_c = \frac{\sum\limits_{k=1}^{m} \rho_k v_k}{\rho_c} \tag{4.153}$$

while

$$\rho_c = \sum_{k=1}^{m} \rho_k \tag{4.154}$$

Generalizing the governing equation (4.117) for this case one should note that liquids will contribute only in the expression of the total acceleration:

$$a_i = \sum_{k=1}^{m} \rho_k \left(\frac{\partial}{\partial t} + v_{k1}\frac{\partial}{\partial x_1}\right)^2 u_i \tag{4.155}$$

where v_{k1} is the projection of the velocity v_k on the axis x_1, but not in the expression of the total potential energy.

Thus, instead of equation (4.117) one obtains:

$$\rho\frac{\partial^2 u_i}{\partial t^2} + \sum_{k=1}^{m} \rho_k \left(\frac{\partial}{\partial t} + v_{k1}\frac{\partial}{\partial x_1}\right)^2 u_i = \sum_{i=1}^{3} a_{ij}\frac{\partial^2 u_j}{\partial x_1^2} \tag{4.156}$$

where a_{ij} are defined by (4.118).

Comparing (4.156) and (4.113) and taking into account the contribution of the additional acceleration term in the characteristic equation:

$$\sum_{k=1}^{m} \rho_k \left(\frac{\partial}{\partial t} + v_{k1}\frac{\partial}{\partial x_1}\right)^2 u_i \rightarrow \sum_{k=1}^{m} \rho_k \left(\widetilde{\lambda} - v_{k1}\right)^2 \qquad (4.157)$$

where $\widetilde{\lambda}$ is the characteristic speed of wave propagation, one arrives at the following expression for $\widetilde{\lambda}$:

$$\rho\widetilde{\lambda}^2 + \sum_{k=1}^{m} \rho_k \left(\widetilde{\lambda} - v_{k1}\right)^2 = \lambda_j \ , \ j = 1, 2, 3, \qquad (4.158)$$

while λ_j are the eigenvalues of the matrix (4.127) given by the equations (4.136)–(4.138).

Hence,

$$\widetilde{\lambda} = \frac{\rho_c}{\rho + \rho_c} v_{c1} \pm \sqrt{\left(\frac{\rho_c}{\rho + \rho_c} v_{c1}\right)^2 - \frac{1}{\rho + \rho_c}\left(\sum_{k=1}^{m} \rho_k v_{k1}^2 + \lambda_j\right)} \qquad (4.159)$$

where v_{c1} is the projection of the mean relative velocity v_c on the axis x.

The conditions of instability for such a porous elastic material are written in the form resembling equations (4.139)–(4.141) when $\widetilde{\lambda}$ becomes imaginary:

$$T_{ii}^0 < -\frac{1}{2}\frac{\partial T_{ij}}{\partial \varepsilon_{ij}} - \frac{\rho_c^2}{\rho + \rho_c} v_{c_i}^2 + \sum_{k=1}^{m} \rho_k v_{k_i}^2, \ i \neq j \qquad (4.160)$$

This formula is simplified if the liquid densities and velocities are identical ($\rho_1 = \rho_2 = \cdots = \rho_c$, $v_1 = v_2 = \cdots = v_c$)

$$T_{ii}^0 < -\frac{1}{2}\frac{\partial T_{ij}}{\partial \varepsilon_{ij}} - \frac{\rho_c^2}{\rho + \rho_c} v_{c_i}^2 \qquad (4.161)$$

Here v_{ii} is the flow velocity in the direction of the stress T_{ii}^0.

4.3.4 Boundaries of Applicability of Classical Models

All the results discussed above were based on formal analysis of mathematical models of elastic materials, and their practical usefulness has to be demonstrated. The most obvious and visual application of these results can be fund in the area of one and two-dimensional models such as strings, membranes, etc. whose states are defined not only by internal geometry (strains), but also by external geometry (shape). As shown in Sect. 4.3.1, in this model, unlimited decrease of the scale of motions may be consistent with the boundedness of stresses and energy. The problem of the shape instability there occurs as a result of any local compression and manifests itself in wrinkling in the course of which the shape looses its smoothness.

One Dimensional Continuum

By one dimensional continuum we will understand a material line which state is characterized by an elongation or contraction ε as a parameter of internal geometry, and by the curvature Ω_3 and twist Ω_1 defining the shape of this line, i.e., representing parameters of external geometry. Such a continuum can be represented, for instance, by an ideally flexible string or filament. Concentrating attention to the instability of its shape and considering only extensible string, one arrives at the following governing equation:

$$\rho_0 \frac{\partial^2 \mathbf{r}}{\partial t^2} = \frac{\partial}{\partial \psi}\left(T \frac{\partial \mathbf{r}}{\partial \psi}\right) + \mathbf{F} + \mathbf{F}', \quad \left|\frac{\partial \mathbf{r}}{\partial \psi}\right| = 1 \tag{4.162}$$

where $\mathbf{r}$ is the radius vector of the string particles, ψ is the coordinate along the string, T is the string tension, $\mathbf{F}$, $\mathbf{F}'$ are the tracing and dead forces respectively.

The parameters characterizing the external geometry of the filament, i.e., its curvature Ω_3 and twist Ω_1 are expressed via the radius vector, respectively:

$$\Omega_3 = \left|\frac{\partial^2 \mathbf{r}}{\partial \psi^2}\right|, \quad \Omega_1 = \frac{\left(\frac{\partial \mathbf{r}}{\partial \psi} \times \frac{\partial^2 \mathbf{r}}{\partial \psi^2}\right) \cdot \frac{\partial^3 \mathbf{r}}{\partial \psi 3}}{\frac{\partial^2 \mathbf{r}}{\partial \psi^2}} \tag{4.163}$$

As shown in the Sect. 4.3.3, Hadamard's instability is associated with the failure of hyperbolicity of the governing dynamical equations of the continuum, i.e., with the appearance of imaginary eigenvalues of the motion (4.127). These eigenvalues are associated with the characteristic speeds of wave propagations, and in particular, with speeds of propagation of discontinuities of the highest order derivatives of the dynamical parameters. We will apply here a well-established technique of determining these characteristic speeds based upon the theory of propagation of discontinuities in hyperbolic equations (Zak 1968).

First of all, as follows from the compatibility condition at the front of the discontinuity:

$$\left[\frac{\partial}{\partial t}\right] = -\lambda \left[\frac{\partial}{\partial \psi}\right] \tag{4.164}$$

where λ is the characteristic speed of wave propagation, and $[x]$ denotes the discontinuity of the parameter x. Obviously:

$$\begin{aligned} [x] &= 0 \quad \text{if } \tfrac{\partial x}{\partial \psi} \neq 0 \\ \left[\tfrac{\partial x}{\partial \psi}\right] &= 0 \quad \text{if } \tfrac{\partial^2 x}{\partial \psi^2} \neq 0, \text{ etc.} \end{aligned} \tag{4.165}$$

Turning to (4.162), one obtains:

$$\left[\frac{\partial \mathbf{r}}{\partial t}\right] = -\lambda \left[\frac{\partial \mathbf{r}}{\partial \psi}\right] = 0, \text{ since } \frac{\partial^2 \mathbf{r}}{\partial t^2} \neq 0 \tag{4.166}$$

Assuming that

$$\mathbf{F} = 0, \text{ while } [\mathbf{F}'] = 0 \text{ even if } \mathbf{F}' \neq 0 \tag{4.167}$$

the following relationship for the discontinuity can be derived from (4.162):

$$\rho_0 \left[\frac{\partial^2 \mathbf{r}}{\partial t^2}\right] = T \left[\frac{\partial^2 \mathbf{r}}{\partial \psi^2}\right] \tag{4.168}$$

and since

$$\frac{\partial^2 \mathbf{r}}{\partial t^2} = \lambda \left[\frac{\partial^2 \mathbf{r}}{\partial \psi^2}\right] \tag{4.169}$$

(4.168) reduces to

$$\left(\lambda^2 - \frac{T}{\rho}\right) \left[\frac{\partial^2 \mathbf{r}}{\partial \psi^2}\right] = 0 \tag{4.170}$$

For non-zero discontinuity:

$$\left[\frac{\partial^2 \mathbf{r}}{\partial \psi^2}\right] \neq 0 \tag{4.171}$$

one arrives at the characteristic speed

$$\lambda_1 = \pm \sqrt{\frac{T}{\rho}} \tag{4.172}$$

which transport the discontinuities of the filament curvature Ω_3. Indeed, as follows from (4.163):

$$[\Omega_3] = \left[\frac{\partial^2 \mathbf{r}}{\partial \psi^2}\right] \tag{4.173}$$

One can easily verify that the same result (4.172) could be obtained from the general formula (4.137) since for a one-dimensional continuum, the shear modulus G vanishes:

$$\frac{\partial T_{12}}{\partial \gamma_{12}} = G \equiv 0 \tag{4.174}$$

For a particular case $\mathbf{F} = 0$, one can obtain the same characteristic speed (4.172) from the general formula (4.138) taking into account that:

$$\frac{\partial T_{13}}{\partial \gamma_{13}} = G \equiv 0 \tag{4.175}$$

However, if $\mathbf{F} \neq 0$ the formula (4.137) cannot be applied directly, and we will turn again to this governing equation (4.162).

First of all, the force $\mathbf{F}$, can be decomposed into the following three components $\mathbf{F}_1$ $\mathbf{F}_2$, and $\mathbf{F}_3$ tangent, normal and binormal to the filament, respectively:

$$F = F_1 \frac{\partial \mathbf{r}}{\partial \psi} + \frac{F_2}{\Omega_3} \frac{\partial^2 \mathbf{r}}{\partial \psi^2} + \frac{F_3}{\Omega_3} \frac{\partial \mathbf{r}}{\partial \psi} \times \frac{\partial^2 \mathbf{r}}{\partial \psi^2} \tag{4.176}$$

Then, differentiating (4.167) with respect to ψ, projecting it into the binormal $b = \frac{\partial \mathbf{r}}{\partial \psi} \times \frac{\partial^2 \mathbf{r}}{\partial \psi^2}$, and assuming that

$$\left[\frac{\partial^2 \mathbf{r}}{\partial \psi}\right] = \Omega_3 \neq 0, \text{ but } [\Omega_3] = 0 \tag{4.177}$$

one arrives at the following characteristic equation

$$\rho_0 \left[\frac{\partial^3 \mathbf{r}}{\partial \psi \partial t^2}\right] \cdot \mathbf{b} = \left(T + \frac{F_2}{\Omega_3}\right) \left[\frac{\partial^3 \mathbf{r}}{\partial \psi^3}\right] \cdot \mathbf{b} \tag{4.178}$$

It was taken into account that

$$\frac{F_3}{\Omega_3} \frac{\partial}{\partial \psi} \left(\frac{\partial \mathbf{r}}{\partial \psi} \times \frac{\partial^2 \mathbf{r}}{\partial \psi}\right) \cdot \left(\frac{\partial \mathbf{r}}{\partial \psi} \times \frac{\partial^2 \mathbf{r}}{\partial \psi}\right) = \frac{F_3}{\Omega_3} \frac{\partial \mathbf{b}}{\partial \psi} \cdot \mathbf{b} = 0 \tag{4.179}$$

Now (4.178) reduces to

$$\left\{\rho_0 \lambda_1^2 - \left(T + \frac{F_2}{\Omega_3}\right)\right\} \left[\frac{\partial^3 \mathbf{r}}{\partial \psi^3}\right] \cdot \left(\frac{\partial \mathbf{r}}{\partial \psi} \times \frac{\partial^2 \mathbf{r}}{\partial \psi^2}\right) = 0 \tag{4.180}$$

but, as follows from (4.163)

$$\left[\frac{\partial^3 \mathbf{r}}{\partial \psi^3}\right] \cdot \left(\frac{\partial \mathbf{r}}{\partial \psi} \times \frac{\partial^2 \mathbf{r}}{\partial \psi^2}\right) = [\Omega_1] \, \Omega_3^2 \tag{4.181}$$

Hence, if the discontinuity of the twist Ω_1 is non-zero, and the curvature Ω_3 at the front of this discontinuity is also non-zero, but continuous ($[\Omega_3] = 0$), the characteristic speed transporting twist discontinuities will be different from (4.172):

$$\lambda_1 \pm \sqrt{\frac{1}{\rho_0}\left(T + \frac{F_2}{\Omega_3}\right)} \neq \lambda \tag{4.182}$$

This new characteristic wave was discovered by Zak (1968). Equation (4.182) becomes more complicated if the forcing $\mathbf{F}$ depends upon the filament velocity (motion of a filament on a rough surface with the static friction effects, or in an ideal fluid with the added mass effects).

As noted above, the system of characteristic speeds (4.172) corresponds to the waves transporting discontinuities of the string curvature Ω_3 or discontinuities of the unit tangent vector $\partial \mathbf{r}/\partial \psi$, i.e., apex points (Fig. 4.5a). The instability associated with the speeds (4.172) occurs if

$$T < 0 \tag{4.183}$$

The inequality (4.183) expresses the well-known fact that a compressed string is unstable because of the loss of stability of the curvature or the unit

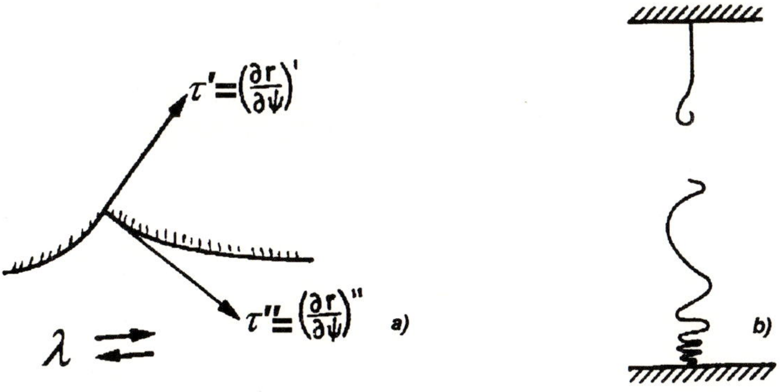

Fig. 4.5.

tangent $\partial \mathbf{r}/\partial\psi$. The shape of such a string cannot be described by smooth functions, and theoretically, the string can be rolled up into a point (Fig. 4.5b).

The second system of characteristic speeds (4.182) corresponds to the waves transporting discontinuities of the string twist Ω or the angle of revolution of the binormal about the tangent to the string. In contrast to (4.172), these speeds depend not only on the tension, but also on the tracing force and curvature. The instability associated with the speeds (4.182) occurs if

$$T < \frac{F_2}{\Omega_3} \tag{4.184}$$

Thus, even a stretched string can be unstable if subjected to the tracing force whose normal component is sufficiently large while the curvature at the point of the application of this force is small (for instance, the point of inflection of the string). In this case the instability of the string shape is caused by the loss of stability of the string twist or the position of the osculating planes (Fig. 4.6a).

It can be shown that elastic strings possess an additional system of waves with characteristic speeds:

$$\lambda_{5,6} = \pm\sqrt{\frac{1}{\rho_0}\frac{\partial T}{\partial \varepsilon}} = \pm\sqrt{\frac{E}{\rho_0}} \tag{4.185}$$

transporting discontinuities of elongations or contractions ε. These speeds can be obtained directly from the general equation (4.136). Usually these waves are not associated with any failure of hyperbolicity.

These results can be generalized on a one-dimensional, ideally flexible pipe within which an inviscid liquid flows, by combining (4.172) and (4.182) with (4.161). Then, the conditions for the failure of hyperbolicity are given

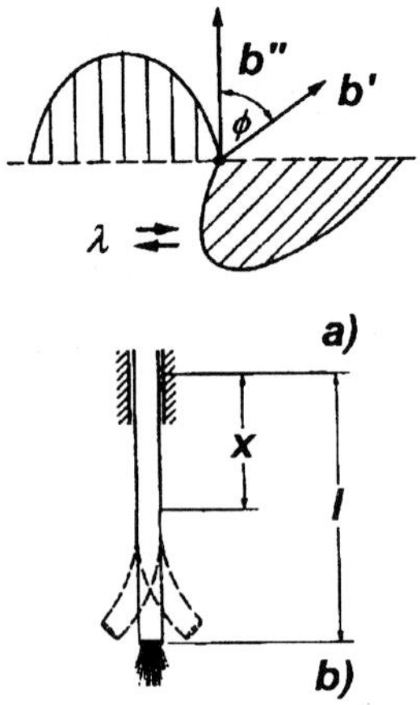

Fig. 4.6.

in the form:

$$T < \frac{\rho\rho^1}{(\rho + \rho^1)} u^2 \tag{4.186}$$

$$T < \frac{F_n}{\Omega} + \frac{\rho\rho^1}{(\rho + \rho^1)^2} u^2 \tag{4.187}$$

where ρ^1, u are the density and relative velocity of the liquid, respectively. This means that a flow within a pipe destabilizes its shape.

In order to illustrate the last results, let us consider a vertical, ideally flexible, inextensible pipe with a free lower end suspended in the gravity field. Assuming that the flow within this pipe has constant velocity u_0, let us define the area of instability (Fig. 4.6b). The tension T referred to the entire pipe's cross-section is given by the following equation

$$T = \rho g \xi (l - x) \tag{4.188}$$

where l is the length of the pipe, x is the coordinate along the length of the pipe, and ξ is the ratio of the area of the cross-section occupied by the pipe's walls relative to the entire cross-sectional area.

Substituting (4.188) into (4.186), one obtains the unstable area of the pipe:

$$l \geq x \geq l - \frac{u_0^2}{g\xi(1+\varepsilon)}, \quad \varepsilon = \frac{\rho}{\rho^1} \tag{4.189}$$

Hence, for the ideal flexible pipe, the free end is always unstable. (Such a phenomenon is well known from the experiments.) In the limit case $u_0 \to 0$, when the pipe can be considered as a string, the unstable area is concentrated around the free end. As shown by Zak (1970, 1983) such an instability manifests itself in an accumulation of energy at the free end (snap of a whip) and will be considered below.

Two Dimensional Continuum Models

By a two-dimensional continuum we understand a material surface which state is characterized by a plain strain tensor ε_{ij} $(i, j = 1.2)$ as a parameter of internal geometry, and coefficients of the second fundamental form β_{ij} $(i, j = 1, 2)$ as parameters of external geometry defining the shape of the surface. Such a continuum can be represented, for instance, by an ideally flexible membrane, film, net, etc.

The governing equations for this continuum can be written in the following form:

$$g^{1/2}\rho_0\frac{\partial^2\mathbf{r}}{\partial t^2} = \frac{\partial}{\partial\psi^{(1)}}\left(T^{11}g^{1/2}\frac{\partial\mathbf{r}}{\partial\psi^{(1)}}\right) + \frac{\partial}{\partial\psi^{(1)}}\left(T^{12}g^{1/2}\frac{\partial\mathbf{r}}{\partial\psi^{(2)}}\right) +$$

$$\frac{\partial}{\partial\psi^{(2)}}\left(T^{12}g^{1/2}\frac{\partial\mathbf{r}}{\partial\psi^{(1)}}\right) + \frac{\partial}{\partial\psi^{(2)}}\left(T^{22}g^{1/2}\frac{\partial\mathbf{r}}{\partial\psi^{(2)}}\right) + \mathbf{F} \qquad (4.190)$$

where $\mathbf{r}$ is the radius vector of the surface point, $\psi^{(1)}$, $\psi^{(2)}$ are the material coordinates, ρ is the initial density, g characterizes the initial internal geometry of the film in semi-geodesical coordinates $\psi_0^{(1)}$, $\psi_0^{(2)}$, $\mathbf{F}$ is the external force, T_{ij} is the stress tensor.

This equation must be complemented by the constitutive and geometrical equations.

$$T_{ij} = f(\varepsilon_{11}, \varepsilon_{22}, ..., \text{etc.}) \qquad (4.191)$$

$$\left|\frac{\partial\mathbf{r}}{\partial\psi^{(1)}}\right| = 1 + 2\varepsilon_{11}, \quad \left|\frac{\partial\mathbf{r}}{\partial\psi^{(2)}}\right| = g + 2\varepsilon_{22}, \quad \frac{\partial\mathbf{r}}{\partial\psi^{(1)}}\cdot\frac{\partial\mathbf{r}}{\partial\psi^{(2)}} = 2\varepsilon_{12}$$

$$g_* = (1 + 2\varepsilon_{11})(g + 2\varepsilon_{22}) - 4\varepsilon_{12}^2 \qquad (4.192)$$

The analysis of these equations, similar to those performed for a string, leads to the following characteristic speeds (Zak 1994a)

$$\widetilde{\lambda}_{1,2} = \pm\sqrt{\frac{1}{\rho_0}\left(T_{ii} + \frac{\partial T_{ii}}{\partial\varepsilon_{ii}}\right)} \qquad (4.193)$$

$$\widetilde{\lambda}_{3,4} = \pm\sqrt{\frac{1}{\rho_0}\left(T_{ii} + \frac{1}{2}\frac{\partial T_{ij}}{\partial\varepsilon_{ij}}\right)}, \quad i \neq j \qquad (4.194)$$

$$\widetilde{\lambda}_{5,6} = \pm\sqrt{\frac{T_{ii}}{\rho_0}} \qquad (4.195)$$

which also can be obtained directly from the general equations (4.136–4.138).

The first two systems correspond to in-plane normal and shear elastic waves transporting discontinuities of strains ε_{ii}, ε_{ij}. The last system corresponds to the transverse waves transporting discontinuities of the coefficients b_{ij} of the second fundamental form [see (4.94) and Fig 4.7a]. Usually

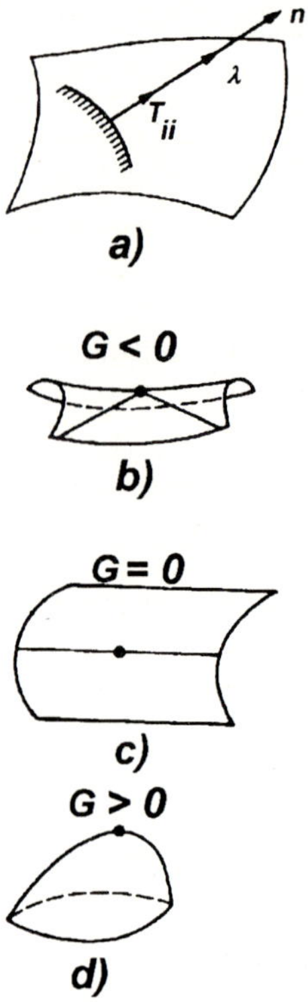

Fig. 4.7.

the failure of hyperbolicity in two dimensional continua is associated with this system of characteristic waves and leads to the instability of the shape if

$$T_{ii} < 0 \tag{4.196}$$

i.e., if a compression occurs in the direction in which the characteristic waves (4.195) are possible.

The last comment is very important because as shown in Sect. 4.3.1, the characteristic waves (4.195) can propagate only in some selected directions which are normal to asymptotic lines. Recall that in contrast to a one dimensional continuum where the shape parameters (curvature and twist) can be changed independently from the elongations, in two dimensional continua, there are some limitations imposed on the changes of the shape in the form of the equations of compatibility with the changes of strains [the Gauss equations (4.195–4.196)]. As follows from (4.103–4.106), at the points of negative Gauss curvature there are two directions of possible shape wave propagations (Fig. 4.6b). At the point of zero Gauss curvature there is only one such direction (Fig. 4.7c). Finally, at the points of positive Gauss curvature the shape discontinuities are impossible (Fig. 4.7d).

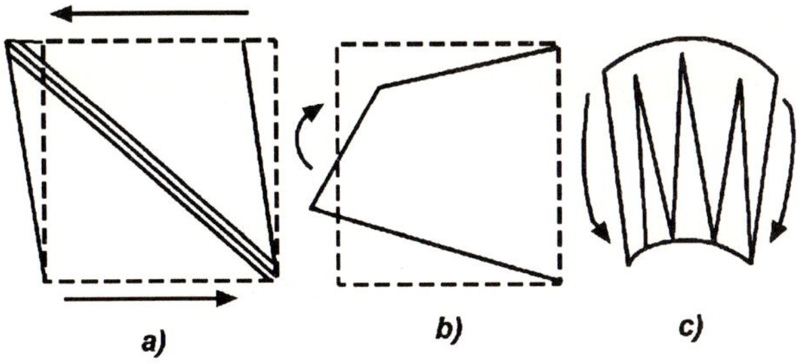

Fig. 4.8.

Thus the instability of the shape defined in terms of the coefficients of the second fundamental form b_{ij} is possible only if a compression occurs in the direction normal to the asymptotic line of the surface at the corresponding point.

Practically the shape instability of a surface is exhibited in formation of wrinkles which coincide with the directions of the positive principal stresses if other principal stresses are negative.

The wrinkles can be observed, for instance, in the course of shearing, twisting, or in-plane bending of a membrane (Fig. 4.8a,b,c).

If both of the principal stresses are negative, then even the lines of wrinkles lose their smoothness, and a membrane can be rolled up into a point.

By combing the formulas (4.195) and (4.161), the above-described results are generalized on inflated two dimensional models:

$$\lambda = \frac{\rho^1}{\rho + \rho^1} u_n \pm \sqrt{\frac{T_n}{\rho + \rho^1} - \frac{\rho \rho^1}{(\rho - \rho^1)^2} u_n^2} \tag{4.197}$$

where u_n is the projection of the velocity of the flow of a fluid on the normal to the front of the wave of a discontinuity, T_n is the tension to the same front and referred to the entire cross-section.

The condition of the failure of hyperbolicity now is given by:

$$T_n < \frac{\rho \rho^1}{\rho + \rho^1} u_n^2 \tag{4.198}$$

As follows from (4.198), a flow within a two dimensional model also destabilizes its shape.

Slightly different criteria of Hadamard's instability can be obtained for liquid films (Zak 1985a,b). A liquid film is understood to be a material surface possessing the following properties: its internal stresses are determined by a plane stress tensor which spherical part is presented by a positive scalar surface tension:

$$\overset{\circ}{T} > 0 \tag{4.199}$$

while the deviatoric part is given by the Newtonian fluid relation:

$$\text{dev } T = 2\mu \text{ dev } e \tag{4.200}$$

Here T is the stress tensor, e is the rate-of-strain tensor, μ is the viscosity of the liquid ($\mu =$ Const). Thus:

$$T = \overset{\circ}{T} E + 2\mu \text{ dev } e \tag{4.201}$$

where E is the unit (plane) tensor.

The equation of the film motion, as well as the continuity equation, are now written in the following invariant form:

$$\varrho \mathbf{a} = \nabla \cdot T + \varrho \mathbf{F} \tag{4.202}$$

$$\frac{\partial \varrho}{\partial t} = \nabla \cdot (\varrho \mathbf{u}) = 0 \tag{4.203}$$

where $\mathbf{a}$ is the acceleration of the liquid, ϱ is the density of the liquid referred to unit surface area, $\mathbf{F}$ is the external force per unit mass, $\mathbf{u}$ is the relative velocity of the liquid with respect to the fixed system of coordinates within the medium.

In contrast to the three-dimensional case, here in the course of film motion the Gaussian curvature of its surface may not be zero, which means that the Eulerian system of coordinates in general cannot be a Cartesian one.

As shown in the differential geometry, the simplest orthogonal system of coordinates at a surface is a semi-geodesical system. Recall that one family of coordinates of such a system is formed by geodesical lines; another family of lines is orthogonal to the latter, but in general the lines of the second family are not geodesical. In the particular case when the Gaussian curvature of the surface is zero, the above system is transformed into the Cartesian system. It is shown that an arbitrary line at the smooth surface can be included in the semi-geodesical system and thereby can define the system uniquely. Such a line can coincide, for instance, with a fixed boundary of the film surface.

The metric relationships in the semi-geodesical system are written in the following form:

$$\left| \frac{\partial \mathbf{r}}{\partial \psi_1} \right| = 1, \quad \left| \frac{\partial \mathbf{r}}{\partial \psi_2} \right| = \sqrt{g}, \quad \frac{\partial \mathbf{r}}{\partial \psi_1} \cdot \frac{\partial \mathbf{r}}{\partial \psi_2} = 0 \tag{4.204}$$

where $\psi_2 =$ Const is the family of geodesical lines, $\psi_1 =$ Const is the family of lines orthogonal to the latter, $\mathbf{r}(\psi_1, \psi_2)$ is the radius-vector as a function of the semi-geodesical coordinates ψ_1, ψ_2. The metric coefficient $\sqrt{g}$ characterizes the internal geometry of the film surface and directly connected with the Gaussian curvature:

$$G = -\frac{1}{\sqrt{g}} \frac{\partial^2 \sqrt{g}}{\partial \psi_2} \tag{4.205}$$

Now the relative velocity $\mathbf{u}$ is given as:

$$\mathbf{u} = u^{(1)} \frac{\partial \mathbf{r}}{\partial \psi_1} + u^{(2)} \frac{\partial \mathbf{r}}{\partial \psi_2} \tag{4.206}$$

where

$$u^{(1)} = \frac{\partial \psi_1}{\partial t} = u_1^0, \; u^{(2)} = \frac{\partial \psi_2}{\partial t} = u_2^0 \sqrt{g} \tag{4.207}$$

i.e., $\dot{u}_i$ is the ith component of the velocity of the liquid with respect to the film points with the fixed coordinates ψ_1, ψ_2 or the physical component of the vector $\mathbf{u}$, while $\mathbf{u}^{(i)}$ is the contravariant components of this velocity.

The translation velocity of the film particles, together with the fixed coordinate system ψ_1, ψ_2, is written in the following form:

$$\dot{\upsilon} = \frac{\partial \mathbf{r}}{\partial t} \tag{4.208}$$

Hence, the absolute velocity and acceleration of the film particles yield, respectively:

$$\upsilon = \frac{\partial \mathbf{r}}{\partial t} + u^{(1)} \frac{\partial \mathbf{r}}{\partial \psi_1} + u^{(2)} \frac{\partial \mathbf{r}}{\partial \psi_2} = \left(\frac{\partial}{\partial t} + u^{(1)} \frac{\partial}{\partial \psi_1} + u^{(2)} \frac{\partial}{\partial \psi_2} \right) \mathbf{r} \tag{4.209}$$

$$\begin{aligned} \mathbf{a} &= \left(\frac{\partial}{\partial t} + u^{(1)} \frac{\partial}{\partial \psi_1} + u^{(2)} \frac{\partial}{\partial \psi_2} \right)^2 \mathbf{r} \\ &+ \left[\left(\frac{\partial}{\partial t} + u^{(1)} \frac{\partial}{\partial \psi_1} + u^{(2)} \frac{\partial}{\partial \psi_2} \right) u^{(1)} \right] \frac{\partial \mathbf{r}}{\partial \psi_1} \\ &+ \left[\left(\frac{\partial}{\partial t} + u^{(1)} \frac{\partial}{\partial \psi_1} + u^{(2)} \frac{\partial}{\partial \psi_2} \right) u^{(2)} \right] \frac{\partial \mathbf{r}}{\partial \psi_1} \end{aligned} \tag{4.210}$$

For convenience, we will introduce a normal $\mathbf{n}$ to the film surface:

$$\mathbf{n} \cdot \left(\frac{\partial \mathbf{r}}{\partial \psi_1} \times \frac{\partial \mathbf{r}}{\partial \psi_2} \right) = 0 \tag{4.211}$$

with the length:

$$|\mathbf{n}| = h(\psi_1, \psi_2) \tag{4.212}$$

where h is the variable thickness of the film.

Then the density referred to the film surface will be variable due to the variable thickness:

$$\varrho = \varrho_0 \widetilde{h},\ \widetilde{h} = \frac{h}{h_0},\ \dot{\varrho} = \frac{\varrho_0}{h_0} = \frac{\varrho}{h} \tag{4.213}$$

where ϱ_0 is the film density corresponding to some fixed thickness h_0, $\dot{\varrho} =$ Const is the density per unit volume.

Obviously, this compressibility is fictitious because physically, the liquid under consideration is incompressible.

The expressions for $\nabla \cdot T$, $\nabla \cdot (\varrho \mathbf{u})$ and e in (4.201–4.202) in the semigeodesic system ψ_1, ψ_2 follow from the vector and tensor analysis in orthogonal coordinates:

$$\begin{aligned}\nabla \cdot T = & \frac{1}{\sqrt{g}}\{\frac{\partial}{\partial \psi_1}\left[\widetilde{h}\sqrt{g}\left(T^{11}\frac{\partial \mathbf{r}}{\partial \psi_1} + T^{12}\frac{\partial \mathbf{r}}{\partial \psi_2}\right)\right] \\ & + \frac{\partial}{\partial \psi_2}\left[\widetilde{h}\sqrt{g}\left(T^{12}\frac{\partial \mathbf{r}}{\partial \psi_1} + T^{22}\frac{\partial \mathbf{r}}{\partial \psi_2}\right)\right]\} \\ & + \frac{1}{\sqrt{g}}\{\frac{\partial}{\partial \psi_1}\left[\sqrt{g}\left(\dot{T}^{11}\frac{\partial \mathbf{r}}{\partial \psi_1} + T^{12}\frac{\partial \mathbf{r}}{\partial \psi_2}\right)\right] \\ & + \frac{\partial}{\partial \psi_2}\left[\sqrt{g}\left(T^{12}\frac{\partial \mathbf{r}}{\partial \psi_1} + T^{22}\frac{\partial \mathbf{r}}{\partial \psi_2}\right)\right]\}\end{aligned} \tag{4.214}$$

(Here the contravariant viscous stresses T^{ij} are treated differently from the contravariant surface tension stresses $\overset{\circ}{T}{}^{ij}$because the latter are applied only at the film surface),

$$\nabla \cdot (\varrho \mathbf{u}) = \frac{\varrho_0}{\sqrt{g}}\left[\frac{\partial}{\partial \psi_1}\left(\widetilde{h}\sqrt{G_{u^{(1)}}}\right) + \frac{\partial}{\partial \psi_2}\left(\widetilde{h}\sqrt{g}u^{(2)}\right)\right]$$

$$\begin{aligned} e_{11} &= \frac{\partial u^{(1)}}{\partial \psi_1},\ e_{22} = \sqrt{g}\left(\frac{\partial u}{\partial \psi_2} + u^{(1)}\frac{\partial \sqrt{g}}{\partial \psi_1}\right) \\ e_{12} &= \frac{g}{2}\frac{\partial}{\partial \psi_1}\left(u^{(2)}\sqrt{g} + \frac{1}{2}\frac{\partial u^{(1)}}{\partial \psi_2}\right)\end{aligned} \tag{4.215}$$

where e_{ij} are the covariant components of the rate-of-strain tensor, and, therefore:

$$\begin{aligned} T^{11} &= 2\mu\frac{\partial u^{(1)}}{\partial \psi_1},\ T^{22} = \frac{2\mu}{g\sqrt{g}}\left(\frac{\partial u^{(2)}}{\partial \psi_2} + u^{(1)}\frac{\partial \sqrt{g}}{\partial \psi_1}\right) \\ T^{12} &= \mu\left[\frac{\partial}{\partial \psi_1}\left(u^{(2)}\sqrt{g}\right) + \frac{1}{g}\frac{\partial u^{(1)}}{\partial \psi_2}\right]\end{aligned} \tag{4.216}$$

while

$$\overset{*}{T}{}^{11} = \frac{2\overset{\circ}{T}}{\widetilde{h}h_0}, \quad \overset{*}{T}{}^{22} = \frac{2\overset{\circ}{T}}{\widetilde{h}h_0 g}, \quad \overset{*}{T}{}^{12} = 0 \tag{4.217}$$

where $\overset{*}{T}{}^{ij}$ are the contravariant components of the spherical tensor of the surface tension stresses, referred to the unit of the film thickness.

Substituting (4.210), (4.211)–(4.217) into (4.202) and (4.203) one arrives at the governing equations for the viscous liquid film motion, referred to the middle surface:

$$\left(\frac{\partial}{\partial t} + u^{(i)}\frac{\partial}{\partial \psi_i}\right)^2 \mathbf{r} + \sum_{i=1}^{2}\left[\left(\frac{\partial}{\partial t} + u^{(i)}\frac{\partial}{\partial \psi_i}\right) u^{(i)}\right]\frac{\partial \mathbf{r}}{\partial \psi_i} - \mathbf{F}$$

$$= \frac{1}{\widetilde{h}\sqrt{g}\dot{e}}\frac{\partial}{\partial \psi_1}\left\{\widetilde{h}\sqrt{g}\left[2\mu\frac{\partial u^{(1)}}{\partial \psi_1}\frac{\partial \mathbf{r}}{\partial \psi_1} + \left(\mu\frac{\partial}{\partial \psi_1}(u^{(2)}\sqrt{g}) + \frac{\mu}{g}\frac{\partial u^{(1)}}{\partial \psi_2}\frac{\partial \mathbf{r}}{\partial \psi_2}\right)\right]\right\}$$

$$+\frac{1}{\widetilde{h}\sqrt{g}\dot{e}}\frac{\partial}{\partial \psi_2}\{\widetilde{h}\sqrt{g}[\left(\mu\frac{\partial}{\partial \psi_1}(u^{(2)}\sqrt{g}) + \frac{\mu}{g}\frac{\partial u^{(1)}}{\partial \psi_2}\right)\frac{\partial \mathbf{r}}{\partial \psi_1}$$

$$+\frac{2\mu}{g\sqrt{g}}(\frac{\partial u^{(2)}}{\partial \psi_2} + u^{(1)}\frac{\partial \sqrt{g}}{\partial \psi_1})\frac{\partial \mathbf{r}}{\partial \psi_2}]\}$$

$$+\frac{2\dot{T}}{h_0\sqrt{g}\dot{e}}[\frac{\partial}{\partial \psi_1}(\frac{\sqrt{g}}{\widetilde{h}}\frac{\partial \mathbf{r}}{\partial \psi_1}) + \frac{\partial}{\partial \psi_2}(\frac{1}{\widetilde{h}\sqrt{g}}\frac{\partial \mathbf{r}}{\partial \psi_2})] \tag{4.218}$$

$$\frac{\partial \widetilde{h}}{\partial t}\frac{1}{\sqrt{g}}\left[\frac{\partial}{\partial \psi_2}\left(\widetilde{h}\sqrt{g}u^{(1)}\right) + \frac{\partial}{\partial \psi_2}\left(\widetilde{h}\sqrt{g}u^{(2)}\right)\right] = 0 \tag{4.219}$$

Equations (4.218–4.219) and (4.204) contain one vector and four scalar unknowns: $\mathbf{r}$, $u^{(1)}$, $u^{(2)}$, $\sqrt{g}$, h and form one vector and four scalar equations, i.e., the system of the governing equation for the viscous liquid film is closed.

Let us select the semi-geodesical coordinates ψ_1, ψ_2 at the film so that at the fixed point M and at the fixed instant of time t_0, the line of discontinuities coincides with the coordinate line ψ_1 = Const. Then

$$\left[\frac{\partial}{\partial \psi_2}\right] = 0 \tag{4.220}$$

Utilizing (4.218–4.219) and (4.204) at the selected point and at the instant t_0 and assuming that $[\partial h/\partial \psi_1] = 0$, one obtains:

$$\left\{(u_1 - \lambda)^2 - \frac{2}{\overset{*}{\rho}}\left(\frac{\overset{\circ}{T}}{\widetilde{h}h_0} + \mu\frac{\partial u_1}{\partial \psi_1}\right)\right\}[b_{11}] = 0 \tag{4.221}$$

If

$$[b_{11}] \neq 0 \tag{4.222}$$

then

$$\lambda = u_1 \pm \sqrt{\frac{2}{\overset{*}{\rho}}\left(\frac{\overset{\circ}{T}}{\tilde{h}h_0} + \mu\frac{\partial u_1}{\partial \psi_1}\right)} \tag{4.223}$$

Thus the discontinuities of the film curvature propagate with the characteristic speed (4.223). As noticed above, some limitations are imposed on the propagation of the characteristic waves in viscous films which are identical to the limitations in membranes and inviscid films. Indeed, referring to the Gauss formula which in semi-geodesical coordinates reduces to:

$$b_{11}b_{22} - b_{12}^2 = \frac{1}{\sqrt{g}}\frac{\partial^2\sqrt{g}}{\partial\psi_1^2}, \quad b_{ij} = \frac{1}{\sqrt{g}}\frac{\partial^2 \mathbf{r}}{\partial\psi_i\partial\psi_j}\cdot\left(\frac{\partial\mathbf{r}}{\partial\psi_1}\times\frac{\partial\mathbf{r}}{\partial\psi_2}\right) \tag{4.224}$$

and taking into account (4.204) and (4.223) one gets:

$$[b_{12}] = 0, \quad [b_{22}] = 0, \quad \left[\frac{\partial^2\sqrt{g}}{\partial\psi_1^2}\right] = 0 \tag{4.225}$$

Hence, as follows from (4.224):

$$[b_{11}]\, b_{22} = 0 \tag{4.226}$$

Thus the instability of the shape defined in terms of the coefficients of the second fundamental form b_{ij} is possible only if a compression occurs in the direction normal to the asymptotic line of the surface at the corresponding point.

Thus, the discontinuities of the coefficient b_{11} are possible only if

$$b_{12} = 0 \tag{4.227}$$

This means that the line of the shape discontinuity of a film must be superimposed with the asymptote to the surface, i.e., with the curve having zero normal section curvature. Consequently, at points of negative Gaussian curvature ($G < 0$), there are two directions of possible shape wave propagation. At points of zero Gaussian curvature ($G = 0$) there is only one direction of possible shape wave propagation. Finally, at points having positive Gaussian curvature ($G > 0$), shape discontinuities are impossible.

These criteria of Hadamard's instability are associated with the imaginary characteristic speed (4.223) transporting the discontinuities of the curvature coefficients of the film surface. However, one should note that the characteristic speed depends on the inplane direction of the corresponding wave propagation. Obviously, the criterion of instability must be associated with a direction along which the viscous compression is maximum.

This direction is defined by the well-known equation:

$$tg\,\dot{\varphi} = \frac{\overset{+}{T} - T'_{11}}{T'_{12}} \tag{4.228}$$

while

$$\overset{+}{T'} = \frac{T'_{11} + T'_{12}}{2} - \frac{1}{2}\sqrt{(T'_{11} - T'_{22}) + 4T^{2\prime}_{12}} \tag{4.229}$$

where $\dot{\varphi}$ is the angle between this principal direction and the inplane tangent to the coordinate ψ_1, $\overset{+}{T'}$ is the minimal normal stress; $T'_{11}, T'_{12}, T'_{22}$ are the physical components of the stress tensor in the local Cartesian basis e_i while

$$e_i = \frac{\partial \mathbf{r}}{\partial \psi_1} = \mathbf{r}_1,\; e_2 = \frac{1}{\sqrt{G}}\frac{\partial \mathbf{r}}{\partial \psi_1} = \frac{1}{\sqrt{G}}\mathbf{r}_2,\; e_1 \cdot e_2 = 0 \tag{4.230}$$

In this basis the criterion of the shape instability is written in the following form:

$$\overset{+}{T'} < 0 \tag{4.231}$$

Returning to the coordinate basis $\mathbf{r}_i$ and taking into account that

$$T'_{11} = \frac{2\overset{\circ}{T}}{h} + T^{11},\; T'_{22} = \frac{2\overset{\circ}{T}}{h} + GT^{22},\; T'_{12} = \sqrt{G}T^{12} \tag{4.232}$$

with reference to (4.216) one arrives at the following criterion of the shape instability for the film:

$$\frac{2\overset{\circ}{T}}{h} < \mu\{\frac{\partial u^{(1)}}{\partial \psi_1} + \frac{1}{\sqrt{G}}(\frac{\partial u^{(2)}}{\partial \psi_2} + u^{(1)}\frac{\partial \sqrt{G}}{\partial \psi_1})$$

$$-\{\left[\frac{\partial u^{(1)}}{\partial \psi_1} - \frac{1}{\sqrt{G}}(\frac{\partial u^{(2)}}{\partial \psi_2} + u^{(1)}\frac{\partial \sqrt{G}}{\partial \psi_1})\right]^2$$

$$+\left[\sqrt{G}\frac{\partial}{\partial \psi_1}(u^{(2)}\sqrt{G}) + \frac{1}{\sqrt{G}}\frac{\partial u^{(1)}}{\partial \psi_2}\right]^2\}^{1/2} \tag{4.233}$$

This inequality leads to the instability of the curvature of the surface. In order to determine possible types of this instability let us note that the inequality (4.233) can lead to such a zone

$$\varphi_1 \leqq \varphi \leqq \varphi_2 \tag{4.234}$$

within which the characteristic speed λ will be imaginary, and consequently, the instability occurs in any direction between the angles φ_1, φ_2. These angles are defined from conditions that the resulting normal stress in this direction are equal to zero:

$$\varphi_{1,2} = \arctan\left[\frac{1}{T'_{22}}\left(-T'_{12} \pm \sqrt{T'_{12} - T'_{11}T'_{22}}\right)\right] \tag{4.235}$$

where T'_{ij} are expressed via $\partial u^{(i)}/\partial\psi$; and $\overset{\circ}{T}$ by means of (4.232) and (4.216).

Three different situations can occur:

a)

$$T'^2_{12} > T'_{11}T'_{22} \tag{4.236}$$

i.e., φ_1, φ_2 are real and there are zones of instability given by (4.234).

b)

$$T'^2_{12} \leqq T'_{11}T'_{22},\ T'_{11} < 0 \tag{4.237}$$

Here the instability can occur in any direction because

c)

$$\overset{+}{T'} = T'_{ii}\,|_{\varphi=\dot{\varphi}} \leqq 0, \text{ and } \overset{+}{T'} = T'_{ii}\,|_{\varphi=\dot{\varphi}} < 0 \tag{4.238}$$

$$T'^2_{12} < T'_{11}T'_{22},\ T'_{11} > 0 \tag{4.239}$$

In this case

$$\overset{+}{T'} > 0 \tag{4.240}$$

and there is no instability at all.

Three Dimensional Continuum Models

In order to illustrate the failure of hyperbolicity for three dimensional continua, let us consider the simplest models, such as a thin rod and a thin shell. Based on the formulas (4.137–4.138), (4.154) one obtains

$$\lambda = \pm\sqrt{\frac{1}{\rho_0}\left[T + \frac{E}{2(\nu+1)}\right]} \tag{4.241}$$

$$T < \frac{E}{2(1+\nu)} \tag{4.242}$$

for a thin rod and

$$\lambda = \pm\sqrt{\frac{1}{\rho_0}\left[T_n + \frac{E}{2(\nu+1)}\right]} \tag{4.243}$$

$$T_n < -\frac{E}{2(1+\nu)} \tag{4.244}$$

for a thin shell, where the denotations here are the same as those of the formulas (4.172), and (4.195), respectively.

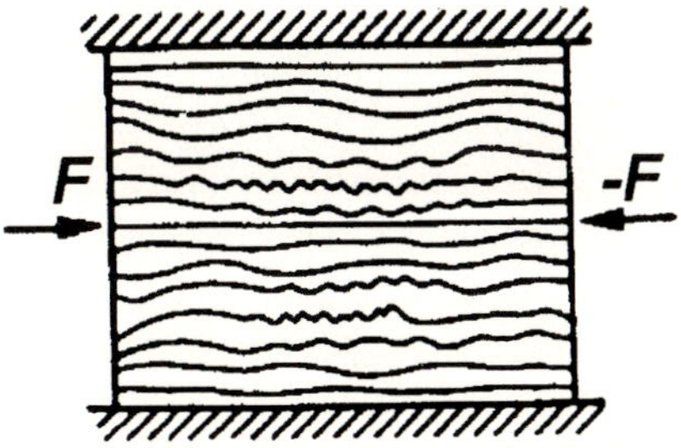

Fig. 4.9.

This means that the same phenomena as described above for a string and membrane emerge in thin rods, and shells, if the compression exceeds the shear modulus (Zak 1983).

But it is easy to conclude, that for such classical (isotropic) elastic materials as steel, the inequalities (4.242), and (4.178) never occur in practice. However, let us consider a laminated elastic material formed by stocking alternatively soft and hard layers of material, while the layers are assumed to be thin. The inequality (4.194) can occur much earlier than yield or failure of the material, and one arrives at instability (wrinkling) of the shapes of layers. This instability was described as a state of microscopic internal collapse of the medium (Fig. 4.9).

By utilizing (4.161), the formulas (4.181–4.244) are generalized for inflated rods and shells:

$$\widetilde{\lambda}_2 = \frac{\rho^1}{\rho+\rho^1}\, u \pm \sqrt{\frac{1}{\rho+\rho^1}\left[T + \frac{E}{2(\nu+1)} - \frac{\rho\rho^1}{\rho+\rho^1}\, u^2\right]} \tag{4.245}$$

$$T < -\left[\frac{E}{2(\nu+1)} - \frac{\rho\rho^1}{\rho+\rho^1}\, u^2\right] \tag{4.246}$$

$$\widetilde{\lambda}_2 = \frac{\rho^1}{\rho+\rho^1}\, u \pm \sqrt{\frac{1}{\rho+\rho^1}\left[T + \frac{E}{2(\nu+1)} - \frac{\rho\rho^1}{\rho+\rho^1}\, u_n^2\right]} \tag{4.247}$$

$$T_n < -\left[\frac{E}{2(\nu+1)} - \frac{\rho\rho^1}{\rho+\rho^1}\, u^2\right] \tag{4.248}$$

Thus, even for classical isotropic materials, the instability can occur if the flow velocity will be large enough.

Now the formula (4.189) is generalized, if one assumes that the pipe resists bending (Fig. 4.6).

$$\ell \le x \le \ell - \frac{\frac{E}{2\rho(v+1)} - u_0^2}{g\xi(1+\varepsilon)} \tag{4.249}$$

Hence, the pipe is stable everywhere, if

$$u_0^2 < \frac{E}{2\rho(\nu+1)} \tag{4.250}$$

So far our attention has been focused on elastic bodies. Let us turn now to an ideal gas. Its bulk coefficient of elasticity E is expressed as:

$$E = -\frac{1}{\varrho}\frac{d\rho}{d(1/\varrho)} = S\frac{d\rho}{d\varrho} \tag{4.251}$$

and the speed of characteristic acoustic wave transporting changes in the density ϱ, and the pressure ρ, is

$$\lambda = \pm\sqrt{\frac{E}{\varrho}} = \pm\sqrt{\frac{d\rho}{d\varrho}} \tag{4.252}$$

In perfect gases

$$\frac{d\rho}{d\varrho} > 0 \tag{4.253}$$

and therefore, Hadamard's instability does not occur. However, it can occur in real gases where the relationship between pressure ρ and the specific volume, $v_n = 1/\varrho$ is expressed by an empirical equation introduced by Van der Waals:

$$\rho = \frac{R_u\theta}{v_n - b} - \frac{a}{v_n^2} \tag{4.254}$$

Here θ is the absolute temperature, R_u is the universal gas constant, and a and b are empirical constants which take into account the finite size of the molecules and the attraction forces between them.

It is easily verifiable that the hyperbolicity of the acoustic equation

$$\frac{\partial^2\varphi}{\partial x^2} - \frac{1}{\lambda^2}\frac{\partial^2\varphi}{\partial t^2} = 0 \tag{4.255}$$

where φ stands for ρ, ϱ, or v_n, and λ is defined by (4.252), fails at the rising branch of the Van der Waals isothermal process (4.254) since there

$$\frac{d\rho}{d\varrho} < 0 \tag{4.256}$$

The instability of gas density and pressure is well known from experiments.

The next example of Hadamard's instability will be concerned with the propagation of gravity waves inside of incompressible fluids. Such waves are due to an inhomogeneity of the fluid caused by the gravitational field. As shown in Landau and Lifshitz (1959), waves whose length is small in comparison with distances over which the gravitational field causes a marked change in density, are governed by the following equations

$$\frac{\partial \boldsymbol{v}}{\partial t} = \frac{g}{\varrho}\left(\frac{\partial\varrho_0}{\partial S_0}\right)_\rho S' - \nabla\left(\frac{\rho'}{\varrho_0}\right), \quad \nabla\cdot\boldsymbol{v} = 0, \quad \frac{\partial S'}{dt} + \boldsymbol{v}\nabla S_0 = 0 \tag{4.257}$$

Here $\boldsymbol{v}$ and $\varrho = \varrho_0 + \varrho'$ are the fluid velocity and density, respectively, g is the free fall acceleration, $S = S_0 + S'$ is the fluid entropy, ϱ_0 and S_0 are the equilibrium values, and ϱ', S' are small dynamical perturbations of ϱ and S, respectively.

The solution of (4.257) can be written in the form of a propagating plane wave:

$$\boldsymbol{v} = \boldsymbol{v}_0 e^{i(\mathbf{k}\cdot\mathbf{r}-\omega t)} \tag{4.258}$$

where ω is the frequency, and $\mathbf{k}$ is the wave vector, while

$$\omega = -\sqrt{\frac{1}{\rho}\left(\frac{\partial \varrho}{\partial S}\right)_\rho g \frac{dS}{dZ} \sin^2\alpha} \tag{4.259}$$

Here the Z axis is vertically upwards, and α is the angle between this axis and the direction of $\mathbf{k}$.

It is interesting to note that although the system (4.257) is not hyperbolic, nevertheless the frequency (4.259) depends upon the direction of the wave vector, but not on its magnitude. This means that for a fixed direction of the wave vector $\mathbf{k}$, the frequency ω does not depend explicitly upon initial boundary conditions, and one can apply the same arguments as those considered for (4.120); in particular, if ω in (4.259) is imaginary, i.e., if

$$\left(\frac{\partial \varrho}{\partial S}\right)_p g \frac{dS}{dZ} > 0 \tag{4.260}$$

Hadamard's instability of solutions to (4.257) occurs. Physically this instability manifests itself in the emergence of convection.

Surface Phenomena

In this item, the conditions of the instability of the shape of a surface separating an elastic material and inviscid liquid flow will be considered.

As shown above, the dynamics of such surfaces, as well as the dynamics of two dimensional continua are characterized by a change in shape, i.e., in the external geometry apart from a change in the internal geometry. However, in contrast to the above mentioned two dimensional continua, the shape of the surface separating two media is defined by their strains, while the governing equations of this surface are coupled with the governing equations of the media playing the role of boundary conditions. Nevertheless some basic dynamical properties of such surfaces can be formally investigated independently from the governing equations of the entire media. These properties depend only on the coefficients of the highest order derivatives, which define the type of governing equations of the separating surface.

Let us consider a free surface of an elastic body streamlined by the flow of an inviscid fluid (Fig. 4.10).

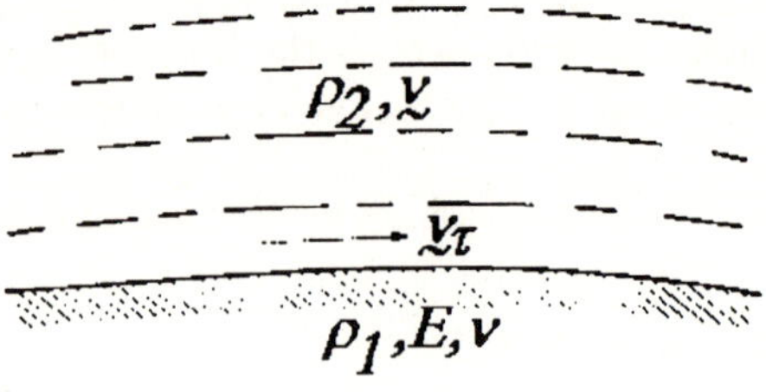

Fig. 4.10.

Applying the principle of virtual work to the volume v_0 containing the inviscid fluid, elastic body, and separating surface, the following is obtained:

$$\int_{V_0} (\rho_1 \mathbf{a} \cdot \delta \mathbf{U}_1 + \rho_2 \mathbf{a}_2 \cdot \delta \mathbf{U}_2 - \delta \Pi)\, dV_0 = 0 \tag{4.261}$$

where ρ_1, ρ_2, $\mathbf{U}_1$, $\mathbf{U}_2$, $\mathbf{a}_1$, $\mathbf{a}_2$ are, respectively, densities, displacements, accelerations of the elastic body, and fluid; and Π is the potential energy of initial stresses and external forces.

The displacements $\mathbf{U}_1$, $\mathbf{U}_2$ are mutually independent in any region which does not contain the separating surface.

Hence, for such volumes, the expression (4.211) leads to the well known governing equations for the elastic body or fluid.

Meanwhile, the above displacements are mutually dependent at the separating surface because of its impenetrability:

$$\mathbf{U}_1^n = \mathbf{U}_1 \cdot n = \mathbf{U}_2 \cdot n = \mathbf{U}_2^n = \mathbf{U}^n \tag{4.262}$$

Consequently, for the variation of the potential energy at the separating surface, it is sufficient to take into account only those terms containing normal displacements and their derivatives:

$$\delta \Pi = \frac{\partial \Pi}{\partial \left(\partial U_1^n / \partial \psi\right)} \delta \frac{\partial U_1^n}{\partial \psi} + \frac{\partial \Pi}{\partial \left(\partial U_2^n / \partial \psi\right)} \delta \frac{\partial U_2^n}{\partial \psi} \tag{4.263}$$

where ψ is the coordinate along an arbitrarily selected direction on the separating surface.

Substituting (4.263) into the integral (4.261) after standard transformations, once can write

$$(\rho_1 \mathbf{a}_1 + \rho_2 \mathbf{a}_2) \cdot \mathbf{n} = \frac{\partial^2 \Pi}{\partial \left(\partial U_1^n / \partial \psi\right)^2} \frac{\partial^2 U_1^n}{\partial \psi_1^2} + \frac{\partial^2 \Pi}{\partial \left(\partial U_2^n / \partial \psi\right)^2} \frac{\partial^2 U_2^n}{\partial \psi^2} + \alpha \tag{4.264}$$

while

$$\mathbf{a}_1 \cdot \mathbf{n} = \frac{\partial^2 U_1^n}{\partial t^2} + \alpha_1 \tag{4.265}$$

$$\mathbf{a}_2 \cdot \mathbf{n} = \left(\frac{\partial}{\partial t} + V_\tau \frac{\partial}{\partial \psi}\right)^2 U_2^n + \alpha_2 \tag{4.266}$$

where V_τ is the projection of the fluid velocity $\mathbf{V}$ on the direction τ, and α, α_1, α_2 are the terms which do not contain the derivatives $\partial^2 U_1^n/\partial\psi^2$, $\partial^2 U_2^n/\partial\psi^2$.

Now taking into account (4.262), (4.265–4.266), the equation (4.264) is transformed to the following form

$$\rho_1 \frac{\partial^2 U^n}{\partial t^2} + \rho_2 \left(\frac{\partial^2 U^n}{\partial t^2} + 2V_\tau \frac{\partial^2 U^n}{\partial t \partial \psi} + V_\tau^2 \frac{\partial^2 U^n}{\partial \psi^2}\right) = \beta \frac{\partial^2 U^n}{\partial \psi^2} + \alpha_3 \tag{4.267}$$

where

$$\beta = \frac{\partial^2 \Pi}{\partial \left(\partial U_1^n/\partial\psi\right)^2} + \frac{\partial^2 \Pi}{\partial \left(\partial U_2^n/\partial\psi\right)^2} \tag{4.268}$$

$$\alpha_3 = \alpha + \alpha_1 + \alpha_2 \tag{4.269}$$

Obviously for the fluid

$$\frac{\partial^2 \Pi}{\partial \left(\partial U_2^n/\partial\psi\right)^2} \equiv 0 \tag{4.270}$$

because the potential energy of shearing deformations in the fluid is zero.

For potential energy of an isotropic elastic body, one can write

$$\delta\Pi = \left[\frac{\partial U_1^n}{\partial \psi} T_{\psi\psi} + \left(1 + \frac{\partial U_1^n}{\partial \xi}\right) T_{\psi\xi}\right] \delta\left(\frac{\partial U_1^n}{\partial \psi}\right) + \alpha_\psi \tag{4.271}$$

where ξ is the coordinate along the normal $\mathbf{n}$ to the separating surface, $T_{\psi\psi}$ are the normal stresses at the separating surface related to the direction τ, $T_{\psi\xi}$ are the shearing stresses related to the directions τ and $\mathbf{n}$ (Fig. 4.11), and α_ψ is the term which does not contain the variation $\delta(\partial U_1^n/\partial\psi)$.

Then

$$\frac{\partial^2 \Pi}{\partial \left(\partial U_1^n/\partial\psi\right)^2} = \frac{\partial}{\partial \left(\partial U_1^n/\partial\psi\right)} \left[\frac{\partial U_1^n}{\partial \psi} T_{\psi\psi} + \left(1 + \frac{\partial U_1^n}{\partial \xi}\right) T_{\psi\xi}\right]$$

$$= T_{\psi\psi} \frac{\partial T_{\psi\xi}}{\partial \left(\partial U_1^n/\partial\psi\right)} \tag{4.272}$$

For small shearing strains

$$\frac{\partial \Pi}{\partial \left(\partial U_1^n/\partial\psi\right)} = \frac{E}{2(1+\nu)} \tag{4.273}$$

where E, ν are Young's modulus and Poisson's ratio.

Hence the formulas (4.272–4.273) give

$$\frac{\partial^2 \Pi}{\partial \left(\partial U_1^n/\partial\psi\right)^2} = T_{\psi\psi} \frac{E}{2(1+\nu)} \tag{4.274}$$

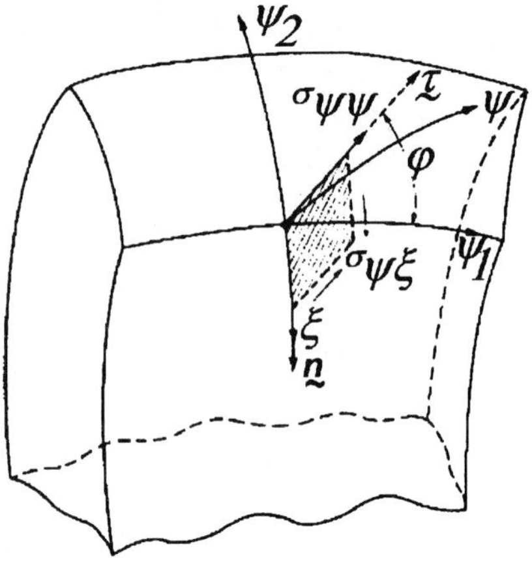

Fig. 4.11.

Now (4.267) is transformed to the final form

$$\rho_1 \frac{\partial^2 U^n}{\partial t^2} + \rho_2 \left(\frac{\partial^2 U^n}{\partial t^2} + 2V_\tau \frac{\partial^2 U^n}{\partial t \partial \psi} + V_\tau^2 \frac{\partial^2 U^n}{\partial \psi^2} \right) = \left\{ T_{\psi\psi} \frac{E}{2\,(1+\nu)} \right\} \frac{\partial^2 U^n}{\partial \psi^2} + \alpha_3 \tag{4.275}$$

Certainly this equation is not complete since, in addition to the unknown U^n, it contains other components of the displacements U_1, U_2 via the term α_3. Hence, in general, (4.194) must be completed by the three dimensional equations of elasticity and of fluid, playing the role of boundary conditions for them.

However, as shown below, there are some basic dynamical properties of free surfaces of elastic bodies which can be qualitatively derived only from (4.275).

Let us derive the characteristic speeds of wave propagation in a free surface of an elastic body along an arbitrarily selected direction τ with the corresponding coordinate φ.

Referring to (4.275) and taking into account the equations of compatibility at the front of the discontinuity [see (4.164)]

$$\left[\frac{\partial}{\partial t} \right] = -\lambda \left[\frac{\partial}{\partial \psi} \right] \tag{4.276}$$

where λ is the characteristic speed of propagation of the discontinuities of the shape of the surface, one gets

$$\left\{ (\rho_1 + \rho_2)\,\lambda^2 - 2\rho_2 V_\tau \lambda + \rho_2 V_\tau^2 - T_{\psi\psi} - \frac{E}{2(\nu+1)} \right\} \left[\frac{\partial^2 U^n}{\partial \psi^2} \right] = [\alpha_3] \tag{4.277}$$

But according to (4.265–4.266) and (4.169), the term α_3 does not contain the derivatives $\partial^2 U^n/\partial t^2$, $\partial^2 U^n/\partial t\partial\psi$, $\partial^2 U^n/\partial\psi^2$ having the discontinuities. Consequently,

$$[\alpha_3] = 0 \tag{4.278}$$

Thus, the equations (4.277–4.278) lead to the following characteristic equations for the speed λ :

$$(\rho_1 + \rho_2)\,\lambda^2 - 2\rho_2 V_\tau \lambda + \rho_2 V_\tau^2 - T_{\psi\psi} - \frac{E}{2(\nu+1)} = 0 \tag{4.279}$$

which defines λ by the formula (Zak 1982)

$$\lambda = \frac{\rho_2}{\rho_1+\rho_2} V_\tau \pm \sqrt{\frac{1}{\rho_1+\rho_2} T_{\psi\psi} + \frac{E}{2(1+\nu)} - \frac{\rho_1\rho_2}{(\rho_1+\rho_2)^2} V_\tau^2} \tag{4.280}$$

Three different types of characteristic waves can be distinguished in association with the formula (4.280):

a) Subcritical waves, if

$$V_\tau^2 < \frac{1}{\rho^2}\left\{T_{\psi\psi} + \frac{E}{2(1+\nu)}\right\} \tag{4.281}$$

In this case, the characteristic speeds λ_1 and λ_2 have different signs, and waves propagate in both direction $\pm\tau$.

b) Supercritical waves, if

$$V_\tau^2 > \frac{1}{\rho^2}\left\{T_{\psi\psi} + \frac{E}{2(1+\nu)}\right\} \tag{4.282}$$

Here, both the characteristic speeds, λ_1, λ_2 have the same sign, and the waves propagate only in the direction of the fluid velocity V_τ.

c) Critical waves, if

$$V_\tau^2 = \frac{1}{\rho^2}\left\{T_{\psi\psi} + \frac{E}{2(1+\nu)}\right\} \tag{4.283}$$

In this case, one of the characteristic waves propagate in the direction of the fluid velocity V_τ, while the other one does not propagate, but forms a stationary motionless wave.

Let us now consider some particular cases:

a) For the free surface of an elastic body when $\rho_2 = 0$, formula (4.280) is simplified:

$$\lambda = \left\{\frac{1}{\rho_1}\left[T_{\psi\psi} + \frac{E}{2(1+\nu)}\right]\right\}^{1/2} \tag{4.284}$$

b) If

$$T_{\psi\psi} \gg \frac{E}{2(1+\nu)},\ \rho_2 = 0 \tag{4.285}$$

the dynamical properties of the free surface become similar to the properties of a film, since

$$\lambda = \pm \left(\frac{T_{\psi\psi}}{\rho_1} \right)^{1/2} \tag{4.286}$$

c) If

$$T_{\psi\psi} \gg \frac{E}{2(1+\nu)}, \text{ but } \rho_2 \neq 0 \tag{4.287}$$

the dynamical properties of the free surface become similar to the properties of a surface separating two inviscid fluids with interfacial tension $T_0 = T_{\psi\psi}$

$$\lambda = \frac{\rho_2}{\rho_1 + \rho_2} V_\tau \pm \sqrt{\frac{T_0}{\rho_1 + \rho_2} - \frac{\rho_1 \rho_2}{(\rho_1 + \rho_2)^2} V_\tau^2} \tag{4.288}$$

Let us make some comments for the special case when

$$\rho_2 = 0, \; T_{\psi\psi} = \frac{E}{2(1+\nu)} \tag{4.289}$$

Introducing for convenience a new variable

$$\overset{*}{T} = T_{\psi\psi} + \frac{E}{2(1+\nu)} \tag{4.290}$$

we will investigate the domain where

$$\overset{*}{T} (\psi < \psi_0) > 0, \; \overset{*}{T} (\psi = \psi_0) = 0 \tag{4.291}$$

For this case, the governing equation (4.275) can be simplified:

$$\rho_1 [\frac{\partial^2 U^n}{\partial t^2}] + T[\rho_1 \frac{\partial^2 U^n}{\partial \psi^2}] = 0 \tag{4.292}$$

Equation (4.292), together with the assumption (4.291) is similar to the equation for the propagation of transverse discontinuities in a string with a free end suspended in a gravity field. Such a problem will be discussed it in detail in the next section. In particular, it will be learned there that if the integral

$$\int_{\psi_0}^{\psi} \frac{dx}{\left\{ \rho_1 \overset{*}{T} (x) \right\}^{1/2}} \tag{4.293}$$

converges, then any infinitesimal transverse disturbance appearing in the neighborhood of the point ψ_0 will be unstable, so that at the point ψ_0 the concentration of the mechanical energy will tend to infinity (the effect of cumulation of energy displaying itself in a snap at the end of the string).

The same kind of energy cumulation phenomenon appears at the free surface of the elastic body if

$$T_{\psi\psi}\ (\psi_0) = \frac{E}{2(1+\nu)} \tag{4.294}$$

$$T_{\psi\psi}\ (\psi < \psi_0) > -\frac{E}{2(1+\nu)} \tag{4.295}$$

and the integral

$$\int_{\psi_0}^{\psi} \frac{dx}{\rho_1 \left\{T_{\psi\psi}\ +\ \frac{E}{2(1+\nu)}\right\}^{1/2}} \tag{4.296}$$

converges.

Recall that all the above results were referred to an arbitrary direction τ selected on the separating surface. To relate these results to the fixed system of coordinates ψ_1, ψ_2 at the surface, the following formula must be used:

$$T_{\psi\psi} = \frac{1}{2}(T_{11} + T_{22}) + \frac{1}{2}(T_{11} - T_{22}) \cos 2\phi + T_{12} \sin 2\phi \tag{4.297}$$

$$V_\tau = V_1 \cos\phi + V_2 \sin\phi \tag{4.298}$$

where T_{11}, T_{22}, T_{12} are the components of the stress tensor in the system ψ_1, ψ_2; V_1, V_2 are the components of the fluid velocity V_τ in the system ψ_1, ψ_2; ϕ is the angle between the above selected direction τ and the unit vector of the coordinate ψ.

Thus, instead of (4.283), one arrives at the more complicated expression for the characteristic speeds which is related to the coordinates ψ_1, ψ_2 :

$$\lambda = \frac{\rho_2}{\rho_1+\rho_2}(V_1 \cos\phi + V_2 \sin\phi)$$

$$\pm\{\frac{1}{\rho_1+\rho_2}[\frac{1}{2}(T_{11}+T_{22}) + \frac{1}{2}(T_{11}-T_{22})\cos 2\phi + T_{12}\sin 2\phi$$

$$+\frac{E}{2(1+\nu)} + \frac{\rho_1\rho_2}{(\rho_1+\rho_2)^2}(V_1\cos\phi + V_2\sin\phi)^2]\} \tag{4.299}$$

This formula shows that the characteristic speeds, as well as the characteristic properties of the free surfaces presented above, depend not only on the location of the point at the surface, but also on the direction τ selected at that point.

Now let us make some remarks about the degree of the arbitrariness in selection of the direction τ along which the characteristic waves can propagate. In other words, let us derive any restrictions which are imposed on the angle in the formula (4.299).

For this purpose, recall that the second derivatives of the transverse displacement U^n can be expressed by the coefficients of the second fundamental form of the surface b_{ij}, so that

$$[\frac{\partial^2 U^n}{\partial \psi_1}] = [b_{11}], \; [\frac{\partial^2 U^n}{\partial \psi_2}] = [b_{22}], \; [\frac{\partial^2 U^n}{\partial \psi_1 \partial \psi_2}] = [b_{12}] \tag{4.300}$$

But formula (4.226) shows that the possible directions of propagations of the waves transporting the discontinuities of the surface shape must coincide with asymptotic lines on the surface. Hence,

$$\phi = \arctan\left\{\frac{1}{b_{11}}\left[-b_{12} \pm (-G)^{1/2}\right]\right\} \tag{4.301}$$

where G is Gauss curvature of the surface. Thus, such waves are impossible at the points of elliptic type.

In conclusion, it is necessary to emphasize that all the above presented surface waves (in contrast to Rayleigh surface waves) are located only at the free surface as a surface of mathematical singularities, although they certainly influence the motion of the internal particles of the elastic body via a term α_3 in the equation (4.277) which couples it with the equations of the corresponding elastic body. However, such an induced motion of internal particles does not possess the characteristic properties.

Referring to the governing equation (4.275), let us study the propagation of high frequency oscillations of the transverse displacements of the free surface of an elastic body. First of all, recall that this equation, being linear with respect to the second order derivatives $\partial^2 U^n/\partial t^2$, $\partial^2 U^n/\partial \psi^2$, $\partial^2 U^n/\partial \psi \partial t$, strictly speaking, is nonlinear with respect to the terms $\partial U^n/\partial t$, $\partial U^n/\partial \psi$, U^n, which is contained in α_3. For small amplitudes of oscillations and their first derivatives, this term can be linearized:

$$\alpha_3 = \alpha_{31}\frac{\partial U^n}{\partial t} + \alpha_{32}\frac{\partial U^n}{\partial \psi} + \alpha_{33}U^n + \alpha_{34} \tag{4.302}$$

where α_{34} does not depend on the terms $\partial U^n/\partial t$, $\partial U^n/\partial \psi$, U^n. For further simplification, all the coefficients of (4.275) can also be linearized with respect to the arbitrarily selected point ψ_0 and instant of time $t_0 = 0$. Then (4.275) is written in the form of a linear differential equation with constant coefficients:

$$\rho_1\frac{\partial^2 U^n}{\partial t^2} + \rho_2(\frac{\partial^2 U^n}{\partial t^2} + 2V_\tau^0\frac{\partial^2 U^n}{\partial t \partial \psi} + V_\tau^{02}\frac{\partial^2 U^n}{\partial \psi^2}) - \left\{T^0_{\psi\psi}\frac{E}{2(1+\nu)}\right\}\frac{\partial^2 U^n}{\partial \psi^2}$$

$$= \alpha^0_{31}\frac{\partial U^n}{\partial t} + \alpha^0_{32}\frac{\partial U^n}{\partial \psi} + \alpha^0_{33}U^n + \alpha^0_{34} \tag{4.303}$$

Recall that this equation is valid only for small amplitudes with small first derivatives, the small area around the above selected point ψ_0, and the small period of time Δt.

Let us derive the solution of (4.303) for the following initial conditions:

$$\left.\begin{aligned} U_0^n &= \frac{1}{\lambda_0} e^{-\lambda_0 \psi i}, \quad \lambda_0 > 0 \\ U_0^\tau &= U \cdot \tau = 0 \end{aligned}\right\} t = 0 \tag{4.304}$$

assuming that λ_0 can be made as large as necessary, i.e.,

$$\lambda_0 > N \to \infty \tag{4.305}$$

Consequently, the initial disturbances can be made as small as necessary, i.e.,

$$U_0^n < \frac{1}{N} \to 0 \tag{4.306}$$

The corresponding solution is written in the form

$$U_*^n = C_1 e^{-\lambda_0(\lambda_1 t - \psi)i} + C_2 e^{-\lambda_0(\lambda_2 t - \psi)i} \tag{4.307}$$

where λ_1, λ_2 are the roots of the characteristic equation

$$(\rho_1 + \rho_2)\lambda^2 - 2\rho_2 V_\tau^0 \lambda + \rho_2 V_\tau^{02} - T_{\psi\psi}^0 - \frac{E}{2(\nu+1)} = \frac{1}{\lambda_0}\alpha_\psi \tag{4.308}$$

Here, the term α_ψ contains the coefficients at the unknown U^n and their first derivatives $\partial U^n/\partial t$, $\partial U^n/\partial \psi$.

The right hand side of this equation tends to zero according to the assumption (4.305), and one arrives at the characteristic equation (4.279). This means that λ_1 and λ_2 in the solution (4.307) coincide with the characteristic speeds of the wave propagation (4.280).

Now let us assume that λ_1, λ_2 are complex, i.e.,

$$\text{Im } \lambda_{1,2} \neq 0 \tag{4.309}$$

Then the solution (4.307) will contain the term

$$\frac{1}{\lambda_0} e^{|\text{Im } \lambda_{1.2}|\Delta t} \sin \lambda_0 \psi \tag{4.310}$$

which leads to infinity, if $\lambda_0 \to \infty$ (within an arbitrarily short period of time Δt and within an infinitesimal area around the point ψ_0).

Hence, one arrives at the following situation

$$|U_*^n| \to \infty \tag{4.311}$$

in spite of the fact that

$$|U_0^n| \to 0 \tag{4.312}$$

However, strictly speaking, because of the utilization of the governing equation (4.303) in linearized form, the condition (4.312) must be weakened:

$$|U_*^n| \nrightarrow 0 \text{ if } |U_0^n| \to 0 \tag{4.313}$$

Turning to the same line of argumentation as those applied for isotropic elastic bodies [see (4.123–4.126)], one concludes that the conditions of Hadamard's instability, and therefore the loss of smoothness of the surface separating the elastic body and the inviscid flow given by the following inequality:

$$T_{\psi\psi} \leq \frac{\rho_1 \rho_2}{(\rho_1 + \rho_2)} V_\tau^2 - \frac{E}{2(1+\nu)} \tag{4.314}$$

because the condition (4.310) now holds.

Hence, the stability of the smooth shape of a free surface decreases as compression $T_{\psi\psi}$ and fluid viscosity increases.

The criterion of wrinkling is simplified, if $\rho_2 = 0$ or $V_\tau = 0$:

$$T_{\psi\psi} \leq -\frac{E}{2(1+\nu)} \tag{4.315}$$

In the particular case ($E = 0$), one arrives at the criterion of wrinkling for the film surface without resistance to bending:

$$T_{\psi\psi} \leq 0 \tag{4.316}$$

In terms of the formula (4.299) the criterion of wrinkling is given by the inequality

$$\frac{1}{2}(T_{11} + T_{22}) + \frac{1}{2}(T_{11} - T_{22}) \cos 2\phi + T_{12} \sin 2\phi$$
$$\leq \frac{\rho_1 \rho_2}{(\rho_1 + \rho_2)^2} (V_1 \cos\phi + V_2 \sin\phi)^2 \tag{4.317}$$

which shows that the criterion of wrinkling depends also on the angle ϕ, i.e., on the direction τ on the surface. It is clear that wrinkles appear in such directions ϕ for which the inequality (4.317) holds. Hence, the contraction of the free surface filaments under compression is divided into two stages. During the first stage, when the compression is smaller than required by the inequality (4.317), the contraction is carried out without the loss of smoothness of the shape (like for the internal filaments). During the second stage, when the compression approaches the limit value.

$$T^*_{\psi\psi} = \frac{\rho_1 \rho_2}{(\rho_1 + \rho_2)} V_\tau^2 - \frac{E}{2(1+\nu)} \tag{4.318}$$

the mechanism of contraction changes, now being carried out by microdeflections of the filament so that its length (measured along all the microdeflections) remains invariable. It is clear that such a mechanism of contraction does not produce an additional resistance. Hence, actually, the compression cannot exceed the critical value $T^*_{\psi\psi}$, and formally, the second stage of contraction is similar to plastic deformation.

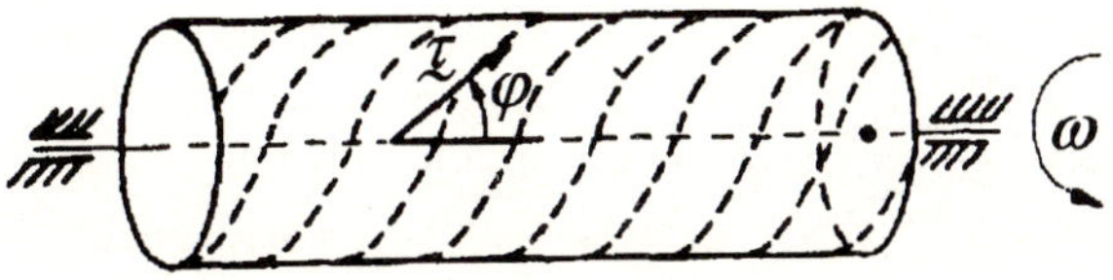

Fig. 4.12.

Let us illustrate the criteria of Hadamard's instability by some examples.

First, consider a cylindrical linear-elastic shaft transmitting some torques (Fig. 4.12). The shearing stresses at its free surface are

$$T_{12} = \frac{Er}{2(\nu+1)} \frac{\alpha}{l} \tag{4.319}$$

where r is the radius of the shaft, α is angle of twist, and l is the length of the shaft.

The maximum compressive stresses appear along the direction forming the angle $\phi = 45°$ with the generator of the cylinder, while

$$\widetilde{T} = T_{12} = \frac{Er}{2(\nu+1)} \frac{\alpha}{l} \tag{4.320}$$

The critical twist angle α^* which leads to the wrinkles of the surface of the shaft is defined by the equation

$$\frac{Er}{2(\nu+1)} \frac{\alpha^*}{l} + \frac{E}{2(\nu+1)} = 0 \tag{4.321}$$

$$\alpha^* = \frac{l}{r} \tag{4.322}$$

Hence, the intensity of wrinkles is proportional to the difference

$$\alpha - \frac{l}{r} \tag{4.323}$$

For the shaft rotating in an inviscid fluid with the angular velocity ω, instead of (4.320), one gets

$$\widetilde{T} = \frac{E}{2(\nu+1)} + \frac{1}{2}\frac{\rho_1\rho_2}{(\rho_1+\rho_2)}\omega^2 r^2 - \frac{1}{2}\left[\frac{\rho_1^2\rho_2^2}{(\rho_1+\rho_2)^2}\omega^2 r + \frac{E^2 r^2}{(\nu+1)^2}\frac{\alpha^{*2}}{l^2}\right]^{1/2} \tag{4.324}$$

Hence, the critical twist angle α^* will be

$$\alpha^* = \frac{1}{r}\left[l - \frac{2\rho_1\rho_2\omega^2 r^2(\nu+1)}{(\rho_1+\rho_2)E}\right]^{1/2} \tag{4.325}$$

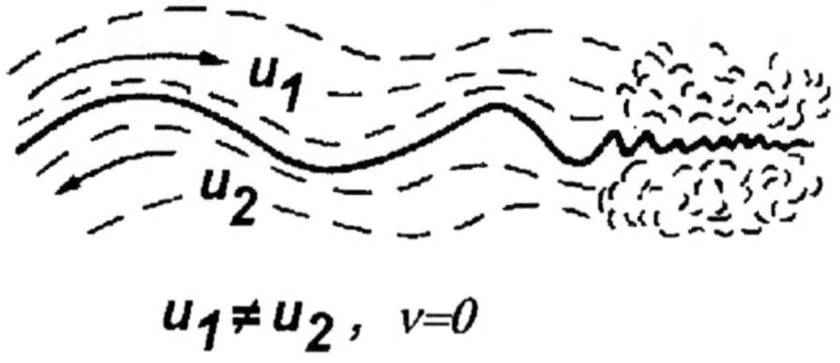

Fig. 4.13.

The angle ϕ which defines the direction of wrinkles now depends on the twist angle α :

$$\phi = \arctan \frac{\frac{1}{2}\frac{\rho_1\rho_2}{(\rho_1+\rho_2)}\omega^2 r^2 - \frac{1}{2}\left[\frac{\rho_1\rho_2}{(\rho_1+\rho_2)}\omega^2 r + \frac{E^2 r^2}{(\nu+1)^2}\frac{\alpha^{*2}}{l^2}\right]^{1/2}}{\frac{Er}{2(\nu+1)}\frac{\alpha}{l}} \tag{4.326}$$

Formula (4.325) shows that if

$$\omega > \frac{(\rho_1+\rho_2)E}{2\rho_1\rho_2 r^2(\nu+1)} \tag{4.327}$$

then wrinkles are induced only by the rotation of the shaft even without any initial stresses.

As a second example, consider a surface of a tangential jump of velocity in an inviscid fluid (Fig. 4.13). Turning to (4.288) and eliminating the surface tension

$$T_0 = 0 \tag{4.328}$$

one obtains the characteristic speed of the surface wave propagation

$$\lambda = \frac{1}{2}\left[(U_2 - U_1) \pm \sqrt{-(U_2 - U_1)^2}\right] = \frac{1}{2}(U_2 - U_1)(1 \pm i) \tag{4.329}$$

which has a non zero imaginary part.

Hence, we arrived at a well known result that in inviscid fluids, tangential jumps of velocities are unstable. However, in view of our previous arguments [see (4.304–4.314)], this is Hadamard's instability. (In fluid mechanics it is called Kelvin-Helmholtz instability.)

As our last example, consider a surface separating a sand and an inviscid flow (Fig. 4.14), presenting sand as an elastic body with the following limitation for the shear modulus at the separating surface

$$G \leq kp \tag{4.330}$$

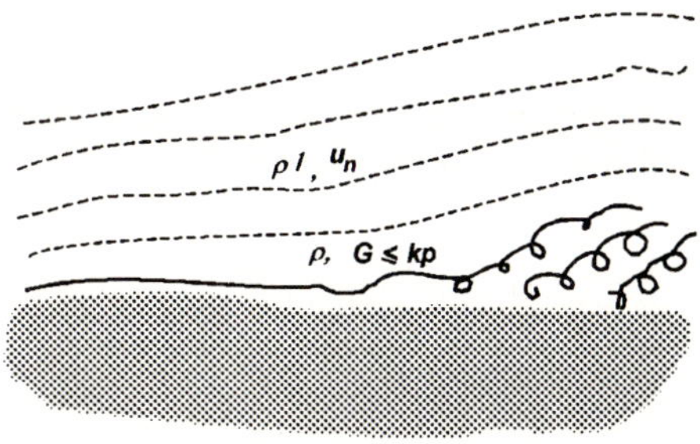

Fig. 4.14.

where p is the pressure of the fluid at this surface and k is the friction coefficient of sand, one derives from (4.314) the condition of the instability (erosion) of sand by the flow (Fig. 4.13).

$$u_n^2 > \frac{\rho + \rho^1}{\rho \rho^1} kp \tag{4.331}$$

4.4 Cumulative Effects

4.4.1 Degenerating Hyperbolic Equations

A cumulative effect can be introduced as a preinstability state which is associated with the change of type of the governing equations from hyperbolic to parabolic when at least one of the characteristic speeds becomes zero. Actually this state represents the boundary for the Hadamard instability, and depending on how the motion approaches this boundary, it may remain stable, or unstable. The simplest example of this type of situation is the governing equation for a vertical ideally flexible inextensible string with a free lower end suspended in a gravity field (Fig. 4.5b). Projecting this equation onto a horizontal direction, one arrives at the governing equations for small transverse motion of the string

$$\frac{\partial^2 x}{\partial t^2} - \frac{T}{\rho} \frac{\partial^2 x}{\partial \psi^2} = 0 \tag{4.332}$$

with the characteristic speeds given by (4.177):

$$\lambda = \pm \sqrt{\frac{T}{\rho}} \tag{4.333}$$

Since the tension of the string T vanish at the free end

$$T = 0 \text{ at } \psi = l \tag{4.334}$$

where l is the length of the string, the characteristic speeds (4.333) vanish too at $\psi = l$, and therefore, (4.332) degenerate into parabolic type at the very end of the string.

As a second example, consider a one dimensional model of the shear wave propagation in a soil column of the height H:

$$\rho\frac{\partial^2 u}{\partial t^2} = \frac{\partial}{\partial x}\left(G\frac{\partial u}{\partial x}\right) \tag{4.335}$$

where ρ is the density, u is the horizontal displacement, G is the shear modulus, t is time, and x is the vertical coordinate with the origin at the surface.

Ignoring the small shear stresses at the surface, the shear modulus can be taken in the following form:

$$G = 0.5\rho g x \tag{4.336}$$

Since

$$G = 0 \text{ at } x = 0 \tag{4.337}$$

(4.334) degenerates into parabolic type at the soil surface.

The same effect can appear in (4.275). Assuming, for simplicity, that the conditions (4.289–4.291) hold, one arrives at (4.292) which becomes parabolic at $\psi = \psi_0$.

For the sake of concreteness, we will investigate the solution to (4.336) subject to the initial and boundary conditions formulated below:

$$u(x,0) = \varphi(x),\ 0 \leq x \leq H \tag{4.338}$$

$$\frac{\partial u}{\partial x}(x,0) = \psi(x)$$

$$u(H,t) = \alpha(t)$$

$$\frac{\partial u}{\partial x}(0,t) = 0 \tag{4.339}$$

Thus, it is assumed that the soil column is fixed at $x = H$ and there is no shear stress at the surface, i.e., at $x = 0$.

One should note that for simplicity in this model all the damping and creep effects are ignored.

4.4.2 Uniqueness of the Solution

Let us assume that there exist two solutions of the problem under consideration: $u'(x,t)$ and $u''(x,t)$; and let us examine the difference:

$$u^*(x,t) = u'(x,t) - u''(x,t) \tag{4.340}$$

The function $u^*(x,t)$ satisfied (4.334) with additional homogeneous conditions:

$$\rho \frac{\partial^2 u^*}{\partial t^2} = \frac{\partial}{\partial x}\left(G \frac{\partial u^*}{\partial x}\right) \tag{4.341}$$

$$u^*(x,0) = 0, \; \frac{\partial u^*}{\partial t}(x,0) = 0 \tag{4.342}$$

$$u^*(H,t) = 0, \frac{\partial u^*}{\partial x}(0,t) = 0 \tag{4.343}$$

For the total energy one gets:

$$E(t) = E(0) = \frac{1}{2} \int_0^H \left\{ G \left(\frac{\partial u^*}{\partial x}\right)^2 + \rho \left(\frac{\partial u^*}{\partial t}\right)^2 \right\} \Bigg|_{t=0} dx = 0 \tag{4.344}$$

If the solution is sought in the open interval

$$0 \leq x \leq H \tag{4.345}$$

which does not include the surface point $x = 0$, then the uniqueness of the solution is obvious.

However, this proof cannot be applied to the closed interval

$$0 \leq x \leq H \tag{4.346}$$

which includes the surface point $x = 0$. Indeed, in this case according to (4.337)

$$G = 0 \text{ at } x = 0 \tag{4.347}$$

and any arbitrarily selected derivative $\partial u^*/\partial x$ at $x = 0$ will satisfy the equality in (4.344).

Thus, for the closed interval (4.346), the uniqueness of the solution can be guaranteed only in the class of functions having continuous derivative $\partial u^*/\partial x$, otherwise the infinite number of different solutions can be offered to satisfy equation (4.334) with the conditions (4.338–4.339) in the interval (4.346). As will be shown in the following, the artificial mathematical restriction about the continuity of the derivative $\partial u^*/\partial x$ excludes such important physical phenomena as cumulative effects (Zak 1970).

From the mathematical point of view the singularity at the point $x = 0$ is associated with the fact that the original equation is hyperbolic in the open interval (4.345), but degenerates into a parabolic equation at the point $x = 0$. The physical meaning of this singularity will be discussed in the following section.

4.4.3 Stability of the Solution

Starting with the initial conditions (4.338) let us assume that

$$\varphi(x)\begin{cases} >0 \text{ for } & 0<\overset{*}{x}_1,\ x<x_2<H \\ =0 \text{ for } & x<x_1, \text{ and } x>x_2 \end{cases} \tag{4.348}$$

$$\psi(x)|_{0=x\le H}=0 \tag{4.349}$$

$$\alpha(t)=0,\ t\ge 0 \tag{4.350}$$

i.e., we consider an initial disturbance in a local interval $(\overset{\circ}{x}_1, \overset{\circ}{x}_2)$ contained within the interval $(0, H)$.

From the differential equations of the characteristics one finds the equations of the characteristics passing through x_1 and x_2:

$$t'=\int_{x_1}^{\overset{*}{x}_1}\frac{d\varepsilon}{\sqrt{G(\varepsilon)/\varphi}},\quad t''=\int_{x_2}^{\overset{*}{x}_2}\frac{d\varepsilon}{\sqrt{G(\varepsilon)/\varphi}},\quad 0<x_1<x_2<H \tag{4.351}$$

Here x_1 and x_2 are the coordinates of the leading and trailing fronts of the discontinuity wave of derivatives $\partial^2u/\partial t^2$ and $\partial^2u/\partial x^2$, where

$$x_1=\overset{*}{x}_1,\ x_2=\overset{*}{x}_2 \text{ at } t=0$$

A singular solution coincident for both characteristics holds for

$$x_1=x_2=0$$

because

$$\left.\frac{dx_1}{dt}\right|_{x_1=0}=\left.\frac{dx_2}{dt}\right|_{x_2=0}=0 \tag{4.352}$$

Two cases may arise:

a) The improper integral

$$\int_x^{x_0}\frac{d\eta}{\sqrt{G(\eta)/\rho}}<\infty \text{ for } x\to 0 \tag{4.353}$$

i.e., converges, which then means that coincidence of the characteristics (4.351) occurs for the finite $t=t^*$.

Then

$$\left|\frac{\partial u}{\partial t}\right|\to\infty \text{ for } t=t^*<\infty \tag{4.354}$$

From the mathematical viewpoint this instability predicts a cumulation of the shear strain energy at the soil surface $x=0$. At the same time it illustrates the ambiguity in the solution which has been remarked in the investigation of equation (4.344).

b) If the improper integral (4.353) diverges the characteristics (4.351) coincide at $t^* \to \infty$ and the cumulation effect does not occur.

For the particular case of soil where the shear modulus is given in the form (4.336) the integral (4.353) converges and the time t^* defining the moment of the formation of the shear strain energy cumulation at the soil surface is

$$t^* = 2\sqrt{\frac{2\overset{*}{x}_2}{g}}$$

In the general case when the shear modulus is a more complicated function of the elevation the cumulative effect occurs if

$$G > \varepsilon_1 x^{2+\varepsilon_2}$$

where $\varepsilon_1, \varepsilon_2$ are arbitrarily small positive constants, because then the integral (4.353) converges (Zak 1983).

4.4.4 Snap of a Whip

The results presented above can be applied to (4.332) describing transverse oscillations of a vertical ideally flexible inextensible string with a free lower end suspended in a gravity field.

The tension of the string due to gravity is given by the following equation:

$$T = \gamma(l - x) \tag{4.355}$$

where γ, l are the specific weight and length of the string.

Referring to the formula (4.333) one concludes that the characteristic speed of transverse displacements tends to zero at the free end:

$$T|_{x=l} = 0, \; \tilde{\lambda}_{3,4} \to 0 \text{ if } x \to l \tag{4.356}$$

In other words, for small transverse displacements of the string, the governing equation is of the hyperbolic type only in the open interval, excluding the end:

$$0 \leq x < l \tag{4.357}$$

As shown in the previous section, in this open interval there exists a unique stable solution. However, in the closed interval, including the end

$$0 \leq x \leq l \tag{4.358}$$

the solution is not unique and there are unstable solutions if the improper integral

$$\int_0^x \frac{d\xi}{[T(\xi)/\rho]^{1/2}} \tag{4.359}$$

converges for $x \to l$.

This result has the very clear physical interpretation: suppose that an isolated transverse wave of small amplitude was generated at the point of suspension (Fig. 4.14). The speed of propagation of the leading front of the transverse wave will be smaller than the speed of the trailing front because the tension decrease from the point of suspension to the free end [see (4.333–4.353)]. Hence the length of the above wave will be decreasing and in some cases (4.359) will tend to zero. Then according to the law of conservation of energy the specific kinetic energy per unit of length will tend to infinity producing a snap (snap of a whip).

It can easily be verified by substituting (4.355) in (4.359) that for the string in the gravity fields the integral (4.360) converges; i.e., the instability in the form of a snap occurs (Zak 1970).

4.4.5 General Case

The same type of instability as a result of cumulation of energy near the boundary of the failure of hyperbolicity can exist in two and three dimensional models in the domains where inequalities (4.148–4.150) are close enough to the corresponding equalities.

As shown in the previous section, the cumulative effect as a pre-instability phenomenon, is associated with the modification of the originally hyperbolic governing equation into parabolic one when the characteristic speed becomes zero. For the sake of concreteness, we will consider a characteristic speed of shear wave propagation in elastic bodies based upon (4.56–4.57).

$$\lambda_{1i} = \pm\sqrt{\frac{1}{\rho}\left(T_{11} + \frac{1}{2}\frac{\partial T_{1i}}{\partial \varepsilon_{1i}}\right)} = 0, \; i = 1, 2, 3 \tag{4.360}$$

Usually the domain (4.360) forms a surface S in an elastic body. This separates the stable zone, where the governing equation is hyperbolic, from the unstable zone, where the governing equation is elliptic.

In this section we will consider the behavior of the solution within the stable zone near the surface S, i.e., a pre-instability effect. As shown in the previous section, there are some situations where small disturbances, occurring near the boundary of stability, propagate to this boundary and grow without limit. General criteria for such a cumulative effect in an elastic material will be given below.

Let us fix an arbitrary point M at the surface S, where the condition (4.49) holds, and select a local Cartesian basis in M such that the axis x_1, is directed toward the zone of stability. Assuming that in the initial state the characteristic speed λ_{1i} is a differentiable function of all its arguments, one can find such small x^* that

$$|\lambda_{1i}| > 0, \left|\frac{\partial \lambda_{1i}}{\partial x_1}\right| < 0 \text{ at } 0 < x_1 < \overset{*}{x}_1 \tag{4.361}$$

Recalling that characteristic speed λ, transport discontinuities $(\partial\varepsilon_{1i}/\partial x_1)$ pose the following initial value problem:

$$\tilde{\varepsilon}_{li}\big|_{t=0} = \begin{cases} \alpha\phi(x_1) > 0 \text{ for } & 0 < \overset{0^1}{x_1} < x_1 < \overset{0^{11}}{x_1} < \overset{*}{x_1} \\ = 0 \text{ for } & x_1 \leq \overset{0'}{x_1} \text{ and } x_1 \geq \overset{0^{11}}{x_1} \end{cases}$$

$$\left.\frac{\partial\tilde{\varepsilon}_{1i}}{dt}\right|_{t=0} = 0 \tag{4.362}$$

while the arbitrary constant multiplier α can be made as small as necessary, and

$$\left.\frac{d\phi}{dx_1}\right|_{x_1 = \overset{0^1}{x_1}} \neq 0 \tag{4.363}$$

$$\left.\frac{d\phi}{dx_1}\right|_{x_1 = \overset{0^{11}}{x_1}} = 0 \tag{4.364}$$

Thus, in the initial state due to (4.362)

$$[\varepsilon_{1i}] = 0 \tag{4.365}$$

but due to equation (4.363)

$$\left[\frac{\partial\varepsilon_{1i}}{\partial x_1}\right] \neq 0 \text{ at } x = \overset{0^1}{x_1} \text{ and } x = \overset{0^{11}}{x_1} \tag{4.366}$$

The initial energy of the disturbed motion

$$E_0 = \int_{\overset{0^1}{x_1}}^{\overset{0^{11}}{x_1}} \tilde{E} dx_1 \tag{4.367}$$

where $\tilde{E}$ is the specific energy per unit of the length along the axis x_1.

Because of the conditions (4.366), the disturbed zone which initially was contained between $\overset{0^1}{x_1}$ and $\overset{0^{11}}{x_1}$ will move with the characteristic speeds $\pm\lambda_{1i}$ in both directions along the x_1 axis. However, our attention will be concentrated at the speed directed toward the boundary of instability, i.e., the point M. Introducing the characteristic speeds λ_{1i}^1 and λ_{1i}^{11} for the leading and trailing fronts, respectively, one arrives at the following differential equations governing the motions of these fronts:

$$\frac{dx_1^1}{dt} = -\lambda_{1i}^1, \quad \frac{dx_1^{11}}{dt} = -\lambda_{1i}^{11} \tag{4.368}$$

while

$$x_1^1\big|_{t=0} = \overset{0^1}{x_1}, \quad x_1^{11}\big|_{t=0} = \overset{0^{11}}{x_1} \tag{4.369}$$

and

$$\lambda_{1i}^{1}=\lambda_{1i}(x_1^1),\ \lambda_{1i}^{11}=\lambda_{1i}(x_1^{11}) \tag{4.370}$$

The solutions to these equations are, respectively:

$$t=\int_{x_1^1}^{\overset{0}{x}{}_1^1}\frac{dx_1}{\lambda_{1i}^{1}}\ \text{and}\ t=\int_{x_1^{11}}^{\overset{0}{x}{}_1^{11}}\frac{dx}{\lambda_{1i}^{11}},\ 0\le x_1^1\le x_1^{11}<\overset{*}{x}_1 \tag{4.371}$$

and can be considered as the characteristics passing through $\overset{0}{x}{}_1^1$ and $\overset{0}{x}{}_1^{11}$.

These solutions express in the implicit form the equations of motions of the leading and trailing fronts:

$$x_1^1=f_1(\overset{0}{x}{}_1^1,t),\ x_1^{11}=f_2(\overset{0}{x}{}_1^{11},t) \tag{4.372}$$

A singular solution coincident for both characteristics holds for

$$x_1^1=x_1^{11}=0 \tag{4.373}$$

because of

$$\lambda_{1i}^{1}=\lambda_{1i}^{11}=0\ \text{at}\ x_1^1=x_1^{11}=0 \tag{4.374}$$

if

$$\int_{\xi_1}^{\xi}\frac{dx_1}{\lambda_{1i}}<\infty\ \text{at}\ \xi\to 0 \tag{4.375}$$

i.e., this improper integral converges. Then the coincidence of the characteristics occurs for a finite time

$$t=t^*<\infty \tag{4.376}$$

and consequently,

$$x_1^{11}-x_1^1\to 0\ \text{for}\ t\to t^* \tag{4.377}$$

Now turning to (4.367) and exploring the principle of the total mechanical energy conservation in an ideal elastic material one obtains

$$E_0=\int_{\overset{0}{x}{}_1^1}^{\overset{0}{x}{}_1^{11}}\widetilde{E}dx_1=\int_{x_1^1}^{x_1^{11}}\widetilde{E}dx_1=E=\text{Const}>0 \tag{4.378}$$

Applying the first mean value theorem to the last integral in (4.378):

$$E=(x_1^{11}-x_1^1)\,\overset{*}{\widetilde{E}},\ \overset{*}{\widetilde{E}}=\widetilde{E}(\overline{x_1}),\ x_1^1\le\overline{x_1}\le x_1^{11} \tag{4.379}$$

with reference to (4.377) one concludes that

$$\widetilde{E} = \frac{E}{x_1^{11} - x_1^1} \to \infty \text{ if } t \to t^* < \infty \tag{4.380}$$

But the specific mechanical energy of the transverse wave in an elastic material is given by the following expression:

$$\widetilde{E} = \frac{1}{2}\rho v_i^2 + \lambda_{1i}\varepsilon_{1i}^2 \tag{4.381}$$

where v_i is the transverse component of the velocity, ε_{1i} is the corresponding strain, and λ_{1i} is given by (4.360).

Taking into account (4.380–4.381), and (4.360), one gets:

$$v_i \to \infty \text{ if } t \to t^* < \infty \tag{4.382}$$

Obviously, the estimate (4.382) does not hold if the improper integral (4.375) diverges, because then any disturbance originated in the interval $0 < x_1 < \overset{*}{x}_1$ will not reach the boundary point M in a finite time period, i.e., the coincidence of the leading and trailing fronts of the wave never occurs.

Thus if the characteristic speed λ_{1i} decreases fast enough to its zero value at the boundary surface S [i.e., the improper integral (4.375) converges] then the cumulative effect occurs at this boundary: the specific mechanical energy, as well as the transverse velocity, tend to infinity.

The result formulated has been obtained under the particular conditions (4.362). However, by using the superposition principle for the governing equation, linearized with respect to the selected point M, and adding the conditions (4.362), which may differ from the zero conditions as little as desired, to the arbitrary initial conditions, one arrives at the same estimates (4.380) and (4.382).

4.4.6 Failure of Lipschitz Conditions

The cumulative effects are accompanied by a very interesting mathematical phenomenon: the failure of Lipschitz conditions for the differential equations of characteristics (4.370–4.371):

$$\frac{ds}{dt} = \dot{s} = \lambda(s) \tag{4.383}$$

Indeed, if the characteristic speed follows (4.332) or (4.334), i.e.,

$$\lambda = \pm\sqrt{\frac{T}{\rho}}, \text{ or } \lambda = \sqrt{\frac{G}{\rho}}, \text{ repectively,} \tag{4.384}$$

then

$$|\frac{\partial\lambda}{\partial s}| = |\frac{1}{2\lambda}\frac{\partial\lambda^2}{\partial s}| \rightarrow 0 \text{ at } s \rightarrow s_0 \tag{4.385}$$

if

$$|\frac{\partial\lambda^2}{\partial S}| > 0 \text{ at } s \rightarrow s_0 \tag{4.386}$$

As follows from (4.336), and (4.355),

$$|\frac{\partial\lambda^2}{\partial s}| = 0.5g > 0, \text{ and } |\frac{\partial\lambda^2}{\partial s}| = g > 0, \text{ respectively}$$

Hence, the loss of the uniqueness of the solution to (4.332) and (4.334) can be formally associated with the failure of the Lipschitz condition at the point where the characteristic speed vanishes. Recall that all classical dynamics as well as the theory of differential equations, are based upon this condition. Nevertheless, without its violation all the cumulative effects (which are observed from experiments) cannot be simulated. Such an inconsistency between the mathematical constraints imposed upon the dynamical models, and the physical nature of motions, will be discussed in detail in the next chapter.

4.5 Lagrangian Turbulence

In this section, the criteria for Hadamard's instability are generalized to continua which are not necessarily elastic. In particular, they are associated with the concept of Lagrangian turbulence (Zak 1986a,b).

Lagrangian turbulence in continua results from chaotic instabilities of trajectories of individualized (marked) particles. In inviscid fluids this phenomenon was introduced by Arnold (1978), who proved that Lagrangian turbulence is not necessarily accompanied by an instability of the velocity field, i.e., by classical (Eulerian) turbulence. Intuitively, Lagrangian turbulence in fluids appears to be natural since changes in particle positions do not cause any restoring forces. It is harder to imagine Lagrangian turbulence in solids; however, as shown in the previous sections, it occurs in elastic bodies under severe compressions (wrinkling phenomenon).

We will discuss the conditions leading Lagrangian turbulence in continua as well as possible mathematical tools for its description in the following section.

4.5.1 Measure of Lagrangian Turbulence

Lagrangian turbulence in continua can be introduced as a failure of Euclidean metric of a material (frozen) system of coordinates. We will describe L-turbulence in terms of a space with torsion S and curvature R. The property of torsion arises when there is an infinitesimal "local slip" in any small neighborhood of the continuum so that an originally closed contour becomes disconnected (a closure failure of displacements). The curvature characterizes local disclinations following from a closure failure of rotations. Using the terminology of Hamiltonian mechanics one can associate R and S with the "angle variables." Indeed, their growth leads to mixing of trajectories rather than to an increase of energy. In order to introduce R and S we will start with the geometry of deformations.

A state of an infinitesimal element of a continuum is characterized by the displacement vector U^i and its gradient $\nabla_j U^i$, while the latter tensor allows an introduction of the metric tensor of a material (frozen) system of coordinates q^i:

$$g_{ij} = \overset{\circ}{g}_{ij} + \nabla_i U_j + \nabla_j U_i + \nabla_i U_k \nabla_j U^k \tag{4.387}$$

in which $\overset{\circ}{g}_{ij}$ is the metric tensor in an initial (displacement-free) state.

The rate of change of $\nabla_j U^i$ from one point to another is characterized by the connection coefficients (Christoffel symbols):

$$\Gamma^p_{ki} = \frac{1}{2} g^{ps} (\partial g_{sk}/\partial q^i + \partial g_{si}/\partial q^k - \partial g_{ik}/\partial q^s \tag{4.388}$$

The existence of U^i and $\nabla_j U^i$ for each infinitesimal element does not yet guarantee their smooth dependence on q^i: such a dependence must be guaranteed by additional (compatibility) conditions imposed on Γ^p_{ki}:

$$R_{rm} = \partial\Gamma^n_{rn}/\partial q^m - \partial\Gamma^n_{rm}/\partial q^n$$

$$+\Gamma^p_{rn}\Gamma^n_{pm} - \Gamma^p_{rm}\Gamma^n_{pn} = 0 \tag{4.389}$$

while

$$\nabla_m g^{ij} R_{ij} = 2\nabla_k R^k_m \tag{4.390}$$

where $R_{rm} = R_{mr}$ is the Ricci (curvature) tensor which must satisfy the Bianchi identity (4.390). Formally one can consider a situation when this tensor is non-zero. It would mean that material coordinates can be introduced in some abstract three dimensional (Riemannian) space with curvature; however, in actual space local material coordinates for two neighboring elements will not match each other so that global material coordinates do not exist. Such a situation can be easily imagined in an inviscid fluid where, for instance, one element may rotate when the rest of the fluid remains motionless (tangential velocity jump). Specifically, a non-zero curvature leads to incompatibility of rotations of two infinitesimal neighboring elements. In this case the vector U^i and the tensor $\nabla_j U^i$ will not be smooth as functions of Cartesian coordinates x_k. However, they will remain smooth as functions of q^i, i.e., the transition from x_k to the q^i will not be smooth. Hence, the Ricci tensor can be used as a measure for Lagrangian turbulence.

This approach can be generalized by introducing an incompatibility not only of rotations, but of displacements too. In this case, the space will be characterized by the curvature R and the torsion S:

$$S^k_{ij} = \Gamma^k_{ij} - \Gamma^k_{ji}$$

However, here we will confine our analysis to the case $S = 0$.

4.5.2 Governing Equations

In this section we express the momentum equations in terms of the Ricci tensor. For simplicity we will confine the analysis to linear stability by linearizing the momentum equations with respect to an initial state in which the material coordinates are Cartesian: $q^i = x_j$. Then (4.387–4.389) reduce to the following:

$$g_{ij} = \delta^i_j + \nabla_i U^k + \nabla_j U^k = \delta^i_j + 2e_{ij} \tag{4.391}$$

$$\Gamma^p_{ki} = \frac{\partial \varepsilon_{pk}}{\partial q^i} + \frac{\partial \varepsilon_{pi}}{\partial q^k} - \frac{\partial \varepsilon_{ki}}{\partial q^p}, \tag{4.392}$$

$$R_{rm} = \frac{\partial^2 \varepsilon_{ii}}{\partial q^r \partial q^m} - \frac{\partial^2 \varepsilon_{nr}}{\partial q^m \partial q^n} - \frac{\partial^2 \varepsilon_{nm}}{\partial q^r \partial q^n} + \frac{\partial^2 \varepsilon_{rm}}{\partial q^n \partial q^n} \tag{4.393}$$

in which e_{ij} is the stress tensor.

We will start with the momentum equation in the following form:

$$\rho_0 \partial^2 U / \partial t^2 = -\nabla p + \nabla \cdot T + F \tag{4.394}$$

in which ρ_0, p, T and F are initial density, pressure, stress deviator (i.e., the traceless part of the stressor), and the mass force, respectively. Exploring (4.391) one can express (4.394) in terms of ε:

$$\rho_0 \partial^2 \varepsilon / \partial t^2 = \frac{1}{2}(\nabla + \nabla^T)(-\nabla p + \nabla \cdot \mathbf{T} + F) \tag{4.395}$$

in which ∇^T is the transposed operator.

Introducing the operation

$$\mathbf{M} = \text{ink}\ \varepsilon = \nabla \times \nabla \times \varepsilon \tag{4.396}$$

which characterizes the incompatibility of reference configurations and recalling that in the linear approximation

$$R^{ij} = M^{ij} \tag{4.397}$$

one obtains the governing equation for the Ricci tensor R in an invariant form

$$\rho_0 \ddot{R} = \frac{1}{2}\left[\nabla \times \nabla \times (\nabla + \nabla^T)(-\nabla \rho + \nabla \cdot T + F)\right] \tag{4.398}$$

or in coordinate form

$$\rho_0 \ddot{R}^{11} = \nabla_2^2 \alpha_{33} + \nabla_3^2 \alpha_{22} - 2\nabla_2 \nabla_3 \alpha_{23}, \tag{4.399}$$

$$\rho_0 \ddot{R}^{12} = -\nabla_1 \nabla_2 \alpha_{33} + \nabla_3 (\nabla_1 \alpha_{23} + \nabla_2 \alpha_{31} - \nabla_3 \alpha_{12}) \tag{4.400}$$

The remaining four equations are obtained by cyclic permutations of the indeces. Here

$$\alpha_{kp} = \frac{1}{2}(\frac{\partial^2 T^{ik}}{\partial q^i \partial q^p} + \frac{\partial T^{lk}}{\partial q^p}\Gamma^i_{li} + T^{lk}\frac{\partial \Gamma^i_{li}}{\partial q^p} + \frac{\partial T^{il}}{\partial q^p}\Gamma^k_{li}$$

$$+T^{il} + \frac{\partial \Gamma^k_{li}}{\partial q^p} + \frac{\partial^2 T^{ip}}{\partial q^i \partial q^k} + \frac{\partial T^{lp}}{\partial q^k}\Gamma^i_{li} + T^{lp}\frac{\partial \Gamma^i_{li}}{\partial q^k}$$

$$\frac{\partial T^{il}}{\partial q^k}\Gamma^p_{li} + T^{il}\frac{\partial \Gamma^p_{li}}{\partial q^k} + \frac{\partial F^k}{\partial q^p} + \frac{\partial F^p}{\partial q^k}$$

$$+2F^j\Gamma^k_{jp} - 2\frac{\partial p}{\partial q_j}\Gamma^k_{jp} - 2\frac{\partial^2 p}{\partial q^k \partial q^p} + O^2(\Gamma^k_{ij}) \tag{4.401}$$

where O^2 contains the second order terms with respect to Γ^k_{ij} and its derivatives.

Obviously, the right hand parts of (4.399–4.400) also depend on R_{ij} and its space derivatives via the space derivatives of the connection coefficients Γ^k_{ij} as well as via the stresses (through constitutive laws). These equations must be complemented by the Bianchi identities of which a linearized version follows from (4.390):

$$\nabla_m R_{ij} = 2\nabla_k R^k_m \tag{4.402}$$

as well as by appropriate boundary conditions.

4.5.3 Criteria for Lagrangian Turbulence

In this section we will investigate the behavior of small perturbations of the Ricci tensor in (4.398). In order to simplify the stability analysis, let us pose the initial conditions in the following form:

$$\varepsilon_{12} = k^{-1} q^1 q^2 \exp(ikq^1)$$

$$\varepsilon_{11} = \varepsilon_{22} = \varepsilon_{33} = \varepsilon_{13} = \varepsilon_{23} = 0, \ \dot{\varepsilon}_{12} = 0$$

$$\text{at } t = 0 \tag{4.403}$$

assuming that $k \to \infty$ and consider (4.398) in a small neighborhood of the surface $q^1 = 0$. Then the initial disturbance of the Ricci tensor is [see (4.403)]

$$\overset{\circ}{R}{}^{33} = k^{-1} \exp(ikq^1) \to 0$$

$$\overset{\circ}{R}{}^{11} = \overset{\circ}{R}{}^{22} = \overset{\circ}{R}{}^{12} = \overset{\circ}{R}{}^{13} = \overset{\circ}{R}{}^{23} = 0 \tag{4.404}$$

and the solution to (4.398) can be sought as

$$\overset{\circ}{R}{}^{23} = k^{-1} \exp[i(kq^1 - \omega t)] \tag{4.405}$$

Thus, the only terms which survive in (4.398) will be those which contain $\partial^4 \varepsilon_{12}/\partial q^1 \partial q^1 \partial q^1 \partial q^2$ since they have the order of 1:

$$\frac{\partial^4 \varepsilon_{12}}{\partial q^1 \partial q^1 \partial q^1 \partial q^2} = \exp[i(kq^1 - \omega t)] \tag{4.406}$$

while

$$\frac{\partial^3 \varepsilon_{12}}{\partial q^1 \partial q^1 \partial q^2} = k^{-1} \exp[i(kq^1 - \omega t)] \to 0$$

$$\text{since } k \to \infty \tag{4.407}$$

After simple calculations, (4.398) reduces to

$$\rho_0 \frac{\partial^2 R^{33}}{\partial t^2} = \nu \frac{\partial^3 R^{33}}{\partial t \partial q^1 \partial q^1} + (G + T^{11}) \frac{\partial^2 R^{33}}{\partial q^1 \partial q^1} \tag{4.408}$$

in which $\nu = \partial T^{12}/\partial \dot{\varepsilon}_{12}$ is the viscosity, and $G = \partial T^{12}/\partial \varepsilon_{12}$ is the elastic shear modulus.

Substituting (4.405) into (4.408), one arrives at the characteristic equation with respect to ω:

$$\rho_0 \omega^2 + i\nu k^2 \omega - (G + T^{11})k^2 = 0 \tag{4.409}$$

whence

$$\omega = \frac{1}{2}[\frac{\nu k^2}{\rho_0} i \pm (-\frac{\nu^2 k^4}{\rho_0^2} + \frac{4(G + T^{11})}{\rho_0} k^2)^{1/2}] \tag{4.410}$$

For an elastic body ($\nu = 0$)

$$\omega = \pm k \left[(1/\rho_0)(G + T^{11})\right]^{1/2} \tag{4.411}$$

and, therefore,

$$\overset{\circ}{R}{}^{33} = k^{-1} \exp\{i[kq^1 \pm [(G + T^{11})/\rho_0]^{1/2} t]\} \to \infty \tag{4.412}$$

although

$$\overset{\circ}{R}{}^{33} \to 0 \text{ at } k \to \infty \text{ if } T^{11} < -G \tag{4.413}$$

Thus, infinitesimal initial disturbances (4.404) lead to an unbounded growth of the solution. Moreover, since

$$k^{-1} e^k \to \infty \text{ at } k \to \infty \tag{4.414}$$

the solution becomes infinitely large within a finite time period (Hadamard instability).

This result was obtained under specially selected initial conditions (4.404), but it can be generalized to include any initial conditions. Indeed, by adding (4.404) to these initial conditions one observes from the preceding argument by superposition that a vanishingly small change in the initial conditions would lead to unboundedly large solutions. Hence, L-turbulence in elasticity occurs if at least one principal value of stress deviators is negative while its absolute value is greater than the shear modulus.

Exactly the same condition leads to a collapse of elasticity under sever compressions (the wrinkling phenomenon). These results are identical to those obtained in the previous section based upon (4.117).

For a viscous fluid ($G = 0$)

$$\text{Im } \omega_1 \to \left|T^{11}\right|/\nu, \ \text{Im } \omega_2 \to -i\infty, \ T^{11} < 0 \tag{4.415}$$

and therefore, the instability (caused by ω_1) occurs if

$$T^{11} < 0 \tag{4.416}$$

Hence, in viscous flows L-turbulence emerges in directions where the viscous stress is negative. From this one concludes that all shear flows are L-turbulent since there always exist directions with negative stress. However, it is important to emphasize that the character of instability in an inviscid fluid is different from the case of elasticity. Indeed, as follows from (4.406), ω_1 is bounded while in (4.411) $\omega \to \infty$ if $k \to \infty$. This is why (4.414) is not applicable to viscous fluids and the solution becomes infinitely large only at $k \to \infty$, i.e., one deals with Liapunov rather than Hadamard type of instability. In this connection, it is interesting to notice that in two dimensional models of liquid films, the condition (4.416) leads to Hadamard instability, i.e., to the wrinkling phenomenon.

For a visco-elastic body

$$\text{Im } \omega_1 \to (T^{11} + G)/\nu \text{ at } k \to \infty \text{ if } T^{11} + G < 0$$

and the instability occurs if the condition (4.413) holds. However, here T includes both the elastic and viscous stresses while the instability is of the Liapunov type.

For an inviscid fluid (4.408) is not applicable since all the derivatives $\partial\Gamma_{ij}^k/\partial q^m$ in (4.401) and, therefore, the right hand side of (4.408) vanish together with T_{ij}. Hence, one has to take into account terms $\partial\Gamma_{ip}^k(F^j - \partial p/\partial q^j)$ in (4.401) which reduces (4.419) and (4.409) to:

$$\rho_0 \partial^2 R^{33}/\partial t^2 = (F^1 - \partial p/\partial q^1)\partial R^{33}/\partial q^1 \tag{4.417}$$

$$\rho_0 \omega^2 = i(\partial p/\partial q^1 - F^1)k \tag{4.418}$$

whence

$$\omega = \pm \left[(1/2\rho_0) \left|\partial p/\partial q^1 - F^1\right| k\right]^{1/2} \tag{4.419}$$

Thus, here L-turbulence occurs if

$$\nabla_i p - F^i \neq 0 \tag{4.420}$$

i.e., in any non-inertial motion, while the instability is of the Hadamard type.

Thus, it has been demonstrated that the curvature (Ricci) tensor of a material (frozen) system of coordinates can be used as a measure of L-turbulence. The conditions sufficient for instability of an initially infinitesimal disturbance of this tensor and, therefore, the criteria for L-turbulence were derived for fluid, elastic and visco-elastic bodies. However, as follows from the derivation, these conditions are only sufficient but not necessary. For necessary conditions one has to consider the full version of (4.399–4.401). In addition, it has to be remembered that all the results are obtained from the linearized equations. This is why the growth of unstable solutions may be bounded and, therefore, instead of (4.412), it should be written: R^{33} does not tend to zero if its initial value vanishes. Obviously, in order to find the quantitative post-instability behavior of R^{33} one has to consider the non-linear version of (4.398) which can be obtained by utilizing (4.389).

It should be emphasized that here the L-turbulence is understood as chaotic instability (with mixing) of trajectories of individualized (physical) particles in actual space in contrast to analytical mechanics where chaos is introduced as an orbital instability of trajectories of a representative (not physical) particle in abstract configuration space. The difference between these two types of chaos has been discussed by Zak (1985).

4.5.4 The Intensification of Vorticity by Extension of Vortex Lines

In this section it will be shown that not only failure of hyperbolicity (Hadamard instability) leads to failure of differentiability of solutions. The example given below demonstrates Liapunov instability accompanied by a decrease in the scale of the motion.

Let us start with the Helmholtz equation for an inviscid incompressible fluid without external forces:

$$\frac{\partial \mathbf{\Omega}}{\partial t} + \mathbf{V}\nabla\mathbf{\Omega} - \mathbf{\Omega}\nabla\mathbf{V} = 0, \ \mathbf{\Omega} = \nabla \times \mathbf{V} \tag{4.421}$$

where $\mathbf{V}$, $\mathbf{\Omega}$ are the velocity and vortex of the fluid, respectively, and rewrite it for small uniform disturbances of vortices Ω', assuming that undisturbed motion is potential:

$$\frac{\partial \mathbf{\Omega}'}{\partial t} - \mathbf{\Omega}'\nabla \overset{*}{\mathbf{V}} = 0 \tag{4.422}$$

where

$$\mathbf{\Omega} = \overset{*}{\mathbf{\Omega}} + \mathbf{\Omega}', \ \mathbf{V} = \overset{*}{\mathbf{V}} + \mathbf{V}'$$

while

$$\overset{*}{\mathbf{\Omega}} = 0$$

$$\mathbf{\Omega}'|_{t=0} = \mathbf{\Omega}_0 = \text{Const}, \ \text{i.e.,} \ \nabla\mathbf{\Omega}'|_{t=0} = 0 \tag{4.423}$$

Now the behavior of the vortices is determined by the eigenvalues of the tensor $\nabla \overset{*}{\mathbf{V}}$. This tensor can be decomposed into the form:

$$\nabla \overset{*}{\mathbf{V}} = \ \text{def} \ \overset{*}{\mathbf{V}} + \mathbf{\Omega} \tag{4.424}$$

But since the initial motion is potential, i.e., $\overset{*}{\mathbf{\Omega}} = 0$, one concludes that $\nabla \overset{*}{\mathbf{V}}$ is a symmetric tensor and all the eigenvalues λ_k are real:

$$\text{Im}\,\lambda_k = 0, \ (k = 1, 2, 3) \tag{4.425}$$

Turning to the condition of incompressibility

$$\nabla \cdot \overset{*}{\mathbf{V}} = 0 \tag{4.426}$$

and taking into account that

$$\nabla \cdot \overset{*}{\mathbf{V}} = I_1 \tag{4.427}$$

where I_1 is the first invariant of the tensor $\nabla \overset{*}{\mathbf{V}}$ one gets:

$$I_1 = \lambda_1 + \lambda_2 + \lambda_3 = 0 \tag{4.428}$$

Let us avoid consideration of the special case when

$$\lambda_1 = 0, \ \lambda_2 = 0, \ \lambda_3 = 0 \tag{4.429}$$

which corresponds to the uniform initial flow. Then, as follows from (4.425) and (4.426), the tensor $\nabla \overset{*}{\mathbf{V}}$ has definite positive and negative characteristic roots, which is sufficient for growth of initially infinitesimal vorticity. But in accordance with Thomson's theorem the circulation of the velocity along an arbitrary contour must be constant:

$$\oint \mathbf{V} \cdot d\mathbf{S} = \text{Const} \tag{4.430}$$

which means that the growth of vortices is carried out not by the growth of velocities but by decrease of the scale of motion as in previous example.

However, one should notice the difference between this instability (which is a Liapunov-type instability) and the Hadamard instability discussed in

Sect. 4.3. As emphasized above, the Hadamard instability does not depend explicitly upon initial or boundary conditions, and an unstable parameter becomes unbounded during finite time period. In contradistinction to this, the instability of solutions to (4.421) may explicitly depend upon both initial and boundary conditions, while an unstable parameter can become unbounded only at $t \to \infty$. The last property follows from the fact that the linearized version of (4.421); i.e., (4.422) is an ordinary differential equation. In addition to this, the growth of the vorticity in the course of the instability may change the condition (4.428), and the linearized version of (4.421) will not be adequate any more.

Despite of all the limitations under which the instability of (4.421) was obtained, the result elucidates the mechanism of vorticity intensification by expansion of vortex lines. More specific results concerning these effects can be obtained for a unidirectional viscous flow:

$$U_x = \alpha x,\; U_r = -\frac{1}{2}\alpha r,\; \alpha = \text{Const},\; \Omega_x = \Omega,\; \Omega_r = \Omega_\varphi = 0 \qquad (4.431)$$

where x, r and φ are cylindrical coordinates.

Turning to the viscous version of (4.421):

$$\frac{\partial \mathbf{\Omega}}{\partial t} + \mathbf{V}\nabla\mathbf{\Omega} - \mathbf{\Omega}\nabla\mathbf{V} - \nu\nabla^2\mathbf{\Omega} = 0 \qquad (4.432)$$

one obtains for the unidirectional flow (4.431)

$$\frac{\partial \Omega}{\partial t} = \frac{\alpha}{2r}\frac{\partial(\Omega r^2)}{\partial r} + \nu\left(\frac{\partial^2\Omega}{\partial r^2} + \frac{1}{r}\frac{\partial\Omega}{\partial r}\right) \qquad (4.433)$$

where ν is the viscosity.

The solution to (4.433) tends to the distribution:

$$\Omega(r) = \Omega_1 \exp\left(-\frac{\alpha r^2}{\nu}\right) \quad \text{at } t \to \infty \qquad (4.434)$$

The solution (4.434) represents a steady flow [playing the role of a static attractor for (4.433)], in which the vorticity is concentrated within a radial distance of order $(\nu/\alpha)^{1/2}$ from the axes of symmetry and in which the intensification of vorticity due to the extension of vortex lines is ultimately balanced by the rate of decrease due to lateral spreading by viscous diffusion. It can be shown that in turbulent flows the integral of Ω^2 (which is a measure of the total amount of vorticity in a unit material volume) reaches a large value (proportional to a positive power of the Reynolds number of the flow) before the loss by viscous motion equals or exceeds the gain by extension of vortex lines.

In order to elucidate physical effects behind the phenomenon of vorticity intensification by extension of vortex lines, let us consider an internal rotation of a cylinder about its axis of symmetry. According to the conservation of the angular momentum:

$$\frac{d}{dt}(I\omega) = 0, \; I = \frac{1}{2}mr^2 \tag{4.435}$$

where m, I, r, and ω are the mass, moment of inertia, radius and angular velocity of cylinder, respectively. Let us assume now that this cylinder is stretched in the axial direction such that its radius decreases

$$r_1 < r \tag{4.436}$$

Then, as follows from (4.435):

$$mr^2\omega = mr_1^2\omega_1 \tag{4.437}$$

and therefore the new angular velocity

$$\omega_1 = \omega\frac{r^2}{r_1} > \omega \tag{4.438}$$

is larger than the original one.

Hence, here the increase of the angular velocity of rotation is caused by stretching the length of the cylinder. Loosely speaking, the same effect occurs when vortex lines are stretched. Indeed, according to the Helmholtz theorem, vortex lines are material lines consisting of the same individual particles, and application of the conservation of the angular momentum to an element of a vortex line leads to the same qualitative results as presented in (4.438).

4.6 Other Types of Instability

There are many different mechanisms leading to decreases in scale of motions via the increase of vorticity in fluids. In this section we will give only a brief description of them.

4.6.1 Thermal Instability

This instability arises when a fluid is heated from below. When the temperature difference across the fluid layer is great enough, the stabilizing effects of viscosity and thermal conductivity are overcome by the destabilizing buoyancy, and the instability occurs in the form of a thermal convection. A particular case of this kind of instability was discussed in Sect. 4.3.4 (4.260).

4.6.2 Centrifugal Instability

Here the instability occurs in a fluid owing to the dynamical effects of rotation or the streamline curvature. For instance, as shown by Rayleigh, an inviscid flow between two rotating coaxial cylinders is unstable if the angular momentum $|r^2\Omega|$ decreases anywhere inside the interval $r_1 < r < r_2$ where Ω is the angular velocity of rotation of the fluid, r_1 and r_2 are the radii of the coaxial cylinders.

It can be demonstrated that in general, centrifugal instability arises from adverse distributions of angular momentum.

4.6.3 Rayleigh-Taylor Instability

This derives from the character of the equilibrium of an incompressible heavy fluid of variable density. For instance, it is shown that in the case of variable density of exponential distribution

$$\rho = \rho_0 e^{\beta Z}, \ \beta = \text{Const} \tag{4.439}$$

where Z is the vertical coordinate, the equilibrium is unstable if

$$\beta > 0 \tag{4.440}$$

i.e., if the heavier layers are above the lighter layers.

4.6.4 Reynolds Instability

This instability results from an imbalance between inertial and viscous forces. It occurs when the Reynolds number (R) exceeds certain critical values which depend upon the type of a flow and its boundary condition. For a particular case of inviscid shear flow ($R \to \infty$) with parallel streamlines. Rayleigh showed that a necessary condition for instability is that the basic velocity profile should have an inflection point.

4.6.5 Mathematical Approach to Analysis of Instabilities

The approach to analysis of these instabilities is based upon linearization of the governing equations with respect to the basic flow. Small disturbances of the flow parameters are sought in the exponential form. For instance, in case of the Reynolds instability of incompressible flows with parallel streamlines this form is:

$$\varphi(x, y, t) = \varphi(y)e^{i(\alpha x - \beta t)}, \ \alpha = \text{Const}, \ \beta = \text{Const}, \ \frac{\beta}{\alpha} = c \tag{4.441}$$

where φ is the stream function, x and y are Cartesian coordinates along and across the flow, respectively. Substitution of (4.441) into the linearized governing (Navier-Stokes) equations leads to an ordinary differential equations with respect to $\varphi(y)$ (called the Orr-Sommerfeld equation):

$$(U-c)(\varphi''-\alpha^2\varphi)-U''\varphi=-\frac{i}{\alpha R}\left(\varphi''''-2\alpha^2\varphi''+\alpha^4\varphi\right) \qquad (4.442)$$

Here U is the velocity of the basic flow, $R=\frac{U_m b}{\nu}$ is the Reynolds number, ν is the fluid viscosity, b is the width of the flow, U_m is the maximum of U.

Since U and U'' are known from the basic flow, while the constants R and α can be treated as given parameters, the only unknowns in (4.442) are $\varphi(y)$ and c. But in addition, the function $\varphi(y)$ must satisfy also the following four boundary conditions:

$$\varphi=0,\ \varphi'=0,\ \text{at } y=0 \text{ and } y\to\infty \qquad (4.443)$$

and this is only possible for some discrete sequence of c. Turning to (4.441), one concludes that the basic flow is unstable if among this sequence c there are such that

$$\operatorname{Im} c>0 \qquad (4.444)$$

It should be emphasized that, in contradistinction to the Hadamard instability, all other types considered above are Liapunov instabilities, so an unbounded growth of vorticity can be detected only at $t\to\infty$. But since the stability analysis is based upon linearized equations, it is always a possibility that an originally unstable flow will "find" an alternative state which is stable. Nevertheless, in many cases the instability may persist, and the scale of motion will decrease gradually. Usually in case of non-zero viscosity, the vorticity is always bounded. However, since our scale of observation is finite, a very small scale of motions can be better interpreted as a loss of differentiability of the motion parameters with respect to spatial coordinates.

5
Stabilization Principle

The flows that occur in Nature must not only obey the equations of fluid dynamics, but also be stable. *–Landau and Lifshitz*

5.1 Instability as Inconsistency Between Models and Reality

5.1.1 General Remarks

It has been demonstrated in the previous chapter, that there are some domains of dynamical parameters where the motion cannot be predicted because of instability of the solutions to the corresponding governing equations. How can it be interpreted? Does it mean that the Newton's laws are not adequate? Or is there something wrong with our mathematical models?

In order to answer these questions, we will discuss some general aspects of the concept of instability, and in particular, a degree to which it is an invariant of motion. We will demonstrate that instability is an attribute of a mathematical model rather than of a physical phenomenon; and consequently, that it depends upon the frame of reference, upon the class of functions in which the motion is described, and upon the way in which the distances between the basic and perturbed solutions is defined.

5.1.2 *Instability Dependence Upon Metrics of Configuration Space*

Let us turn to orbital instability discussed in the Sect. 4.2. The metric of configuration space where the finite-degree-of-freedom dynamical system with N generalized coordinates $q^i (i = 1, 2 \cdots N)$ is represented by a unit-mass particle, was defined by (4.17–4.18). Now there are at least two possible ways to define the distance between the basic and disturbed trajectories. Following Synge (1926), we will consider the distance in a kinematical and kinematico-statistical sense. In the first case the corresponding points on the trajectories are those for which time t has the same value. In the second case the correspondence between points on the basic trajectory C and a disturbed trajectory C^* is established by the condition that P (a point on C) should be the foot of the geodesic perpendicular let fall from P^* (a point on C^*) on C, i.e., here every point of the disturbed curve is adjacent to the undisturbed curve (regardless of the position of the moving particle at the instant t). As shown by Synge, both definitions of stability are invariant with respect to coordinate transformations, and in both cases the stability implies that the corresponding distance between the curves C and C^* remains permanently small.

It is obvious that stability in the kinematical sense implies stability in the kinematico-statical sense, but the converse is not true. Indeed, consider the motion of a particle of unit mass on a plane under the influence of a force system derivable from a potential:

$$\Pi = -x + \frac{1}{2}y^2 \tag{5.1}$$

Writing down the equations of motion and solving them, we get:

$$x = \frac{1}{2}t^2 + At + B \tag{5.2}$$

$$y = C \sin (t + \alpha) \tag{5.3}$$

where A, B, C and D are constants of integration.

Let the undisturbed motion be:

$$x = \frac{1}{2}t^2 + t \tag{5.4}$$

$$y = 0 \tag{5.5}$$

The motion is clearly unstable in the kinematical sense. However, from the viewpoint of stability in the kinematico-statical sense, the distance between corresponding points:

$$PP^* = y = C \sin (t + D) \tag{5.6}$$

remains permanently small if C is small. Hence, there is stability in the kinematico-statical sense.

Thus, the same motion can be stable in one sense, and unstable in another, depending upon the way in which the distance between the trajectories is defined.

It should be noted that in both cases, the metric of configuration space was the same [see (4.17–4.18)]. However, as shown by Synge (1926), for conservative systems, one can introduce a configuration space with another metric.

$$g_{mn} = (E - \Pi)\alpha_{mn} \tag{5.7}$$

where α_{mn} are expressed by (4.17), and E is the total energy.

The system of motion trajectories here consists of all the geodesics of the manifold. The correspondence between points on the trajectories is fixed by the condition that the arc O^*P^* should be equal to the arc OP, where O and O^* are arbitrarily selected origins on the basic trajectory and any disturbed one, respectively.

As shown by Synge, the problem of stability here (which is called stability in the action sense) is that of the convergence of geodesics in Riemannian space. If two geodesics pass through adjacent points in nearly parallel directions, the distance between points on the geodesics equidistant from the respective initial points is either permanently small or not. If not, there is instability. It appears that stability in the action sense may not be equivalent to stability in the kinematico-statical sense for distances which change the total energy E.

Turning to the example (5.1), let us take the initial point O at the origin of coordinates and the initial point O^* on the y axis. Then the disturbance being infinitesimal, the (action) distance between corresponding points is:

$$PP^* = (E - \Pi)^{1/2}\, y = 2^{-1/2}(t + 1)\, C \sin(t + D) \tag{5.8}$$

Hence, the motion is unstable in the action sense.

5.1.3 *Instability Dependence Upon Frame of Reference*

Dynamical instability depends not only upon the metric in which the distances between trajectories are defined, but also upon the frame of reference in which the motion is described. Such a dependence was already noted in Sect. 4.3. In this section we will introduce and discuss an example which illustrates the dependence of criteria of hydrodynamics stability and the onset of turbulence upon the frame of reference.

The linear theory of hydrodynamic stability is based upon Eulerian representation of fluid motions in which the frame of reference is chosen *a priori*. Strictly speaking, such a representation provides criteria of stability for the velocity field rather than the fluid motion. The difference between these two types of stability was illustrated by Arnold (1988), who introduced flows with stable velocity fields and unstable trajectories (Lagrangian turbulence). If the classical (Eulerian) turbulence is associated with the instability of streamlines then it is reasonable to study this instability in a streamline frame of reference in which streamlines form a family of initially unknown Eulerian coordinates, while the remaining two Lagrangian coordinates are found from the compatibility conditions. Such a frame of reference is completely defined by the motion, and therefore, it contains a minimum of arbitrarily chosen parameters (Zak 1990b).

First of all, we will show that criteria of stability in this frame of reference do not necessarily coincide with the classical criteria which are derived from the Orr-Sommerfeld equation. For this purpose, we will introduce a small disturbance velocity field for incompressible plane flow in Cartesian coordinates x, y:

$$V_x = \phi'(y)e^{i(\alpha x-\beta t)}, \quad V_y = -i\alpha\phi(y)e^{i(\alpha x-\beta t)}, \quad \alpha\,\beta = \text{Const} \tag{5.9}$$

where the prime denotes differentiation.

The angle θ between streamlines and the $\dot{x}$-direction is

$$\theta = \frac{i\alpha\phi}{\dot{V}}\,e^{i(\alpha x-\beta t)} \tag{5.10}$$

in which $V(y)$ is the velocity profile of the basic flow. The orthogonal streamline coordinates ξ, ζ are found from the system:

$$\frac{\partial x}{\partial \xi} = H_1\cos\theta\,, \quad \frac{\partial x}{\partial \zeta} = -H_2\sin\theta\,, \quad \frac{\partial y}{\partial \xi} = H_1\sin\theta\,, \quad \frac{\partial y}{\partial \zeta} = H_2\cos\theta\,, \tag{5.11}$$

where H_1 and H_2 are the Lame coefficients defined by the compatibility conditions ($\partial^2 x/\partial\xi\partial\zeta = \partial^2 x/\partial\zeta\partial\xi$, etc.)

$$\left(\frac{\partial\theta}{\partial x} + \frac{\partial\theta}{\partial y}\tan\theta\right)H_1 = \tan\theta\frac{\partial H_1}{\partial x} - \frac{\partial H_1}{\partial y} \tag{5.12}$$

and

$$\left(-\frac{\partial\theta}{\partial y} + \frac{\partial\theta}{\partial x}\tan\theta\right)H_2 = \tan\theta\frac{\partial H_2}{\partial y} - \frac{\partial H_2}{\partial x} \tag{5.13}$$

As follows from (5.10–5.13), the coordinate transformation

$$x = x(\xi, \zeta, t)\,, \quad y = y(\xi, \zeta, t) \tag{5.14}$$

in general will depend on time. Hence, for the stream function one obtains

$$\psi = \phi(y)\,e^{i(\alpha x-\beta t)} = \phi\,[y(\xi,\ \zeta,\ t]e^{i[\alpha x(\xi,\zeta,t)-\alpha t]} \tag{5.15}$$

i.e.

$$\left.\frac{\partial\psi}{\partial t}\right|_{x,y=\text{Const}} \neq \left.\frac{\partial\psi}{\partial t}\right|_{\xi,\zeta=\text{Const}} \tag{5.16}$$

In other words, the stability criteria in frames x, y and ξ, ζ are not necessarily the same.

This preliminary conclusion provides motivation to analyze criteria of hydrodynamic stability in streamline coordinates.

Confining our investigation to a plane incompressible inviscid flow one derives the momentum equations in streamline coordinates from the Lagrange equation:

$$\frac{d}{dt}\frac{\partial W}{\partial \xi} - \frac{\partial W}{\partial \xi} = \frac{1}{\rho}\frac{\partial p}{\partial \xi}, \quad - \frac{\partial W}{\partial \zeta} = -\frac{1}{\rho}\frac{\partial p}{\partial \zeta} \tag{5.17}$$

in which the kinetic energy

$$W = \frac{1}{2} H_1^2 \xi^2 \tag{5.18}$$

and the velocity

$$V = V_1 = H_1 \xi, \quad V_2 = H_2 \xi = 0 \tag{5.19}$$

while p and ρ are pressure and density, respectively.

The momentum equations read

$$H_1 \frac{\partial V}{\partial t} + V\left(\frac{\partial H_1}{\partial t} + \frac{\partial V}{\partial \xi}\right) = -\frac{1}{\rho}\frac{\partial p}{\partial \xi} \tag{5.20}$$

and

$$-\frac{V^2}{H_1 H_2}\frac{\partial H_1}{\partial \xi} = \frac{1}{\rho}\frac{\partial p}{\partial \zeta} \tag{5.21}$$

The continuity equation follows from the condition

$$\operatorname{div} V = 0, \quad \text{i.e.} \ \frac{\partial}{\partial \xi}(V H_2) = 0 \tag{5.22}$$

Equations (5.11–5.13) are completed by the compatibility (Lame's) equation

$$\frac{\partial}{\partial \xi}\left(\frac{1}{H_1}\frac{\partial H_2}{\partial \xi}\right) + \frac{\partial}{\partial \zeta}\left(\frac{1}{H_2}\frac{\partial H_1}{\partial \zeta}\right) = 0 \tag{5.23}$$

Linearizing these equations with respect to an unperturbed shear flow:

$$\overset{*}{V} = \overset{*}{V}(y) \tag{5.24}$$

and taking into account that for this flow the streamline coordinates coincide with the Cartesian coordinates:

$$\xi = x, \quad \zeta = y, \quad \widetilde{H}_1 = \widetilde{H}_2 = 1 \tag{5.25}$$

one obtains after eliminating the pressure.

$$\{\frac{\partial^2 \tilde{V}}{\partial t \partial \zeta}\frac{\partial^2 \tilde{H}_1}{\partial t \partial \zeta} + \breve{V}(\zeta)\frac{\partial^2 \tilde{V}}{\partial \xi \partial \zeta} + \breve{V}^2(\zeta)\frac{\partial^2 \tilde{H}_1}{\partial \xi \partial \zeta} = 0 \tag{5.26}$$

$$\frac{\partial \tilde{V}}{\partial \xi} + \tilde{V}(\zeta)\frac{\partial \tilde{H}_2}{\partial \xi} = 0 \tag{5.27}$$

$$\frac{\partial^2 \tilde{H}_2}{\partial \xi^2} + \frac{\partial^2 \tilde{H}_1}{\partial \zeta^2} = 0 \tag{5.28}$$

where $\tilde{V}, \tilde{H}_1$, and $\tilde{H}_2$ are small perturbations of V, H_1, and H_2, respectively.

If the solution for $\tilde{V}$ is assumed to be of the form

$$\tilde{V} = \dot{V}(\zeta)\phi'(\zeta)\, e^{i(\alpha\xi-\beta t)}, \quad \alpha,\ \beta = \text{Const} \tag{5.29}$$

then, as follows from equations (5.27–5.28),

$$\tilde{H}_2 = -\,\phi'(\zeta)\, e^{i(\alpha\xi-\beta t)}, \quad -\frac{\partial \tilde{H}_1}{\partial \zeta} = -\alpha^2\phi(\zeta)\, e^{i(\alpha\xi-\beta t)} \tag{5.30}$$

Substituting the values (5.29–5.30) into (5.26) one arrives at the governing equation for $\phi(\xi)$:

$$\phi'' - \frac{c\dot{V}'(\zeta)}{\dot{V}(\zeta)[\dot{V}(\zeta) - c]}\phi' - \alpha^2\phi = 0, \quad c = \frac{\beta}{\alpha} \tag{5.31}$$

which is different from the Orr-Sommerfeld equation.

If the basic flow $\dot{V}(y)$ is bounded by rigid walls:

$$y = y_1, \quad y = y_2 \tag{5.32}$$

then the streamlines at $\eta = y_1$ and $\zeta = y_2$ must coincide with these walls, i.e.

$$\frac{\partial \boldsymbol{\tau}}{\partial \xi} = -\frac{1}{H_2}\frac{\partial H_1}{\partial \zeta}\mathbf{n} = 0 \text{ at } y = y_1 \text{ and } y = y_2 \tag{5.33}$$

in which $\boldsymbol{\tau}$ and $\mathbf{n}$ are the unit tangent and the unit normal vectors to the streamlines.

Hence

$$\frac{\partial H_1}{\partial \zeta} = 0 \text{ at } y = y_1 \text{ and } y = y_2 \tag{5.34}$$

and therefore, with references to (5.30),

$$\phi(\zeta_1 = y_1) = 0, \quad \phi(\zeta_2 = y_2) = 0 \tag{5.35}$$

These equations express the boundary conditions for (5.31). In order to show that the stability criteria in streamline coordinates are different from those given by the Orr-Sommerfeld equations let us select a special velocity profile $\mathring{V}(y)$ such that the coefficient of ϕ' in (5.31) reduces to a constant. Obviously, such a profile must satisfy the first-order differential equation

$$\frac{\mathring{V}'}{\mathring{V}(\mathring{V} - c)} = y = \text{Const} \quad \text{Im}\,\gamma = 0 \tag{5.36}$$

and consequently

$$\overset{\circ}{V} = \frac{c}{1 - e^{c\gamma y}} \tag{5.37}$$

while (5.31) for this profile reduces to

$$\phi'' - c\gamma\phi' - \alpha^2\phi = 0 \tag{5.38}$$

Its general solution is

$$\phi = C_1 \exp(\lambda_1 y) + C_2 \exp(\lambda_2 y) \tag{5.39}$$

where

$$\lambda_{1,2} = \frac{cy}{2} + \sqrt{\frac{c^2\gamma^2}{4} + \alpha^2} \tag{5.40}$$

Substitution of the boundary conditions (5.35) into (5.39) leads to a system of homogeneous equations:

$$C_1 \exp(\lambda_1 y_1) + C_2 \exp(\lambda_2 y_1) = 0, \quad C_1 \exp(\lambda_1 y_2) + C_2 \exp(\lambda_2 y_2) = 0 \tag{5.41}$$

and for a non-trivial solution

$$\det \begin{pmatrix} \exp(\lambda_1 y_1) & \exp(\lambda_2 y_1) \\ \exp(\lambda_1 y_2) & \exp(\lambda_2 y_2) \end{pmatrix}$$

$$= \exp(\lambda_1 y_1 + \lambda_2 y_2) - \exp(\lambda_1 y_2 + \lambda_2 y_1) = 0 \tag{5.42}$$

i.e., $\lambda_1 = \lambda_2$; or, with reference to (5.40),

$$c = \pm\, i\, 2\, \alpha / y = \pm\, c_0 i \tag{5.43}$$

Since α and γ are real, c is imaginary, and therefore, solutions (5.30) are unstable for any y_1 and y_2.

Now we will show that the Orr-Sommerfeld equation predicts stability for the same profile. Indeed, substituting c from (5.43) into (5.37) and separating the real part of the velocity profile, one obtains

$$\mathrm{Re}\, \dot{V} = \pm \operatorname{cotan}(2\alpha y) \tag{5.44}$$

This profile has only one inflection point (at $y = \pi/\, 4\alpha$). Consequently, according to the point-of-inflection criterion proved by Tollmien, any profile of the form (5.44) which does include the inflection point, i.e.

$$0 \le y_1 \le y \le y_2 < \frac{\pi}{4\alpha} \tag{5.45}$$

is stable.

It is important to emphasize that these two different results regarding the same velocity profile are not mutually exclusive: the first is related to the stability of the fluid motion referred to streamline coordinates, while the second is related to the stability of the velocity field. But which one of these approaches is actually related to the onset of turbulence? The dynamics of fluid motion, and in particular, the stability of streamlines, is directly related to the onset of turbulence inasmuch as the stability of particle trajectories is directly related to the onset of Lagrangian turbulence. At the same time, the stability of velocity fields is indirectly related to the onset of turbulence. This is why the linearized version of the classical theory of stability cannot explain the instability of plane Couette flows. In this connection it is worth noting that by an appropriate selection of α, y_1 and y_2 in (5.44–5.45), the velocity profile, (5.45), can be made as close as necessary to a straight line, thereby predicting the instability of any flow which is arbitrarily close to the Couette flow.

5.1.4 Instability Dependence Upon Class of Functions

The properties of solutions to differential equations, such as existence, uniqueness, and stability, have a mathematical meaning only if they are referred to a certain class of functions. For instance, as shown previously in Sect. 4.4 (4.332) and (4.334), we have a unique stable solution in an open interval (4.345) in the class of bounded functions, while in a closed interval, (4.346), the uniqueness and stability are not guaranteed. Most of the results concerning the properties of solutions to differential equations require differentiability (up to a certain order) of the functions describing the solutions. However, the mathematical restrictions imposed upon the class of functions which guarantee the existence of an unique and stable solution, do not necessarily lead to the best representation of the corresponding physical phenomenon. Indeed, turning again to (4.332) and (4.334), one notices that the unique and stable solution (4.345) does not describe a cumulation effect (a snap of a whip) which is well pronounced in experiments. At the same time, an unstable solution in a closed interval (4.346) gives a qualitative description of this effect. Hence, pure mathematical restrictions imposed upon the solutions are not always consistent with the physical nature of motions. In this context, the long-term instability in classical dynamics discussed in the Chap. 4, can be interpreted as a discrepancy between these mathematical restrictions and physical reality. This means that unpredictability in classical dynamics is a price paid for mathematical "convenience" in dealing with dynamical models. Therefore, the concept of unpredictability in dynamics should be put as unpredictability in a selected class of functions, or in a selected metrics of configuration space, or in a selected frame of reference.

Now the following problem can be posed. How to select an appropriate mathematical representation of a physical phenomena? The answer to this question will be discussed below.

5.2 Postinstability Models in Continua

5.2.1 *Concept of Multivaluedness of Displacements*

The formal mathematical basis for the models with multivalued displacement fields will be presented in the next sections. Here we will start with some physical interpretations of such models based on the technique of the separation of motions of different scales (Zak 1982a,b,c).

As shown in Chap. 4, Sect. 3, the failure of hyperbolicity in continua leads to the following type of solution to the corresponding governing equations:

$$\mathbf{u} = \mathbf{u}_1(\mathbf{r}, t) + \mathbf{u}_2(\mathbf{r}, t) \sin \mathbf{k} \cdot \mathbf{r}, \ \ \mathbf{k} \to \infty, \ \ \mathbf{u}_2 \to 0 \tag{5.46}$$

where $\mathbf{u}$ is the displacement, $\mathbf{u}$ and $\mathbf{u}_1, \mathbf{u}_2$ are smooth functions characterized by a distance scale l upon which their changes are negligible, and $\mathbf{k}$ is the wave vector in the direction of the imaginary characteristic speed, i.e., $\mathbf{k} \gg 1/l$.

According to (4.120), (4.125-4.126) the function $\mathbf{u}_2$ can be presented as;

$$\mathbf{u}_2 = \dot{\mathbf{u}}_2 \exp(k\sqrt{-\lambda^2})t, \ \ \dot{\mathbf{u}}_2 = \mathbf{u}_2|_{t=0} \sim \frac{1}{k}, \ \ \lambda = \frac{\overset{\circ}{\lambda}}{\varrho_0} \tag{5.47}$$

where ϱ_0 is the density, and λ is the characteristic speed of wave propagation in the $\mathbf{k}$ direction.

The expression for $\overset{\circ}{\lambda}{}^2$ follows from (4.132–4.133):

$$\overset{\circ}{\lambda}{}^2 = T^{ii} + \frac{1}{2}\frac{\partial T^{ii}}{\partial \varepsilon_{ij}} \ \ (i \neq j) \tag{5.48}$$

for transverse waves and

$$\overset{\circ}{\lambda}{}^2 = T^{ii} + \frac{\partial T^{ii}}{\partial \varepsilon_{ii}} \tag{5.49}$$

for longitudinal waves.

In the course of the instability λ becomes imaginary, and therefore, the displacement $\mathbf{u}_2$ start growing exponentially. However, one should notice that (5.46) was derived under the assumption that the displacement gradients, i.e., $\nabla(\mathbf{u}_2 \sin \mathbf{k} \cdot \mathbf{r})$, are small. This is why the period of time in (5.47) should be limited by

$$t \leqq t^* \sim \sqrt{\frac{1}{k^2(-\lambda^2)}} \tag{5.50}$$

Then

$$\nabla(\mathbf{u}_2 \sin \mathbf{k} \cdot \mathbf{r}) \sim \frac{1}{kl} \ll 1 \tag{5.51}$$

while

$$\nabla^2(\mathbf{u}_2 \sin \mathbf{k} \cdot \mathbf{r}) \sim \frac{1}{l} \tag{5.52}$$

Now integrating (5.46) over the infinitesimal interval $2\pi/k$ and taking into account that

$$\frac{k}{2\pi}\int_0^{2\pi/k} \mathbf{u}_1 dx = \mathbf{u}_1, \quad \frac{k}{2\pi}\int_0^{2\pi/k} \mathbf{u}_2 \sin \mathbf{k} \cdot \mathbf{r} \cdot dx = 0 \quad \left(x = \frac{\mathbf{k} \cdot \mathbf{r}}{k}\right) \tag{5.53}$$

one obtains

$$\frac{k}{2\pi}\int_0^{2\pi/k} \mathbf{u} dx \simeq \mathbf{u}_1, \quad \frac{k}{2\pi}\int_0^{2\pi/k} \mathbf{u} \sin \mathbf{k} \cdot \mathbf{r} \cdot dx \simeq \frac{1}{2}\mathbf{u}_2 \tag{5.54}$$

This means that the smooth displacement $\mathbf{u}_1$ practically remains unchanged during the infinitesimal interval $2\pi/k \ll l$ while the fluctuating displacement $\mathbf{u}_2$ has vanishingly small mean value and do not effect $\mathbf{u}_1$. In other words, due to the difference in scales these two displacements can be treated as geometrically independent, and therefore the displacement field becomes double-valued.

Actually, in the course of the instability there occurs infinite (and even uncounted) set of fluctuating terms $\mathbf{u}_2(\xi)$ (where ξ is the parameter of the multivaluedness), and consequently, the displacement field can be treated as multivalued. It is worth emphasizing that the multivaluedness should be considered only as a convenient mathematical tool for description of non-smooth states of continua.

Thus we will reformulate the original model starting from the moment when the instability occurred and introduce the multivalued displacement field:

$$\grave{\mathbf{u}} = \mathbf{u} + \mathbf{u}_r(\xi), \; 0 \leqq \xi \leqq 1, \; \xi = \xi_1, \xi_2, \xi_3 \tag{5.55}$$

where ξ_1, ξ_2, ξ_3 are uncounted sets of numbers of particles superimposed at the same point of space, $\mathbf{u}$ and $\mathbf{u}_r(\xi)$ are smooth and the fluctuating displacements, while

$$\int_0^1 \mathbf{u}_r(\xi) d\xi = 0 \tag{5.56}$$

In the next section we will introduce generalized models of continua based upon the concept of the multivaluedness of displacement and velocity fields.

5.2.2 *Generalized Models of Continua*

Definition

A postinstability in continua can be defined in the following terms:

$$v(\mathbf{r}_2) - v(\mathbf{r}_1) \nrightarrow 0 \text{ if } \mathbf{r}_2 - \mathbf{r}_1 \rightarrow 0 \tag{5.57}$$

where v is the velocity, and $\mathbf{r}_1, \mathbf{r}_2$ are the position vectors of the points in space. Obviously the condition (5.57) cannot be simulated by smooth functions; but it can be properly simulated if (5.57) is replaced by the following:

$$v(\mathbf{r}_2) - v(\mathbf{r}_1) \neq 0 \text{ if } \mathbf{r}_2 - \mathbf{r}_1 = 0 \tag{5.58}$$

Enlarging the class of functions in this manner implies that several individual particles with different velocities can appear at the same point in space. From the mechanical point of view, it corresponds to the relaxation of the principle of impenetrability, with utilization of a technique of multivalued functions.

Generally speaking, there are other possibilities of formally enlarging the class of differentiable functions describing postinstability motions of continuous media. But, the technique of multivalued functions seems reasonable also from the dynamics point of view, because in the course of instability of a motion the solutions tend to be multivalued (see examples in Sect. 5.2.1). However, as will be shown below, the above multivaluedness can appear only in very special cases, the conditions for which have to be found. Certainly, one should note that the multivaluedness mentioned above is an attribute of a mathematical (but not physical) model.

Kinematic Concepts

Let us consider the vector function:

$$v = v(\mathbf{r}, t) \tag{5.59}$$

We shall consider any mathematical limitations imposed on (5.59) as a constraint, and the medium with such a constraint is said to be a constrained medium. For example, the requirement of differentiability of the function (5.59) with respect to the space coordinates may be considered as a constraint. Relaxing the differentiability of the function (5.59) may be carried out, for instance, by means of the introduction of the above mentioned multivalued function:

$$v = v(\mathbf{r}, t, \xi^{(1)}), \ 0 \leqq \xi^{(1)} \leqq 1$$

$$\xi^{(1)} = \xi_1^{(1)}, \xi_2^{(1)}, \xi_3^{(1)} \tag{5.60}$$

where $\xi_1^{(1)}, \xi_2^{(1)}, \xi_3^{(1)}$ are uncounted sets of numbers of particles superimposed at the same point of space. Thus the function (5.60) simulates a medium as a mixture of ξ components, which are kinematically different (Fig. 5.1a).

For further transformations, the following decomposition of the velocity will be reasonable:

$$v = v_c + v_r(\xi^{(1)}), \ v_c = \int_{\xi^{(1)}} v(\xi^{(1)}) d\xi^{(1)} \tag{5.61}$$

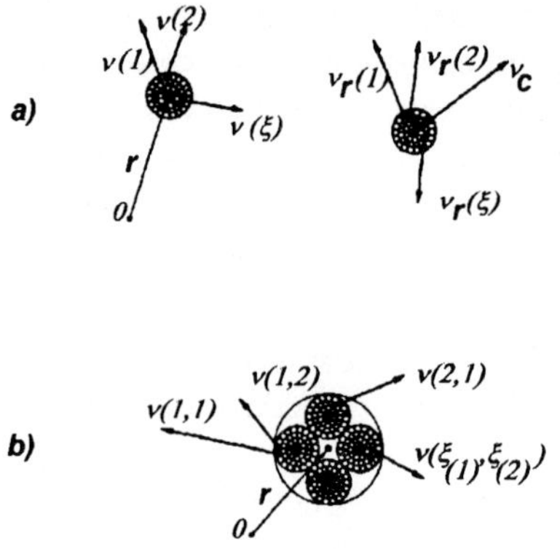

Fig. 5.1.

Here, $\boldsymbol{v}_c$ is the velocity of the "center of inertia" of the set of particles superimposed at the same point of space (an analog of the classical velocity), while $\boldsymbol{v}_r(\xi^{(1)})$ are the relative velocities, or pulsations, with respect to the above-introduced "center of inertia."

For the first step, the multivalued function (5.60) can be considered as differentiable "as many times as necessary" for any fixed $\xi^{(1)}_*$. Otherwise, a secondary multivaluedness can be introduced (Fig. 5.1):

$$\boldsymbol{v} = \boldsymbol{v}(\mathbf{r}, t, \xi^{(1)}, \xi^{(2)}),\ 0 \leqq \xi^{(1)} \leqq 1,\ 0 \leqq \xi^{(2)} \leqq 1 \text{ etc.} \tag{5.62}$$

Hence the requirement regarding differentiability of as a single-valued function has the following mathematical formulation:

$$\delta_{\xi^{(1)}} \overset{+}{\boldsymbol{v}} = 0, \text{ or } \delta_{\xi^{(1)}} \nabla \overset{+}{\boldsymbol{v}} = 0 \tag{5.63}$$

where $\delta_{\xi^{(1)}}$ is the symbol of $\xi^{(1)}$ variation, $\overset{+}{\boldsymbol{v}}$ is the vector of virtual velocity of the particles, and $\nabla \overset{+}{\boldsymbol{v}}$ is the gradient of the vector $\overset{+}{\boldsymbol{v}}$.

Obviously, the equality

$$\delta_{\xi^{(2)}} \nabla \overset{+}{\boldsymbol{v}} = 0 \tag{5.64}$$

expresses the fact that the function (5.62) is multivalued and differentiable for any fixed $\xi^{(1)}_*$, but does not have a secondary multivaluedness, etc.

Dynamical Concepts

Utilizing the principle of virtual work:

$$\int_V \int_{\xi^{(2)}} (\varrho \mathbf{a} - F) \cdot \delta_{\xi^{(1)}} \overset{+}{\boldsymbol{v}}\, d\xi^{(1)} d\mathbf{V} = 0 \tag{5.65}$$

where $\mathbf{V}$ is an arbitrary volume in space occupied by a medium, $\overset{+}{\boldsymbol{v}}$ is a virtual velocity, $\mathbf{a}$ is the acceleration, and ϱ is the density; let us multiply the second equality in (5.63) by an arbitrary tensor $(T)^T$ having the dimension of stresses and playing the role of Lagrange multiplier:

$$(T)^T \cdot \cdot \delta_{\xi^{(1)}} \nabla \overset{+}{\boldsymbol{v}} = 0 \qquad (5.66)$$

where $(T)^T$ denotes the transpose of T.

Obviously, the tensor T has the same degree of multivaluedness as the tensor $\nabla \overset{+}{\boldsymbol{v}}$; i.e.,

$$T = T(\xi^{(1)}) \qquad (5.67)$$

Taking into account the identities:

$$\nabla \cdot [(A)^T \cdot \mathbf{r}] = \mathbf{r} \cdot \nabla \cdot (A)^T + (A)^T \cdot \cdot (\nabla \mathbf{r})^T \qquad (5.68)$$

where A and $\mathbf{r}$ are arbitrary tensor and vector, then integrating (5.66) over the volume $\mathbf{V}$ and parameter of multivaluedness $\xi^{(1)}$, and adding term by term to the equality (5.65) one obtains:

$$\int_V \int_{\xi^{(1)}} (\varrho \boldsymbol{v} - \mathbf{F} - \nabla \cdot T) \cdot \delta_{\xi^{(1)}} \overset{+}{\boldsymbol{v}} \, d\xi^{(1)} d\mathbf{V}$$

$$+ \int_S \int_{\xi^{(1)}} (\mathbf{F} - T \cdot \mathbf{n}) \cdot \delta_{\xi^{(1)}} \overset{+}{\boldsymbol{v}} \, d\xi^{(1)} ds = \mathbf{0} \qquad (5.69)$$

whence because of independence of variations $\delta_{\xi^{(1)}} \overset{+}{\boldsymbol{v}}$:

$$\varrho \mathbf{a} = \nabla \cdot T + \mathbf{F} \qquad (5.70)$$

$$T \cdot \mathbf{n} = \mathbf{F}_S \qquad (5.71)$$

Here S is the surface bounding the volume V, $\mathbf{n}$ is the unit normal to this surface and $\mathbf{F}$, $\mathbf{F}_S$ are the volume, and surface external forces. Remembering that here the acceleration $\mathbf{a}$ is single-valued i.e., $\mathbf{a}$ does not depend on $\xi^{(1)}$, and decomposing the multivalued tensor $T(\xi^{(1)})$ in the manner of the formula (5.61):

$$\dot{T} = T_c + T_r(\xi^{(1)}), \; T_c = \int_{\xi^{(1)}} T(\xi^{(1)}) d\xi^{(1)} \qquad (5.72)$$

let us integrate (5.70–5.71) over $\xi^{(1)}$:

$$\varrho \mathbf{a} = \nabla \cdot T_c + \mathbf{F} \qquad (5.73)$$

$$T_c \cdot \mathbf{n} = \mathbf{F}_S \tag{5.74}$$

These are classical governing equations of a continuous media motion with the corresponding boundary conditions. The stress tensor T plays the role of a reaction of the constraint (5.63). In other words, the requirement of differentiability "as many times as necessary," or single-valuedness of velocities, is equivalent to the possibility of introducing a stress tensor.

Subtracting (5.68), (5.74) from (5.70), (5.71) respectively leads to additional equations:

$$\nabla \cdot T_r(\xi^{(1)}) = 0, \; T_r(\xi^{(1)}) \cdot \mathbf{n} = 0 \tag{5.75}$$

which are not coupled with (5.73–5.74) and therefore can be studied separately.

Equations (5.75) give the conditions of equilibrium of the "micro-stresses" $T_r(\xi^{(1)})$, which appear as constraints providing the absence of the multivaluedness of velocities.

Now let us derive the equations of motion of the medium with the constraint (5.64); i.e., for the medium where the velocities can have the first degree of multivaluedness:

$$\boldsymbol{v} = \boldsymbol{v}(\xi^{(1)}) \tag{5.76}$$

while all the derivatives with respect to space coordinates $\nabla \boldsymbol{v}, \boldsymbol{\nabla}^2 \boldsymbol{v}$ etc. exist at any fixed $\xi_*^{(1)}$. Such a medium can be understood, for instance, as a continuum beyond the limits of stability described in Sect. 4.3.

The principle of virtual work (5.65) and the equation of constraint (5.66) are rewritten in the form:

$$\int_V \int_{\xi^{(1)}} \int_{\xi^{(2)}} (\varrho \mathbf{a} - F) \cdot \delta_{\xi^{(2)}} \boldsymbol{v} \, d\xi^{(2)} d\xi^{(1)} d\mathbf{V} = 0 \tag{5.77}$$

$$(T)^T \cdot \cdot \delta_{\xi^{(1)}} \nabla \overset{+}{\boldsymbol{v}} = 0 \tag{5.78}$$

where

$$\mathbf{a} = \mathbf{a}\xi^{(1)} \tag{5.79}$$

but

$$T = T(\xi^{(1)}, \xi^{(2)}) \tag{5.80}$$

Then instead of (5.69) one obtains:

$$\int_V \int_{\xi^{(1)}} \int_{\xi^{(2)}} (\varrho \mathbf{a} - \mathbf{F} - \boldsymbol{\nabla} \cdot \mathbf{T}) \cdot \delta_{\xi^{(2)}} \overset{+}{\boldsymbol{v}} \, d\xi^{(2)} d\xi^{(1)} d\mathbf{V}$$

$$+ \int_S \int_{\xi^{(1)}} \int_{\xi^{(2)}} (\mathbf{F} - \mathbf{T} \cdot \mathbf{n}) \cdot \delta_{\xi^{(2)}} \overset{+}{\boldsymbol{v}} \, d\xi^{(2)} d\xi^{(1)} d\mathbf{V} \tag{5.81}$$

whence because of independence of variations $\overset{+}{v}\ d\xi^{(2)}$:

$$\varrho \mathbf{a}(\xi^{(1)}) = \nabla \cdot T(\xi^{(1)}, \xi^{(2)}) + \mathbf{F} \tag{5.82}$$

$$T(\xi^{(1)}, \xi^{(2)}) \cdot \mathbf{n} = \mathbf{F}_S \tag{5.83}$$

Now decomposing:

$$T = T_c(\xi^{(1)}) + T_r(\xi^{(1)}, \xi^{(2)}),\ T_c(\xi^{(1)}) = \int_{\xi^{(2)}} T(\xi^{(1)}, \xi^{(2)}) d\xi^{(2)} \tag{5.84}$$

and integrating (5.82–5.83) over the $\xi^{(2)}$ one obtains the governing equations:

$$\varrho \mathbf{a}(\xi^{(1)}) = \nabla \cdot T_c(\xi^{(1)}) + \mathbf{F} \tag{5.85}$$

$$T_c(\xi^{(1)}) \cdot \mathbf{n} = \mathbf{F}_S \tag{5.86}$$

The analog of (5.75) is derived similarly:

$$\nabla \cdot T_r(\xi^{(1)}, \xi^{(2)}) = 0,\ T_r(\xi^{(1)}, \xi^{(2)}) = 0 \tag{5.87}$$

Exploring this decomposition (5.61) and introducing an additional decomposition

$$T_c(\xi^{(1)}) = T_{cc} + T_{cr}(\xi^{(1)}),\ T_{cc} = \int_{\xi^{(1)}} F(\xi^{(1)}) d\xi^{(1)} \tag{5.88}$$

$$\mathbf{F} = \mathbf{F}_c + \mathbf{F}_r,\ \mathbf{F}_c = \int_{\xi^{(1)}} \mathbf{F}(\xi^{(1)}) d\xi^{(1)} \tag{5.89}$$

the equation of motion (5.86) is written in the final form:

$$\varrho \left(\frac{\partial \boldsymbol{v}_c}{\partial t} + \boldsymbol{v}_c \nabla \boldsymbol{v}_c - \int_{\xi^{(1)}} \boldsymbol{v}_r \nabla \boldsymbol{v}_r d\xi^{(1)} \right) = \nabla \cdot T_{cc} + \mathbf{F}_c \tag{5.90}$$

$$\varrho \left(\frac{\partial \boldsymbol{v}_c}{\partial t} + \boldsymbol{v}_c \nabla \boldsymbol{v}_r + \boldsymbol{v}_r \nabla \boldsymbol{v}_c + \boldsymbol{v}_r \nabla \boldsymbol{v}_c - \int_{\xi^{(1)}} \boldsymbol{v}_r \nabla \boldsymbol{v}_r d\xi^{(1)} \right) = \nabla \cdot T_{cr} + \mathbf{F}_r \tag{5.91}$$

These two vector equations contain two vector unknowns $\boldsymbol{v}_c, \boldsymbol{v}_r(\xi^{(1)})$ which are coupled nonlinearly. In other words, the behavior of the "classical" component of velocities $\boldsymbol{v}_c$ depends on the behavior of multivalued pulsations $\boldsymbol{v}_r(\xi^{(1)})$ and vice versa.

The system (5.90–5.91) must be completed by the corresponding constitutive equation, which characterizes a new state of the fluid:

$$T_c(\xi^{(1)}) = f[\text{def}\ \boldsymbol{v}(\xi^{(1)})] \tag{5.92}$$

$$T_{cc} = \int\limits_{\xi^{(1)}} f[\text{def } \boldsymbol{v}(\xi^{(1)})] d\xi^{(1)} \tag{5.93}$$

$$T_{cr} = f[\text{def } \boldsymbol{v}(\xi^{(1)})] - \int\limits_{\xi^{(1)}} f[\text{def } \boldsymbol{v}(\xi^{(1)})] d\xi^{(1)} \tag{5.94}$$

Clearly, these constitutive equations must satisfy the general laws which were formulated by Noll because for any multivalued model at any fixed parameter of the multivaluedness $\xi^{(1)}$ all the functions are differentiable "as many times as necessary" and all the vectors $\nabla \cdot \boldsymbol{v}$, $\nabla \times \boldsymbol{v}$, and tensors $\nabla \boldsymbol{v}$, $\nabla^2 \boldsymbol{v}$, etc., exist. Thus, for the domains where the classical model of fluid is unstable, the enlarged model given by (5.90–5.91), together with new constitutive equations (5.93–5.94) can be applied.

Obviously, it can happen that in some domains of parameters this enlarged model also becomes unstable. Then the model with the second degree of multivaluedness must be introduced. Such a model can be formally derived from the previous model by replacing in (5.90–5.94):

$$\boldsymbol{v}_c \to \boldsymbol{v}_c(\xi^{(1)}), \ \boldsymbol{v}_r \to \boldsymbol{v}_r(\xi^{(1)}, \xi^{(2)})$$

$$T_{cc} \to T_{cc}(\xi^{(1)}), \ T_{cr}(\xi^{(1)}) \to T_{cr}(\xi^{(1)}, \xi^{(2)}) \tag{5.95}$$

The process can be carried out many times.

5.2.3 Applications to the Model of an Inviscid Fluid

Governing Equation

By definition, in an inviscid fluid the shearing stresses are zero; i.e.,

$$\text{dev } T = 0 \tag{5.96}$$

Taking into account that the tensors $\nabla \boldsymbol{v}$, T can be decomposed:

$$\nabla \boldsymbol{v} = \frac{1}{3}(\nabla \cdot \boldsymbol{v}) E + \text{ dev } (\nabla \boldsymbol{v}) \tag{5.97}$$

$$T = -\frac{1}{3} p E + \text{ dev } T \tag{5.98}$$

where p is the spherical part of the tensor T (pressure), (5.66) is also decomposed:

$$p \delta_{\xi^{(1)}} (\nabla \cdot \overset{+}{\boldsymbol{v}}) = 0 \tag{5.99}$$

$$\text{dev } (T)^T \cdot \cdot \delta_{\xi^{(1)}} \text{ dev } \nabla \overset{+}{\boldsymbol{v}} = 0 \tag{5.100}$$

But in accordance with (5.96)

$$\text{dev } (T)^T = 0 \tag{5.101}$$

Then as follows from (5.100),

$$\delta_{\xi^{(1)}} \operatorname{dev} \nabla \overset{+}{v} \neq 0 \tag{5.102}$$

$$v = v(\xi^{(1)}) \tag{5.103}$$

But the same transformations can be done for any constraint:

$$\delta_{\xi^{(1)}} \nabla \overset{+}{v} \neq 0, \ \ (i = 2, 3, ..., \text{etc.}) \tag{5.104}$$

$$v = v(\xi^{(1)}, \xi^{(2)}, ..., \xi^{(n)} \text{ etc.}) \tag{5.105}$$

In other words, as follows from the definition (5.103), the field of velocities in an inviscid fluid can be multivalued. Moreover, the degree of this multivaluedness tends to infinity.

It is important to emphasize that here the appearance of multivaluedness is not caused by instability, but follows from the principle of release from constraints, according to which the absence of reactions (shearing stresses) must be compensated by introducing additional degrees of freedom [see (5.100)]. From this point of view the behavior of an inviscid fluid is not fully described by the Euler's equations. This is why the classical theory of an inviscid fluid abounds with paradoxes.

The subsequent discussion will be confined to the model with the first degree of multivaluedness, i.e., when

$$v = f(\xi^{(1)} = \xi) \tag{5.106}$$

The enlarged Euler's equations for an incompressible fluid are written in the following form:

$$\varrho \left(\frac{\partial v_c}{\partial t} + v_c \nabla v_c + \int_\xi v_r \nabla v_r d\xi \right) = -\nabla p_c + \mathbf{F}_c \tag{5.107}$$

$$\nabla \cdot v_c = 0 \tag{5.108}$$

Hence, because of nonlinear structure of the convective component of fluid acceleration, the pulsation velocities $\bar{v}_r$ (representing motions on the micro-scale) contribute to the motion of the original scale. It is easy to show that (5.107) coincides with the Reynolds equation describing turbulent motions. Indeed, the Reynolds decomposition of the velocity field is based upon the time averaging of the total velocity, and formally it coincides with the decompositions (5.61), while the rules of time averaging and the integration over ξ are also formally the same. Denoting

$$\rho \int_\xi v_r \nabla v_r d\xi = \overline{\rho v_r \nabla v_r} \tag{5.109}$$

one arrives at the Reynolds equations for modelling turbulent motions.

Thus a more general view of the class of functions in which motions of fluids are described, leads us to a remarkable result: it turns out that the Euler's equations for inviscid fluid are not adequate unless they are written in the Reynolds form.

Let us now introduce a model of atomized inviscid incompressible fluid postulating that

$$\nabla \boldsymbol{v}_c \geqq 0 \tag{5.110}$$

Then we can write:

$$\delta_\xi \nabla \cdot \boldsymbol{v}_c \neq 0 \tag{5.111}$$

but

$$p\delta_\xi \nabla \cdot \boldsymbol{v}_c = 0 \tag{5.112}$$

That is

$$p = 0 \tag{5.113}$$

and the governing equations are given by:

$$\varrho\left(\frac{\partial \boldsymbol{v}_c}{\partial t} + \boldsymbol{v}_c \nabla \boldsymbol{v}_c + \int_\xi \boldsymbol{v}_r \nabla \boldsymbol{v}_r d\xi\right) = \mathbf{F}_c \tag{5.114}$$

$$\frac{\partial \varrho}{\partial t} + \nabla \cdot (\varrho \boldsymbol{v}_c) = 0 \tag{5.115}$$

These equations hold true until

$$\varrho < \varrho_0 \tag{5.116}$$

where ϱ_0 is the density before atomization.

Both of the systems (5.107–5.108) and (5.114–5.115) can be combined in one generalized model:

$$\widetilde{\varrho}\left(\frac{\partial \boldsymbol{v}_c}{\partial t} + \boldsymbol{v}_c \nabla \boldsymbol{v}_c + \int_\xi \boldsymbol{v}_r \nabla \boldsymbol{v}_r d\xi\right) = -\nabla p_c + \mathbf{F}_c \tag{5.117}$$

$$\frac{\partial \widetilde{\varrho}}{\partial t} + \nabla \cdot (\widetilde{\varrho} \boldsymbol{v}_c) = 0 \tag{5.118}$$

where

$$p = \frac{1}{2}(p_c + |p_c|) \tag{5.119}$$

$$\widetilde{\varrho} = \varrho_0 + \frac{1}{2}(\varrho - \varrho_0 + |\varrho - \varrho_0|) \tag{5.120}$$

One should recall that the inviscid model of a fluid is an idealization: actually it describes a high Reynolds number fluid motions where the viscous forces can be ignored in comparison to the forces of inertia. But as follows from the theory of hydrodynamics stability, such motions are unstable, and as a result of this, the multivaluedness of velocity field occurs in the same way as it was described in this section. Hence the same effect of the multivaluedness of velocity field follows from the principle of release of constraints (for zero viscosity $\nu = 0$) and from the hydrodynamics stability (for small viscosity $\nu \to 0$).

Application to Lagrangian Dynamics

In this section we will return to finite dimensional dynamical systems which can be represented by the Lagrange equations (4.57) and introduce models describing postinstability motions of such systems. As in section 4.2, we will consider a mechanical system with N degrees of freedom and the kinetic energy

$$W = a_{ij}\dot{q}^i\dot{q}^j \tag{5.121}$$

in which $\dot{q}^i$, and $\dot{q}^j$ are the generalized coordinates and velocities respectively, and introduce an N dimensional (abstract) space with the metric:

$$ds^2 = a_{sk}dq^s dq^k, \ \dot{s}^2 = 2W \tag{5.122}$$

Then the equations of motion

$$q^i = q^i(t) \tag{5.123}$$

satisfy the following differential equations

$$\ddot{q}^\alpha + \Gamma^\alpha_{\beta\gamma}\dot{q}^\beta\dot{q}^\gamma = Q^\alpha \tag{5.124}$$

where Q^α is the force vector, $\Gamma^\alpha_{\beta\gamma}$ are the Christoffel symbols:

$$\Gamma^\ell_{sk} = \frac{1}{2}a^{\ell p}\left(\frac{\partial a_{sp}}{\partial q^k} + \frac{\partial a_{kp}}{\partial \dot{q}^s} - \frac{\partial a_{sk}}{\partial q^p}\right),$$

$$a^{\alpha\beta}a_{\beta\gamma} = \delta^\alpha_\gamma = \begin{cases} 0 \text{ if } & \alpha \neq \gamma \\ 1 \text{ if } & \alpha = \gamma \end{cases} \tag{5.125}$$

Equation (5.124) can be interpreted as a parametrical equation of the trajectory C of a representing point M with the contravariant coordinates q^α (Fig. 5.2). The unit tangent vector $\tau = \nu_0$ to this trajectory is defined as:

$$\tau^\alpha = \nu^\alpha_0 = \frac{dq^\alpha}{ds} = \frac{1}{\sqrt{2W}}\dot{q}^\alpha, \ a_{mn}\nu^m_0\nu^n_0 = 1 \tag{5.126}$$

while the unit normals $\nu_1, \nu_2, ..., \nu_{N-1}$ are given by the Frenet equations:

$$\frac{d\nu^i_p}{ds} + \Gamma^i_{kq}\nu^q_p\frac{dq^k}{ds} = -\chi_p\nu^i_{p-1} + \chi_{p+1}\nu^i_{p+1} \tag{5.127}$$

where $\chi_1, \chi_2, ..., \chi_{N-1}$ are the curvatures of the trajectory, and s is the arc coordinate along this trajectory.

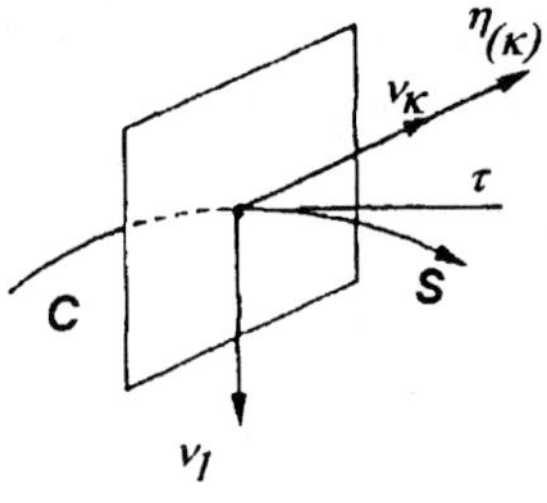

Fig. 5.2.

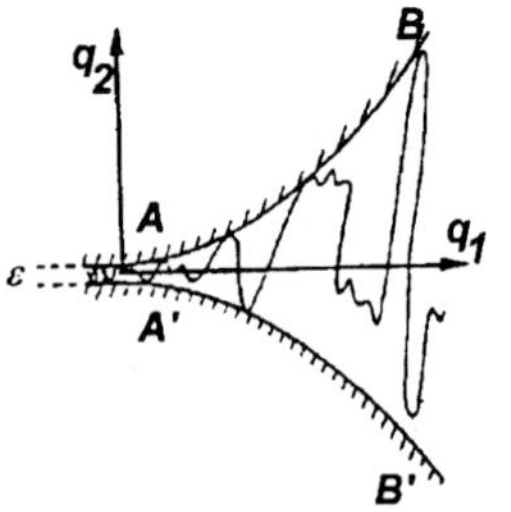

Fig. 5.3.

The principle normal ν_1 is coplanar with the tangent ν_0, and the force vector Q. The rest of the curvatures as well as the directions of the rest of the normals are defined by (5.126).

For simplicity, we will confine ourselves to the particular case when

$$\Gamma^{\ell}_{sk} = \text{Const} \tag{5.128}$$

In case of orbital instability of (5.124), their solution becomes chaotic. It means that in configuration space, two different trajectories which may be initially indistinguishable (because of finite scale of observation), diverge exponentially, so that a real trajectory can "fill up" all the space between these exponentially diverging trajectories (Fig. 5.3). In other words, in the domain of exponential instability, each trajectory "multiplies," and therefore, the predicted trajectory becomes multivalued, so that velocities can be considered as random variables:

$$\dot{q}^i = \dot{q}^i(t, \xi), \; 0 \leqq \xi \leqq 1 \tag{5.129}$$

where $\dot{q}$ for a fixed t is a function on a probability space, and ξ is a point in the probability space.

Let us refer the original equations of motion (5.124) to a non-inertial frame of reference which rapidly oscillates with respect to the original inertial frame of reference.

Then the absolute velocity of $\dot{q}$ can be decomposed into the relative velocity $\dot{q}_1$ and the transport velocity $\dot{q}_2 = 2\dot{q}_{2(0)} \cos\omega t$

$$\dot{q} = \dot{q}_1 + 2\dot{q}_{2(0)} \cos\omega t, \quad \omega \to \infty \tag{5.130}$$

while $\dot{q}_1$ and $\dot{q}_2^{\circ}$ are "slow" functions of time in the sense that

$$\omega \gg \frac{1}{\tau} \tag{5.131}$$

where τ is the time scale upon which the changes of q_1 and $\dot{q}_{2(0)}$ can be ignored. Then

$$q \tilde{=} q_1 \text{ since } \int_0^{t \gg \tau} \dot{q}_{2(0)} \cos\omega t dt \simeq \frac{1}{\omega} \dot{q}_{2(0)} \sin\omega t \to 0 \text{ if } \omega \to \infty \tag{5.132}$$

In other words, a fast oscillating velocity practically does not change the displacements.

Taking into account that

$$\frac{\omega}{2\pi} \int_0^{2\pi/\omega} \dot{q}_1 dt \simeq \dot{q}_1$$

$$\int_0^{2\pi/\omega} \dot{q}_{2(0)} \sin\omega t dt = 0$$

$$\int_0^{2\pi/\omega} \dot{q}_{2(0)}^2 \cos^2 \omega t dt = \frac{1}{2} \dot{q}_{2(0)}^2 \tag{5.133}$$

one can transform the system

$$\dot{x}^i = a_j^i + b_{jm}^i x^i x^m, \quad i = 1, 2, ..., n \tag{5.134}$$

into the following form:

$$\dot{\bar{x}}^i = a_j^i \bar{x}^j + b_{jm}^i \bar{x}^i \bar{x}^m + b_{mj}^i \overline{x^j x^m}, \quad i = 1, 2, ..., n \tag{5.135}$$

where $\overline{x}^i$ and $x^i x^j$ are means and double correlations of x^i as random variables, respectively.

As will be shown below, the transition from (5.134) to (5.135) is identical to the Reynolds transformation: i.e., applied to the Navier-Stokes equations, it leads to the Reynolds equations, and therefore the last term in (5.135) (which is a contribution of inertial forces due to fast oscillations of the frame of reference) can be identified with the Reynolds stresses. From a mathematical viewpoint, this transformation is interpretable as enlarging the class of smooth functions to multivalued ones. Indeed, as follows from (5.131), for any arbitrarily small interval Δt, there always exists such a large frequency $\omega > \Delta t/2\pi$ that within this interval the velocity $\dot{q}$ runs through all its values, and actually the velocity field becomes multivalued.

Clearly (5.135) result from time averaging. In case of applicability of the ergodic hypothesis, the same equations can be obtained from ensemble averaging. However, formally the averaging procedure can be introduced axiomatically based upon Reynolds conditions:

$$\overline{a+b} = \overline{a} + \overline{b}, \ \overline{ka} = k\overline{a}, \ \overline{k} = k \ (k = \text{Const})$$

$$\overline{\frac{\partial a}{\partial l}} = \frac{\partial \overline{a}}{\partial l}, \ \overline{\overline{a}b} = \overline{a}\overline{b}$$

This leads to the identity

$$\overline{ab} = \overline{a}\overline{b} + \overline{a'b'}$$

where $a = \overline{a} + a'$ and $b = \overline{b} + b'$.

Let us consider again the mechanical system (5.121–5.128). Substituting the decomposition (5.130) into (5.124), one obtains

$$\ddot{q}_1^\alpha + \Gamma^\alpha_{\beta\delta}\dot{q}_1^\beta\dot{q}_1^\gamma + \Gamma^\alpha_{\beta\delta}\dot{q}^\beta_{2(0)}\dot{q}^\alpha_{2(0)} = Q^\alpha \tag{5.136}$$

Here the terms

$$Q^\alpha_{(i)} = -\Gamma^\alpha_{\beta\gamma}\dot{q}^\beta_{2(0)}\dot{q}^\gamma_{2(0)} \tag{5.137}$$

represent inertia forces caused by the transport motion of the frame of reference.

Applying a velocity decomposition similar to (5.130),

$$\mathbf{v} = \overline{\mathbf{v}} + 2\tilde{\mathbf{v}}\cos\omega t, \ \omega \to \infty \tag{5.138}$$

to the momentum equation for a continuum in Eulerian representation

$$\rho\left(\frac{\partial\overline{\mathbf{v}}}{\partial t} + \overline{\mathbf{v}\nabla\mathbf{v}}\right) = \nabla\cdot\sigma \tag{5.139}$$

where σ is the stress tensor, one obtains

$$\rho\left(\frac{\partial\overline{\mathbf{v}}}{\partial t} + \overline{\mathbf{v}}\nabla\overline{\mathbf{v}}\right) = \nabla\cdot(\sigma + \tilde{\sigma}) \tag{5.140}$$

in which $\tilde{\sigma}$ is the Reynolds stress tensor with the components

$$\tilde{\sigma}_{ij} = -\overline{\rho\tilde{v}_i\tilde{v}_j} \tag{5.141}$$

In terms of the Reynolds equations $\overline{\mathbf{v}}$ and $\tilde{\mathbf{v}}$ represent the mean velocity and the amplitude of fast velocity fluctuations, respectively.

The most significant advantage of the Reynolds-type equations (5.135), (5.136), and (5.140) is that they are explicitly expressed via the physically reproducible parameters $\overline{x^i}$, $\overline{x^i x^j}$ which describe for instance, a mean velocity profile in turbulent motions, or a power spectrum of chaotic attractors. However, as a price for this, these equations require a closure since the number of unknowns is larger than the number of equations. Actually the closure problem existed for almost 100 years, since the Reynolds equations were derived. In the next sections, based upon the stabilization principle introduced by Zak (1985a; 1986a,b; 1990b), this problem will be discussed.

Postinstability Models of Flexible Bodies

The introduction of the concept of multivaluedness of the velocity field is not the only way to enlarge the class of functions in which motions of dynamical systems are described. We will illustrate this by turning to models of ideal flexible bodies such as strings and membranes. For simplicity, let us start with a model of an inextensible string.

The condition of inextensibility can be expressed in the form of a geometrical constraint:

$$\left|\frac{\partial \mathbf{r}}{\partial \psi}\right| = 1 \tag{5.142}$$

i.e.,

$$\delta \left|\frac{\partial \mathbf{r}}{\partial \psi}\right| = 0 \tag{5.143}$$

Multiplying (5.128) by T, one arrives at the condition

$$T\delta \left|\frac{\partial \mathbf{r}}{\partial \psi}\right| = 0 \tag{5.144}$$

where the Lagrangian multiplier T is the string tension.

However, strictly speaking, the condition of inextensibility restricts only the elongation of the string element, but not its contraction. Indeed, admitting non-differentiable configurations of the string (see Fig. 5.4) one concludes that the constraint (5.142) can be weakened to the following :

$$\left|\frac{\partial \mathbf{r}}{\partial \psi}\right| \leq 1 \tag{5.145}$$

and consequently, instead of (5.143–5.144) one has respectively:

$$\delta_\xi \left|\frac{\partial \mathbf{r}}{\partial \psi}\right| \begin{cases} = 0 \text{ if } & \left|\frac{\partial \mathbf{r}}{\partial \psi}\right| = 1 \\ \neq 0 \text{ if } & \left|\frac{\partial \mathbf{r}}{\partial \psi}\right| < 1 \end{cases} \tag{5.146}$$

and

$$T_{11} \begin{cases} > 0 \text{ if } & \left|\frac{\partial \mathbf{r}}{\partial \psi}\right| = 1 \\ = 0 \text{ if } & \left|\frac{\partial \mathbf{r}}{\partial \psi}\right| < 1 \end{cases} \tag{5.147}$$

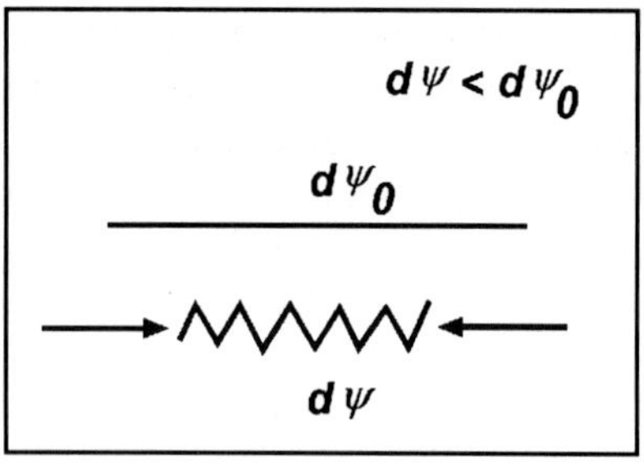

Fig. 5.4.

This means that a string can never be compressed, by definition, if its non-differential configurations are admissible. This demonstrates again the dependence of the concept of instability upon a class of fluctuations in which the motion is described. Indeed, as shown in Sect. 4.3.2, a compressed string ($T_{11} < 0$) is unstable in the class of differentiable functions. At the same time, the enlarging the class of smooth functions by admitting non-differential configurations of the string, eliminates the unstable domains, by definition, since, as follows from the principle of release of constraints (5.132), $T \geqslant 0$.

The same approach can be applied to membranes. In the next section we will derive and analyze the models for postinstability behavior of strings and membranes in connection with the stabilization principle which will be introduced below.

5.3 Stabilization Principle and the Closure Problem

5.3.1 *General Remarks*

Revisiting the dynamical system (5.107–5.108) which describe motions in the class of multivalued functions, one notes that the systems are not complete in the sense that the number of unknowns is larger than the number of equations. In particular, the vector which expresses the bulk contribution of the "micro-scale" motions into the averaged motion, represents excessive unknowns. Such an incompleteness creates a closure problem. This problem was first identified in connection with the Reynolds equations describing turbulent motion. The problem of turbulence arose almost a hundred years ago as a result of the discrepancy between theoretical fluid mechanics and experiments. However, in spite of considerable research activity, there is no general approach to prediction of turbulence based upon theoretical models. Most of the efforts were directed toward finding a "physical" law which would couple the Reynolds stresses with the rate-of-strain of the average motion, and thereby, would represent additional equations required for the closure of the Reynolds equations. For instance, Prandtl introduced the mixing length assumption

$$\tau = \rho \ell^2 \left| \frac{\partial u}{\partial y} \right| \frac{\partial u}{\partial y} \tag{5.148}$$

for the two dimensional version of the Reynolds equation

$$\frac{\partial u}{\partial t} + u\frac{\partial u}{\partial x} + v\frac{\partial u}{\partial y} = \frac{1}{\rho}\frac{d\tau}{\partial y}, \quad \tau = -\rho\overline{u_r v_r} \tag{5.149}$$

Here u, v and u_r are the mean and fluctuation velocity projections on the Cartesian coordinates x and y respectively, τ is the shear component of the Reynolds stress, and ℓ is a so-called mixing length which is supposed to be found from experiment.

By exploiting the closure (5.148), Prandtl solved several problems of two dimensional theory of turbulence: he found a mean velocity profile of an axisymmetrical turbulent flow in a pipe, and he described the smoothing out of a velocity discontinuity, etc., with all of his solutions being sufficiently close to experimental results.

However, the same closure (5.148) failed to provide satisfactory solutions in many other cases, which means that the closure (5.148) can not be considered as a "physical" law. But does any "physical" law of the type (5.148) exist, in principle? And is such a law necessary for the closure? Indeed, as shown in the previous section, the Reynolds stresses can be interpreted as a contribution of the inertia forces of a rapid oscillating frame of reference, while this frame of reference can be chosen arbitrarily! However, such an interpretation leads to another question: is it possible to find such a frame of reference which provides the "best" representation of the motion? Obviously, in this representation, the motion must be stable, and therefore, the restoration of stability of the originally unstable motion can be chosen as the main criterion for selection of the frame of reference, and therefore, of the Reynolds stresses. From the mathematical viewpoint, it means that if the original motion is unstable in the class of smooth functions, this instability can be eliminated by enlarging the class of functions. The simplest example of this is the admission of non-differentiable configurations of an ideal flexible string introduced by the constraint (5.74) which automatically eliminates the Hadamard instability. From this viewpoint, the Prandtl closure (5.148) can be treated as a feedback which stabilizes an originally unstable laminar flow. Indeed, turning, for instance, to a plane Poiseuille flow with the parabolic velocity profile, one arrives at its instability if the Reynolds number is larger than $R_{\text{cr}} \cong 5772$. Experiments show that a new steady turbulent profile is not parabolic any more: it is very flat near the center and is very steep near the walls. The same profile follows from the Prandtl solution based upon the closure (5.148). But since this profile can be experimentally observed, it must be stable, and this stabilization is carried out by the "feedback" (5.148).

5.3.2 Formulation of the Stabilization Principle

Based upon remarks made in the previous section, we will formulate the following stabilization principle. Consider a dynamical model which in some domain of its parameters becomes unstable in the class of differentiable functions, i.e., instability leads to an unbounded growth of ignorable variables. As noted earlier, this means that the corresponding physical phenomena cannot be adequately described in the class of differentiable functions, and the original model must be modified. The modification of the model should be based upon the enlarging of the original class of functions in such a way that the instability is eliminated (Zak 1985a,b; 1986a,b; 1987; 1990b; 1994a).

In the next three sections, the stabilization principle will be applied to the prediction of postinstability behavior of flexible bodies (strings, films), of fluids (turbulent motions), and of finite dimensional dynamical systems (chaos).

5.4 Applications of Stabilization Principle

5.4.1 Postinstability Motions of Flexible Bodies

We will start with applications of the stabilization principle to predictions of postinstability motions of one and two dimensional continua, i.e., to flexible bodies such as strings and films, since for these models, the application of this principle requires the simplest strategy.

Postinstability Motions of Strings

A one dimensional continuum which can be represented by an inextensible string is described by (4.162). Its motion becomes unpredictable if the string tension is negative:

$$T < 0 \tag{5.150}$$

since then the solutions to (4.162) become unstable in the class of differentiable functions [see (4.183)].

Enlarging the class of differentiable functions by admitting non-differentiable string configurations (see Fig. 5.4) was carried out by introducing the unilateral constraint (5.145) instead of the constraint (5.142). This leads to the following governing equation instead of (4.162) (at $\mathbf{F} = 0$):

$$\rho \frac{\partial^2 \mathbf{r}}{dt^2} = \frac{\partial}{\partial \psi}(\tilde{T}\frac{\partial \mathbf{r}}{\partial \psi}) + \mathbf{F}' \tag{5.151}$$

$$\tilde{T} = \begin{cases} T \text{ if } & T \geqslant 0 \\ 0 \text{ if } & T < 0 \end{cases} \tag{5.152}$$

$$\left|\frac{\partial \mathbf{r}}{\partial \psi}\right| \begin{cases} = 1 \text{ if } & T \geqslant 0 \\ < 1 \text{ if } & T < 0 \end{cases} \tag{5.153}$$

$$\rho = \begin{cases} \rho_0 \text{ if } & T \geqslant 0 \\ \frac{\rho_0}{|\frac{\partial \mathbf{r}}{\partial \psi}|} \text{ if } & T < 0 \end{cases}, \quad \frac{d\psi}{d\psi_0} = \left|\frac{\partial \mathbf{r}}{\partial \psi}\right| \tag{5.154}$$

Here $\mathbf{F}$ and $\mathbf{F}'$ are the tracing and dead forces, respectively.

As follows from these equations, in the domain $T = 0$, the length of the string element $|\partial \mathbf{r}/\partial \psi|$ as well as the string linear density ρ become new variables which are found from (5.151–5.154). Indeed, at $T < 0$:

$$\frac{\partial^2 \mathbf{r}}{\partial t^2} = \mathbf{F}' \tag{5.155}$$

and the string motion degenerates into a flow of free material particles:

$$\mathbf{r} = \mathbf{r}_0 + (\frac{\partial \mathbf{r}}{\partial t})_0 t + \int_0^t \mathbf{F}' dt \tag{5.156}$$

so

$$\frac{\partial \mathbf{r}}{\partial \psi} = \frac{\partial \mathbf{r}_0}{\partial \psi} + \frac{\partial}{\partial \psi}(\frac{\partial \mathbf{r}}{\partial t})_0 t + \int_0^t \frac{\partial \mathbf{F}'}{\partial \psi} dt \tag{5.157}$$

Here

$$\mathbf{r}_0 = \mathbf{r}(t=0), \quad \frac{\partial \mathbf{r}}{\partial t}(t=0) \tag{5.158}$$

and $\mathbf{F}'$ are the dead (non-tracing) forces. Hence, all the parameters of the contracted string are found from the solution (5.156).

Obviously, the Hadamard instability of the original governing equations (4.162) is eliminated from (5.151–5.154).

So far we consider the case when tracing force $\mathbf{F}$ is zero. However, as follows from (4.182), in the presence of tracing forces, there is another possibility of Hadamard's instability if

$$T < -\frac{F_2}{\Omega_3} \tag{5.159}$$

where F_2 is the normal component of $\mathbf{F}$, and Ω_3 is the string curvature.

Applying the stabilization principle, one has to set:

$$T \geqslant \max\,(0, -\frac{F_2}{\Omega_3}) = \hat{T} \tag{5.160}$$

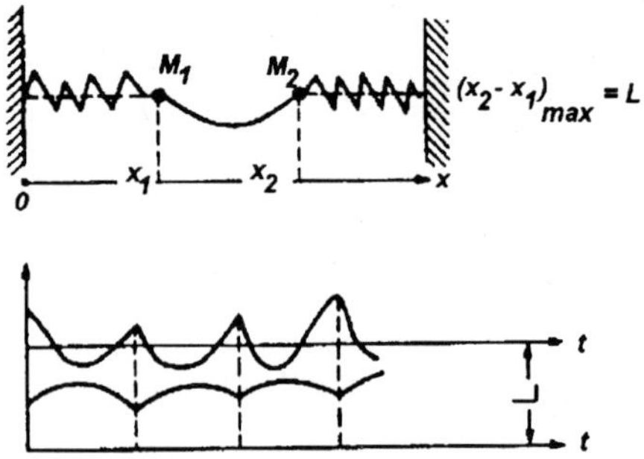

Fig. 5.5.

i.e.,

$$\tilde{T} = \begin{cases} T \text{ if } & T \geqslant \hat{T} \\ 0 \text{ if } & T < \hat{T} \end{cases} \tag{5.161}$$

Thus, in the general case, the system of governing equations for a string is:

$$\rho \frac{\partial^2 \mathbf{r}}{\partial t^2} = \frac{\partial}{\partial \psi}(\tilde{T} \frac{\partial \mathbf{r}}{\partial \psi}) + \mathbf{F} + \mathbf{F}' \tag{5.162}$$

$$\rho = \begin{cases} \rho_0 \text{ if } & T \geqslant \hat{T} \\ \frac{\rho_0}{|\frac{\partial \mathbf{r}}{\partial \psi}|} \text{ if } & T < 0 \end{cases} \tag{5.163}$$

while $\hat{T}$ and $\tilde{T}$ are defined by (5.160) and (5.161) respectively.

The following examples will illustrate the postinstability behavior of strings (Zak 1979a,b).

Example 1

Let us consider the motion of a string between two elastic elements (Fig. 5.5) while neglecting the mass of the film in comparison with the masses of the element. The equation of motion can be written in the following form:

$$\ddot{x}_1 + \frac{c_1 + c_2}{m_1 + m_2} x_1 = 0, \; \ddot{x}_2 + \frac{c_1 + c_2}{m_1 + m_2} x_2 = 0 \text{ if } x_2 - x_1 = l \tag{5.164}$$

$$\ddot{x}_1 + \frac{c_1}{m_1} x = 0, \; \ddot{x}_2 + \frac{c_2}{m_2} x_2 = 0 \text{ if } x_2 - x_1 < l \tag{5.165}$$

where c_1, c_2 are stiffness, m_1, m_2 are masses of the elastic elements. It is easy to show that the first type of motion is unstable. Indeed, imparting the infinitesimal decrement Δx_2 to the coordinate x_2 we arrive at the second type of motion. However, if according to the equation (5.165) the equality

$$x_2 - x_1 = l \tag{5.166}$$

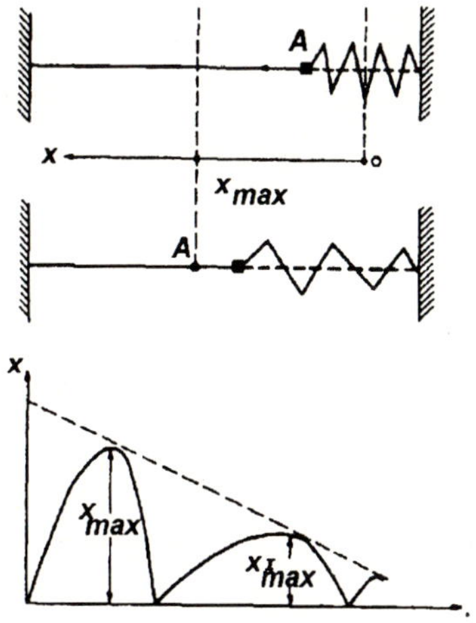

Fig. 5.6.

is restored and consequently

$$(\dot{x}_2 - \dot{x}_1)|_{t=-0} \geqslant 0 \tag{5.167}$$

as a result of reflection we have:

$$(x_2 - \dot{x}_1)|_{t=+0} = -\,(\dot{x}_2 - \dot{x}_1)|_{t=-0} \leqslant 0 \tag{5.168}$$

Thus the condition (5.166) is violated and we return to the second type of motion again. The exception takes place only if

$$(\dot{x}_2 - \dot{x}_1)|_{t=-0} = 0 = (\dot{x}_2 - \dot{x}_1)|_{t=+0} \tag{5.169}$$

But practically, this case is not realistic.

Example 2

Proceeding from the previous example let us take into account the string mass and confine ourselves to only one elastic element with the mass m_0 (Fig. 5.6). According to (4.172) at $T = 0$, any disturbance can propagate only with the velocity of motion of individual particles of the film whose coordinates coincide with the coordinate of the point A. Thus we come across the problem of the motion of a particle with variable mass. This mass can be expressed by the formula:

$$m = \rho x \tag{5.170}$$

where x is the current coordinate of the point A. The differential equation for such motion can be written as follows:

$$\frac{d}{dt}[(m_0 + \rho x)\dot{x}] = -cx$$

or

$$(m_0 + \rho x)\ddot{x} = -cx - \rho\dot{x}^2 \tag{5.171}$$

where c is the stiffness of the elastic element. Introducing a new variable:

$$z = \dot{x}^2 \tag{5.172}$$

we get a linear equation

$$\frac{1}{2}(m_0 + \rho x)\frac{dz}{dx} + \rho z + cx = 0$$

Hence

$$v = \dot{x} = \frac{\left\{\frac{m_0^2}{\rho^2}v_0^2 - \frac{c}{\rho}\left(\frac{m_0}{\rho^2}x^2 + \frac{2}{3}x^3\right)\right\}^{1/2}}{\frac{m_0}{\rho} + x} \tag{5.173}$$

where $v_0 = v|_{x=0}$, and

$$t = \int_0^x \frac{\left(\frac{m_0}{\rho} + x\right)dx}{\left\{\frac{m_0^2}{\rho^2}v_0^2 - \frac{c}{\rho}\left(\frac{m_0}{\rho^2}x^2 + \frac{2}{3}x^3\right)\right\}^{1/2}} \tag{5.174}$$

The amplitude $x_{\max}$ of the first oscillation is defined by means of the equation

$$x_{\max}^3 + \frac{3m_0}{2\rho}x_{\max}^2 = \frac{3m_0^2}{2c\rho}v_0^2 \tag{5.175}$$

This amplitude corresponds to the time

$$t^* = \int_0^x \frac{\left(\frac{m_0}{\rho} + x\right)dx}{\left\{\frac{m_2}{\rho^2}v_0^2 - \frac{c}{\rho}\left(\frac{m_0}{\rho}x^2 + \frac{2}{3}x^3\right)\right\}^{1/2}} \tag{5.176}$$

The starting equation (5.171) is valid only if $t \leqslant t^*$. After this period (Fig. 5.6) the point A does not move but the string particles "flow out" from it. Now, instead of (5.171), (5.173–5.174) we have:

$$[m_0 + \rho(x_{\max} - x)]\ddot{x} = -cx + \rho\dot{x}^2 \tag{5.177}$$

$$v = \frac{\left\{\frac{c}{\rho}\left[\left(\frac{m_0}{\rho} + x_{\max}\right)(x_{\max}^2 - x^2) - \frac{2}{3}(x_{\max}^3 - x^3)\right]\right\}^{1/2}}{\frac{m_0}{\rho} + x_{\max} - x} \tag{5.178}$$

If $x = 0$ we get

$$v_1 = \frac{\left\{\frac{c}{\rho}\left[\left(\frac{m_0}{\rho}x_{\max}^2 + \frac{1}{3}x_{\max}^3\right)\right]\right\}^{1/2}}{\frac{m_0}{\rho} + x_{\max}} = \frac{\frac{m_0}{\rho}\left(v_0^2 - \frac{\rho c}{3m_0^2}x_{\max}^3\right)^{1/2}}{\frac{m_0}{\rho} + x_{\max}} \tag{5.179}$$

At this moment as a result of reflection from the point 0 which corresponds to the natural length of the film, the velocity v_1, changes its sign. At the same moment the value v_1, is changed too, since before shock the moving mass was $(m_0 + \rho x_{\max})$ and after the shock only the mass of the elastic body begins to move. Hence, after reflection we arrive at the starting problem, but the new initial velocity $v_0^{(1)}$ is defined by the formula:

$$v_0^{(1)} = \left(v_0^2 - \frac{\rho c}{3m_0^2} x_{\max}^3\right)^{1/2} < v_0 \tag{5.180}$$

For the new amplitude instead of (5.175) we obtain:

$$x_{\max}^{(1)^3} + \frac{3m_0}{2\rho} x_{\max}^{(1)^2} = \frac{3m_0^2}{2c\rho}\left(v_0^2 - \frac{\rho c}{3m_0^2} x_{\max}^3\right) \tag{5.181}$$

Thus the following recurrent formulas can be written:

$$v^{(n)} = \left\{(v_0^{(n-1)})^2 - \frac{\rho c}{3m_0^2}(x_{\max}^{(n-1)})^3\right\}^{1/2} < v^{(n-1)} \tag{5.182}$$

$$(x_{\max}^{(n)})^3 + \frac{3m_0}{2\rho}(x_{\max}^{(n)})^2 = \frac{3m_0^2}{2c\rho}[(v_0^{(n-1)})^2 - \frac{\rho c}{3m_0^2}(x_{\max}^{(n-1)})^3] \tag{5.183}$$

$$(x_{\max}^{(n-1)}) < (x_{\max}^{(n)}), \ n = 1, 2, \ldots \tag{5.184}$$

Hence we obtained damped oscillations which are accompanied by the sequence of decreasing shocks (Fig. 5.6). The loss of energy in these oscillations is a result of a nonelastic process of string contraction.

5.4.2 Postinstability Motions of Films

Governing Equations

A two dimensional continuum which can be represented by a film or a membrane, is described by (4.190). As shown in Sect. 4.3.2, its motion becomes unpredictable if at least one of the principle stresses becomes negative.

$$\overset{+}{T}_{11} < 0 \text{ or } \overset{+}{T}_{22} < 0 \tag{5.185}$$

since then the solutions to (4.190) become unstable in the class of differentiable functions.

Application of the stabilization principle requires that in the governing equations of a film the principle stresses must be non-negative:

$$\overset{+}{T}_{11} \geqslant 0, \ \overset{+}{T}_{22} \geqslant 0 \tag{5.186}$$

As in the case of a string, this requirement necessitates the enlarging of the class of functions by admitting non differential film configurations (for instance, the film surface may be covered by wrinkles). For inextensible films this enlargement is carried out by the following unilateral constraints:

$$\left|\frac{\partial \mathbf{r}}{\partial q_i}\right| \begin{cases} = g_{ii}^{1/2} \text{ if } \overset{+}{T}_{ii} \geqslant 0 \\ < g_{ii}^{1/2} \text{ if } \overset{+}{T}_{ii} < 0 \end{cases}, \; i = 1, 2 \tag{5.187}$$

where q_i are the coordinates coinciding with the directions of principle stresses, and g_{ii} are the coefficients of the first quadratic dorm of the film surface.

Since the formulations of the stabilization principle here is significantly simplified if the concept of principle stresses is exploited, we will introduce a system of coordinates q_1, q_2, which coincides with the principle directions of the stresses:

$$\left|\frac{\partial \mathbf{r}}{\partial q_1}\right|^2 = g_{11}, \quad \left|\frac{\partial \mathbf{r}}{\partial q_2}\right|^2 = g_{22}, \quad \frac{\partial \mathbf{r}}{\partial q_1} \cdot \frac{\partial \mathbf{r}}{\partial q_2} = 0 \tag{5.188}$$

where g_{11} and g_{22} are the new unknowns.

Introducing the coefficients of coordinates transformation

$$a_{ij} = \frac{\partial \psi_i}{\partial q_j} \tag{5.189}$$

and comparing (5.188) and (4.192), we have:

$$\frac{\partial \mathbf{r}}{\partial q_1} = a_{11}\frac{\partial \mathbf{r}}{\partial \psi_1} + a_{21}\frac{\partial \mathbf{r}}{\partial \psi_2}, \quad \frac{\partial \mathbf{r}}{\partial q_2} = a_{12}\frac{\partial \mathbf{r}}{\partial \psi_1} + a_{22}\frac{\partial \mathbf{r}}{\partial \psi_2} \tag{5.190}$$

$$g_{11} = a_{11}^2(1 + 2\varepsilon_{11}) + a_{21}^2(g + 2\varepsilon_{22}) + 4a_{11}a_{21}\varepsilon_{12} \tag{5.191}$$

$$g_{22} = a_{12}^2(1 + 2\varepsilon_{11}) + a_{22}^2(g + 2\varepsilon_{22}) + 4a_{12}a_{22}\varepsilon_{12} \tag{5.192}$$

$$0 = a_{11}a_{12}(1 + 2\varepsilon_{11}) + a_{21}a_{22}(g + 2\varepsilon_{22}) + 2(a_{11}a_{22} + a_{21}a_{12})\varepsilon_{12} \tag{5.193}$$

Besides this, from (5.189) the equations of compatibility follow:

$$\frac{\partial a_{11}}{\partial q_2} = \frac{\partial a_{12}}{\partial q_1}, \quad \frac{\partial a_{22}}{\partial q_1} = \frac{\partial a_{21}}{\partial q_2} \tag{5.194}$$

It will be more convenient now to rewrite the components of deformation in the new system q_1, q_2:

$$\tilde{\varepsilon}_{11} = a_{11}^2\varepsilon_{11} + 2a_{11}a_{21}\varepsilon_{12} + a_{21}^2\varepsilon_{22} \tag{5.195}$$

$$\tilde{\varepsilon}_{22} = a_{12}^2\varepsilon_{11} + 2a_{12}a_{22}\varepsilon_{12} + a_{22}^2\varepsilon_{22} \tag{5.196}$$

$$\tilde{\varepsilon}_{12} = a_{11}a_{12}\varepsilon_{11} + (a_{11}a_{22} + a_{12}a_{21})\varepsilon_{12} + a_{21}a_{22}\varepsilon_{22} \tag{5.197}$$

Then instead of (5.191–5.193) we have [see (4.204)]:

$$g_{11} = a_{11}^2 + a_{21}^2 g + 2\tilde{\varepsilon}_{11} \tag{5.198}$$

$$g_{22} = a_{12}^2 + a_{22}^2 g + 2\widetilde{\varepsilon}_{22} \tag{5.199}$$

$$0 = a_{11}a_{12} + a_{21}a_{22}g + 2\widetilde{\varepsilon}_{12} \tag{5.200}$$

Taking into account the fact that the system of coordinates q_1, q_2 is not a material system, and introducing the relative velocities v_1, v_2 of the individual particles with respect to these coordinates, we set

$$v_1 = \frac{\partial q_1}{\partial t}, \quad v_2 = \frac{\partial q_2}{\partial t} \tag{5.201}$$

Then keeping in mind the equality (5.189), we arrive at the equations of compatibility:

$$\frac{\partial a_{11}}{\partial t} = \frac{\partial}{\partial q_1}(a_{11}v_1 + a_{12}v_2), \quad \frac{\partial a_{22}}{\partial t} = \frac{\partial}{\partial q_2}(a_{21}v_1 + a_{22}v_2) \tag{5.202}$$

Thus we have introduced 8 new unknowns g_{11}, g_{22}, a_{11}, a_{12}, a_{21}, a_{22}, u_1, u_2, and 7 new equations (5.194), (5.198–5.200), (5.202). The additional new equation has to express the fact that the system q_1, q_2, corresponds to the principle directions of stresses. Hence:

$$\overset{+}{T}_{12} = 0 \tag{5.203}$$

In new coordinates instead of (4.190) we can obtain:

$$\begin{aligned}(g_{11}g_{22})^{1/2}\rho_0[(\frac{\partial}{\partial t} + v_1\frac{\partial}{\partial q_1} + v_2\frac{\partial}{\partial q_2})^2\mathbf{r} + (\frac{\partial v_1}{\partial t} + v_1\frac{\partial v_1}{\partial q_1} + v_2\frac{\partial v_1}{\partial q_2})\frac{\partial \mathbf{r}}{\partial q_1}\\ (\frac{\partial v_2}{\partial t} + v_1\frac{\partial v_2}{\partial q_1} + v_2\frac{\partial v_2}{\partial q_2})\frac{\partial \mathbf{r}}{\partial q_1}] - \mathbf{F}\\ = \frac{\partial}{\partial q_i}\left\{\overset{+}{T}_{11}\,(g_{11}g_{22})^{1/2}\frac{\partial \mathbf{r}}{\partial q_1}\right\} + \frac{\partial}{\partial q_2}\left\{\overset{+}{T}_{22}\,(g_{11}g_{22})^{1/2}\frac{\partial \mathbf{r}}{\partial q_2}\right\}\end{aligned} \tag{5.204}$$

The system (5.194), (5.198–5.200), (5.202) is closed. For an elastic film the equations must be complemented by constitutive relationships which, for a linear elastic material can be written in the form:

$$\widetilde{\varepsilon}_{11} = \frac{1}{E}(\overset{+}{T}_{11} - \mu\overset{+}{T}_{22}), \quad \widetilde{\varepsilon}_{22} = \frac{1}{E}(\overset{+}{T}_{22} - \mu\overset{+}{T}_{11}), \quad \widetilde{\varepsilon}_{12} = \frac{(1+\mu)}{E}\overset{+}{T}_{12} = 0 \tag{5.205}$$

where $\widetilde{\varepsilon}_{ij}$ are the principle strains, $\overset{+}{T}_{ij}$are the principle stresses, E is the modulus of elasticity, and μ is the Poisson ratio. In this system the condition of instability are given the form (5.185). If one or both of the inequalities (5.185) hold, we have to set there:

$$\overset{+}{T}_{11} = 0 \text{ or } \overset{+}{T}_{22} = 0 \tag{5.206}$$

and cancel the correspondent constitutive equation in (5.205) for the case of an elastic film. Thus the system remains closed and defines the unknowns $v_1, \widetilde{\varepsilon}_{11}$, or $v_2, \widetilde{\varepsilon}_{22}$ which now can be utilized for the description of wrinkles. For instance, if

$$\overset{+}{T}_{22} = 0 \tag{5.207}$$

the relative height of wrinkles with respect to the initial thickness of film is expressed by the following equality:

$$\Delta\widetilde{h} = \max\left(1, \frac{1}{1 + \widetilde{\varepsilon}_{22} + \frac{\mu \overset{+}{T}_{11}}{E}}\right) \tag{5.208}$$

If both of the principle stresses are zero

$$T_{11} = T_{22} = 0 \tag{5.209}$$

we arrive at the equation:

$$\frac{\partial^2 \mathbf{r}}{\partial t^2} = \frac{\mathbf{F}}{\rho_0 (g)^{1/2}} \tag{5.210}$$

The solution of this equation defines deformations by means of (4.192) and the density ρ by means of the formula:

$$\rho = \frac{\rho_0}{[(1 + 2\varepsilon_{11})(g + \varepsilon_{22}) - 4\varepsilon_{12}^2]^{1/2}} \tag{5.211}$$

The equation (5.210) is valid until

$$\widetilde{\varepsilon}_{11} < 0, \ \widetilde{\varepsilon}_{22} < 0 \tag{5.212}$$

where $\widetilde{\varepsilon}_{22}, \widetilde{\varepsilon}_{22}$ are the principle deformations.

Several examples discussed below will illustrate the application of the generalized model of a film to statics and dynamics of postinstability states.

Statics of WrinklingFilms

The governing equation for the equilibrium of a film with possible regions of wrinkling can be written in the following form (Zak 1982b):

$$\frac{\partial}{\partial q_1}\left[\frac{1}{2}(T_{11} + |T_{11}|)\left|\frac{\partial \mathbf{r}}{\partial q_1}\right| \left|\frac{\partial \mathbf{r}}{\partial q_1}\right| \frac{\partial \mathbf{r}}{\partial q_1}\right]$$

$$+ \frac{\partial}{\partial q_2}\left[\frac{1}{2}(T_{22} + |T_{22}|)\left|\frac{\partial \mathbf{r}}{\partial q_1}\right| \left|\frac{\partial \mathbf{r}}{\partial q_2}\right| \frac{\partial \mathbf{r}}{\partial q_2}\right] + \mathbf{F} = 0 \tag{5.213}$$

where $\mathbf{r}$ is the radius-vector of the film points, $\mathbf{F}$ is the vector of external forces, q_1, q_2 are the orthogonal (non-material) coordinates which coincide with the directions of the principles stresses T_{11} and T_{22}. Thus,

$$\frac{\partial \mathbf{r}}{\partial q_1} \cdot \frac{\partial \mathbf{r}}{\partial q_2} = 0 \tag{5.214}$$

while the strains in this system of coordinates are given by

$$2\varepsilon_{11} = \left| \frac{\partial \mathbf{r}}{\partial q_1} \right|^2 - \alpha_{11}^2 - \alpha_{21}^2 g_0 \tag{5.215}$$

$$2\varepsilon_{22} = \left| \frac{\partial \mathbf{r}}{\partial q_2} \right|^2 - \alpha_{12}^2 - \alpha_{22}^2 g_0 \tag{5.216}$$

$$2\varepsilon_{12} = -\alpha_{11}\alpha_{12} - \alpha_{21}\alpha_{22} g_0 \tag{5.217}$$

$$\frac{\partial \alpha_{11}}{\partial q_2} = \frac{\partial \alpha_{12}}{\partial q_1}, \ \frac{\partial \alpha_{22}}{\partial q_1} = \frac{\partial \alpha_{21}}{\partial q_2}, \ \left(\alpha_{ij} = \frac{\partial \psi_i}{\partial q_j} \right) \tag{5.218}$$

$$\left| \frac{\partial \mathbf{r}}{\partial \psi_1^0} \right| = 1, \ \left| \frac{\partial \mathbf{r}}{\partial \psi_2^0} \right|^2 = g_0, \ \frac{\partial \mathbf{r}}{\partial \psi_1^0} \cdot \frac{\partial \mathbf{r}}{\partial \psi_2^0} = 0 \tag{5.219}$$

where ψ_1, ψ_2 are the material (frozen) coordinates, and ψ_1^0, ψ_2^0 are the initial state of the material coordinates which are chosen as semi-geodesical coordinates. We will consider here only wrinkling regions of films where

$$T_{11} > 0, \ T_{22} = 0 \tag{5.220}$$

Therefore instead of (5.213) the governing equations now are written in the following simplified form:

$$\frac{\partial}{\partial q_1} \left(T_{11} \left| \frac{\partial \mathbf{r}}{\partial q_1} \right| \left| \frac{\partial \mathbf{r}}{\partial q_2} \right| \frac{\partial \mathbf{r}}{\partial q_1} \right) + \mathbf{F} = 0 \tag{5.221}$$

The introduction of a fictitious stress.

$$T^* = T_{11} \left| \frac{\partial \mathbf{r}}{\partial q_1} \right| \left| \frac{\partial \mathbf{r}}{\partial q_2} \right| \tag{5.222}$$

reduces the governing equation to the final form:

$$\frac{\partial}{\partial q_1} (T^* \frac{\partial \mathbf{r}}{\partial q_1}) + \mathbf{F} = 0 \tag{5.223}$$

which formally coincides with the governing equation for a string.

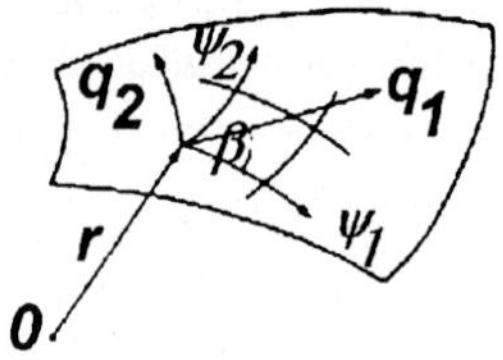

Fig. 5.7.

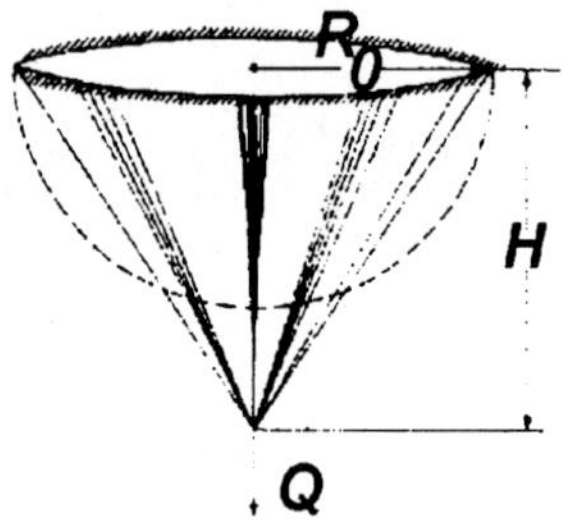

Fig. 5.8.

Example 1. Equilibrium without External Forces

If

$$\mathbf{F} = 0 \tag{5.224}$$

then (5.223) has the first integral

$$T^* \frac{\partial \mathbf{r}}{\partial q_1} = \text{Const} \tag{5.225}$$

Consequently, wrinkles form a family of straight lines along which the product (Fig. 5.7)

$$T^* \left| \frac{\partial \mathbf{r}}{\partial q_1} \right| = T_{11} \left| \frac{\partial \mathbf{r}}{\partial q_1} \right|^2 \left| \frac{\partial \mathbf{r}}{\partial q_2} \right| = \text{Const} \tag{5.226}$$

These general results in some special cases allow us to find the equilibrium state of a film and to investigate some singular effects without involving all the equations (5.215–5.226).

Stretching of a Semi-spherical Film. Let us consider a semi-spherical film which is symmetrically stretched by the forces **Q** as shown in Fig. 5.8.

Clearly the equilibrium shape of the film cannot have positive Gaussian curvature. Indeed, taking the scalar product of (5.153) and the unit normal **n** to the surface of the film, one obtains

$$(T_{11} + |T_{11}|)\, b_{11} + (T_{22} + |T_{22}|)\, b_{22} = 0 \tag{5.227}$$

where

$$b_{11} = \frac{\partial^2 \mathbf{r}}{\partial q_1^2} \cdot \mathbf{n},\ b_{22} = \frac{\partial^2 \mathbf{r}}{\partial q_2^2} \cdot \mathbf{n} \tag{5.228}$$

are the coefficients of the second fundamental form of the film surface.

Hence,

$$b_{11} b_{22} \leqslant 0 \text{ if } T_{11} > 0 \tag{5.229}$$

i.e.,

$$G = \frac{b_{11} b_{22} - b_{12}^2}{\left|\frac{\partial \mathbf{r}}{\partial q_1}\right| \left|\frac{\partial \mathbf{r}}{\partial q_2}\right|} \leq 0 \tag{5.230}$$

where G is the Gaussian curvature.

Thus, stretching of a semi-spherical film even by infinitesimal force $\mathbf{Q}$ leads to buckling (snapping) in the course of which the sphere is transformed into a surface with a non-positive Gaussian curvature. In order to find this surface, first of all one should exclude from consideration the possibility that for the post-buckling state both of the principal stresses are positive:

$$T_{11} > 0,\ T_{22} > 0 \tag{5.231}$$

Indeed, in this case the following conditions must be true:

$$\left|\frac{\partial \mathbf{r}^1}{\partial q_1}\right| \left|\frac{\partial \mathbf{r}^1}{\partial q_2}\right| \geq \left|\frac{\partial \mathbf{r}}{\partial q_1}\right| \left|\frac{\partial \mathbf{r}}{\partial q_2}\right| \tag{5.232}$$

where $\mathbf{r}^1 = \mathbf{r}^1(q_1\ q_2)$ is the equation of the post-buckling surface. In other words the area of the film cannot be decreased if all the stresses are positive. But taking into account that by symmetry the post-buckling film forms a surface of revolution. one can write the Gaussian formula for the pre- and post-buckling cases as

$$G = \frac{1}{R}\frac{d^2 R}{dS^2} = R_0 > 0,\ G' = \frac{1}{R^1}\frac{d^2 R^1}{dS^2} < 0 \tag{5.233}$$

where R, R^1 are the radii of revolution for the pre- and post-buckling surfaces, R_0 is the radius of the sphere, S is the arc coordinate of the meridian curve, i.e.,

$$\frac{1}{R^1}\frac{d^2 R^1}{dS^2} > \frac{1}{R}\frac{d^2 R}{dS^2} \tag{5.234}$$

while

$$R^1 = R \text{ if } S = 0 \text{ and } S = \frac{1}{2}\pi R_0 \tag{5.235}$$

Clearly,

$$R^1 < R,\ 0 < S < \frac{1}{2}\pi R_0 \tag{5.236}$$

Hence, the area of the film surface after buckling is decreased which contradicts (5.232), and consequently, the assumption (5.231).

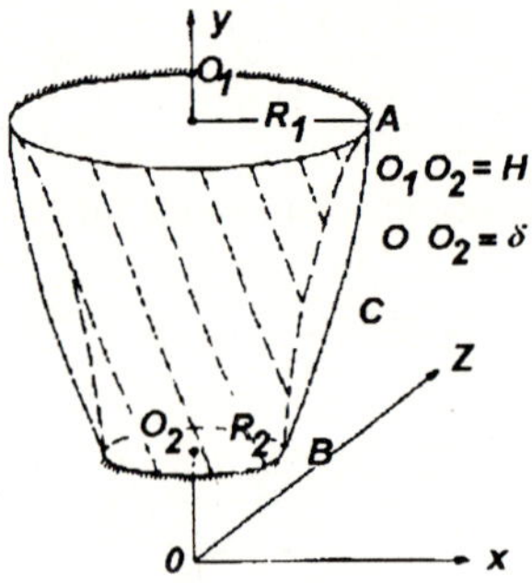

Fig. 5.9.

Thus, the post-buckling state is characterized by (5.220). Now exploring (5.225) one concludes that the post-buckling film forms a lined surface which by symmetry must be a cone of revolution. In the course of transformation of the semi-sphere into the cone all the parallels are contracted forming straight-lined wrinkles. Because of the positive tension

$$T_{11} > 0 \tag{5.237}$$

the length of the wrinkle cannot be smaller than its pre-buckling length, i.e.,

$$H = R_0 \sqrt{(\frac{\pi}{2})^2 - 1} \tag{5.238}$$

where H is the height of the cone (Fig. 5.8).

The contractions (wrinkling) of parallels are given by the formula

$$\varepsilon_{22} = \frac{2\theta}{\pi} - \sin\theta < 0 \text{ if } \theta < \frac{\pi}{2} \tag{5.239}$$

where $\pi/2 - \theta$ is the latitude of the parallel.

Hence, infinitesimal stretching of a semi-spherical film leads to a jump in the shape via snapping which is accompanied by wrinkling. As a result the sphere is transformed into a cone whose geometry is defined irrespective of the elastic properties of the film.

Torsion of a Convex Film of Revolution. Let us consider a film which forms a convex surface of revolution bounded by two undeformable circumferences with radii R_1 and R_2 (Fig. 5.9). We will assume that this surface is generated by the rotation of the curve C

$$x = \psi_1(S), \; y = \psi_2(S) \tag{5.240}$$

about the axis of symmetry y. Then the arc coordinate S and the angle γ of rotation of the curve C about the axis y form the semi-geodesical system

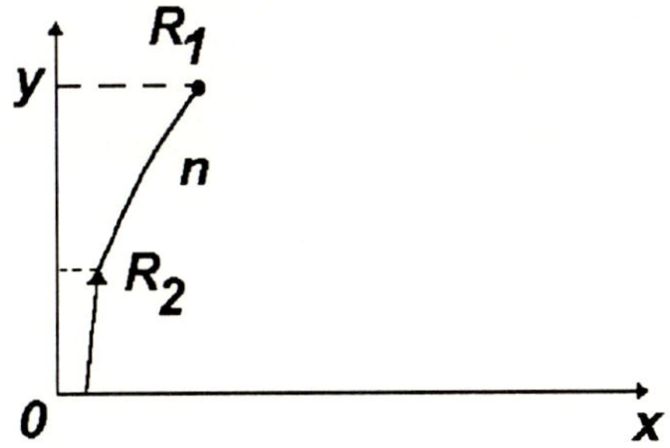

Fig. 5.10.

of meridians and parallels

$$\psi_1^0 = S,\ \psi_2^0 = \gamma \tag{5.241}$$

with the following coefficients of the first fundamental form:

$$g_{11} = 1,\ g_{12} = 0,\ g_{22} = G = [\psi_1(S)]^2 \tag{5.242}$$

Now let us assume that the lower circumference is fixed, while the upper circumference is turned through the torsion angle α about the axis y. Arguing the same way as in the previous case, one can conclude that even infinitesimal torsion leads to a snapping of the convex surface into a lined (wrinkled) surface of revolution, i.e. into a hyperboloid of revolution of one sheet, irrespective of the initial shape of the film. Consequently, the initial curve C is transformed into the corresponding hyperbola h. Clearly, the larger the difference is between the curves C and h the greater is the effect of snapping-through. As in the previous case such a buckling is accompanied by a jump in the intensity of wrinkling because of the contraction of parallels.

For infinitesimal torsion the post-buckled state of the film can be defined without involving of the whole system of (5.215–5.219). Indeed, let us write the equation of the hyperbola h (Fig. 5.10) as

$$Ax_2 - By^2 = 1,\ A > 0,\ B > 0 \tag{5.243}$$

where

$$A = \frac{\Delta_1(\delta)}{\Delta(\delta)},\ B = \frac{\Delta_2(\delta)}{\Delta(\delta)} \tag{5.244}$$

$$\Delta(\delta) = \begin{vmatrix} \delta^2 & -R_2^2 \\ (\delta + H)^2 & -R_1^2 \end{vmatrix},\ \Delta_1(\delta) = \begin{vmatrix} 1 & -R_2^2 \\ 1 & R_1^2 \end{vmatrix},$$

$$\Delta_2(\delta) = \begin{vmatrix} \delta^2 & 1 \\ (\delta + H) & 1 \end{vmatrix} \tag{5.245}$$

and H is the height of the body of revolution formed by the film.

The unknown parameter δ can he found by taking into account that for small torsion the lengths of the curves C and h must be equal, i.e.,

$$l_c = \int_{R_2}^{R_1} \left(1 + \frac{1}{B}\frac{A^2x^2}{Ax^2 - 1}\right)^{1/2} dx$$

$$= \left(\frac{1}{A}\right)^{1/2} \left\{ E\left[\left(1 + \frac{A}{B}\right)^{1/2}, \psi_1\right] - E\left[\left(1 + \frac{A}{B}\right)^{1/2}, \psi_2\right]\right\} = \gamma(\delta) \tag{5.246}$$

where E is the elliptical function of the second kind in Legendre's form, and ψ_1, ψ_2 are given by

$$\psi_1 = \arcsin(R_1\sqrt{A}), \quad \psi_2 = \arcsin(R_2\sqrt{A}) \tag{5.247}$$

Hence

$$\delta = \gamma^{-1}(\ell_c) \tag{5.248}$$

and for the convenience of the further transformations the parameters A, B in (5.181) will be considered as known, being defined by R_1, R_2, H, l_c, via the formulas (5.244–5.248). Now the configuration of the film after small torsion is found as a hyperboloid of one sheet formed by rotation of the hyperbola (5.243) about the axis y:

$$Ax^2 - By^2 + Az^2 = 1 \tag{5.249}$$

The wrinkles form two families of straight lines corresponding to the left- and right-hand torsion:

$$z + y = \frac{\lambda_1}{A^{1/2}}(1 + xB^{1/2}), \quad \lambda_1(z - y) = \frac{1}{A^{1/2}}(1 - xB^{1/2}) \tag{5.250}$$

$$(z + y) = \frac{\lambda_2}{A^{1/2}}(1 - xB^{1/2}), \quad \lambda_2(z - y) = \frac{1}{A^{1/2}}(1 + xB^{1/2}) \tag{5.251}$$

The angle β_1, between wrinkles and parallels is given by

$$\beta_1 = \arccos\left[\frac{A}{(A + B)(1 + By^2)}\right]^{1/2} \tag{5.252}$$

Stress Singularities. The integral (5.226) shows that the real tension T can tend to infinity at the points where

$$\left|\frac{\partial \mathbf{r}}{\partial q_2}\right| \to 0 \tag{5.253}$$

Clearly, such a situation occurs at the top of a cone where wrinkles converge. Hence at such points the tension T_{11} becomes unbounded under a finite load.

Example 2. Equilibrium with External Forces Normal to the Surface of a Film

Let us assume now that external forces are normal to the surface of a film, i.e.,

$$\mathbf{F} = F\mathbf{n}, \left(\mathbf{n} \cdot \frac{\partial \mathbf{r}}{\partial q_1} = 0, \ \mathbf{n} \cdot \frac{\partial \mathbf{r}}{\partial q_2} = 0\right) \tag{5.254}$$

Such a case covers inflated films, films loaded by an inviscid fluid, and films stretched on a smooth rigid surface.

Taking the scalar product of the (5.223) and the vector $\partial \mathbf{r}/\partial q_2$ and keeping in mind (5.214) and (5.254), one arrives at

$$\frac{\partial^2 \mathbf{r}}{\partial q_1^2} \cdot \frac{\partial \mathbf{r}}{\partial q_2} = 0 \tag{5.255}$$

which shows that wrinkles form a family of geodesical lines on the surface of the film:

$$q_2 = \text{Const} \tag{5.256}$$

This means that, as in previous case, the coordinates q_1, q_2 form a semi-geodesical system.

Taking the scalar product of the (5.223) and the vector $\partial \mathbf{r}/\partial q_1$, and keeping in mind (5.254), one arrives at the same integral

$$T^* \left|\frac{\partial \mathbf{r}}{\partial q_1}\right| = \text{Const}, \text{ or } T_{11} \left|\frac{\partial \mathbf{r}}{\partial q_1}\right|^2 \left|\frac{\partial \mathbf{r}}{\partial q_2}\right| = \text{Const} \tag{5.257}$$

along a wrinkle as in the previous case [see (5.226)]. This means that the tension T_{11} becomes unbounded at the points where wrinkles converge, irrespective of the rate of loading.

Taking the scalar product of the (5.223) and the normal $\mathbf{n}$ to the film, one obtains

$$T^* b_{11} = -F \tag{5.258}$$

i.e.,

$$\text{sgn } b_{11} = -\text{sgn } F \tag{5.259}$$

The following example illustrates applications of the results given above.

Small Torsion of a Film of Revolution. Let us consider a film which is stretched on a rigid smooth surface of revolution and bounded by two undeformable circumferences with radii R_1 and R_2 (Fig. 5.11), and assume that the lower circumference is fixed, while the upper circumference is turned through the torsion angle α about the axis y. As shown above, the wrinkles (if they occur) will form a family of geodesical lines. Thus, for a spherical film wrinkles will coincide with inclined great circles, while for a cylindrical film they will form a family of helices with constant slopes. In the last case the complete solution of the problem can be obtained if the angle of

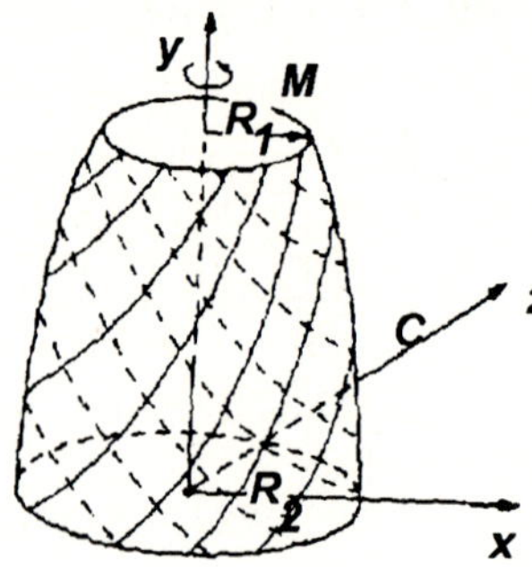

Fig. 5.11.

the torsion is small. Indeed, it is easy to verify that the constant principal strains

$$\varepsilon_{11} = \frac{\alpha}{2}, \ \varepsilon_{22} = -\frac{\alpha}{2}, \ \varepsilon_{12} = 0 \tag{5.260}$$

satisfy (5.215–5.217) together with the boundary conditions, if one takes into account that for the cylinder

$$\left|\frac{\partial \mathbf{r}}{\partial q_1}\right| = \left|\frac{\partial \mathbf{r}}{\partial q_2}\right| = g_0 = 1 \tag{5.261}$$

Then the coefficients a_{ij} are given by the following formulas:

$$a_{11} = a_{22} = a_{21} = -a_{12} = \frac{\sqrt{2}}{2}(1+\alpha) \text{ if } \alpha > 0 \tag{5.262}$$

$$a_{11} = a_{22} = a_{12} = -a_{21} = \frac{\sqrt{2}}{2}(1+\alpha) \text{ if } \alpha < 0 \tag{5.263}$$

Here the initial semi-geodesical system is formed by the family of generators ψ_1, and parallels ψ_2. Thus, in the case of left-hand torsion, wrinkles q_2 coincide with the family of left-hand helices intersecting the cylinder generators at the angle of 45°, while the coordinates q_1, form the orthogonal right-hand family of helices, and vice versa. Hence, the directions of wrinkles become unstable, jumping from –45° to 45° when the torsion angle changes sign. This phenomenon as a nonlinear singularity is similar to the phenomenon of snapping discussed above.

Clearly, the pressure between the film and rigid surface generated by the torsion is defined by (5.258), (5.261):

$$p = \frac{1}{2} T_{11} R_0^{-1} \tag{5.264}$$

where R_0 is the radius of the cylinder.

Example 3. Equilibrium in a Potential Field

Let us assume that external mass forces form a potential field:

$$\mathbf{F} = \rho_0 G_0^{1/2} \nabla u \tag{5.265}$$

Then from (5.223)

$$\frac{\partial}{\partial q_1}\left(T^* \frac{\partial \mathbf{r}}{\partial q_1}\right) + \rho_0 G_0^{1/2} \nabla u = 0 \tag{5.266}$$

where ρ_0 is the mass per unit area.

Taking the scalar product of this equation and the vector $\partial\, \mathbf{r}/\partial q_2$, one obtains

$$T^* \frac{\partial^2 \mathbf{r}}{\partial q_1^2} \cdot \frac{\partial \mathbf{r}}{\partial q_2} = \rho_0 G_0^{1/2} \frac{\partial u}{\partial q_2} \tag{5.267}$$

i.e.,

$$\frac{\partial^2 \mathbf{r}}{\partial q_1^2} \cdot \frac{\partial \mathbf{r}}{\partial q_2} = 0, \text{ if } \frac{\partial u}{\partial q_2} = 0 \tag{5.268}$$

Hence, wrinkles form a family of geodesical lines, if the lines q_2 form equipotential surfaces $u = \text{Const}$.

Taking the scalar product of this equation and the vector $\partial\, \mathbf{r}/\partial q_1$, and introducing the notations

$$\left|\frac{\partial \mathbf{r}}{\partial q_1}\right|^2 = g_{11}, \quad \left|\frac{\partial \mathbf{r}}{\partial q_2}\right|^2 = g_{22} \tag{5.269}$$

one arrives at the following equation:

$$\frac{\partial T^*}{\partial q_1} + \rho_0 (g_{11}^{-1} g_0)^{1/2} \frac{\partial u}{\partial q_1} = 0 \tag{5.270}$$

For small strains when the change in area of a film can be ignored,

$$g_{11} g_{22} \sim \overset{0}{g}_{11} \overset{0}{g}_{22} = g_0^{1/2}(q_1,\ q_2) \tag{5.271}$$

Equation (5.270) leads to the following integral:

$$T_{11} g_0^{1/2} = \text{Const} - \rho_0 g_0^{1/2}(q_1,\ q_2) \frac{\partial u}{\partial q_1} dq_1 \tag{5.272}$$

In the particular case when

$$\overset{0}{g}_{11} \overset{0}{g}_{22} = g_0 = \text{Const} \tag{5.273}$$

this integral simplifies to

$$T_{11} + \rho_0 u = \text{Const} \tag{5.274}$$

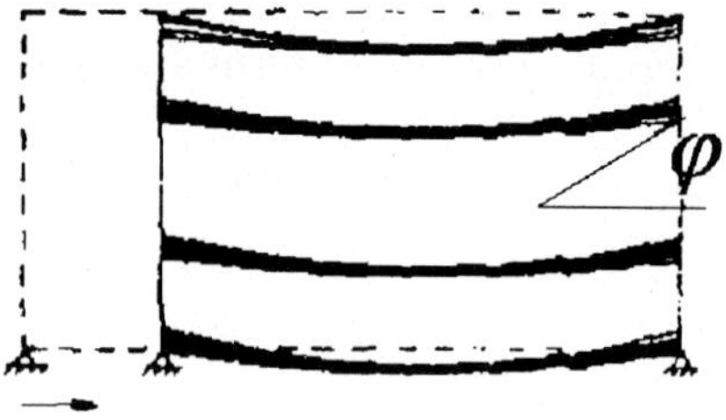

Fig. 5.12.

or with reference to (5.222)

$$T^* + \rho_0 u = \text{Const} \tag{5.275}$$

i.e., the tension T^* is constant at equipotential surfaces.

Let us assume that

$$\mathbf{F} = \mathbf{F}^0 = \text{Const} \tag{5.276}$$

Then taking the vector product of (5.223) and $\mathbf{F}^0$, one obtains

$$T^* \frac{\partial \mathbf{r}}{\partial q_1} \times \mathbf{F}^0 = \text{Const} \tag{5.277}$$

i.e., wrinkles lie in the planes parallel to the forces F.

Similarly,

$$T^* \frac{\partial \mathbf{r}}{\partial q_1} \times \mathbf{i} = \text{Const} \tag{5.278}$$

where

$$\mathbf{i} \cdot \mathbf{F}^0 = 0, \ \mathbf{i} \cdot \frac{\partial \mathbf{r}}{\partial q_1} \times \frac{\partial^2 \mathbf{r}}{\partial q_1^2} = 0 \tag{5.279}$$

i.e., the projection of the product $T^* \, \partial \mathbf{r}/\partial q_1$, on the direction $\mathbf{i}$ orthogonal to $\mathbf{F}^0$ if the plane of the wrinkle is constant.

Vertical Plane Film in the Gravity Field. In order to illustrate the results given above we will consider a vertical plane film in the gravity field, assuming that initially the film was stretched between two rigid vertical supports, but then these supports were moved so that the distance between them became smaller. Ignoring the extensibility of the film let us define the shape of those wrinkles which appear (Fig. 5.12). Clearly, the wrinkles form a family of parallel curves intersecting the supports while the length of each of them is equal to the initial horizontal length of the film, i.e.,

$$\left| \frac{\partial \mathbf{r}}{\partial q_1} \right| = g_{11}^{1/2} = 1 \tag{5.280}$$

where q_1, is the arc coordinate along a wrinkle. Because of the rigidity of the supports the film cannot be contracted in the vertical direction. Hence, each element is subjected only to shearing so that its area is given by

$$(g_{11}g_{22})^{1/2} = \cos\psi \tag{5.281}$$

where ψ is the angle between the tangent of the wrinkle and the horizontal line. Consequently,

$$\sqrt{g_{22}} = \cos\psi \tag{5.282}$$

Projection of the governing equation (5.266) onto the tangent and normal to a wrinkle gives

$$\frac{\partial T^*}{\partial q_1} = g\rho_0 \sin\psi \tag{5.283}$$

and

$$T^* \frac{d\psi}{dq_1} = g\rho_0 \cos\psi \tag{5.284}$$

where

$$\nabla u = -g\mathbf{j} \tag{5.285}$$

with $\mathbf{j}$ being the unit vertical vector, while

$$\frac{d\psi}{dq_1} = \Omega$$

is the curvature of a wrinkle.

Clearly,

$$\frac{dT^*}{T^*} = \tan\psi d\psi$$

and

$$T^* = \frac{C}{\cos\psi} \tag{5.286}$$

Then as follows from (5.284)

$$\frac{d\psi}{dq_1} = \frac{g}{C_1}\cos^2\psi \tag{5.287}$$

and

$$q_1 = \frac{C_1}{g}\frac{\tan\ \psi}{\rho_0} + C_2 \tag{5.288}$$

This is the equation of a wrinkle in natural form while the arbitrary constants C_1, C_2 are defined from the boundary conditions.

Formula (5.287) shows that the wrinkles form a family of catenaries.

Dynamics of Wrinkling Films

Singular Effects

Several singular effects are associated with the discontinuity in the transformation to the principal stress directions which define the directions of wrinkles. Indeed, these directions are given by the formula:

$$\tan\phi = \frac{\overset{+}{T}_{11} - T_{xx}}{T_{xy}} \tag{5.289}$$

where $\overset{+}{T}_{11}$, is the principal stress, T_{xx}, T_{xy} are the normal and shear stress in a Cartesian system x, y; ϕ is the angle between the directions of stresses $\overset{+}{T}_{11}$, T_{xx}.

Clearly, the directions of wrinkles (the angle ϕ) are changed by a jump (from $-45°$ to $+45°$) if

$$T_{xy} = 0 \tag{5.290}$$

This singularity explains the jump at the force deflection characteristic in the course of shearing of a film stretched between two parallel supports, (Fig. 5.13a,c). The same effect leads to the snap-through of a convex film of revolution into the hyperboloid of revolution of one sheet after infinitesimal torsion (Fig. 5.13d), or to a jump from left to right helix geodesical lines of wrinkles of a cylindrical film subject to a torsion with respect to its axis (Fig. 5.13b).

Shocks and Shock Waves

The effect of shocks is generated by discontinuity of the stress-strain dependence at $T_{ii} = 0$, in the course of transition of the film from the state of contraction to the state of elongation (Fig. 5.13f).

But besides this, a fundamental nonlinear phenomenon of classical shock wave formation occurs within the domain of wrinkling. Formally such a possibility comes from the nonlinear terms of the convective part of acceleration in (5.204). These terms appear because wrinkles move with respect to fixed individual particles. The mechanism of formation of shock waves here is similar to the same phenomenon in a continuous flow of free particles having different initial velocities [see (3.23) and Fig. 3.5 and Fig. 5.4e]. In film these shock waves appear in the form of lines of condensation of the mass (thick wrinkles) and accompanied by the dissipation of energy due to non elastic collisions between the particles in the course of wrinkling (Zak 1983).

For a quantitative description of shock wave formation let us consider the simplest free medium – a one dimensional continuous flow of free particles. Introducing an Eulerian space coordinate x one arrives at the following

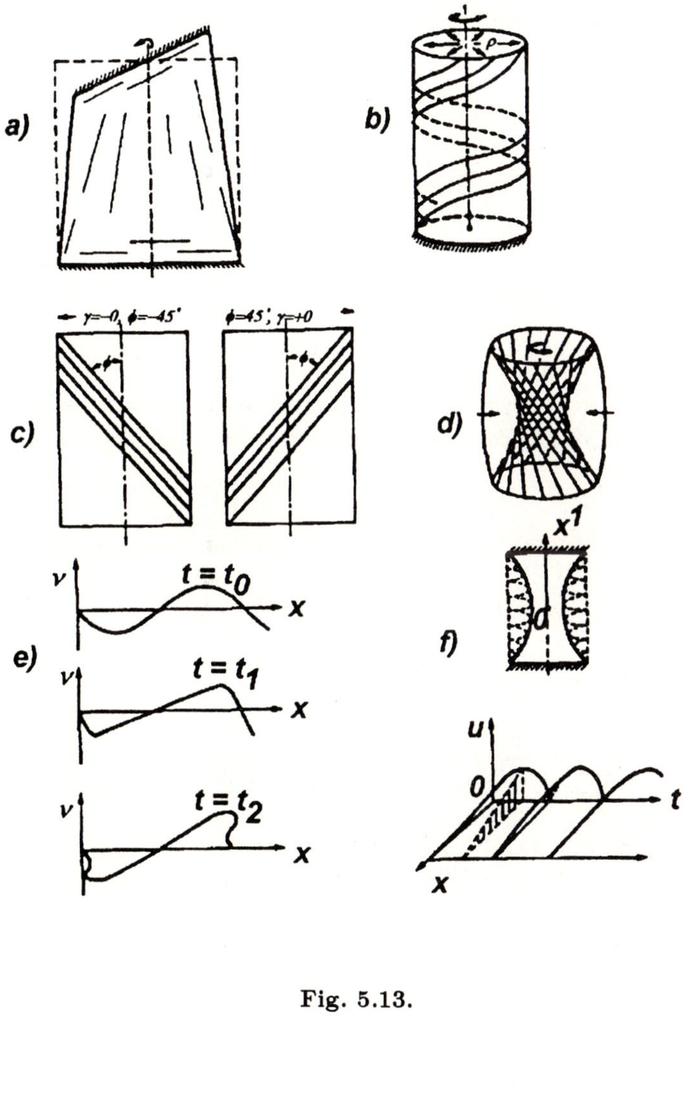

Fig. 5.13.

governing equation of motion:

$$\frac{\partial v}{\partial t}+v\frac{dv}{\partial x}=0 \tag{5.291}$$

and the continuity equation:

$$\frac{\partial \rho}{\partial t}+\rho\frac{\partial v}{\partial x}+v\frac{\partial \rho}{\partial x}=0 \tag{5.292}$$

where v, ρ are the velocity and linear density, respectively. Let us assume now that the initial distribution of the velocity is given in the following form:

$$v|_{t=0}=v_0\sin\frac{x}{\ell_0} \tag{5.293}$$

where v_0, ℓ_0, can be made as small as necessary, i.e.,

$$v_0\to 0,\ \ell_0\to 0 \tag{5.294}$$

The solution to (5.291) with the initial condition (5.293) can be written in the following implicit form:

$$v=v_0\sin\left[\frac{1}{\ell_0}(x-vt)\right] \tag{5.295}$$

Obviously,

$$v=0 \text{ if } x=\ell_0\pi k\ (k=1,2,...,\text{etc.}) \tag{5.296}$$

$$v_{\max}=v_0 \text{ if } x-v_0t=\frac{\ell_0}{2}\pi k\ (k=1,2,...,\text{etc.}) \tag{5.297}$$

$$\frac{\partial v}{\partial x}=\frac{\frac{v_0}{\ell_0}\cos\left[\frac{1}{\ell_0}(x-vt)\right]}{1+\frac{v_0}{\ell_0}t\cos\left[\frac{1}{\ell_0}(x-vt)\right]} \tag{5.298}$$

$$\frac{\partial v}{\partial x}=0 \text{ if } v=v_0 \tag{5.299}$$

$$\frac{\partial v}{\partial x}=\frac{\frac{v_0}{\ell_0}}{1+t\frac{v_0}{\ell_0}} \text{ if } v=0 \tag{5.300}$$

$$\frac{\partial v}{\partial x}\to\pm\infty \text{ if } 1+\frac{v_0}{\ell_0}t\cos\left[\frac{1}{\ell_0}(x-vt)\right]=0 \tag{5.301}$$

The last equation in (5.301) is transformed:

$$\frac{v_0}{\ell_0}\cos\left[\frac{1}{\ell_0}(x-vt)\right]=-\frac{1}{t} \tag{5.302}$$

while, as follows from (5.295):

$$\frac{v_0}{\ell_0}\sin\left[\frac{1}{\ell_0}(x-vt)\right]=\frac{v}{\ell_0} \tag{5.303}$$

whence:

$$\frac{1}{t^2} + \frac{v^2}{\ell_0^2} = \frac{v_0^2}{\ell_0^2} \tag{5.304}$$

and the velocity v is eliminated from (5.302):

$$\frac{v_0}{\ell_0} \cos \frac{1}{\ell_0} \left(x - \ell_0 \sqrt{\frac{v_0^2}{\ell_0^2} t^2 - 1} \right) = \frac{1}{t} \tag{5.305}$$

The roots of (5.305):

$$x_i = f_i(t) \tag{5.306}$$

describe the motions of singular points where in accordance with (5.301) the slope of the curve $v(x)$ becomes vertical (Fig. 5.13e).

Apparently, there are no real roots in the interval:

$$t < \frac{\ell_0}{v_0} \tag{5.307}$$

but there is infinite number of roots for:

$$t > \frac{\ell_0}{v_0} \tag{5.308}$$

Indeed, in this case

$$x_i = \ell_0 \left[\arccos \left(-\frac{\ell_0}{v_0 t} \right) + \sqrt{\frac{v_0^2}{\ell_0^2} t^2 - 1} \right] \tag{5.309}$$

Obviously, the distances between these singular points can be made as small as necessary by decreasing the constant ℓ_0 in the initial condition (5.293).

It follows from the continuity equation (5.292) that the singular points can be considered as points of mass condensation because:

$$\frac{\partial \rho}{\partial t} \to \pm\infty \text{ if } \frac{\partial v}{\partial x} \to \pm\infty \tag{5.310}$$

At the same time in accordance with (5.300):

$$\frac{\partial \rho}{\partial t} \to 0 \text{ if } t \to \infty \text{ at } v = 0 \tag{5.311}$$

i.e., the motionless "holes" in the continuum are formed. It follows from (5.296) that their coordinates are given by

$$x = \ell_0 \pi k \ (k = 1, 2, ..., \text{etc.}) \tag{5.312}$$

i.e., the distances between the "holes" can be made as small as necessary by decreasing the constant ℓ_0 in the initial condition (5.293).

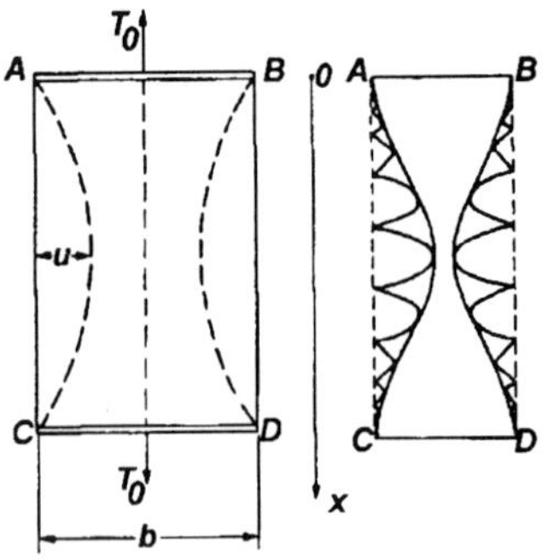

Fig. 5.14.

However, in contrast to the singular points of mass condensation, which are formed within finite periods of time, the "holes" are formed during infinite periods of time.

Thus, this example illustrates the fact that a free medium with differentiable (single-valued) initial conditions is inclined to degenerate into a countable set of "thick" particles, or points of superposition of different particles, and "holes," the distances between which can be vanishingly small.

At the same time the example gives a quantitative picture of formation of thick wrinkles as shock waves in the simplified post-instability models.

Singular Thick Wrinkles

The behavior of singular thick wrinkles as shock waves are governed by (5.204) written in the integral form. In the simplest case this wrinkle can be considered as a string with a variable mass (due to absorption or emission of film particles):

$$u\frac{\partial^2 u}{\partial t^2} + \left(\frac{\partial u}{\partial t}\right)^2 = a^2 u \frac{\partial^2 u}{\partial x^2}, \quad a^2 = \frac{T}{\rho_0} \tag{5.313}$$

where u is the in-plane displacement of the wrinkle, T is the tension along the wrinkle, ρ_0 is the initial film density (Fig. 5.13f).

The most interesting effects (snaps) occurs at the moment when the wrinkle vanishes, i.e., when

$$u \to 0, \quad \frac{\partial u}{\partial t} \to \infty \tag{5.314}$$

The quantitative analysis of (5.313) will be given below in connection with the behavior of films in solar arrays (Zak 1980).

Example 1. Let us consider an inextensible film of a solar array which is stretched between two parallel supports AB and CD as shown in Fig. 5.14.

Assume that the film is attached to a rigid support AB and wound on a drum CD which provides a constant tension, T_0. Let the free edge AC deform as shown in Fig. 5.14. Assume this deformation u to be small such that the elongation along AC can be neglected analogous to string theory. Since in the initial state the film is subjected to tension only in the vertical direction of Fig. 5.14 and such deformation leads to contraction or compression of its transverse elements, it can result in the instability of the film shape. As follows from (4.195) the contraction cannot propagate in space in front of the source of the disturbance, and can only coincide with the disturbance. Hence the deforming edge AC, as a singular boundary wrinkle, will absorb new particles of the contracting film and its motion can be described by the governing equation of a string with a variable mass:

$$\rho\frac{\partial^2 u}{\partial t^2} - T_0^*\frac{\partial^2 u}{\partial x^2} = F_R \tag{5.315}$$

where ρ_0 is the initial film density, ρ the density of the moving wrinkle, T_0 the film tension, u the transverse in-plane displacement, x the coordinate of the film length, t the time, and F_R the reactive force produced by the variable mass.

From the theory of bodies with variable mass:

$$F_R = \frac{d\rho}{dt}v \tag{5.316}$$

where v is the velocity of the particles being absorbed with respect to the moving mass. It is clear that

$$v = -\frac{\partial u}{\partial t} \tag{5.317}$$

$$\rho = \rho_0 u \tag{5.318}$$

Consequently

$$F_R = -\rho_0(\frac{\partial u}{\partial t})^2 \tag{5.319}$$

From the original equations (5.315) and (5.319) the following can be derived:

$$u\frac{\partial^2 u}{\partial t^2} + (\frac{\partial u}{\partial t})^2 = a^2 u\frac{\partial^2 u}{\partial x^2}$$

$$a^2 = \frac{T_0}{\rho_0} \tag{5.320}$$

The solution of (5.320) can be expressed in the following form:

$$u = \theta X,\ \theta = \theta(t),\ X = X(x) \tag{5.321}$$

Separating variables,

$$X'' + \lambda X = 0 \tag{5.322}$$

$$\theta\ddot{\theta} + \dot{\theta}^2 + a^2\lambda\theta^2 = 0 \tag{5.323}$$

where the constant λ must be determined from the boundary conditions. Equation (5.323) together with the boundary conditions

$$X(0) = 0, \ X(t) = 0 \tag{5.324}$$

leads to the following eigenvalues:

$$\lambda_n = (\pi n/l)^2, \ n = 1, 2, ... \tag{5.325}$$

Thus

$$X_n = \sin(\pi n/l)x \tag{5.326}$$

substituting (5.325) into (5.323):

$$\theta_n\ddot{\theta}_n + \dot{\theta}_n^2 + \alpha_n^2\theta_n^2 = 0$$

$$\alpha_n = \frac{T_0}{\rho_0}(\frac{\pi n}{t})^2 \tag{5.327}$$

writing this equation in the form:

$$(\ddot{\theta}_n^2) + 2\alpha_n^2\theta_n^2 = 0 \tag{5.328}$$

the following solution can be obtained:

$$\theta_n = \sqrt{\left|A_n \sin\alpha_n\sqrt{2}t + B_n \cos\alpha_n\sqrt{2}t\right|} \tag{5.329}$$

where A_n and B_n are constants.

Thus any function

$$u_n = \sin(\pi n/\ell)x \cdot \sqrt{\left|A_n \sin\alpha_n\sqrt{2}t + B_n \cos\alpha_n\sqrt{2}t\right|} \tag{5.330}$$

is a solution of the original (5.315). However, because of the nonlinearity of this equation the superposition of (5.330) is, generally speaking, not necessarily a solution. This means that in the general case when the initial deformation cannot be represented in the form:

$$u_n(t=0) = C\sin(\pi n/\ell)x, \ (n = 1, 2, ...) \tag{5.331}$$

where C is a constant, the method of separation of variables is not applicable.

Let us consider (5.330) at $n = 1$. Setting the initial conditions in terms of the impulse $I_0 \sin\pi/l\ x$ and the kinetic energy $W_0 \sin\pi/l\ x$ of the initial disturbance in the form of a shock, we can define constants A_1 and B_1 as,

$$A_1 = \sqrt{2}I_0/\alpha_1, \ B_1 = I_0^4/4W_0^2 \tag{5.332}$$

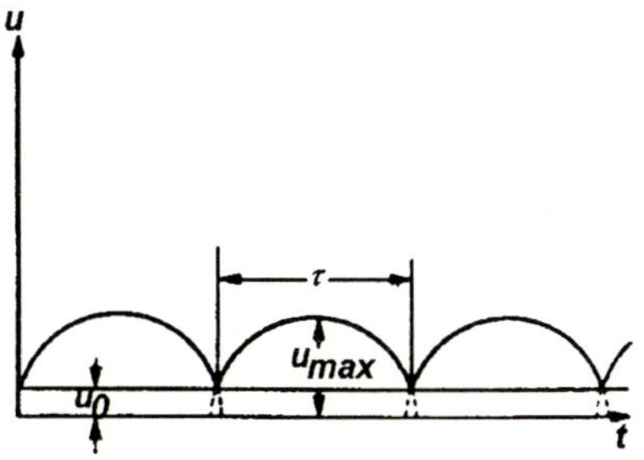

Fig. 5.15.

Consequently,

$$u_{t=0} = I_0^2/2W_0, \ \dot{u}_{t=0} = 2W_0/I_0 \tag{5.333}$$

This means that at the time of the initial shock neither the displacement nor the velocity are equal to zero. In other words, in the course of the initial shock both the displacement and the velocity change abruptly:

$$\triangle u = I_0^2/2W_0, \ \ \triangle \dot{u} = 2W_0/I_0 \tag{5.334}$$

This effect follows from the fact that the mass of the wrinkle tends to zero if $u \to 0$ and the original equation (5.315) has a singularity. The subsequent motion can be described by the expression:

$$u = \sin\frac{\pi}{\ell}x \cdot \sqrt{\left|\frac{\sqrt{2}I_0}{\alpha_1}\sin\sqrt{2}\alpha_1 t + \frac{I_0^4}{4W_0^2}\cos\sqrt{2}\alpha_1 t\right|} \tag{5.335}$$

This solution possesses some features typical of harmonic oscillations, shown in Fig. 5.15. These are:

a) The amplitude of the oscillation at any fixed x is constant and defined only by the initial conditions:

$$u_{\max} = \sqrt[4]{\frac{2I_0^2}{\alpha_1^2} + \frac{I_0^4}{4W_0^2}}\sin\frac{\pi}{\ell}x \tag{5.336}$$

b) The period of oscillation (interval between two shocks) is also constant:

$$\tau = \arccos\frac{I_0^3\alpha_1 - 4\sqrt{2}W_0^2}{I_0^3 + 4\sqrt{2}W_0^2} \tag{5.337}$$

But at the same time the solution has some nonlinear features:

a) The period of oscillation according to (5.337) depends upon the initial conditions and consequently, on the amplitude.

b) At the time

$$t^* = t + \tau K, \ (K = 1, 2, ...) \tag{5.338}$$

there are shocks with velocities:

$$\dot{u} = \frac{4W_0}{I_0} \sin \frac{\pi}{\ell} x \tag{5.339}$$

as a result of a reflection from the line defined by

$$u = \frac{I_0^2}{2W_0} \sin \frac{\pi}{\ell} x \tag{5.340}$$

c) At all times the zone

$$u \leq \frac{I_0^2}{2W_0} \sin \frac{\pi}{\ell} x \tag{5.341}$$

approaches but never reaches the film width b and the line defined by (5.340) forms the edge wrinkle with mass

$$m = \rho_0 \frac{I_0^2}{2W_0} \sin \frac{\pi}{\ell} x \tag{5.342}$$

d) The harmonic frequencies ν after expansion of the solution in Fourier series will be multiples of the main frequency:

$$\nu = \pi / \tau \tag{5.343}$$

which depends upon the initial conditions according to (5.337).

Some remarks about the general case of the initial conditions can be made. First of all, note that if (5.330) is the particular solution of (5.320), then the function:

$$u = \sum_n \sin \frac{\pi n}{\ell} x \cdot \sqrt{\left| \sum_n \left(A_n \sin \alpha_n \sqrt{2} t + B_n \cos \alpha_n \sqrt{2} t \right) \right|}$$

$$u_{max} > I_0 / 2W_0 \tag{5.344}$$

will also be a solution of the above equation. Applying the initial conditions one obtains:

$$u(x,0) = \sum_n \sin \frac{\pi n}{\ell} x \sqrt{\left| \sum_n B_n \right|} \tag{5.345}$$

$$\dot{u}(x,0) = \sum_n \sin \frac{\pi n}{\ell} x \left(\sum_n \alpha_n B_n / \sqrt{2 \left| \sum_n B_n \right|} \right) \tag{5.346}$$

Hence if the given initial conditions

$$u(x,0) = f(x), u(\dot{x},0) = \psi(x) \tag{5.347}$$

can be approximated in the form of (5.345–5.346), the solution of (5.320) can be written in the form of (5.344). Note that the approximations of (5.345–5.346) differ from the Fourier series expansion, therefore strictly speaking. it is not an exact solution for the general case.

The solutions of (5.330), (5.335), or (5.344) describe the symmetric oscillations of the edges AC and BD with respect to the centerline. The time of the reflections t^* corresponds to the time when the distance between the edges is maximum. The asymmetric oscillations differ from the above case only by the time of occurrence and by the location of the shocks. For instance, if the oscillations of the film edges are described by:

$$u_i = \sin\frac{\pi n_i}{\ell}x \cdot \sqrt{\left|A_i \sin\sqrt{2}\alpha_i t + B_n \cos\alpha_n\sqrt{2}\alpha_i t\right|} \tag{5.348}$$

$$u_i \geq I_0^{(i)}/2W_0^{(i)}, \quad (i = 1, 2, ...) \tag{5.349}$$

the shock will appear at the time when

$$u_2 - u_1 = b - \sum_{i=1}^{2} \frac{I_0^{(i)}}{2W_0^{(i)}} \tag{5.350}$$

New coefficients A' and B' after the shock are easily defined from the balance of the impulse and the kinetic energy before and after the shocks.

In the particular case when

$$u_2 - u_1 = b \tag{5.351}$$

the oscillations are without contraction. Instead of (5.320) one now obtains:

$$\frac{\partial^2 u}{\partial t^2} = a^2 \frac{\partial^2 u}{\partial x^2} \tag{5.352}$$

i.e., the classical equation for transverse oscillations.

Note that in the general case when the initial conditions are different from those (5.331) and every particle has its own timing for the shocks, it is possible to introduce the speed of the shock propagation along the length of the film:

$$\lambda' = \sqrt{T_0/\rho_0} = a \tag{5.353}$$

Thus in this case we arrive at additional longitudinal shocks oscillating with the period.

$$T' = 2\ell/a \tag{5.354}$$

In order to investigate longitudinal oscillations in films of a rotating solar sail, consider an inextensible film of a solar sail rotating with angular velocity ω (Fig. 5.16). Assume that the film is attached along the line AA and the end is free; the film can be considered, with a sufficient degree of

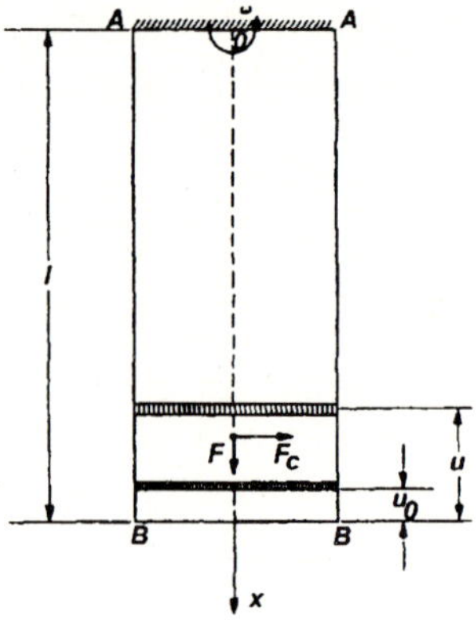

Fig. 5.16.

accuracy, as uniformly stretched along its length and hence can be treated as an inextensible flexible filament.

Let the free end BB be deformed uniformly with respect to the film width. This deformation will lead to a contraction of the film elements. This contraction leads to a wrinkle formation as a result of an instability of the film shape which cannot propagate along the film ahead of the source of the disturbance. In the course of this deformation or crumpling the line BB will absorb new film particles and move toward the fixed edge AA. Invoking the theory of particles with variable mass the following governing equation is obtained.

$$\frac{d}{dt}\left(m\frac{du}{dt}\right) = F, \; u \geq 0 \tag{5.355}$$

where m is the variable mass of the moving "wrinkle," u the displacement of the above "wrinkle," and F the external force. Introducing the linear density of the film

$$m = \rho u, \; F = \rho\omega^2 u(\ell - u) \tag{5.356}$$

where t is the length of the film. Thus instead of (5.315) one obtains

$$\frac{d}{dt}\left(u\frac{du}{dt}\right) = -\omega^2 u(\ell - u)$$

or

$$u\ddot{u} + \dot{u}^2 + \omega^2 u(\ell - u) = 0 \tag{5.357}$$

introducing a new variable:

$$\dot{u}^2 = Z \tag{5.358}$$

the following equation is obtained

$$\frac{dZ}{du} + 2\frac{Z}{u} + 2\omega^2 u(\ell - u) = 0 \tag{5.359}$$

Hence

$$\dot{u} = \sqrt{(C/u^2) - \omega^2(2/3u\ell - 1/2u^2)} \tag{5.360}$$

where C = an arbitrary constant.

Taking into account that

$$\dot{u} = 0 \text{ if } u = u_{\max} \tag{5.361}$$

$$\dot{u} = \frac{\omega}{\widetilde{u}}\sqrt{2/3\ell u_{\max}(1-\widetilde{u}^3) - 1/2u_{\max}^2(1-\widetilde{u}^4)} \tag{5.362}$$

where

$$\widetilde{u} = u/u_{\max} \tag{5.363}$$

equation (5.362) defines the solution u in terms of elliptic functions:

$$t = \frac{1}{\omega}\int_0^{\widetilde{u}} \frac{\widetilde{u}d\widetilde{u}}{\sqrt{2/3lu_{\max}(1-\widetilde{u}^3) - 1/2u_{\max}^2(1-\widetilde{u}^4)}} \tag{5.364}$$

It follows from (5.362) that the solution has singularities at $u = 0$:

$$\dot{u} \to \infty \text{ if } u \to 0 \tag{5.365}$$

which express the reflections from the free end. Following the argument used previously one arrives at a similar situation: at the time of the initial shock the displacement and velocity are defined by (5.333), the subsequent jumps in velocity at the time of reflections from the point

$$u = I_0^2/2W_0 \tag{5.366}$$

are defined by the formula:

$$\Delta\dot{u} = 2W_0/I_0 \tag{5.367}$$

In general the solution to (5.364) looks like the solution to (5.335) for a fixed x shown in Fig. 5.15. The interval between shocks can be evaluated by means of (5.364) by considering the inequalities:

$$I_0^2/2W_0 \ll u_{\max} \ll t \tag{5.368}$$

$$\tau = \frac{\sqrt{u_{\max}}}{\omega\sqrt{2/3\ell}}\int_0^1 \frac{\widetilde{u}d\widetilde{u}}{\sqrt{1-\widetilde{u}^3}} = \frac{1}{\omega\sqrt{2/3u_{\max}}} \cdot \frac{1}{\pi}\frac{\sqrt{3}}{4\sqrt{4}}\left[\Gamma(2/3)\right]^2$$
$$= \frac{0.89}{\omega}\sqrt{\frac{u_{\max}}{\ell}} \tag{5.369}$$

Thus the frequencies of harmonic oscillation which result from the expansion in Fourier series are multipliers of the frequency $\nu = 2\pi/\tau$ and hence depend upon the initial conditions.

The longitudinal oscillations defined above in the film of a rotating solar sail lead to additional Coriolis forces, F_c.

$$F_c = \pm 2\rho u \dot{u} \omega \tag{5.370}$$

which act in plane and are orthogonal to the longitudinal axis of the film. These forces lead to transverse in-plane oscillations of the film.

Following a similar argument as above it is observed that contrary to the previous discussion wherein the film tension was constant, the film tension for the transverse in-plane oscillations of a blade is variable given by

$$T = (\rho_0/2)\omega^2 \ell^2 (1 - \widetilde{x}^2), \ \widetilde{x} = x/\ell \tag{5.371}$$

where x is the longitudinal coordinate with origin 0 at the center of rotation. Therefore, instead of (5.320) the governing equation now becomes

$$u\frac{\partial^2 u}{\partial t^2} + (\frac{\partial u}{\partial t})^2 = u\frac{\partial}{\partial x}(\frac{T}{\rho_0}\frac{\partial u}{\partial x}) \tag{5.372}$$

or

$$u\ddot{u} + \dot{u}^2 = u\left[1/2\omega^2\ell^2(1-\widetilde{x}^2)u'' - \omega^2\ell^2\widetilde{x}u'\right] \tag{5.373}$$

Separation of variables leads to (5.328) and (5.329) with $a = \omega l$ while the eigenvalues α_n are defined by the Legendre equations:

$$\left(\widetilde{x}^2 - 1\right) X'' + 2\widetilde{x}X' - \lambda_n X = 0, \ \alpha_n = \omega^2\ell^2\lambda_n \tag{5.374}$$

instead of (5.316). Assuming that the function $X(\widetilde{x})$ is bounded within the interval $0 < \widetilde{x} < 1$ we arrive at the sequence of the eigenvalues:

$$\lambda_n = n(n+2), \ (n = 1, 2, ...)$$

and normalized eigenfunctions:

$$X_n = \sqrt{n + 1/2}P_n(\widetilde{x}) \tag{5.375}$$

where $P_n(\widetilde{x})$ are the Legendre polynomials. The solution to (5.373) can be written in the form similar to (5.344):

$$u = \sum_n \left[\sqrt{n+1/2}P_n(\widetilde{x})\right]$$

$$\cdot\sqrt{\left|\sum_n \left(A_n \sin\alpha_n\sqrt{2}t + B_n\cos\alpha_n\sqrt{2}t\right)\right|}$$

$$u_{\max} \geq I_0/2W_0 \tag{5.376}$$

If one of the eigenfrequencies of (5.376) is equal to any eigenfrequency of (5.364), after expanding the above solution in Fourier series, the resonance which arises between the longitudinal and the transverse in-plane oscillations is caused by the Coriolis force.

In the general case the velocity of shock propagation along the length of the film can be obtained from:

$$\lambda' = \frac{\sqrt{2}}{2}\omega\ell\sqrt{1-\widetilde{x}^2} \tag{5.377}$$

The equations of motion for the shock along the film is the solution to the following differential equation:

$$\lambda' = \pm\ell\frac{d\widetilde{x}_*}{dt} = \pm\frac{\sqrt{2}}{2}\omega\ell\sqrt{1-\widetilde{x}_*^2} \tag{5.378}$$

where $\widetilde{x}_*$ is the coordinate of the moving shock with initial value $\widetilde{x}_*^0$; consequently:

$$\widetilde{x}_*^2 = \pm\tanh(\frac{\sqrt{2}}{2}\omega t) + \widetilde{x}_*^0 \tag{5.379}$$

Thus the shock waves in this case move aperiodically with $\tau \to \infty$. In other words, any shock moving along the film is being retarded and does not experience a reflection from the end of the film.

Next, the effect of damping on the longitudinal oscillations of the film of a rotating solar sail will be examined. In Fig. 5.16 assume that the film is set in motion by a radial shock applied at the fixed end and directed outward to the free end. Such a shock will be distributed instantaneously and uniformly along the entire length of the film due to the assumed inextensibility of the film. This leads to a uniformly distributed velocity of reflection v_0:

$$v_0 = I_0/\rho\ell \tag{5.380}$$

where I_0 is the initial impulse. The governing equation for the moving film with decreasing length can be written in the following form:

$$\rho_0 x\widetilde{x} = \rho_0\omega^2(x^2/2) \text{ or } \widetilde{x} = 1/2\omega^2 x \tag{5.381}$$

where the x coordinate of the film tip describes the translational motion of the entire film. Obviously, the solution for the first period of motion ($\dot{x} < 0$) is:

$$x = \ell\cosh\frac{\omega}{2}t - \frac{2v_0}{\omega\ell}\sinh\frac{\omega}{2}t \tag{5.382}$$

This solution is valid until

$$t = t^* = \frac{2v_0}{\omega\ell}\text{ arctan h (when } \dot{x} = 0) \tag{5.383}$$

At this time the length of the film will be minimum

$$x_{\min} = \ell\sqrt{1-\frac{4v_0^2}{\ell^2\omega^2}},\ v_0 < \frac{\omega\ell}{2} \tag{5.384}$$

while the length $(\ell - x_{\min})$ will be absorbed by the wrinkle in the root of the film. The governing equation for the moving film with increasing length has another form:

$$\rho_0 x\ddot{x} + \rho_0 \dot{x}^2 = \rho_0(\omega^2/2)x^2$$

or

$$(\ddot{x}^2) = 1/2\omega^2\ell^2 \tag{5.385}$$

The difference between (5.381) and (5.385) comes from the fact that during the first period the relative velocity of the separating particles is zero, but the relative velocity of the separating particles during the second period is not zero. The solution of (5.385) is:

$$x = \ell\sqrt{1 - \frac{4v_0^2}{\ell^2\omega^2}}\sqrt{\cosh\frac{\omega}{2}t}, \ x \leq t \tag{5.386}$$

The solution is valid until:

$$t \leq t^{**} = \frac{2}{\omega}\operatorname{arccos h}\frac{1}{1 - \frac{4v_0^2}{\ell^2\omega^2}} \ (\text{when } x = \ell) \tag{5.387}$$

At this time the velocity is determined by the expression:

$$v_0^{(1)} = v_0\sqrt{1/2(1 - \frac{2v_0^2}{\ell^2\omega^2})} < v_0 \tag{5.388}$$

The period of such a complete oscillation is:

$$\tau = t^* + t^{**}$$

$$= \frac{2}{\omega}\left[\operatorname{arctan h}\frac{2v_0}{\omega\ell} + \operatorname{arccos h}\frac{1}{1 - \frac{4v_0^2}{l^2\omega^2}}\right] \tag{5.389}$$

This reflection from the unwrinkled state of the film at the time $t^* = t^{**}$ results in the same state which existed at the initial $t = 0$ but with another velocity $v_0^{(1)} < v_0$.

Continuing this process, the sequence of the reflection velocities is given by:

$$v_0 > v_0^{(1)} > v_0^{(2)} \ldots > v_0^{(n)} \tag{5.390}$$

which can be rewritten using the recurrence formula:

$$v_0^{(n)} = v_0^{(n-1)}\sqrt{1/2\left(1 - \frac{2v_0^{(n-1)2}}{\ell^2\omega^2}\right)} \tag{5.391}$$

The intensity of the shocks at the time of their reflection can be calculated using

$$I^{(n)} = v_0^{(n)}/v_0 \ I_0 < I^{(n-1)} \tag{5.392}$$

Thus the shock intensity is decreasing with time. The periods of subsequent oscillations are evaluated by

$$\tau^{(n)} = \frac{2}{\pi}\left[\operatorname{arctan h}\frac{2v_0^{(n)}}{\omega\ell} + \operatorname{arccos h}\frac{1}{1-\frac{4v_0^{(n)2}}{\ell^2\omega^2}}\right] < \tau^{(n-1)} \tag{5.393}$$

Hence the frequency of oscillation is increasing with time.

The amplitudes of the oscillations are given by the expression:

$$A_n = \ell - x_{\min}^{(n)} = \ell\left[1-\sqrt{\left(1-\frac{2v_0^{(n)}}{\ell^2\omega^2}\right)^2}\right] \leq A_{n-1} \tag{5.394}$$

Consequently the amplitudes are decreasing with time and the damping can be evaluated from

$$\gamma = \left[1-\sqrt{1-(2v_0^{(n)}/l\omega)^2}\right] / \left[1-\sqrt{1-(2v_0^{(n-1)}/l\omega)^2}\right] \tag{5.395}$$

Note that the cause for the apparent damping lies in the loss of kinetic energy during the absorption of film particles by wrinkles at the root without reflection. This is an example of absolute inelastic shock.

The above is a nonlinear phenomenon. The nonlinearity is expressed in the dependence of the period of oscillation and of the damping upon amplitude.

Example 2. In the previous example our attention was focused on the behavior of a single wrinkle. In this example we will capture the averaged effect of wrinkle formation, and analyze its contribution into film vibrations (Zak 1982a,b,c).

We recall that wrinkling results from the instability of a smooth shape of a film having no resistance to bending if compression emerges, while the directions of the wrinkles are orthogonal to the directions of the compression. In contrast to elastic elongation, the contraction of a compressed film is non-elastic being carried out by micro-deflections of filaments (wrinkles) without additional resistance.

Thus, constitutive equations for an elastic film in terms of principle stresses $\widetilde{\sigma}^{ii}$ and strains $\widetilde{\varepsilon}_{ii}$,

$$\widetilde{\sigma}^{ii} = f(\widetilde{\varepsilon}_{11}, \widetilde{\varepsilon}_{22}), \ (i = 1, 2) \tag{5.396}$$

are valid only for positive stresses while negative stresses must be replaced by zero.

Clearly, such strong nonlinearities are essential even for very small strains where usual geometrical and physical nonlinearities can be ignored. This example presents a modified form of the governing equations for wrinkling films and approximate analytical solutions to the simplified equations of nonlinear inplane vibrations. The additional "constitutive" equation for wrinkling film is formulated in the following terms: the principal stresses cannot be negative, i.e.

$$\tilde{\sigma}^{11}, \tilde{\sigma}^{22} \geq 0 \tag{5.397}$$

which means that

$$\tilde{\sigma}^{11} \geq 0, \; \tilde{\sigma}^{22} \geq 0, \; \sqrt{\sigma^{11}\sigma^{22}} - \sigma^{12} \geq 0 \tag{5.398}$$

These inequalities will be automatically satisfied if one introduces new stresses $\overset{*}{\sigma}_{ij}$ instead of σ_{ij} through

$$\overset{*}{\sigma}{}^{11} = \frac{1}{2}\left(\sigma^{11} + |\sigma^{11}|\right) \tag{5.399}$$

$$\overset{*}{\sigma}{}^{22} = \frac{1}{2}\left(\sigma^{22} + |\sigma^{22}|\right) \tag{5.400}$$

$$\overset{*}{\sigma}{}^{12} = \frac{1}{2}[\sigma^{12} + \sqrt{(\sigma^{11} + |\sigma^{11}|)(\sigma^{22} + |\sigma^{22}|)} - \left|\sigma^{12} - \sqrt{(\sigma^{11} + |\sigma^{11}|)(\sigma^{22} + |\sigma^{22}|)}\right|] \tag{5.401}$$

Now standard governing equations of a film are generalized over the region of wrinkling by replacing the stresses according to (5.399–5.401). For instance, invoking (4.195) one obtains:

$$g^{1/2}\rho_0 \frac{\partial^2 \mathbf{r}}{\partial t^2} = \frac{\partial}{\partial \psi^{(1)}}\left(\overset{*}{\sigma}{}^{11} g^{1/2} \frac{\partial \mathbf{r}}{\partial \psi^{(1)}}\right) + \frac{\partial}{\partial \psi^{(1)}}\left(\overset{*}{\sigma}{}^{12} g^{1/2} \frac{\partial \mathbf{r}}{\partial \psi^{(2)}}\right) + \frac{\partial}{\partial \psi^{(2)}}\left(\overset{*}{\sigma}{}^{12} g^{1/2} \frac{\partial \mathbf{r}}{\partial \psi^{(1)}}\right) + \frac{\partial}{\partial \psi^{(2)}}\left(\overset{*}{\sigma}{}^{22} g^{1/2} \frac{\partial \mathbf{r}}{\partial \psi^{(2)}}\right) + \mathbf{F} \tag{5.402}$$

where $\mathbf{r}$ is the radius-vector of the film point, $\psi^{(1)}$, $\psi^{(2)}$ are the material coordinates, ρ is the density, g characterizes the initial internal geometry of the film, $\mathbf{F}$ is the external force, and $\overset{*}{\sigma}_{ij}$ are given by (5.399–5.401).

This equation must be complemented by the constitutive equations

$$\sigma^{ij} = f(\varepsilon_{11}, \varepsilon_{12}, \varepsilon_{22}) \tag{5.403}$$

which with (5.399–5.401) are written in terms of $\overset{*}{\sigma}$ as

$$\overset{*}{\sigma}{}^{11} = \frac{1}{2}\left[f(\varepsilon_{11}, \varepsilon_{12}, \varepsilon_{22}) + |f(\varepsilon_{11}, \varepsilon_{12}, \varepsilon_{22})|\right], \text{ etc.,} \tag{5.404}$$

Hence, by means of the constitutive equations (5.404), a solution of the system will automatically satisfy the inequalities (5.398).

Thus, the system (5.402–5.403) possesses the geometrical and physical nonlinearities which usually can be ignored for very small perturbations.

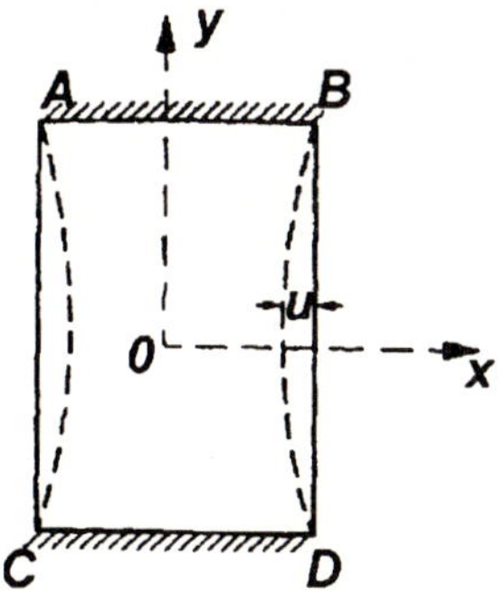

Fig. 5.17.

But the equations (5.399–5.401) contribute a new type of nonlinearity which occurs in systems with unilateral constraints. Such a nonlinearity cannot be ignored even for very small perturbations if the unperturbed state is sufficiently close to the boundary of wrinkling [see (5.289)].

Let us pose now the problem of small transverse inplane vibrations of a plane film stretched between two horizontal parallel supports AB and CD with a stationary tension σ^0_{yy} as shown in Fig. 5.17.

Taking into account that

$$\psi^{(1)} = x,\ \psi^{(2)} = y,\ G = 1,\ r_x = u,\ r_y = v$$

$$\sigma^0_{yy} = \text{Const},\ \sigma_{yy} = \sigma^0_{yy} + \widetilde{\sigma}_{yy},\ \widetilde{\sigma}_{yy}(t) \ll \sigma^0_{yy} \tag{5.405}$$

we find after the standard approximations that

$$g \simeq 1,\ \frac{\partial}{\partial y}\left(\sigma^{yy} g^{1/2} \frac{\partial u}{\partial y}\right) \simeq \sigma^0_{yy} \frac{\partial^2 u}{\partial y^2}$$

and (5.402) is simplified to

$$\rho_0 \frac{\partial^2 u}{\partial t^2} = \frac{\partial \overset{*}{\sigma}_{xx}}{\partial x} + \frac{\partial \overset{*}{\sigma}_{xy}}{\partial y} + \sigma^0_{yy} \frac{\partial^2 u}{\partial x^2} + F_x,\ \rho_0 \frac{\partial^2 \overset{*}{\sigma}}{\partial t^2} = \frac{\partial \overset{*}{\sigma}_{xy}}{\partial x} + \frac{\partial \widetilde{\sigma}_{yy}}{\partial y} \tag{5.406}$$

The linear variant of constitutive equations are

$$\sigma_{xx} = \frac{E}{1-\mu^2}\left(\frac{\partial u}{\partial x} + \mu \frac{\partial v}{\partial y}\right),$$

$$\sigma_{xy} = \frac{E}{2(1-\mu)}\left(\frac{\partial u}{\partial y} + \frac{\partial v}{\partial x}\right),\ \widetilde{\sigma}_{yy} = \frac{E}{1-\mu^2}\left(\frac{\partial v}{\partial y} + \mu \frac{\partial u}{\partial x}\right) \tag{5.407}$$

where E is Young's modulus and μ is Poisson's ratio. Equations (5.399–5.400) are expressed in terms of displacements as

$$\overset{*}{\sigma}_{xx} = \frac{E}{2(1-\mu^2)}\left(\frac{\partial u}{\partial x} + \mu \frac{\partial v}{\partial y}\right)\left(1 + \frac{\partial u/\partial x + \mu\ \partial v/\partial y}{|\partial u/\partial x + \mu\ \partial v/\partial y|}\right),\ \overset{*}{\sigma}_{yy} = \widetilde{\sigma}_{yy} \tag{5.408}$$

In order to simplify (5.401), let us turn to the third inequality in (5.398) which can be rewritten in the form

$$\frac{E}{2(1-\mu)}\left(\frac{\partial u}{\partial y}+\frac{\partial v}{\partial x}\right) \leq \left[\frac{\sigma_{yy}^0 E}{2(1-\mu^2)}\frac{\partial u}{\partial x}\left(1+\frac{\partial u/\partial x+\mu\,\partial v/\partial y}{|\partial u/\partial x+\mu\,\partial v/\partial y|}\right)\right] \tag{5.409}$$

This condition can be satisfied for very small deflections,

$$\left(\frac{\partial v}{\partial x}\right)^2,\ \left(\frac{\partial u}{\partial y}\right)^2 \ll \left|\frac{\partial u}{\partial x}+\mu\frac{\partial v}{\partial y}\right| \tag{5.410}$$

if

$$\overset{*}{\sigma}_{xy}= \frac{E}{4(1-\mu)}\left(\frac{\partial u}{\partial y}+\frac{\partial v}{\partial x}\right)\left(1+\frac{\partial u/\partial x+\mu\,\partial v/\partial y}{|\partial u/\partial x+\mu\,\partial v/\partial y|}\right) \tag{5.411}$$

Indeed, if

$$\frac{\partial u}{\partial x}+\mu\frac{\partial v}{\partial y}\geq 0 \tag{5.412}$$

then in accordance with (5.409–5.411)

$$\overset{*}{\sigma}_{xy}=\sigma_{xy}<\sqrt{\overset{*}{\sigma}_{xx}\overset{*}{\sigma}_{yy}}$$

but if

$$\frac{\partial u}{\partial x}+\mu\frac{\partial v}{\partial y}<0 \tag{5.413}$$

then

$$\overset{*}{\sigma}_{xy}=\sqrt{\overset{*}{\sigma}_{xx}\overset{*}{\sigma}_{yy}}=0 \tag{5.414}$$

Substituting (5.408) and (5.411) into (5.406), one arrives at the final form of the governing equation of small transverse inplane vibrations:

$$\rho_0\frac{\partial^2 u}{\partial t^2}=\left[\frac{E}{2(1-\mu^2)}\frac{\partial^2 u}{\partial x^2}+\frac{E}{4(1+\mu)}\frac{\partial^2 u}{\partial y^2}+\frac{E}{4(1-\mu)}\frac{\partial^2 v}{\partial x\partial y}\right]$$
$$\cdot\left(1+\frac{\partial u/\partial x+\mu\,\partial v/\partial y}{|\partial u/\partial x+\mu\,\partial v/\partial y|}\right)+\sigma_{yy}^0\frac{\partial^2 u}{\partial y^2}+F_x \tag{5.415}$$

$$\rho_0\frac{\partial^2 v}{\partial t^2}=\left[\frac{E}{2(1-\mu^2)}\frac{\partial^2 v}{\partial x^2}+\frac{E}{4(1+\mu)}\frac{\partial^2 v}{\partial x^2}+\frac{E}{4(1-\mu)}\frac{\partial^2 u}{\partial x\partial y}\right]$$
$$\cdot\left(1+\frac{\partial u/\partial x+\mu\,\partial v/\partial y}{|\partial u/\partial x+\mu\,\partial v/\partial y|}\right) \tag{5.416}$$

For better physical interpretation of the results, in this section only horizontal vibrations will be investigated ($v\to 0$). This is why we start with criteria for the decoupling of (5.415) and (5.416) first. Obviously (5.415) and (5.416) are always decoupled in a state of wrinkling because then

$$1 + \frac{\partial u/\partial x + \mu\ \partial v/\partial y}{|\partial u/\partial x + \mu\ \partial v/\partial y|} = 0$$

For a non-wrinkled state the criteria for decoupling can be written in the form

$$\mu \left|\frac{\partial v}{\partial y}\right| \ll \left|\frac{\partial u}{\partial x}\right|, \quad \left|\frac{\partial^2 v}{\partial x\ \partial y}\right| \ll \frac{2}{1+\mu}\left|\frac{\partial^2 u}{\partial x^2}\right| \tag{5.417}$$

We will assume that the horizontal vibrations can be found in the form

$$u = A_0 \cos\frac{\pi}{b}x \sin\frac{\pi}{l}y\,[\alpha_0 + \alpha_1 \sin(\omega_0 t + \varphi_0)]$$

where α_0, A_0, ω_0, φ_0, b, l are constants, while initial conditions for the vertical vibrations are homogeneous,

$$v|_{t=0} = 0, \quad \left.\frac{\partial v}{\partial t}\right|_{t=0} = 0$$

Then one obtains the following evaluations from (5.416):

$$\left|\frac{\partial^2 v}{\partial t^2}\right| \sim \frac{A_0}{bl}\frac{E}{2(1-\mu)\rho_0}, \quad |v| \sim \frac{1}{\omega_0^2}\frac{A_0}{bl}\frac{E}{2(1-\mu)\rho_0}$$

$$\left|\frac{\partial^2 v}{\partial x\ \partial y}\right| \sim \frac{1}{\omega_0^2}\frac{A_0}{b^2 l^2}\frac{E}{2(1-\mu)\rho_0}$$

while

$$\left|\frac{\partial u}{\partial x}\right| \sim \frac{A_0}{b}, \quad \left|\frac{\partial^2 u}{\partial x^2}\right| \sim \frac{A_0}{b^2}, \quad \left|\frac{\partial v}{\partial y}\right| \sim \frac{1}{\omega_0^2}\frac{A_0}{bl^2}\frac{E}{2(1-\mu)\rho_0}$$

It then follows from inequalities (5.417) that

$$\omega_0^2 \gg \frac{\mu E}{2(1-\mu)\rho_0 l^2}, \quad \omega_0^2 \gg \frac{(1+\mu)E}{4\rho_0 l^2(1-\mu)} \tag{5.418}$$

Thus the contribution of the vertical vibrations to the equation for the horizontal vibrations can be ignored if the frequency of the horizontal vibrations is large in the sense of inequalities (5.417).

We will start the analyses with free vibrations. Assuming that the film is unbounded in the horizontal direction, we will consider the uncoupled form of (5.415) for the horizontal vibrations,

$$\rho_0\frac{\partial^2 u}{\partial t^2} = \left[\frac{E}{2(1-\mu^2)}\frac{\partial^2 u}{\partial x^2} + \frac{E}{4(1+\mu)}\frac{\partial^2 u}{\partial y^2}\right]\left(\frac{\partial u/\partial x}{|\partial u/\partial x|} + 1\right)$$

$$+\sigma_{yy}^0\frac{\partial^2 u}{\partial y^2} + F_x \tag{5.419}$$

and set the initial conditions in the form

$$u|_{t=0} = A_0 \cos \frac{\pi}{b} x \sin \lambda_y y \tag{5.420}$$

$$\left.\frac{\partial u}{\partial t}\right|_{t=0} = 0 \tag{5.421}$$

Obviously, the solution to this problem will be accurate enough in such a region of the parameters b, λ_y that the inequalities (5.417) hold.

The constant λ_y is defined from the boundary conditions

$$u = 0, \ y = 0 \text{ and } y = l \tag{5.422}$$

where l is the vertical length of the film and $2b$ is the wave length of the initial deflection. Thus,

$$\lambda_y = \frac{\pi n}{l}, \ (n = 1, 2, ...) \tag{5.423}$$

We will restrict attention to the case $n = 1$ and $b \leq 1$. The equations (5.420–5.421) become

$$u|_{t=0} = A_0 \cos \frac{\pi}{b} x \sin \frac{\pi}{l} y \tag{5.424}$$

$$\left.\frac{\partial u}{\partial t}\right|_{t=0} = 0 \tag{5.425}$$

We seek an approximate solution to (5.420) with the boundary conditions (5.422) and initial conditions (5.424–5.425) in the form

$$u = A_0 \cos \frac{\pi}{b} x \sin \frac{\pi}{l} y \left[\alpha_0 + \alpha_1 \sin\left(\omega_0 t + \varphi_0\right)\right] \tag{5.426}$$

where α_0, A_0, ω_0, φ_0, are constants.

First let us linearize the terms

$$\frac{\partial^2 u}{\partial x^2} \frac{\partial u/\partial x}{|\partial u/\partial x|}, \ \frac{\partial^2 u}{\partial y^2} \frac{\partial u/\partial x}{|\partial u/\partial x|} \tag{5.427}$$

by applying the method of equivalent linearization. Calculation of the coefficients

$$q_0 = \frac{1}{2\pi} \int_0^{2\pi} \frac{\alpha_0 + \alpha_1 \sin\psi}{|\alpha_0 + \alpha_1 \sin\psi|} \cdot (\alpha_0 + \alpha_1 \sin\psi) d\psi$$

$$= \frac{1}{2\pi} \int_0^{2\pi} |\alpha_0 + \alpha_1 \sin\psi| \, d\psi = \alpha_0 \left[1 + 4\frac{\alpha_1}{\alpha_0}\left(1 - \frac{\alpha_0^2}{\alpha_1^2}\right)^{1/2}\right] \quad \text{if } \alpha_0 < \alpha_1 \tag{5.428}$$

$$q_1 = \frac{1}{\pi \alpha_1} \int_0^{2\pi} |\alpha_0 + \alpha_1 \sin\psi| \sin\psi d\psi =$$

$$-\left[\frac{2}{\psi}\arcsin\frac{\alpha_0}{\alpha_1}+\frac{2}{\pi}\frac{\alpha_0}{\alpha_1}\left(1-\frac{\alpha_0^2}{\alpha_1^2}\right)^{1/2}\right] \tag{5.429}$$

$$q_2=\frac{1}{\pi\alpha_1}\int_0^{2\pi}|\alpha_0+\alpha_1\sin\psi|\cos\psi d\psi=0 \tag{5.430}$$

leads to the approximations

$$\frac{\partial^2 u}{\partial x^2}\cdot\frac{\partial u/\partial x}{|\partial u/\partial x|}\simeq -A_0\frac{\pi^2}{b^2}\cos\frac{\pi}{b}x\sin\frac{\pi}{y}\left[q_0+q_1\alpha_1\sin\left(\omega_0 t+\varphi_0\right)\right] \tag{5.431}$$

$$\frac{\partial^2 u}{\partial y^2}\cdot\frac{\partial u/\partial x}{|\partial u/\partial x|}\simeq -A_0\frac{\pi^2}{l^2}\cos\frac{\pi}{b}x\sin\frac{\pi}{y}\left[q_0+q_1\alpha_1\sin\left(\omega_0 t+\varphi_0\right)\right] \tag{5.432}$$

Now substituting (5.427) into (5.419) and taking into account (5.431–5.432) one arrives at the following equations:

$$\left\{\frac{E}{2(1-\mu^2)}\cdot A_0\frac{\pi^2}{b^2}+\left[\sigma_{yy}^0+\frac{E}{4(1-\mu)}\right]A_0\frac{\pi^2}{l^2}\right\}\alpha_0$$

$$=\left\{\frac{-E}{2(1-\mu^2)}A_0\frac{\pi^2}{b^2}-\frac{E}{4(1-\mu)}A_0\frac{\pi^2}{l^2}\right\}q_0 \tag{5.433}$$

$$\rho_0\omega^2=\frac{E\pi^2}{2(1-\mu^2)b^2}(1+q_1)+\frac{E\pi^2}{4(1-\mu)l^2}(1+q_1)\frac{\pi^2}{l^2}\sigma_{yy}^0 \tag{5.434}$$

whence,

$$\omega_0=\left\{\frac{\frac{\pi^2}{l^2}\sigma_{yy}^0+\gamma_0\left[1-\frac{2}{\pi}\arcsin\frac{\alpha_0}{\alpha_1}-\frac{2}{\pi}\frac{\alpha_0}{\alpha_1}\left(1-\frac{\alpha_0^2}{\alpha_1^2}\right)^{1/2}\right]}{\rho_0}\right\}^{1/2} \tag{5.435}$$

$$\frac{\alpha_1}{\alpha_0}\left(1-\frac{\alpha_0^2}{\alpha_1^2}\right)^{1/2}=\frac{1}{4}\gamma_1 \tag{5.436}$$

where

$$\gamma_0=\pi^2E\left[\frac{1}{2(1-\mu^2)b^2}+\frac{1}{4(1-\mu)l^2}\right] \tag{5.437}$$

$$\gamma_1=-2-\frac{\pi^2}{l^2}\frac{\sigma_{yy}^0}{\gamma_0}<0 \tag{5.438}$$

From the initial conditions (5.424–5.425),

$$\alpha_0+\alpha_1\sin\varphi_0=1 \tag{5.439}$$

$$A_0\omega_0\alpha_1\cos\varphi_0=0 \tag{5.440}$$

whence,

$$\left(\frac{1-\alpha_0}{\alpha_1}\right)^2 = 1 \tag{5.441}$$

$$\varphi_0 = \frac{\pi}{2} \tag{5.442}$$

Equations (5.435–5.436), and (5.441–5.442) define the unknowns ω_0, α_0, α_1, and φ_0.

Indeed, as follows from (5.436),

$$\frac{\alpha_1}{\alpha_0} = -\left(1+\frac{\gamma_1^2}{16}\right)^{1/2} \tag{5.443}$$

Hence,

$$\omega_0 = \{\frac{\pi^2}{l^2}\sigma_{yy}^0 + \gamma_0[1 - \frac{2}{\pi}\arcsin\left(1+\frac{\gamma_1^2}{16}\right)^{-1/2} - \frac{1}{2\pi}\left(1+\frac{\gamma_1^2}{16}\right)^{-1}\gamma_1]/\rho\}^{1/2} \tag{5.444}$$

$$-8.472 < \alpha_0 = \frac{1}{1-\left(1+\frac{\gamma_1^2}{16}\right)^{1/2}} < 0 \tag{5.445}$$

$$9.472 > \alpha_1 = \left[1 - \frac{1}{1-\left(1+\frac{\gamma_1^2}{16}\right)^{1/2}}\right] > 1 \tag{5.446}$$

As follows from (5.444), (5.436–5.438), the inequalities (5.417) hold because of the condition $b \leq 1$.

Thus the transverse inplane free oscillations of a wrinkling film are accompanied by the following nonlinear effects:

a) The center of the vibration is shifted from the position of equilibrium towards the domain of contraction of the film. Indeed, according to (5.445), (5.437–5.438) this shift satisfies the inequalities

$$-8.472 < \alpha_0 < 0 \tag{5.447}$$

if

$$0 < \sigma_{yy}^0 < \infty \tag{5.448}$$

b) The amplitude of the vibrations is greater than the initial deflection from the position of the equilibrium, because, as follows from (5.446), (5.442), (5.438),

$$9.472 > \alpha_1 > 1 \tag{5.449}$$

if

$$0 < \sigma_{yy}^0 < \infty$$

c) The frequency of the vibrations is less than the frequency ω' of the corresponding model without wrinkling. Indeed,

$$\omega' = \left\{ \frac{\frac{\pi^2}{l^2}\sigma_{yy}^0 + \pi^2 E\left[\frac{1}{2(1-\mu^2)b^2} + \frac{1}{4(1-\mu)l^2}\right]}{\rho_0} \right\}^{1/2} > \left(\frac{\frac{\pi^2}{l^2}\sigma_{yy}^0 + 2\gamma_0}{\rho_0} \right)^{1/2} \tag{5.450}$$

but

$$1 < \left[1 + \frac{2}{\pi} \arcsin\left(1 + \frac{\gamma_1^2}{16}\right)^{-1/2} - \frac{\gamma_1}{2\pi}\left(1 + \frac{\gamma_1^2}{16}\right)^{-1} \right] < 1.959 \tag{5.451}$$

if

$$0 < \sigma_{yy}^0 < \infty$$

Consequently

$$\omega_0 < \omega' \tag{5.452}$$

Clearly, all of these nonlinear effects are decreasing with the increase of the pretension σ_{yy}^0.

In conclusion of this example, let us turn to forced vibrations. Assuming that the external force in (5.419) is given by

$$F_x = B_0 \cos\frac{\pi}{b}x \sin\frac{\pi}{l}y \sin\omega_* t \tag{5.453}$$

we will seek the forced vibrations in the form

$$u = \cos\frac{\pi}{b}x \sin\frac{\pi}{l}y(\alpha_0^* + \alpha_1^* \sin\omega_* t) \tag{5.454}$$

Substituting this expression into (5.419) and taking into account (5.428–5.432), one arrives at the formulas

$$\frac{\alpha_1^*}{\alpha_0^*} = -\left(1 + \frac{\gamma_1^2}{16}\right)^{1/2} \tag{5.455}$$

$$\alpha_1^* = \frac{B_0}{\omega_*^2 - \omega_0^2} \tag{5.456}$$

where γ_1 and ω_0 are given by (5.438) and (5.444) respectively. Thus, the symmetrical external forces induce non-symmetrical vibrations. As the formulas (5.455–5.456) show, the shift α_0 is always negative and in the course of the resonance ($\omega_* \to \omega_0$) it becomes unbounded as well as the amplitude α_1^*.

5.4.3 Prediction of Chaotic Motions

The strategy for application of the stabilization principle to predict chaotic motions is more sophisticated than those considered in the previous section.

Inertial Motions

In order to clarify the main idea of this strategy, let us turn to the inertial motion of a particle M of unit mass on a smooth pseudosphere S having constant negative curvature (5.376) (Zak 1985a,b; 1994a).

As shown there, the orbital instability, and therefore the chaotic behavior of the particle M can be eliminated by the elastic force (4.47):

$$F = \alpha^2 \varepsilon, \ \alpha^2 = \text{Const} > -2WG, \ G < 0 \tag{5.457}$$

proportional to the normal deviation ε from the geodesic trajectory, which is applied to the particle M. But such a force can appear as an inertial force if the motion of the particle M is referred to an appropriate non-inertial system of coordinates.

Indeed, so far the motion has been referred to an inertial system of coordinates, q_1, q_2, where q_1 is the coordinate along the geodesic meridians, and q_2 is the coordinate along the parallels. Let us introduce now a frame of reference which rotates about the axis of symmetry of the pseudosphere (Fig. 4.1a) with the rapidly oscillating transport velocity:

$$\dot{\varepsilon} = 2\dot{\varepsilon}_0 \cos \omega t, \ \omega \to \infty \tag{5.458}$$

so that the component of the resultant velocity along the meridians and parallels are respectively:

$$v_1 = \dot{q}_1, \ v_2 = \dot{q}_2 + 2\dot{\varepsilon}_0 \cos \omega t \tag{5.459}$$

Since (5.459) has the same structure as (5.130), the Lagrangian of the motion of the particle M relative to the new (non-inertial) frame of reference can be written [with taking into account (5.132)] in the following form [see (4.12)]

$$L^* = \dot{q}_1^2 - \frac{1}{G_0} e^{-2\sqrt{G_0} q_1} \left(\dot{q}_2^2 + \dot{\varepsilon}_0^2 \right) \tag{5.460}$$

The last term in (5.460) represents the contribution of the inertia forces in the new frame of reference. So far, the transport velocity $\dot{\varepsilon}_0$ was not specified, and therefore, the Lagrangian (5.460) has the same element of arbitrariness as the governing equations (5.136) describing chaotic motions [see (5.137)].

Now, based upon the stabilization principle, we are going to specify the transport motion in such a way that the original orbital instability of the inertial motion of the particle M is eliminated. Turning to the condition (4.49), one obtains:

$$\frac{\partial^2 L}{\partial \varepsilon^2} \geqslant -2WG_0 \tag{5.461}$$

where $W = \frac{1}{2} m v_0^2$ is the kinetic energy of the particle.

This condition can be satisfied if the transport velocity $\dot{\varepsilon}_0$ is coupled with the normal deviation ε as following:

$$-\frac{1}{G_0}e^{-2\sqrt{G_0}q_1}\dot{\varepsilon}_0^2 = -WG_0\varepsilon^2 \tag{5.462}$$

As follows from (4.52) in this limit case the Liapunov exponent of the relative motion in the new (non-inertial) frame of reference will be zero:

$$\sigma = \sqrt{-G_0 - \frac{\alpha^2}{W}} = 0,\ \alpha^2 = \frac{\partial^2 L}{\partial \varepsilon^2} \tag{5.463}$$

and the trajectories of perturbed motions do not diverge. The normal deviation from the trajectory of the relative motion (in case of zero perturbed velocity $\dot{\varepsilon}_0$) can be written in the following from:

$$\varepsilon = \varepsilon_0 = \text{Const},\ \varepsilon_0 = \varepsilon(t = 0) \tag{5.464}$$

which means that in the new frame of reference an initial error ε_0 does not grow – it remains constant.

The relative motion along the trajectory is described by the differential equation following from the Lagrangian (5.460) which takes the following form [after substituting (5.462)]

$$L = \dot{q}_1^2 - \frac{1}{G_0}e^{-2\sqrt{G_0}q_1}\dot{q}_2 - WG_0\varepsilon^2 \tag{5.465}$$

i.e.,

$$\ddot{q}_1 - \frac{2\sqrt{-G_0}}{G_0}e^{-2\sqrt{-G_0}q_1}\dot{q}_2 = 0 \tag{5.466}$$

But the original (undisturbed) motion was directed along the meridians, i.e., $\dot{q}_2 \equiv 0$. Consequently,

$$\ddot{q}_1 = 0,\ \dot{q}_1 = v_0 = \text{Const} \tag{5.467}$$

i.e., the relative motion along the trajectory remain unchanged.

Returning to the original (inertial) system one obtains the resultant velocity by summing the relative and transport velocities:

$$v_\tau = \tilde{v}_0 \tag{5.468}$$

$$v_\varepsilon = m\tilde{v}_0^2 G_0 \varepsilon_0 \cos\omega t,\ (\omega \to \infty) \tag{5.469}$$

in which v_τ, and v_ε, are the velocity components parallel and normal to the undisturbed (geodesic) trajectory, respectively.

The equations of the disturbed motion in the original frame of reference are

$$\sigma = \tilde{v}_0 t \tag{5.470}$$

$$\varepsilon = \varepsilon_0 + (1/\omega)m\widetilde{v}_0^2 G_0 \varepsilon_0 \sin\omega t, \ (\omega \to \infty) \tag{5.471}$$

in which σ is the arc coordinate along the undisturbed (geodesic) trajectory.

It is worth noting that the velocity $\widetilde{v}_0$ is different from the original velocity v_0. Indeed, the total kinetic energy of the particle now consists of the kinetic energy of the motion along the trajectory and the kinetic energy of transverse fluctuations expressed by (5.462), i.e.,

$$\frac{v_0^2}{2} = \frac{\widetilde{v}_0^2}{2} + \frac{v_0^2}{2}(q_2^0)^2 \, |G_0| \tag{5.472}$$

whence

$$\overline{v}_0 = v_0[1 - (q_2^0)^2 \, |G_0|]^{1/2} < v_0 \tag{5.473}$$

Obviously the mean or averaged motion represents a macroscopic view of the particle behavior extracted from the microscopic world, while the irreversibility of this motion is manifested by the loss of the initial kinetic energy to microscopic fluctuations.

It should be emphasized that the decomposition of the motion into regular and fluctuation components was enforced by the stabilization principle as a supplement to Newtonian mechanics [see (5.462)], while without this principle any theory where dynamical instability can occur is incomplete.

As follows from (5.468–5.471) the motion in the original frame of reference is stable in the sense that the current deviations of displacements and velocities do not exceed their initial values. However, the displacement-time function (5.471) is not differentiable because its derivative (5.469) is multivalued. Indeed, for any arbitrarily small interval Δt there always exists such a large frequency $\omega > \Delta t/2\pi$ that within this interval the velocity (5.469) runs through all its values. In other words, one arrived at stability in the class of non-differentiable functions. (The mathematical meaning of this result will be discussed below.)

Thus, chaotic motion of a particle on a smooth pseudosphere is represented by the "mean" motion (5.470) along the undisturbed geodesic trajectory [with the constant velocity (5.468) and the fluctuation motion (5.471)] normal to this trajectory. The "amplitude" of these fluctuations is vanishingly small, but the velocity "amplitude" is finite. It is worth emphasizing that this amplitude is proportional to the Gaussian curvature of the surface S, i.e., to the degree of the orbital instability. Therefore, it can be associated with the measure of the uncertainty in the description of the motion.

It is worth mentioning that both "mean" and "fluctuation" components representing the originally chaotic motion are stable. This is why they are not sensitive to initial uncertainties and are fully reproducible. In other words, such a representation of the original chaotic motion is deterministic.

One should note that the condition $\omega \to \infty$ is a mathematical idealization. Practically, ω is finite:

$$\omega \gg 1/T \tag{5.474}$$

where T is a time scale over which changes of the parameters of the motion are negligible. In the same sense the concepts of differentiability and multivaluedness have to be understood. Indeed, the multivaluedness of the functions (5.469–5.470) means that the time interval between two different values of these functions is smaller than the scale of observation T of the examined motion, and there fore, these values can be associated with "almost" the same argument.

Thus, chaos in solutions to the governing equations of orbitally unstable mechanical systems can be eliminated by an appropriate selection of the frame of reference.

What is the mathematical meaning of this procedure? First, it is necessary to distinguish the chaos as a characteristic of solutions to certain mathematical equations from the chaos as a characteristic of motions of some mechanical systems. The first type of chaos cannot be eliminated since it is an inherent property of the mathematical equation. However, chaos as a characteristic of mechanical motions can be eliminated by changing the way of their description, i.e., by referring these equations to such a frame of reference which provides the best "view" of the motion.

Second, as discussed above, the concept of stability is related to a certain class of functions, or a type of space: the same solution can be stable in one space and unstable in another, depending on the definition of the "distance" between two solutions. Indeed, if the distance between the solutions in (5.471) is defined as

$$\rho = \sum_{k=0}^{n} \max \left| \varepsilon_2^{(k)}(t) - \varepsilon_1^{(k)}(t) \right| \tag{5.475}$$

then the solution (5.471) is stable for $n = 0, 1$, but it is unstable for $n = 2, 3$, ..., since its derivatives $\varepsilon^{(2)}$, $\varepsilon^{(3)}$,..., etc., are unbounded. In other words, the concept of stability as well as chaos is an attribute of a mathematical model rather than of a physical phenomenon.

Hence, from a formal mathematical point of view, the occurrence of chaos in description of mechanical motions means only that these motions cannot be properly described by smooth functions if the scale of observation is finite.

It is important to emphasize that, strictly speaking, the representation of chaotic motion in the form (5.470–5.471) is not unique. Indeed, as follows from (5.461), one can "over-stabilize" the motion by choosing the inequality

$$\frac{\partial^2 L}{\partial \varepsilon^2} > -2WG_0 \tag{5.476}$$

instead of the equality in (5.462).

Such overstabilized representations will be also "correct," although they will not be "optimal." Indeed, as noted above, the amplitude of the fluctuations is proportional to the degree of the original instability. This means that in overstabilized representations, the contribution of the fluctuations (i.e., the contribution of the "Reynolds stresses") will be more pronounced, and this will increase the uncertainty in the description of the motion.

One can note that the application of the stabilization principle to representation of chaotic motions in Lagrangian dynamics can be linked to a control problem. Indeed, we are introducing additional rapidly fluctuating forces (coming from non-inertial motions of the frame of reference) which are coupled with the parameters of motions in such a way that the original instability is eliminated.

In the particular case of an inertial motion of a particle M on a pseudosphere, the rate of divergence of the trajectories was constant [see (4.9)], which means that local and global Liapunov exponents are the same. This is why by eliminating the positive local Liapunov exponent, we "automatically" eliminate the global one. In general case, the situation is more complex: the local Liapunov exponents depend upon the position of the system, and by eliminating all the local positive Liapunov exponents one overstabilizes the motion. Indeed, non-positive global Liapunov exponents can exist even if the local ones are positive in some domain of space where the motion can occur. As we will see later, the elimination of global Liapunov exponents is a much harder problem, and this is why in many practical situations we will confine ourselves with the easier problem of elimination of local exponents, i.e., with the overstabilized representations. The overstabilization problem will be discussed later [see (5.5)] in connection with a variational formulation of the stabilization principle.

Potential Motions

Based upon (5.136), for potential motions the governing equations can be written in the following form:

$$\ddot{q}^{\alpha} + \Gamma^{\alpha}_{\beta\gamma}\dot{q}^{\beta}\dot{q}^{\gamma} = -\frac{\partial \Pi}{\partial q^{\alpha}} + Q^{\alpha}_{(i)}, \quad \frac{\partial \Pi}{\partial q^{\alpha}} = -Q^{\alpha} \tag{5.477}$$

where Π is the potential energy of the dynamical system, and $Q^{\alpha}_{(i)}$ are the inertia forces or the "Reynolds stresses" caused by the rapidly oscillating transport motion of the frame of reference [see (5.137)].

For simplicity we will confine ourselves to a two-dimensional dynamical system assuming that $\alpha = 1, 2$. Following the same strategy as those applied to inertial motions, let us couple the inertia forces with parameters of the dynamical system in such a way that the original orbital instability (if it occurs) is eliminated. For this purpose, first we will represent these forces in the form:

$$Q^{\alpha}_{(i)} = -\frac{\partial \Pi_{(i)}}{\partial q^{\alpha}} \tag{5.478}$$

where $\Pi_{(i)}$ is a fictitious potential energy equivalent to the kinetic energy of the fluctuations. Then, turning to the criteria of local orbital stability

(4.45), one finds this potential energy $\Pi_{(i)}$ and consequently, the inertial forces $Q^{\alpha}_{(i)}$ from the condition that the original local orbital instability is eliminated:

$$G + 3\left[\frac{\nabla\left(\Pi + \Pi_{(i)}\right)\cdot \mathbf{n}}{2W}\right]^2 + \frac{1}{2W}\left[\frac{\partial^2\left(\Pi + \Pi_{(i)}\right)}{\partial q^i \partial q^j} - \Gamma^k_{ij}\frac{\partial\left(\Pi + \Pi_{(i)}\right)}{\partial q^k}\right]$$

$$n^i n^j = 0,\ i, j = 1, 2 \tag{5.479}$$

Here W, G, and Γ^k_{ij} are defined by the parameters of the dynamical system (5.477) via (4.17), (4.19–4.20), respectively, and n^i are the contravariant components of the unit normal $\mathbf{n}$ to the trajectory of the basic motion.

Equation (5.479) contains only one unknown $\Pi_{(i)}$, which can be found from it, and this will define the inertial forces, or the "Reynolds stresses" (5.478).

It should be noted that unlike the case of the inertial motion of a particle on a pseudosphere, here the Gaussian curvature G, as well as the gradients of the potential energy Π, are not constants, and consequently, the local Liapunov exponents may be different from the global ones. This means that the condition (5.479) eliminates local positive exponents, and therefore, the solution to (5.477) and (5.479) represents an overstabilized motion. Obviously, elimination of only global positive Liapunov exponents would lead to solutions with fewer uncertainties while some local exponents in certain domains of the phase space may even remain positive. However, the strategy for elimination of global positive exponents is more sophisticated, and it can be implemented only numerically (see Sect. 5.5).

It is worth noting that (5.479) is simplified to the following:

$$G = \frac{1}{2W}\left[\frac{\partial^2\left(\Pi + \Pi_{(i)}\right)}{\partial q^i \partial q^j}\right] n^i n^j = 0 \tag{5.480}$$

if the basic motion is characterized by zero potential forces

$$\frac{\partial \Pi}{\partial q^i} = 0 \tag{5.481}$$

It may occur, for instance, when the dynamical system is in a relative equilibrium with respect to a moving frame (Zak 1994a).

Example: Bending Oscillations of a Rotating Rod

Let us consider a rotating structure carrying a pin-ended rod AB simply supported, as shown in Fig. 5.18 (Zak 1987). Assuming that the rotation is not controlled, and therefore, the structure is conservative, we will analyze the dynamical interaction between the rigid rotation and one of the in-plane bending modes of the rod. Thus, we will introduce two-dimensional

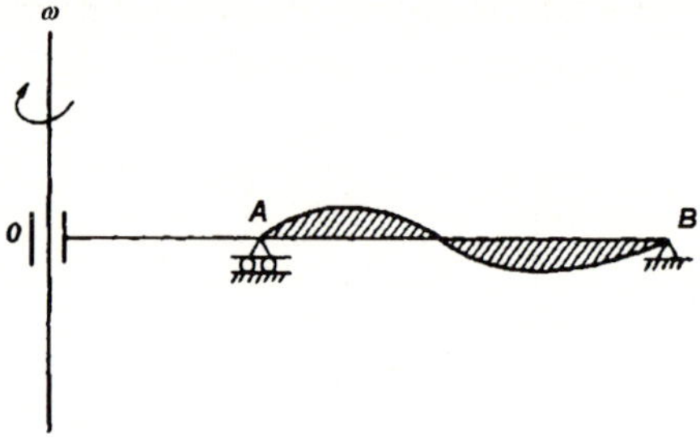

Fig. 5.18.

configuration subspaces with the coordinates q^1 and $q^2_{(n)}$, where

$$q^1 = \phi_1, \; q^2_{(n)} = u_0^{(n)}, \; n = 1, 2, ..., \text{etc.} \tag{5.482}$$

Here ϕ is the angle of rotation, and $u_0^{(n)}$ is the maximum transverse deflection of the rod AB, while the in-plane nth bending mode under consideration is presented as

$$u^{(n)} = u_0^{(n)} \sin \frac{\pi n}{l} x, \; n = 1, 2, ..., \text{etc.} \tag{5.483}$$

in which x is the coordinate along the rod and l is the rod length. The contribution of the coordinates q^1 and q^2 to the kinetic energy of the system is expressed as

$$W = \frac{1}{2}\rho \int_0^l \{u^2\dot\phi^2 + \dot u^2 + 2(R-x)\dot\phi\dot u + \dot\phi^2 \left[R(l-x) - \frac{1}{2}(l^2 - x^2)\right] u'^2$$

$$+\dot\phi^2 (R-x)^2\} dx, \; R > l \tag{5.484}$$

in which ρ is the rod density, R is the radius of rotation of the point A and $u' = du/dx$.

The potential energy of the system depends only on the bending modes:

$$\Pi = \frac{\pi^4 n^4}{4l^3} EI u_0^{(n)^2}, \; \text{i.e.,} \; \frac{\partial \Pi^2}{\partial q^2 \partial q^2} = \frac{\pi^4 n^4}{2l^3} EI \tag{5.485}$$

where EI is the flexural rigidity of the rod.

The metric coefficients defined by (5.484) are

$$a_{11} = \rho \left\{ Rl(R-l) + \frac{l^3}{3} + u_0^2 \left[\frac{l}{2} + \frac{\pi^2 n^2}{4}\left(R - \frac{2}{3}l\right)\right]\right\}$$

$$a_{12} = \rho l (R - \frac{l}{2}), \; a_{22} = \rho l \tag{5.486}$$

while the Gaussian curvature, calculated using the coefficients a_{ij} and (4.19–4.20) is:

$$G_{(n)} = -\frac{1}{2\rho l^4}[12l + \pi^2 n^2(3R - 2l)] < 0 \tag{5.487}$$

Let us assume that the undisturbed motion of the system is characterized by no deflections and constant rotation:

$$u = 0,\ \phi = \omega = \text{Const at } t = 0 \tag{5.488}$$

Then the undisturbed kinetic energy of the system follows from equation (5.482):

$$W = \frac{1}{2}\rho(R^2 l - Rl^2 + \frac{l^3}{3})\omega_0^2 \tag{5.489}$$

The unit tangent to the undisturbed trajectory is defined by its contravariant components:

$$\tau^i = \frac{dq^i}{dt}\frac{1}{\sqrt{2W_0}},\ \text{i.e.,}\ \tau^2 = 0,\ \tau^1 = \frac{\sqrt{2}}{\sqrt{\rho\left(R^2 l - Rl^2 + \frac{l}{3}\right)^3}} \tag{5.490}$$

The contravariant components of the unit normal to the undistributed trajectory are found from the following conditions:

$$a_{11}\nu^1\nu^1 + a_{12}\nu^1\nu^2 + a_{22}\nu^2\nu^2 = 1$$

$$a_{11}\tau^1\nu^1 + a_{12}(\tau^1\nu^2 + \tau^2\nu^1) + a_{22}\tau^2\nu^2 = 0 \tag{5.491}$$

whence

$$\nu^2\nu^2 = \frac{1}{a_{22}} = \frac{1}{\rho l} \tag{5.492}$$

Substituting equations (5.485) (5.487) and (5.492) into inequality (4.45) and taking into account (5.481) one arrives at the conditions for chaotic bending oscillations of the rotating rod:

$$\frac{\rho}{2}\left(\frac{R^2}{l^2} - \frac{R}{l} + \frac{1}{3}\right)\left[12 + \pi n^2\left(3\frac{R}{l} - 2\right)\right]\omega_0^2 > \frac{\pi^4 n^4}{2l^2}EI \tag{5.493}$$

As follows from this condition, lower modes are more sensitive to chaotic instability than higher modes since the right hand side of inequality (5.493) grows faster than the left hand side when n increases.

It should be emphasized that the condition (5.493) was derived for small deflections of the rod and, therefore, the applications of this condition can be justified if all the deflections are bounded by arresting devices, so that

$$|u_0^n| \ll l \tag{5.494}$$

Under this restriction, the mean motion will be sufficiently close to the undisturbed motion and therefore, (5.485), (5.487), (5.489) and (5.492) can be substituted into the closure (5.480) which defines the Reynolds force $\widetilde{Q}^2_{(i)}$ and the potential $\widetilde{\Pi}_{(i)}$

$$\frac{\partial^2 \widetilde{\Pi}_{(i)}}{\partial q^2 \partial q^2} = A, \ \widetilde{Q}^2_{(i)} = -\frac{\partial \widetilde{\Pi}_{(i)}}{\partial q^2} = -Aq^2, \ \widetilde{\Pi}_{(i)} = \frac{1}{2} q^2 q^2 \tag{5.495}$$

in which

$$A = \frac{\rho}{2}\left(\frac{R^2}{l^2} - \frac{R}{l} + \frac{1}{3}\right)\left[12 + \pi n^2 \left(3\frac{R}{l} - 2\right)\right]\omega_0^2 - \frac{\pi^4 n^4}{2l^2} EI \tag{5.496}$$

For finite deflections the mean motion does not necessarily coincide with the undisturbed motion. and therefore, (5.477) and (5.478) are coupled.

General Case

When motions of a dynamical system are not potential, in many cases it is more convenient to represent them in the form of a system of first order differential equations (5.134):

$$\dot{x}^i = a^i_j x^j + b^i_{jm} x^j x^m, \ i = 1, 2, ..., n \tag{5.497}$$

which can be represented in the Reynolds form (5.135):

$$\dot{\overline{x}}^i = a^i_j \overline{x}^j + b^i_{jm} \overline{x}^j \overline{x}^m + b^i_{jm} \overline{\tilde{x}^j \tilde{x}^m}, \ i = 1, 2, ..., n \tag{5.498}$$

with the additional terms $b^i_{jm} \overline{\tilde{x}^j \tilde{x}^m}$ representing the Reynolds stresses.

The Closure Problem

Because of additional unknowns $\overline{\tilde{x}^j \tilde{x}^m}$ in equations (5.498) the closure problem arises. Analogously we will seek additional coupling between the mean motion and the fluctuations:

$$\overline{\tilde{x}^j \tilde{x}^m} = a^{jm}_l \overline{x}^j + a^{jm}_{ln} \overline{x}^l \overline{x}^n, ..., \text{etc.} \tag{5.499}$$

based upon the stabilization principle the application of which will be clarified below.

Firstly, we recall that the solutions to equations (5.497) are chaotic, and consequently, some of the Liapunov exponents of equations (5.497) are positive:

$$\lambda^+_m > 0, \ m = 1, 2, ..., S \tag{5.500}$$

Secondly, we are looking for a decomposition (5.497) in which the mean motion is periodic, rather than chaotic. Hence, the fluctuations should he coupled with the mean motion such that all positive Liapunov exponents become zero, while the rest of the exponents are unchanged. Indeed, in this case the mean motion is a regular motion which is the "closest" to the original chaotic motion. Since the Liapunov exponents for the system (5.498–5.499) depend on the "feedback" coefficients a_l^{jm}, $a_{l\ n}^{jm}$, etc., the closure can now be formulated as follows:

$$\left.\begin{array}{ll} \lambda_m^{+}(a_l^{jm}, a_{l\ n}^{jm}, \ldots) = 0, & i = 1, 2, \ldots, S \\ \lambda_i^{\circ}(a_l^{jm}, a_{l\ n}^{jm}, \ldots) = \lambda_i^{0}(0, 0, \ldots), & i = 1, 2, \ldots, S_0 \\ \lambda_i^{-}(a_l^{jm}, a_{l\ n}^{jm}, \ldots) = \lambda_i^{-}(0, 0, \ldots), & i = 1, 2, \ldots, S_- \end{array}\right\} \tag{5.501}$$

in which λ^+, λ^0 and λ^- are positive, zero and negative Liapunov exponents, respectively. Obviously, those coefficients a_l^{jm} which do not appear in equation (5.501), must be zero.

Thus, the system (5.498–5.499), (5.501) is closed. It defines the regular mean motion and fluctuations which represent the original chaotic motion. Since all the Liapunov exponents for this system are not positive, the solution is stable and predictable in the sense that small changes in the initial conditions cause small changes in both the mean motion and the fluctuations (Zak 1994a).

In the next subsection the application of this approach to the Lorenz strange attractor is illustrated.

Higher Order Approximations

The Reynolds decomposition of the variables x^i in (5.497) generates not only pair correlations $\overline{x^i x^j}$, but also correlations of higher orders, such as triple correlations $\overline{x^i x^j x^k}$, quadruple correlations $\overline{x^i x^j x^k x^m}$, etc.

Indeed, multiplying equations (5.497) by x^i and averaging and combining the results, one obtain the governing equations for the pair correlations $\overline{x^i x^k}$:

$$\dot{\overline{x^i x^k}} = a_j^i \overline{x^j x^k} + a_j^k \overline{x^j x^i} + b_{jm}^i(\overline{x^k x^j x^m} + \overline{x^k x^j}\,\overline{x^m} + \overline{x^k x^m}\,\overline{x^j})$$

$$+ b_{jm}(\overline{x^i x^j x^m} + \overline{x^i x^j}\,\overline{x^m} + \overline{x^i x^m}\,\overline{x^j}) \tag{5.502}$$

which contain nine additional triple correlations $\overline{x^i x^j x^k}$.

Similar equations for the triple correlations will contain all the quadruple correlations etc. In general, one arrives at an infinite hierarchy of equations which are open, since any first N equations relate $(N + 1)$ correlations.

From this viewpoint all the closures discussed above can be considered as first-order approximations which defined only the mean components of the chaotic motions. In order to define both the mean motion and the four correlations one should consider the Reynolds equation (5.498) together with equation (5.502). In this case the evolution of the double correlations is already prescribed by equations (5.502), and consequently, the stabilizing feedback must now couple the triple correlations with the mean and pair correlation components:

$$\overline{x^k x^j x^m} = F(\overline{x^k}, \overline{x^l x^m}) \tag{5.503}$$

The system (5.498), (5.502–5.503) will define periodic mean and pair correlation components.

The strategy for finding higher order correlations will be illustrated in Sect. 5.5.

Computational Strategy

As follows from the above, the closure, i.e. the stabilizing feedback between the Reynolds stresses and the mean components of the motion, can be written in the explicit form only if the criteria for the onset of chaos are formulated explicitly. Since such a situation is an exception rather than a rule, we develop below a computational strategy which allows one to find the closure regardless of the complexity of the original equations.

We will demonstrate this strategy using equations (5.497). The same strategy will be suitable for the Navier-Stokes equations since after an appropriate discretization technique they reduce to the form (5.497).

Turning to (5.498), which follow from equations (5.497) as a result of the Reynolds decomposition, let us linearize them with respect to the original (laminar) state x_0^i :

$$\dot{\overline{x}} = (a_j^i + 2b_{jm}^i \overline{x}_0^m)\overline{x}^j, \text{ with } \overline{x^i x^j} = 0 \text{ at } \overline{x}^i = \overline{x}_0^i \tag{5.504}$$

Introducing small "laminar" disturbances in the form

$$\overline{x}^i = \overline{x}_*^i \exp(\lambda_0 t) \tag{5.505}$$

one arrives at a truncated analog of the Orr-Sommerfeld equations [see (4.424)].

$$\lambda^0 \delta_j^i = (a_j^i + 2b_{jm}^i \overline{x}_0^m)\overline{x}^j \tag{5.506}$$

where the local eigenvalues of equations (5.498)

$$\lambda^0 = \lambda_1^0, \lambda_2^0, ..., \lambda_n^0 \tag{5.507}$$

are the roots or the characteristic equation

$$\det(\lambda^0 \delta_j^i - a_j^i - 2b_{jm}^i \overline{x}_0^m) = 0 \tag{5.508}$$

Applying the same procedure to the second-order Reynolds equations (5.502), one obtains, instead of (5.506),

$$\frac{1}{2}\lambda^0 \delta_k^i = (a_j^i + 2b_{jm}^i \overline{x}_0^m)\overline{x^j x^k} \tag{5.509}$$

and, therefore, the local eigenvalues of (5.502) are twice as large as those for (5.498), i.e. instead of equation (5.505):

$$\overline{x^i x^k} = \sim \exp(2\lambda_0 t) \tag{5.510}$$

If the original "laminar" state x_0^i is unstable, i.e. there are λ_i^0 with positive real parts in (5.507)

$$\operatorname{Re} \lambda_i^0 > 0 \tag{5.511}$$

then the pair correlations (5.510) will grow much faster than the mean motion disturbances (5.505), and one can assume that these correlations will be large enough to stabilize (5.498) while the mean motion will remain sufficiently close to its original state x_0^i. This property makes possible the following computational strategy.

Let us seek a closure to (5.498) in the neighborhood of the original laminar state x_0^i in the form

$$b_{jm}^i \overline{x^j x^m} = C_j^i \overline{x}^j \tag{5.512}$$

in which C_j^i are to be found.

Substituting (5.512) into (5.498) and linearizing them with respect to the original "laminar" state x_0^i, one obtains

$$\dot{\overline{x}}^i = (a_j^i + 2b_{jm}^i \overline{x}_0^m + C_j^i) \overline{x}^j \tag{5.513}$$

while the eigenvalues for this equation follows from

$$\det(\lambda^0 \delta_j^i - a_j^i - 2b_{jm}^i \overline{x}_0^m - C_j^i) = 0 \tag{5.514}$$

The sought coefficients C_j^i must be selected such that

$$\operatorname{Re} \lambda_i = \frac{1}{2} \left(\operatorname{Re} \overset{\circ}{\lambda}_i - \left| \operatorname{Re} \overset{\circ}{\lambda}_i \right| \right) \tag{5.515}$$

Indeed, in this case all the positive real parts of the local eigenvalues causing the instability of the "laminar flow" become zero, while the rest of these eigenvalues remain unchanged.

In order to find C_j^i from condition (5.515), we diagonalize the matrix

$$a_j^i + 2b_{jm}^i \overline{x}_0^m = \{F_{ij}\} \tag{5.516}$$

such that

$$\theta^{-1} F \theta = [\lambda_1, ..., \lambda_n] \tag{5.517}$$

Then the matrix of the sought coefficients

$$C_j^i = \{C_{ij}\} \tag{5.518}$$

is found to be

$$\{C_{ij}\} = \frac{1}{2}\theta[C_1, C_2, ..., C_n]\theta^{-1} \tag{5.519}$$

in which

$$C_i = \frac{1}{2}\left(\text{Re}\ \lambda_i - \left|\text{Re}\ \lambda_i\right|\right) \tag{5.520}$$

Substituting equation (5.519) into (5.513), one obtains a linearized governing equation for the turbulent or chaotic motion at the very beginning of the transition from the laminar motion. Selecting a small time step Δt_1 one can find the next state $\overline{x}_1^i$:

$$\overline{x}_1^i = \overline{x}_0 + \dot{\overline{x}}_0\ \Delta t_1$$

Repeating this procedure for $\overline{x}_1^i$, Δt_2, $\overline{x}_2^i$, etc. one arrives at the evolution of the turbulence, or chaos. The process ends when the solution approaches a regular (static or periodic) attractor whose existence is assumed (Zak 1994a).

Example 1: the Lorenz Attractor

As an example, we will illustrate prediction of the probabilistic structure of the Lorenz attractor by using the stabilization principle.

Applying the Reynolds transformation to the Lorenz attractor:

$$\begin{aligned} \dot{x} &= -\sigma x + \sigma y \\ \dot{y} &= -xz + rx - y \\ \dot{z} &= xy - bz \end{aligned} \tag{5.521}$$

one obtains

$$\begin{aligned} \dot{\overline{x}} &= -\sigma\overline{x} + \sigma\overline{y} \\ \dot{\overline{y}} &= r\overline{x} - \overline{y} - \overline{x}\overline{z} - \overline{xz} \\ \dot{\overline{z}} &= -b\overline{z} + \overline{x}\overline{y} + \overline{xy} \end{aligned} \tag{5.522}$$

where $\overline{x}$, $\overline{y}$, $\overline{z}$ are the mean values of x, y, and z, while $\overline{xz}$ and $\overline{xy}$ are double correlations representing the Reynolds "stresses."

As extra variables, these double correlations must be found from the condition that they suppress the positive Liapunov exponent down to zero. In this case both the mean and the double-correlations components of the motion will be represented by periodic attractors, i.e. in a fully deterministic way.

Numerical implementation of this strategy performed for $\sigma = 10$, $r = 28$, and $b = 8/3$ leads to the following results (Zak et al. 1996): Figure 5.19 represents the original chaotic attractor as a solution to (5.521). In Figs. 5.20–5.22 this attractor is decomposed into two deterministic (periodic) motions: the mean motion (Fig. 5.20) and the double correlations, i.e. the Reynold's stresses (Fig. 5.21–5.22). In order to find all the double correlations, one should exploit the system for triple-correlations which

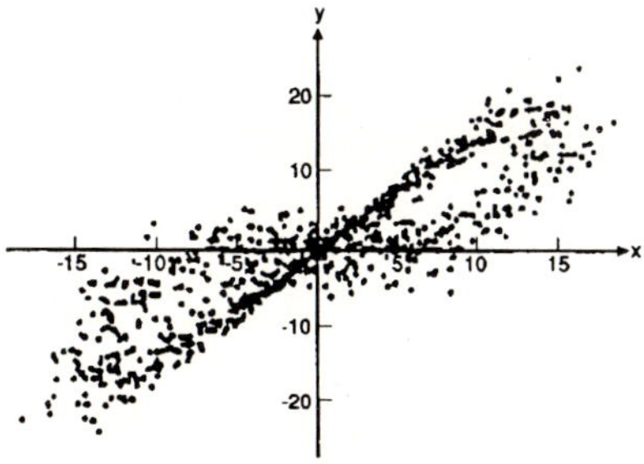

Fig. 5.19.

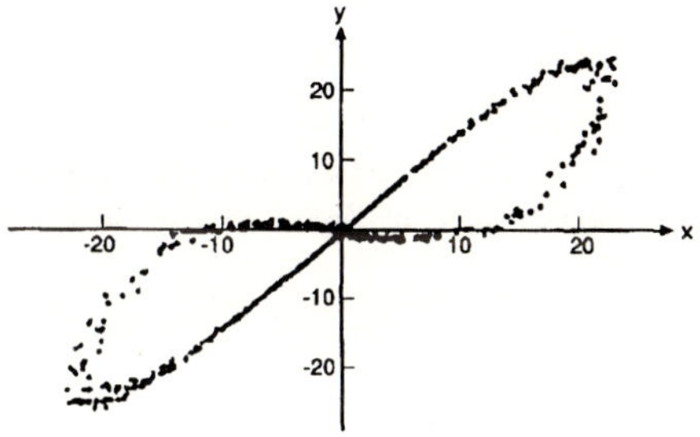

Fig. 5.20.

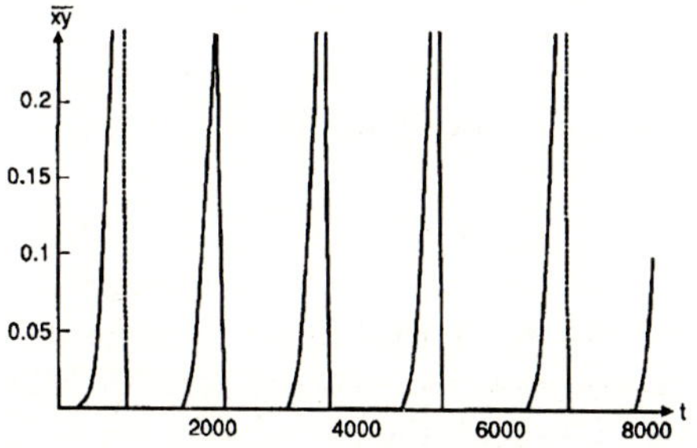

Fig. 5.21.

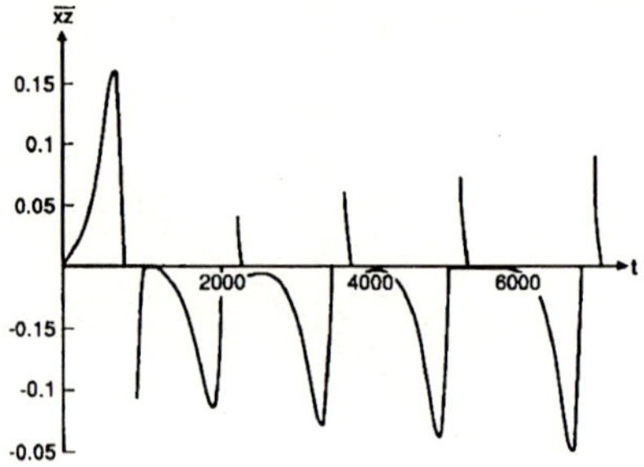

Fig. 5.22.

can be obtained in a straightforward way from (5.521). In this system all the triple-correlations, as extra variables, must be found from stabilization principle in a similar way. By continuing this process, one can find the probabilistic structure of the solution to the Lorenz equations (5.521) to a required accuracy.

Example 2: Charged Particle

The second example illustrates prediction of the mean flow in chaotic behavior of a charged particle in a uniform magnetic field:

$$\dot{v}_x = \frac{x}{r^3} - v_y, \;\; \dot{v}_y = \frac{y}{r^3} - v_x, \;\; \dot{v}_z = \frac{z}{r^3},$$

$$r^2 = x^2 + y^2 + z^2, \;\; \dot{x} = v_y, \;\; \dot{z} = v_z \tag{5.523}$$

The system is chaotic if, for instance,

$$x = 1.5, \;\; y = 0, \;\; z = 4.0; \;\; v_x = v_z = 0.01 \text{ at } t = 0 \tag{5.524}$$

Figure 5.23 demonstrates a chaotic region in the space x, y, z . Application of the stabilization principle leads to the multiperiodic (but not chaotic) mean flow plotted in Fig. 5.24 (Zak and Meyers 1995).

5.4.4 Predictions of Turbulent Motions

Introduction

The problem of turbulence arose almost a hundred years ago as a result of a discrepancy between theoretical fluid dynamics and applied problems. However, in spite of considerable research activity, there is no general deductive theory of high Reynolds number turbulence: the direct utilization of the Navier-Stokes equations leads to unstable and unpredictable (chaotic) solutions. The central problem of turbulence theory is to find suitable methods of converting the infinite hierarchy of equations following from the averaging process of Navier-Stokes equations into a closed set.

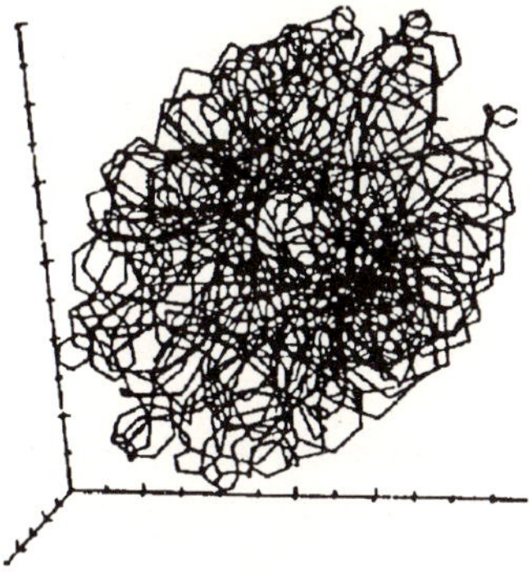

Fig. 5.23.

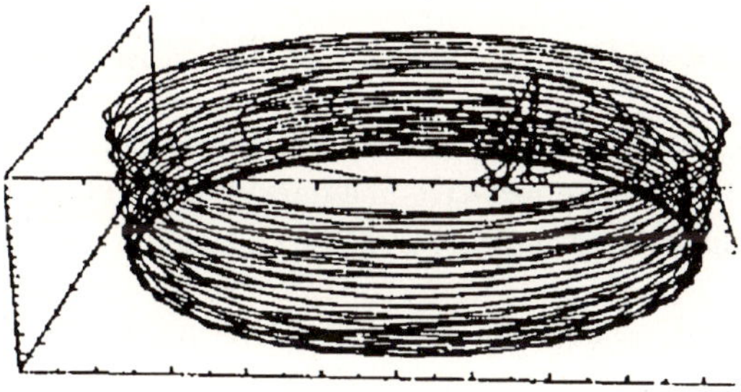

Fig. 5.24.

This section is based upon a different interpretation of the closure problem. The guiding principles of this interpretation can be illustrated by the following example. Let us consider a Couette flow between two parallel flat plates which are displaced relative to each other. The flow remains laminar as long as the Reynolds number $R < 1500$ and the velocity distribution is then linear. For $R > 1500$ this flow becomes unstable, and therefore, practically it can never be observed: any small disturbances (which always exist) are amplified, and being driven by the mechanism of the instability, grow until a new (turbulent) stable state is approached. The turbulent velocity profiles are very flat near the centre and become very steep near the walls, while with increasing Reynolds number, the curvature of these profiles becomes more pronounced. The stability here should be understood in a sense that the flow is reproducible, while a small change of R leads to small change of the flow characteristics.

Prandtl obtained the mean velocity profile analytically based upon the Reynolds equations and additional rheological (the mixing length) assumption which couples the mean velocity v^k and the fluctuation velocities $\tilde{v}^k$:

$$f(v^k, \overline{\tilde{v}^i\tilde{v}^k} = 0 \tag{5.525}$$

However, the same equation (5.525) can be interpreted as a "feedback" which stabilizes the originally unstable laminar flow by driving it to a new turbulent state. In this section the closure in turbulence is based upon the stabilization principle formulated and applied in the previous section of this chapter.

Thus, as follows from the above example, the closure in turbulence can be related to the stabilization effect of the fluctuation velocities. Obviously, the fluctuations can grow only as long as the instability persists, and consequently, the instability of the original laminar flow must be replaced by a neutral stability of the new turbulent flow. This stabilization principle allows one to formulate the closure problem as follows:

Let the original laminar flow described by the Navier-Stokes equations

$$\rho(\partial v^k/\partial t + v^i \nabla_i v^k) = -\nabla_k p + \nu \nabla^2 v^k, \; \nabla_i v^i = 0 \tag{5.526}$$

be unstable, i.e., some of the eigenvalues of the corresponding Orr-Sommerfeld equation (4.424) have negative imaginary parts. Then, the fluctuation velocity $\tilde{v}^k$ in the corresponding Reynolds equation

$$\rho(\partial v^k/\partial t + v^i \nabla_i v^k) = -\nabla_k p + \nu \nabla^2 v^k - \rho \partial_i \, \overline{\tilde{v}^i \tilde{v}^k} \tag{5.527}$$

must be coupled with the mean velocity v^k by a feedback

$$f(v^k, \nabla_i v^k, \ldots \overline{\tilde{v}^i \tilde{v}^k}, \ldots) = 0 \tag{5.528}$$

which is found from the condition that all these negative imaginary parts vanish, and therefore, the solution possesses a neutral stability.

The Reynolds equation (5.527) can be derived using the decomposition (5.130) and taking into account (5.132–5.133). Indeed, the solution to the Navier-Stokes equation (NSE) for high Reynolds numbers ($R > R_{\text{cr}}$) is sought as a sum of smooth and non-smooth components

$$\dot{v} = v + \tilde{v} f(\omega t), \; \omega \to \infty \tag{5.529}$$

where v and $\tilde{v}$ represent the mean and fluctuation velocities, respectively, and f is a periodic function. Substituting (5.529) in NSE and averaging them over the vanishingly small period $2\pi/\omega$ yields

$$\frac{\partial v}{\partial t} + v\nabla v - \nu \nabla^2 v + \frac{1}{\rho}\nabla p = -\overline{\tilde{v}\nabla\tilde{v}} \tag{5.530}$$

$$\nabla \cdot v = 0 \tag{5.531}$$

Equation (5.530) can be interpreted as NSE written in a non-inertial frame of reference which oscillates with the transport velocity $\tilde{v}f(\omega t)$. Therefore, the vector $-\overline{\tilde{v}\nabla\tilde{v}}$ represents the contribution of the corresponding inertial forces, while v is interpreted as a relative velocity. Such a transition will be useful if v is smooth, i.e., if solutions to (5.530–5.531) are stable. For this purpose $\tilde{v}$ should be appropriately coupled with v by a "feedback"

$$\Phi(v, \overline{\tilde{v}\nabla\tilde{v}}, \text{ etc.}) = 0 \tag{5.532}$$

implying that (5.530–5.532) are stable. From the physical standpoint it means that we are looking for such a frame of reference which provides the best "view" of the motion. Hence, in this interpretation the closure in turbulence is not a rheological or statistical problem: it rather resembles the problem of control theory. The same line of argumentation was already applied in the previous section of this chapter to Lagrangian dynamics.

The usefulness of this approach is illustrated by a qualitative analysis of the turbulence between two rotating concentric cylinders (Zak 1986a,b). For vanishing viscosity ($\nu \to 0$) the motion is unstable if $d^2\Pi/dr^2 < 0$ (the Rayleigh criterion), where Π is the potential energy due to small radial displacement of the fluid with respect to its unperturbed configuration. Thus instability is eliminated if some additional potential field $\tilde{\Pi}$ is applied such that $d^2(\Pi + \tilde{\Pi})/d\tilde{r}^2 = 0$. The potential $\tilde{\Pi}$ can be caused, for instance, by a rapidly oscillating force field: $\tilde{\Pi} = 1/2\rho\tilde{v}^2$; but the same effect is generated by the inertia forces in a rapidly oscillating frame of reference. Hence, the feedback (5.532) now can be written as

$$\frac{d^2}{dr^2}\left(\Pi + \frac{\rho\tilde{v}^2}{2}\right) = 0$$

or

$$\frac{u^2}{r} + \frac{1}{2}\frac{d}{dr}(\tilde{v}^2) = C_1^2 = \text{Const} \tag{5.533}$$

where u is the circumferential component of the mean velocity v, and $\tilde{v}$ is the radial component of the fluctuation, and equations (5.530–5.531) for this example reduce to

$$\nu\left(u'' + \frac{u'}{r} - \frac{u}{r^2}\right) = 0$$

i.e.,

$$u = ar + \frac{b}{r} \tag{5.534}$$

Hence, as follows from (5.533) and no-slip boundary conditions

$$\tilde{v}^2 = \frac{1}{2}a^2(r_1^2 - r^2) - 2ab\ln\frac{r}{r_1} - \frac{1}{2}b^2\left(\frac{1}{r^2} - \frac{1}{r_1^2}\right)$$

$$+C(r - r_1) \tag{5.535}$$

while

$$C = \frac{1}{2}b^2\left(\frac{1}{r_1^2} - \frac{1}{r_2^2}\right) + 2ab\ln\frac{r_2}{r_1} - \frac{1}{2}a^2(r_2^2 - r_1^2)$$

$$a = \frac{\omega_2 r_2^2 - \omega_1 r_1^2}{r_2^2 - r_1^2}, \quad b = \frac{(\omega_1 - \omega_2)\, r_1^2 r_2^2}{r_2^2 - r_1^2}$$

in which $r_1, r_2, \omega_1, \omega_2$ are the inner and outer radii and corresponding angular velocities.

Since the cylinders are of infinite length ($l \to \infty$), the problem is two dimensional and the radial fluctuation velocity (5.535) can be interpreted as a "projection" of the Taylor vortices onto the cross-sectional plane for the limit case $l \to \infty$, $v \to 0$.

It is worth mentioning that the solution is obtained directly from NSE without any additional physical assumptions. Clearly, this solution gives a deterministic description of turbulence because it is reproducible: small changes in initial conditions lead to small changes in both mean and fluctuation velocities. The same strategy was applied to chaos in classical dynamics in the previous section.

In the general case the feedback (5.532), can be sought as

$$\overline{\tilde{v}\nabla\tilde{v}} = (1/R - 1/R_{\,\mathrm{cr}})\nabla^2 v \tag{5.536}$$

which leads to "stabilized" NSE in a dimensionless form

$$S\frac{dv}{dt} + v\nabla v + E\nabla p = \frac{1}{R_{\,\mathrm{cr}}}\nabla^2 v$$

$$R_{\mathrm{cr}} = f(v, r, \ldots) \tag{5.537}$$

Here the critical Reynolds number R_{cr} is defined by the stability criteria of laminar flows, namely, from the condition that $R_{\mathrm{cr}}(\alpha) = \text{Const}$ for all the originally unstable frequencies α.

It should be stressed again that the solutions obtained by using the stabilization principle are not unique: they depend upon the degree of overstabilization. The "optimal" solution, i.e., the solution with the minimum Reynolds stress, and therefore, with minimum uncertainties, would be obtained in case of elimination of the global instability (i.e., the global Liapunov exponents). The elimination of local instability leads to stronger overstabilization, and therefore, to less optimal solutions, although computationally the last strategy is much simpler.

Since NSE can be reduced to a system of ordinary differential equations using discretization techniques (such as finite difference, finite elements, spectral representations, etc.), the computational strategy of application the stabilization principle described in Sect. 5 of this chapter is valid for predictions of turbulent motions. This is why in this section our attention will be concentrated on closures following from analytical formulations of instability of the type (5.533) and (5.536).

Example 1. Smoothing out of a Velocity Discontinuity

Introduction. The meaningfulness of the principle formulated above will be illustrated by the consideration of the smoothing out of a velocity discontinuity which was first treated by Prandtl, who closed the Reynolds equation:

$$\frac{\partial u}{\partial t}+u\frac{\partial u}{\partial x}+v\frac{\partial u}{\partial y}=\frac{1}{\rho}\frac{\partial \tau}{\partial y},\quad \tau=-\rho\overline{\widetilde{u}\widetilde{v}} \tag{5.538}$$

by using the mixing length theory:

$$\tau=\rho l^2\left|\frac{\partial u}{\partial y}\right|\frac{\partial u}{\partial y} \tag{5.539}$$

in which u and v are the mean and fluctuation velocity projections on the coordinates x and y, respectively; τ is the turbulent shearing stress, and l is the mixing length which is supposed to be found from experiments. By assuming that l is proportional to the width of the mixing zone b, Prandtl obtained the following solution for the mean horizontal velocity:

$$u(y,t)=u_{+}+u_{-}(\frac{2}{3}\widetilde{y}-\frac{1}{3}\widetilde{y}^3) \tag{5.540}$$

while

$$u_{+}=\frac{1}{2}(u_1+u_2),\ u_{-}=\frac{1}{2}(u_1-u_2)$$

$$\widetilde{y}=y/b,\ b=\frac{3}{2}\beta^2u_{-}t \tag{5.541}$$

in which u_1 and u_2 are the horizontal velocities before the mixing, and β is an experimental coefficient.

In this section, we will obtain the analog of (5.540) without any experimental coefficients (Zak 1986a,b).

Posedness of the Problem. Consider the basic flow of incompressible inviscid fluid of the same density ρ in two horizontal parallel streams of widths H and velocities $\mathring{u}_1,\mathring{u}_2$ $(\mathring{u}_1>\mathring{u}_2)$.

For potential motion the Reynolds equation reduces to

$$\frac{\partial \phi_i}{\partial t}+\frac{1}{2}\left(\frac{\partial \phi_i}{\partial x}\right)^2=\frac{p_i}{\rho}+\text{Const}$$

$$|y|\leqslant H,\ i=1,2 \tag{5.542}$$

in which x is parallel and y is perpendicular to the streams, ϕ_i are the mean velocity potentials, p_i are the pressures which include the Reynolds stresses p_i'':

$$p_i=p_i'+p_i'',\ p_i''=\frac{1}{2}\rho\left(\overline{\widetilde{u}_i^2+\widetilde{v}_i^2}\right),\ i=1,2 \tag{5.543}$$

In addition ϕ_i must satisfy the Laplace equation

$$\nabla^2 \phi_i = 0 \tag{5.544}$$

and the boundary conditions

$$\partial \phi_i / \partial y = 0 \text{ at } |y| = H \tag{5.545}$$

The Closure Problem. Linear stability analysis of this system leads to the following characteristic equation:

$$(\lambda - \overset{\circ}{u}_1)^2 - (\lambda - \overset{\circ}{u}_2)^2 = (p^*/\rho a k)\tanh(kH) \tag{5.546}$$

in which a and $2\pi/k$ are the amplitude and the wavelength of a small perturbation, respectively ($a \ll H$), λ is the speed of wave propagation, and p^* is the Reynolds stress discontinuity:

$$p^* = p_1'' - p_2'' \tag{5.547}$$

Assuming that the basic flow is laminar ($p^* = 0$) one arrives at the well-known solution of (5.546):

$$\lambda = \overset{\circ}{u}_+ \pm i\, \overset{\circ}{u}_- \tag{5.548}$$

which shows that this flow is unstable because of negative imaginary part of λ. Due to this instability small disturbances (which always exist) will grow until their contribution to the Reynolds stress p^* stabilizes the system by eliminating the imaginary parts in (5.548). Hence, p^* must be selected such that

$$\lambda = \overset{\circ}{u}_+ \tag{5.549}$$

and therefore

$$p^* = \frac{2\rho a k}{\tanh(kH)} \overset{\circ}{u}{}_-^2 \tag{5.550}$$

i.e., the Reynolds stress discontinuity is proportional to the square of the velocity discontinuity.

Equation (5.550) can be considered as a feedback (5.528) which makes the system (5.542–5.545), (5.550) closed.

The solution to this system is

$$\phi_1 = \overset{\circ}{u}_1 x - \frac{a\, \overset{\circ}{u}_-}{\sinh(kH)} \cosh[k(y+H)]$$

$$\times \sin[k(x - \overset{\circ}{u}_+ t)], \; 0 \geqslant y \geqslant -H \tag{5.551}$$

$$\phi_2 = \overset{\circ}{u}_2 x - \frac{a\, \overset{\circ}{u}_-}{\sinh(kH)} \cosh[k(y-H)]$$

$$\times \sin[k(x - \overset{\circ}{u}_+ t)], \ 0 \leqslant y \leqslant H \tag{5.552}$$

$$p_1'' = \frac{ak \overset{\circ}{u}{}_-^2 \sinh[k(y+H)]}{\sinh(kH)\tanh(kH)}$$

$$\times \{\cos[k(x - \overset{\circ}{u}_+ t)] + |\cos[k(x - u_+ t]|\}$$

$$0 \geqslant y \geqslant -H \tag{5.553}$$

$$p_2'' = -\frac{ak \overset{\circ}{u}{}_-^2 \sinh[k(y-H)]}{\sinh(kH)\tan(kH)}$$

$$\times \{\cos[k(x - \overset{\circ}{u}_+ t)] - |\cos[k(x - u_+ t]|\}$$

$$0 \leqslant y \leqslant H \tag{5.554}$$

The Smoothing out of the Velocity Discontinuity. Let us consider the case when the system is conservative, and therefore, due to fluctuations the velocity discontinuity is smoothing out. In order to obtain a closed form solution we will assume that

$$ak \ll 1 \tag{5.555}$$

i.e., that the disturbances are given in the form of shallow waves with amplitude much smaller than the wavelength.

As will be shown below, in this case the smoothing out process is much slower than the oscillations with the frequency $k \overset{\circ}{u}_+$. Differentiating with respect to x the momentum equations (5.542), then substituting them into the solution (5.551–5.552) and averaging the resulting expressions over the period $2\pi/k \overset{\circ}{u}_+$, one obtains at $y = 0$:

$$du_1/dt = -A(u_1 - u_2)^2, \ du_2/dt = A(u_1 - u_2)^2$$

$$u_1|_{t=0} = \overset{\circ}{u}_1, \ \ u_2|_{t=0} = \overset{\circ}{u}_2 \tag{5.556}$$

in which

$$A = (2/\pi) ak^2 \coth(kH) \tag{5.557}$$

Now the averaging procedure can be justified. Indeed, the time derivative characterizing the smoothing out process has the order:

$$du/dt_1 \sim ak^2 \overset{\circ}{u}{}^2 \tag{5.558}$$

while the time derivative characterizing oscillations with the frequency ku_+ is evaluated as

$$du/dt_2 \sim k \overset{\circ}{u}{}^2 \tag{5.559}$$

and therefore

$$du/dt_2 \gg du/dt_1 \text{ if } ak \ll 1 \tag{5.560}$$

The solution to (5.556) is written as

$$u_1 = \overset{\circ}{u}_+ - \frac{1}{1/\overset{\circ}{u}_- + 4At}$$

$$u_2 = \overset{\circ}{u}_+ + \frac{1}{1/\overset{\circ}{u}_- + 4At} \tag{5.561}$$

It is easy to conclude from this solution that

$$u_1 + u_2 = \overset{\circ}{u}_1 + \overset{\circ}{u}_2 = \text{Const}$$

and

$$u_1, \ u_2 \to \overset{\circ}{u}_+ \quad \text{at } t \to \infty \tag{5.562}$$

For the mean horizontal velocity profiles one obtains:

$$\hat{u}_1 = \overset{\circ}{u}_1 - \frac{\sinh[k(y-H)]}{\sinh(kH)} \times \left(\overset{\circ}{u}_- - \frac{1}{1/\overset{\circ}{u}_- + [(8ak^2/\pi)\coth(kH)]t} \right)$$

$$0 \geqslant y \geqslant -H \tag{5.563}$$

$$\hat{u}_2 = \overset{\circ}{u}_2 + \frac{\sinh[k(y+H)]}{\sinh(kH)} \times \left(\overset{\circ}{u}_- - \frac{1}{1/\overset{\circ}{u}_- + [(8ak^2/\pi)\coth(kH)]t} \right)$$

$$0 \leqslant y \leqslant H \tag{5.564}$$

The evolution of the mean horizontal velocity field in the course of the smoothing out of the velocity discontinuity described by (5.563–5.564) has the same character as those obtained from the Prandtl solution (5.540), (Fig. 5.25). However, it is worth mentioning again, that (5.563–5.564) do not depend upon any experimental coefficients as (5.540) does.

The decay of the kinetic energy of the fluctuation velocities can be found by the substitution of the discontinuity $(u_1 - u_2)$ from (5.561) into (5.553) and (5.554). After averaging these equations over the period $2\pi/k\ \overset{\circ}{u}_+$, one gets:

$$\left(\overline{\tilde{u}_1^2 + \tilde{v}_1^2}\right) = \frac{2ak \ \sinh[k(y+H)]}{\pi \ \sinh(kH) \ \tanh(kH)} \times \left(\frac{1}{1/\overset{\circ}{u}_- + [(8ak^2/\pi)\coth(kH)]t} \right)^2$$

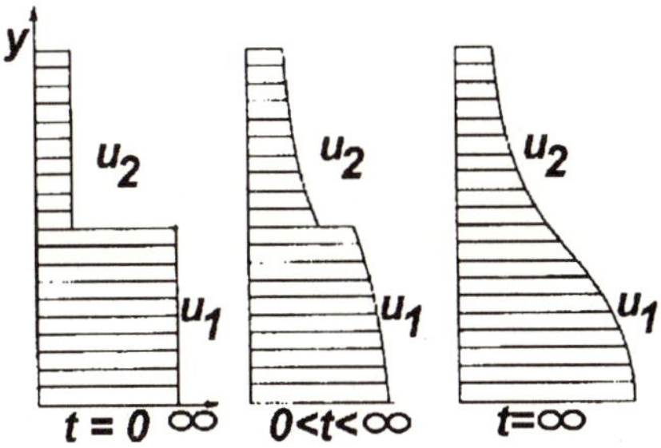

Fig. 5.25.

$$0 \geqslant y \geqslant -H \tag{5.565}$$

$$\left(\widetilde{u}_2^2 + \widetilde{v}_2^2\right) = \frac{2ak \ \sinh[k(y-H)]}{\pi \ \sinh(kH) \ \tanh(kH)}$$

$$\times \left(\frac{1}{1/\overset{\circ}{u}_- + [(8ak^2/\pi)\coth(kH)]t} \right)^2$$

$$0 \leqslant y \leqslant -H \tag{5.566}$$

Example 2 Analysis of Turbulence in Shear Flows

Introduction. If we continue our study to flows whose instability can be found from linear analysis (plane Poiseuille flow , boundary layers), then the closure problem can be formulated as follows: let the original laminar flow described by the Navier-Stokes equations be unstable, i.e. some of the eigenvalues for the corresponding Orr-Sommerfeld equation have positive imaginary parts. Then, the closure is found from the condition that all these positive imaginary parts vanish, and therefore, the solution possesses a neutral stability. However, the closure (5.532) can be written in the explicit form only if the criteria for the onset of instability are formulated explicitly. Since such a situation is an exception rather than a rule, one can apply a step-by-step strategy proposed in Sect. 5. This strategy is based upon the fact that the Reynolds stress disturbances grow much faster than the mean motion disturbances (5.510). Hence, one can assume that these stresses will be large enough to stabilize the mean flow which is still sufficiently close to its original unperturbed state. But the Reynolds stresses being substituted in the Reynolds equations will change the mean velocity profile, and consequently, the conditions of instability. These new conditions, in turn, will change the Reynolds stresses etc. By choosing the iteration steps to be sufficiently small, one can obtain acceptable accuracy. In this example the first step approximation will be applied to a plane Poiseuille flow (Zak 1988).

Formulation of the Problem. Let us consider a plane shear flow with a dimensionless velocity profile:

$$\overline{U} = \overline{U}(y), \; 0 \leqslant y \leqslant 1 \tag{5.567}$$

with boundaries

$$y_1 = 0, \; y_2 = 1 \tag{5.568}$$

and the x coordinate being along the axis of symmetry. The stream function representing a single oscillation of the disturbance is assumed to be of the form

$$\psi(x, y, t) = \varphi(y) e^{i(ax - \beta t)} \tag{5.569}$$

The function $\varphi(y)$ must satisfy the Orr-Sommerfeld equation:

$$(U - C)(D^2 - \alpha^2)\varphi - U''\varphi = (i\alpha \, R)^{-1}(D^2 - \alpha^2)^2\varphi \tag{5.570}$$

in which α and β are constants, R is the Reynolds number and

$$C = \frac{\beta}{\alpha} = C_r + C_i \tag{5.571}$$

and

$$D\varphi = \frac{d\varphi}{dy} = \varphi' \tag{5.572}$$

Equation (5.570) should be solved subject to the boundary conditions, which in the case of a symmetric flow between rigid walls are

$$\varphi = D\varphi = 0 \text{ at } y = y_2, \; D\varphi = D^3\varphi = 0 \text{ at } y = y_1 \tag{5.573}$$

We will start with the velocity profile characterized by the critical Reynolds number:

$$R = R_{\text{cr}} \tag{5.574}$$

Any increase in velocity when

$$R^* > R_{\text{cr}} \tag{5.575}$$

leads to instability of the laminar flow and to transition to a new turbulent flow.

We will concentrate our attention on the situation when the increase in the Reynolds number is sufficiently small,

$$\frac{R^* - R_{\text{cr}}}{R_{\text{cr}}} \ll 1 \tag{5.576}$$

In this case we will be able to formulate a linearized version of the closure (5.532) explicitly based upon the conditions of the instability of the Orr-Sommerfeld equation written for $R = R_{\text{cr}}$ and to obtain the mean velocity profile and Reynolds stress for the corresponding turbulent flow.

Generalized Orr-Sommerfeld Equation. In order to apply the stabilization principle and formulate the closure problem we have to incorporate the Reynolds stresses into the Orr-Sommerfeld equation. For this purpose let us start with the Reynolds equations for a plane shear flow expressed in terms of small perturbations:

$$\frac{\partial \widetilde{U}}{\partial t}+U\frac{\partial \widetilde{U}}{\partial x}+\widetilde{V}\frac{dU}{dy}+\frac{1}{\rho}\frac{\partial \widetilde{P}}{\partial x}=\nu\nabla^2\widetilde{U}+\frac{\partial \widetilde{\tau}}{\partial y} \tag{5.577}$$

$$\frac{\partial \widetilde{V}}{\partial t}+U\frac{\partial \widetilde{V}}{\partial x}+\frac{1}{\rho}\frac{\partial \widetilde{P}}{\partial y}=\nu\nabla^2\widetilde{V}+\frac{\partial \widetilde{\tau}}{\partial x} \tag{5.578}$$

and

$$\frac{\partial \widetilde{U}}{\partial x}+\frac{\partial \widetilde{V}}{\partial y}=0 \tag{5.579}$$

using the boundary layer approximation. Here $U(y)$ is the mean velocity profile, and $\widetilde{U}, \widetilde{V}$, and $\widetilde{P}$ are small velocity and pressure perturbations, ν is the kinematic viscosity and $\widetilde{\tau}$ is the shearing Reynolds stress which is sought in the form

$$\widetilde{\tau}=\widetilde{\tau}(y)e^{i(\alpha x-\beta t)} \tag{5.580}$$

Introducing (5.569) and (5.580) into (5.577–5.579) we obtain, after the elimination of pressure, the generalized Orr-Sommerfeld equation in dimensionless form:

$$(\overline{U}-C)(D^2-\alpha^2)\varphi-\overline{U}''\varphi-(i\alpha\ R)^{-1}(D^2-\alpha^2)^2\varphi=-\frac{i}{\alpha}(D^2+\alpha^2)\tau \tag{5.581}$$

in which

$$\tau=\frac{\widetilde{\tau}}{\rho U_{\max}^2} \tag{5.582}$$

It contains an additional term on the right hand side: the Reynolds stress disturbance, as yet unknown.

The Closure Problem. Returning to our problem, let us apply (5.581) to the case when

$$R=R^*,\ U=U(y) \tag{5.583}$$

Substituting (5.583) into (5.581), one obtains

$$(U-C)(D^2-\alpha^2)\varphi-\overline{U}''\varphi-(i\alpha\ R)^{-1}(D^2-\alpha^2)^2\varphi=-\frac{i}{\alpha}(D^2+\alpha^2)\tau \tag{5.584}$$

With zero Reynolds stress ($\tau=0$), (5.584) would have eigenvalues with positive imaginary parts since $R^*>R_{\mathrm{cr}}$. These positive imaginary parts of the eigenvalues would vanish if R^* is replaced by R_{cr}. Hence, according to the stabilization principle, the Reynolds stresses should be selected such that (5.584) is converted to (5.570) at $R = R_{\text{cr}}$, i.e.

$$-\frac{i}{\alpha}(D^2 + \alpha^2)\tau + (i\alpha\, R^*)^{-1}(D^2 - \alpha^2)^2\varphi = (i\alpha\, R_{\text{cr}})^{-1}(D^2 - \alpha^2)^2\varphi$$

or

$$(D^2 + \alpha^2)\tau = \left(\frac{1}{R_{\text{cr}}} - \frac{1}{R^*}\right)(D^2 - \alpha^2)^2\varphi \tag{5.585}$$

Equation (5.585) relates the disturbance of the mean flow velocity and the Reynolds stress τ. It allows us to reproduce a linearized version of the closure (5.573):

$$\overline{\tau}'' + \alpha^2\overline{\tau} = \left(\frac{1}{R_{\text{cr}}} - \frac{1}{R^*}\right)\left(\overline{\psi}''\,'' - 2\alpha^2\overline{\psi}'' + \alpha^4\overline{\psi}\right) \tag{5.586}$$

in which $\overline{\tau}$ and $\overline{\psi}$ are the dimensionless Reynolds stress and the stream function characterizing the unperturbed flow (for instance, $\psi = -\partial U/\partial X$). Indeed, after perturbing (5.586) and substituting equations (5.569) and (5.580), one returns to (5.585).

It is important to emphasize that (5.586) is not universal closure: it contains two numbers (R_{cr} and α) which characterize a particular laminar flow. Here R_{cr} is the smallest value of the Reynolds number below which all initially imparted disturbances decay, whereas above this value those disturbances which are characterized by α [see (5.569) and (5.580)] are amplified. Both of these numbers can be found from (5.570) as a result of classical analysis of hydrodynamics stability performed for a particular laminar flow. One should recall that closure (5.586) implies a small increment of the Reynolds number over its critical value [see (5.576)]. For large increments the procedure must be performed by steps: for each new mean velocity profile (which is sufficiently close to the previous one) the new R'_{cr} and α' are supposed to be found from the solution of the eigenvalue problem for the Orr-Sommerfeld equation. Substituting R'_{cr} and α' into the closure (5.586) and solving it together with the corresponding Reynolds equations, one finds the mean velocity profile and the Reynolds stress for the next increase of the Reynolds number $R^{*\prime}$, etc.

Plane Poiseuille Flow. In this section we will apply the approach developed above to a plane Poiseuille flow with the velocity profile

$$\overline{U}^0(y) = 1 - y^2 \tag{5.587}$$

and

$$R_{\text{cr}} = 5772.2, \ \alpha = 1.021 \tag{5.588}$$

As a new (supercritical) Reynolds number we will take

$$R^* = 6000 \tag{5.589}$$

The closure (5.586) should be considered together with the governing equation for the unidirectional mean flow:

$$\nu\overline{U}'' + \overline{\tau}' = C = \text{Const} \tag{5.590}$$

or

$$\nu\overline{U}' + \overline{\tau} = \overline{C}_1 y + \overline{C}_2 \tag{5.591}$$

The constants $\overline{C}_1$, and $\overline{C}_2$, can be found from the condition

$$\overline{\tau} = 0 \text{ at } y = 1 \text{ and } y = 0 \tag{5.592}$$

expressing the fact that the Reynolds stress vanishes at the rigid wall and in the middle of the flow.

Hence,

$$\overline{C}_2 = 0 \tag{5.593}$$

since $\overline{U}' = 0$ at $y = 0$ and

$$\overline{C}_1 = \nu\overline{U}_1' \ (\overline{U}_1' = \overline{U}' \text{ at } y = 1) \tag{5.594}$$

Thus

$$\overline{\tau} = \nu(U_1' y - U')$$

or in dimensionless form,

$$\overline{\tau} = \frac{1}{R^*}(\overline{U}_1' y - \overline{U}') \tag{5.595}$$

Substituting (5.595) into the closure (5.586) one obtains the governing equation for the mean velocity profile in terms of the stream function $\overline{\psi}$ while $\overline{U} = \partial\psi/\partial y$:

$$\frac{1}{R_{\text{cr}}}\overline{\psi}''\,'' - \alpha^2\left(\frac{2}{R_{\text{cr}}} - \frac{1}{R^*}\right)\psi'' - \alpha^4\left(\frac{1}{R_{\text{cr}}} - \frac{1}{R^*}\right)\psi = \frac{\alpha^2}{R^*}\overline{\psi}_1'' y \tag{5.596}$$

in which $\overline{\psi}_1'' = \overline{\psi}''$ at $y = y_1$.

Without loss of generality it can be set

$$\psi|_{y=0} = 0 \tag{5.597}$$

Since at the rigid wall $\overline{U} = 0$, one obtains

$$\psi_1' = 0 \tag{5.598}$$

In the middle of the flow due to symmetry:

$$U_0' = 0, \text{ i.e., } \psi_0'' = 0 \tag{5.599}$$

Finally, the flux of the turbulent flow should be the same as the flux of the original (unstable) laminar flow:

$$\psi_1 = \int_0^1 (1 - y^2) dy = \frac{2}{3} \tag{5.600}$$

These four (non-homogeneous) boundary conditions (5.597–5.600) allow one to find four arbitrary constants appearing as a result of integration of (5.596). After substituting the numbers [see (5.588–5.589)] one arrives at the following differential equation:

$$\overline{\psi}''\,'' - 1.08202\overline{\varphi}_1''y - 0.04124\overline{\varphi} = 1.044\overline{\psi}_1''y \tag{5.601}$$

whence

$$\overline{\psi} = C_1 \sin 0.19199y + C_2 \cos 0.19199y + C_3 \sinh 0.19199y$$

$$+ C_4 \cosh 0.19199y - 25.3152\psi_1''y \tag{5.602}$$

Applying conditions (5.597) and (5.599) one finds that

$$C_2 = C_4 = 0 \tag{5.603}$$

Taking into account that

$$\overline{\psi}_1'' = -0.00703C_1 + 0.00712C_3 \tag{5.604}$$

one obtains

$$\overline{\psi} = C_1 \sin 0.19199y + C_3 \sinh 0.19199y + (0.17797C_1 - 0.18024C_3)y$$

Now applying conditions (5.598) and (5.600) one arrives at the following solution:

$$\overline{\psi} = 11.278 \sin 0.19199y - 270.11 \sinh 0.19199y + 50.692y \tag{5.605}$$

and therefore,

$$\overline{U} = 2.1653 \cos 0.19199y - 51.8584 \cosh 0.19199y + 50.692 \tag{5.606}$$

Substituting solution (5.605) into (5.595) one obtains the Reynolds stress profile:

$$R^*\tau = 0.41572 \sin 0.19199y + 9.9563 \sinh 0.19199y - 2.00259y \tag{5.607}$$

Analysis of the Solution. We will start with the comparison of the original laminar velocity profile (5.587) and the mean velocity profile (5.605). Both of them envelop the same area, i.e., the fluxes of the original laminar and post-instability turbulent flows are the same. However, the maximum turbulent mean velocity is smaller than the maximum of the original laminar flow:

$$\overline{U}_{\max} = 0.9989 < \overline{U}_{\max} = 1 \tag{5.608}$$

Also,

$$\left|\overline{U}_0''\right| = 1.99132 < \left|\overline{U}_0^{\circ\prime\prime}\right| = 2 \tag{5.609}$$

At the same time

$$\left|\overline{U}_1'\right| = 1.99132 > \left|\overline{U}_1'\right| = 2 \tag{5.610}$$

Hence, the turbulent mean velocity profile is more flat at the centre and more steep at the walls in comparison with the corresponding laminar flow. This property is typical for turbulent flows.

Turning to the Reynolds stress profile (5.606) one finds that the maximum of the stress module $|\tau|$ is shifted toward the wall:

$$y^* = 0.58 \tag{5.611}$$

which expresses the well-known wall effect.

Finally, the pressure gradient

$$\frac{\partial \overline{p}}{\partial x} = \frac{1}{R^*}\overline{U}_0'' + \overline{\tau}_0' \tag{5.612}$$

for the new turbulent flow is greater than for the original laminar flow:

$$\left|\frac{\partial \overline{p}}{\partial x}\right| = \frac{2.002586}{R^*} > \frac{2}{R^*} = \left|\frac{\partial \overline{p}^\circ}{\partial x}\right| \tag{5.613}$$

Thus, despite the fact that the Reynolds number R^* slightly exceeds the critical value R_{cr}, all the typical features of turbulent flows are clearly pronounced in the solution obtained above.

Thus, it has been demonstrated again that the closure in turbulence theory is based upon the principle of stabilization of the original laminar flow by fluctuation velocities. We will stress again the mathematical meaning of this procedure.

It is well known that the concept of stability is related to a certain class of functions: a solution which is unstable in a class of smooth functions can be stable in an enlarged (non-smooth) class of functions. Reynolds enlarged the class of smooth functions by introducing the field of fluctuation velocities which generated additional (Reynolds) stresses in the Navier-Stokes equations. Now it is reasonable to extend this procedure by choosing these Reynolds stresses such that they eliminate the original instability, i.e., by applying the stabilization principle.

It should be emphasized that one cannot expect that the solution of the type (5.563–5.566) or (5.606) will describe all the peculiarities of turbulent motion; it will rather extract the most essential properties of the motion, i.e., such properties which are reproducible. and therefore, have certain physical meaning. Description of finer details of turbulent motions will require further enlarging the class of functions which may require exploitation of higher order approximations [see (5.502–5.503)]. However, all new closures must be based upon the stabilization principle which will provide the reproducibility of the turbulence structure.

It should be expected that more accurate results could be obtained by elimination of global instability. However, since global instability is very unlikely to be formulated analytically, only numerical procedures can be actually useful for this purpose.

5.5 Stabilization Using Minimum Entropy

In this section we will introduce a variational formulation of the stabilization principle which gives the most general approach to the closure problem, and in particular, minimizes uncertainties in the postinstability models caused, for instance, by overstabilization.

Let us start with an autonomous first order nonlinear system:

$$\dot{x}_i = f_i(x), \quad x = x_1, ..., x_n \tag{5.614}$$

and introduce the error vector ε

$$\dot{\varepsilon} = f_i(x + \varepsilon), \quad \varepsilon_0 = \varepsilon \text{ at } t = 0 \tag{5.615}$$

with the norm $||\varepsilon||$.

The Liapunov exponent along a solution

$$x = x(t) \tag{5.616}$$

in the direction defined by initial error vector ε_0 is:

$$\sigma[\varepsilon_0, x(t)] = \lim_{\substack{t\to\infty \\ ||\varepsilon_0||\to 0}} \frac{1}{t} \ln \frac{||\varepsilon||}{||\varepsilon_0||} \tag{5.617}$$

To circumvent the overflow, one can choose a small fixed interval τ and renormalize ε to a norm of unity. Then:

$$\sigma_n = \frac{1}{n\tau} \sum_{i=1}^{n} \ln d_i \tag{5.618}$$

The values

$$d_k = ||\varepsilon_{k-1}(\tau)||, \quad \varepsilon_k(0) = \frac{\varepsilon_{k-1}(\tau)}{d_k} \tag{5.619}$$

are iteratively computed. Here $\varepsilon_k(\tau)$ is obtained by integrating (5.615) with the initial value $\varepsilon_k(0)$ along the trajectory from

$$x(k\tau) \text{ to } x[(k+1),\tau]$$

It can be shown that the Liapunov exponent

$$\sigma = \lim_{n\to\infty} \sigma_n \tag{5.620}$$

always converges to a single value.

The interval τ can be chosen as:

$$\tau \sim \tau_0 N \tag{5.621}$$

where τ_0 is the integration time-step with N-bits precision.

Equation (5.621) results from the fact that for a chaotic motion, the initial condition is lost completely after a critical number of the iterations τ given by (5.621).

The objective of the stabilization is to suppress all positive Liapunov exponents by zeros:

$$\sigma = \lim_{n\to\infty} \sigma_n = 0 \tag{5.622}$$

if, without stabilization

$$\sigma > 0 \tag{5.623}$$

However, there are many different stabilization strategies which lead to the same result (5.622), since the global Liapunov exponent is defined only at

$$t \to \infty$$

Let us turn to the origin of the stabilization principle. Actually it states that the fluctuation can grow only until the instability exists, and they stop growing as soon as the instability is eliminated. Since the fluctuations represent a measure of uncertainty, one can reinterpret the stabilization principle as the principle of minimum uncertainty, or a minimum of entropy subject to the constraint (5.622).

For the probability density $\varphi(x)$, the principle is formulated as:

$$H = -\int_{-\infty}^{\infty} f(x)\log_2 f(x)dx \to \text{min at } \sigma = 0 \tag{5.624}$$

or if (5.614) is represented in the Reynolds form

$$\dot{\overline{x}} = f(\overline{x}) + R \tag{5.625}$$

where R is the Reynolds force, then

$$E = \frac{1}{n\tau}\sum_{i=1}^{n} \|R_i\| \to \text{min at } \sigma = 0 \tag{5.626}$$

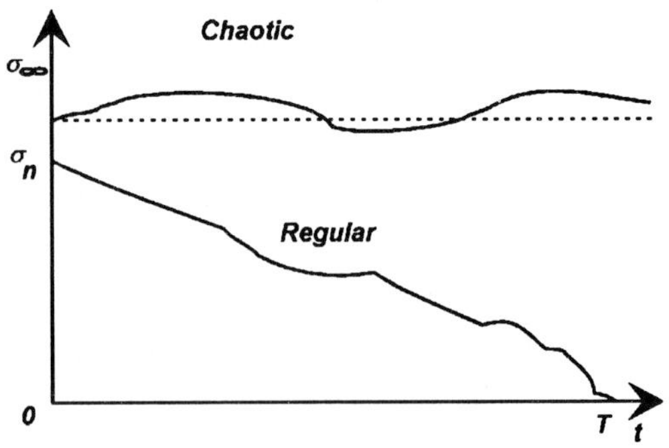

Fig. 5.26.

The formulation (5.626) can be linked to the Gauss principle of minimum constraints.

Before formulating the numerical strategy, note that as follows from computational results, for chaotic motions positive Liapunov exponents [see (5.618)] approach their limits as in Fig. 5.26, while for regular motions which are the "closest" to chaotic, they drop linearly to zero. Here t is the total time of computations. Based upon this qualitative property, one can formulate the computational strategy as follows:

$$\dot{x}' = \begin{cases} f(x) \text{ if } & \sigma_n \leq \sigma_0^{(i)} \\ 0 \text{ if } & \sigma_n > \sigma_0^{(i)} \end{cases} \tag{5.627}$$

[By $\dot{x}'$ we denote those equations which are controllable, i.e., which are nonlinear. Linear equations do not contain the Reynolds forces and they remain in the form (5.614)].

Here σ_n is given by (5.618), and

$$\sigma_0^{(i)} = \frac{t - T}{\overset{*}{T}{}^{(i)} - T} \overset{*}{\sigma}{}^{(i)}, \quad \overset{*}{\sigma}{}^{(i)} = \sigma(\overset{*}{T}{}^{(i)})$$

(see Fig. 5.27). Thus, the control strategy is defined by the parameter $\overset{*}{T}$, i.e., by the time when the linear control of the solution started. The solution should be computed for several $\overset{*}{T}$ (independently), and the one which has the minimum uncertainty (5.626) is the optimal one. These solutions can be computed simultaneously if parallel processing is applied. For further optimization, one can use a nonlinear control between $\overset{*}{T}$ and T using minimization of the uncertainty (5.626) with respect to additional parameters

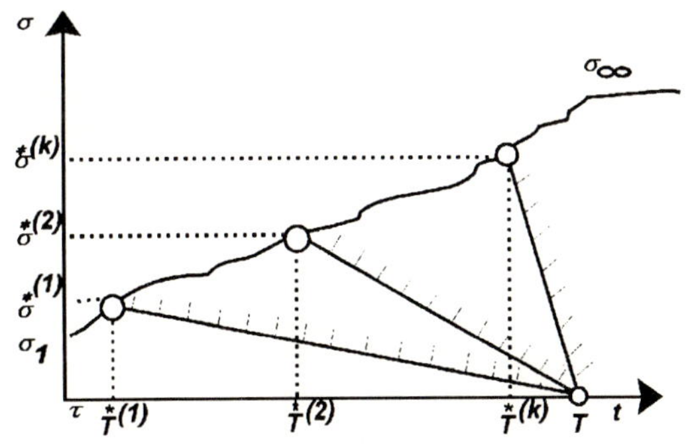

Fig. 5.27.

characterizing the nonlinear curve

$$\sigma_0^{(i)} = F(\overset{*}{T}^{(i)}, a^{(i)}, ...) \tag{5.628}$$

The strategy for the higher order moments now can be developed as following

Consider a dynamical system with quadratic nonlinearities:

$$\dot{x}_i = a_{ij}x_j + a_{ijk}x_jx_k, ... \; i = 1, 2, ..., n \tag{5.629}$$

Due to uncertainties in initial conditions amplified by instabilities, the solution becomes multivalued

$$x_i = \varphi(\xi_1, ..., \xi_n), \;\; 0 \leq \xi_i \leq 1 \tag{5.630}$$

where φ can be considered as a joint distribution of possible values of the vector x_i.

The decomposition:

$$x_i = \overline{x}_i + \widetilde{x}_i, \; \overline{x}_i = \int_0^1 \varphi(x_i)d\xi_1, ..., d\xi_n \tag{5.631}$$

transforms (5.629) to the following:

$$\dot{\overline{x}} = a_{ij}\overline{x}_j + a_{ijk}\overline{x}_j\overline{x}_k + a_{ijk}\overline{x}_{jk}, \; \overline{x}_{jk} = \int_0^1 \varphi x_j x_k d\xi_1, ..., d\xi_n \tag{5.632}$$

where $\overline{x}_i$ and $\overline{x}_{jk}$ are the vectors of expectations and the dispersion tensor for the original (random) vector x_i [compare with (5.498)].

As follows from (5.632), the tensor $\overline{x}_{jk}$ must be symmetric

$$\overline{x}_{jk} = \overline{x}_{kj} \tag{5.633}$$

and semipositive (since $\overline{x}_{ii} \geqslant 0$), i.e., all its diagonal minors must be semipositive

$$\det{}_i \begin{pmatrix} x_{11} & \cdot & \cdot \\ \cdot & x_{ii} & \cdot \\ \cdot & \cdot & x_{nn} \end{pmatrix} \geqslant 0, \; i = 1, 2, ..., n \tag{5.634}$$

The stabilization principle imposes m additional constraints upon the dispersion tensor via the Reynolds forces

$$a_{ijk}\overline{x}_{jk} = \beta_i, \; i = 1, 2, ..., m, \; m \leqslant n \tag{5.635}$$

For $n > 1$, the constraints (5.633–5.635) are not sufficient to uniquely define the dispersion tensor. Hence, we can exploit such an arbitrariness to our advantage by requiring that the tensor introduces minimum uncertainty, i.e., invariant

$$I_1^{(2)} = \sum_{i=1}^{n} \overline{x}_{ii} \to \min \tag{5.636}$$

subject to the constraints (5.633–5.635).

Equations (5.633–5.636) uniquely define the dispersion tensor $\overline{x}_{ij}$. Multiplying (5.629) by x_k and averaging and combining the results, we obtain the following

$$\dot{\overline{x}}_{ik} = a_{ij}\, \overline{x}_{jk} + a_{jk}\overline{x}_{ij} + a_{ijm}(\overline{x}_{kjm} + \overline{x}_{kj}\overline{x}_m + \overline{x}_{km}\overline{x}_j + \overline{x}_{jm}\overline{x}_k)$$
$$+ a_{jmk}(\overline{x}_{ijm} + \overline{x}_{ij}\overline{x}_m + \overline{x}_{im}\overline{x}_j + \overline{x}_{jm}\overline{x}_i) \tag{5.637}$$

which express the third order correlations via the second ones.

The third order correlations form a symmetric tensor of the third rank $\overline{x}_{ijk}$ while

$$x_{ijk} = x_{kji} = x_{jik} =, ..., \text{etc.} \tag{5.638}$$

For $n = 3$, six equations (5.637) contain up to ten different components of the tensor $\overline{x}_{ijk}$. Again this freedom can be exploited to our advantage by applying the same minimum uncertainty principle. For this one can introduce the following invariant for minimization:

$$I_1^{(3)} = \left| \sum_i x_{iij} \right| = \left(\sum_i x_{ii1} \right)^2 + \left(\sum_i x_{ii2} \right)^2 + \left(\sum_i x_{ii3} \right)^2 \to \min \tag{5.639}$$

subject to constraints (5.637–5.638).

Consider a general nonlinear system instead of (5.629):

$$\dot{x} = f(x), \; x = x_1, x_2, ..., x_n \tag{5.640}$$

Rewrite (5.640) in an identical form:

$$\dot{x} = f(\overline{x}) + f(x) - f(\overline{x}) \tag{5.641}$$

since

$$f(x) = f(\overline{x}) + \left.\frac{\partial f}{\partial x^2}\right|_{x=\overline{x}} (x - \overline{x}) + \frac{1}{2}\frac{\partial^2 f}{\partial x^2}(x - \overline{x})^2 +, ..., \text{etc.} \tag{5.642}$$

Equation (5.641), after averaging, reduces to

$$\dot{\overline{x}} = f(x) = f(\overline{x}) + \frac{1}{2}\left.\frac{\partial^2 f}{\partial x^2}\right|_{x=\overline{x}} (x - \overline{x})^2 + \frac{1}{2}\frac{\partial^3 f}{\partial x^3}(x - \overline{x})3 +, ..., \text{etc.} \tag{5.643}$$

Thus, in general, (5.641) contains all the higher order moments. The strategy of finding these moments can be the same with the only complication being that now all the equations (5.633–5.639) are coupled.

5.5.1 *Numerical Example*

Consider the Lorenz attractor:

$$\begin{array}{rl|cl} \dot{x}_1 = & -\sigma x_1 + \sigma x_2 & & \dot{\overline{x}}_1 = -\sigma\overline{x}_1 + \sigma\overline{x}_2 \\ \dot{x}_2 = & -x_1x_3 + rx_1 - x_2 & \rightarrow & \dot{\overline{x}}_2 = -\overline{x}_1\overline{x}_3 + r\overline{x}_1 - \overline{x}_2 - \overline{x}_{13} \\ \dot{x}_3 = & x_1x_2 - bx_3 & & \dot{\overline{x}}_3 = x_1x_2 - bx_3 + \overline{x}_{12} \end{array} \tag{5.644}$$

Now the constraints (5.633) are:

$$\overline{x}_{12} = \overline{x}_{21},\ \overline{x}_{13} = \overline{x}_{31},\ \overline{x}_{23} = \overline{x}_{32} \tag{5.645}$$

the constraints (5.634) are:

$$\overline{x}_{12}\overline{x}_{22} - \overline{x}_{12}^2 \geqslant 0,\ \overline{x}_{11}\overline{x}_{33} - \overline{x}_{13}^2 \geqslant 0,\ \overline{x}_{22}\overline{x}_{33} - \overline{x}_{23}^2 \geqslant 0 \tag{5.646}$$

and the constraints (5.635) are:

$$\overline{x}_{13} = \beta_{13},\ \overline{x}_{12} = \beta_{12} \tag{5.647}$$

As follows from (5.646)

$$\overline{x}_{22} = \frac{\overline{x}_{12}}{\overline{x}_{11}},\ \overline{x}_{33} = \frac{\overline{x}_{13}^2}{\overline{x}_{11}}$$

Hence

$$\overline{x}_{11} + \overline{x}_{22} + \overline{x}_{33} = \overline{x}_{11} + \frac{\beta_{12}}{\overline{x}_{12}} + \frac{\beta_{13}}{\overline{x}_{11}} = E$$

and therefore

$$E \rightarrow \min,\ \text{if } 1 - \frac{\beta_{12}^2}{\overline{x}_{11}^2} - \frac{\beta_{13}^2}{\overline{x}_{11}^2} = 0,$$

i.e.,

$$\overline{x}_{11} = \sqrt{\beta_{12}^2 + \beta_{13}^2}, \ \overline{x}_{33} = \frac{\beta_{13}^2}{\sqrt{\beta_{12}^2 + \beta_{13}^2}},$$

$$\overline{x}_{22} = \frac{\beta_{12}^2}{\sqrt{\beta_{12}^2 + \beta_{13}^2}}, \ \overline{x}_{23} = \frac{\beta_{12}\beta_{13}}{\sqrt{\beta_{12}^2 + \beta_{13}^2}} \tag{5.648}$$

Thus, as shown in Example 1 [see (5.521–5.522)], the application of the stabilization principle in the form (5.501) allows one to find only two components of the velocity correlations, $\overline{x}_{12}$ and $\overline{x}_{13}$, i.e., those which are explicitly enter the governing equations (5.644). Then, based upon the variational formulation of the stabilization principle, the rest of the components are found from (5.648) as a result of the minimization of uncertainties.

Part II

Terminal (Non-Lipschitz) Dynamics in Biology and Physics

6
Biological Randomness and Non-Lipschitz Dynamics

Pluritas non est ponenda sine necessitate. –*William of Occam*

6.1 Introduction

In the kind of curious confusion which is often found in historical anecdotes, the above statement has come down to us as "Occam's Razor." It is curious because if Occam was not in the context it has come to be known; namely, the "Law of Parsimony": that being should not be multiplied without reason (Thorburn 1915). The original intent of Occam was to set forth a rule of methodology – not a rule of existence. And yet, modern science seems to have confused the two. Even Newton himself quoted this dictum in several editions to his *Principia.*

The original fervent interest in deterministic chaos was in part due to the finding that very complicated patterns could be developed out of relatively simple equations – a finding that seemed to confirm the veracity of the dictum. But in the decades since it has become evident that the reality is not so simple. Indeed, *pluritas est ponenda sine necessitate* – being is multiplied without [apparent] necessity. The simple kinds of chaos have been supplanted by various forms. Furthermore, a new area, termed "complexity theory" has developed which attempts to define the putative underlying rules for organization.

Part of the difficulty in dealing with time honored dictums (in this case, it can be traced to Aristotle in his *Physics*, Book I, chapter vi), is their apparent truthfulness and logic. The key word here is logic. For although logic and reason are the tools scientists use, the object of their inquiry – Nature – often fails to be bound by these rules. A careful reading of the history of sciences over the last century and half reveal just as many failures as successes – or at least paradoxes. Consider the wave/particle duality, among others. It might be fair to say that Nature is not logical – or, at least, it is alogical.

What has become even more interesting is the revelation that the previously unwanted, messy quantity, "noise," has become an important factor in various dynamics. And it is not the pseudorandom noise of an algorithm, or calculated probabilities of distributions, but the material noise of the real world.

Nowhere better can this be seen, than in the workings of biological processes. From the first attempts to apply physical laws to the dynamics of physiology, it has been a slow road to the appreciation that their refinement includes significant alterations to traditional descriptions of physical laws. Consider, for example, the area of cardiovascular rheology. The laws of hydrodynamics have been applied to circulation physiology, but in most respects they have had to be refined each step of the way because of the complications provided by the different composition of blood, as well as the dynamic elasticity of the blood vessels themselves.

It is not surprising, then, that some of the more robust uses of deterministic dynamics have occurred in areas which are characterized by some sort of fluid dynamics. For example, changes in renal tubular pressure regulation appear to be associated with strange attractors under pathologic circumstances (Holstein-Rathlou 1993). Also, mixing in terminal acini of the lungs may be characterized by chaotic dynamics (Tsuda et al. 1995).

The picture is not so clear with other systems. Heart rate variability has long been touted as a chaotic system, but several careful investigations seriously put this proposition into question (Zbilut et al. 1991, Zbilut et al. 1995, Zbilut et al. 1996a; Kanters et al. 1994, Kanters et al. 1996). Given the need for beat-to-beat adaptability, it is questionable that a deterministic system could provide the necessary flexibility. Studies purporting control of "cardiac chaos" such as ventricular fibrillation by adjusting unstable fixed points have not yet demonstrated unequivocal proof of deterministic chaos. Indeed, it can be argued that the waves seen in ventricular fibrillation are aperiodic or quasiperiodic, and that the adaptive control methods are a variation of methods suggested by Kapitsa and Landau and Lifschitz (1959). Certainly, similar concerns can be made for claims of chaotic dynamics with breathing (Webber and Zbilut 1996), as well as the electroencephalogram (Pritchard et al. 1995).

In the case of the EEG a problem lies with the fact that it is a surface representation of millions of signals of neurons, which are, in effect, modified by passage through the body, and processed by the acquisition system.

Unlike the ECG, the EEG is not associated with a discrete event such as muscular contraction.

An interesting fact is that those systems which are discrete (e.g., heartbeat, a scroll wave in excitable media), often have a singularity. Such systems have been extensively studied by Winfree (1987), although primarily from the viewpoint of phase dynamics and related topology. Indeed, it might be argued that such systems cannot be dominated by classical deterministic chaos. Instead, the singularity allows for a unique "punctuation" to the dynamics, where according to non-Lipschitz dynamics, control is easily achieved. This is not to say that there has been much progress in the understanding of the fundamental processes underlying these phenomena; however, an important factor is often missing in these simulations, namely the electromechanical feedback of a living tissue which constantly alters the involved ionic conductances (Lab 1996).

6.1.1 Adaptability

What was novel and interesting has now become a regular activity: the search for fractal dimensions, positive Liapunov exponents, and fuzzy phase plane plots. That this should be so, is not surprising. Chaotic characterizations often have the ability to reduce the myriad complex dynamics of interconnected systems into one relatively simple conceptualization; so that what seems untenably difficult to understand becomes instead a relatively simple process of low dimensional strange attractors. With the advent of experimental algorithms to measure dimensions and exponents, as well as computers fast and cheap enough so as to be available to almost any laboratory, it is a commonplace to suggest that chaotic dynamics abound in physiological systems.

Amidst this activity has been a growing realization among many dynamicists, that the algorithms themselves have numerous drawbacks, including questions of noise level, and stationarity which prohibit their uncritical application. Even if these drawbacks were resolved, a more fundamental problem lies in their application: by definition, chaotic systems are fully determined by initial conditions, and they are unpredictable only because of our lack of precision. Whereas such conditions can be fully realizable in the physical world, they are difficult to demonstrate in complex biological organisms. Certainly, one of the main features of living organisms is that they are adaptive throughout a range of time scales. To suggest that biologic dynamics are fully characterized by attractors, strange or otherwise, would limit this adaptability.

It may be argued that these objections can be countered by the existence of "control parameters," which can retune the system. As yet, however, there has been no demonstration of where such a control apparatus can exist physically. Furthermore, given the tremendous amounts of noise in biological systems, and the extreme sensitivity to initial conditions of chaotic systems, the energy expended to run adaptive controllers would be considerable. Certainly, it has even been pointed out that many of the so-called negative feedback systems do not exist when subjected to experimental verification. Additionally, the stability of such systems would come into question, since often infinitesimal differences in control parameters can result in enormously different effects. Consider, for example, the implications for the conduction system of the heart, which must work relatively flawlessly for decades.

Indeed, one of the main problems of chaos theory with respect to biological systems is its lack of ability to explain the singular points of physiological processes where repetitive, stereotypical (orbital) behavior is encountered. This is to say, what determines whether a neuron fires, the heart beats, or the lung inflates. Certainly, the qualitative features of the firing can be well modeled, but the pauses between events cannot. Even the use of stochastic differential equations encounter considerable difficulty, especially when their time dependence is considered. This is not even to consider the separate issue of noise characterization: do biological organisms harbor different kinds of noise generators? All of these points suggest that nonlinear chaotic dynamics fail to describe many biological systems in one of their most important points; namely, adaptability.

An alternative opinion is that the main difficulty is not the nonlinear dynamics per se, but in one of the time honored conditions for differential equations. Specifically, it is pointed out that uniqueness criteria are required for the solution of differential equations. In fact there is no absolute need to require such conditions. Certainly, uniqueness makes things much more neat and tidy from a mathematical standpoint, but it might be counterproductive to adaptability. It is also noted that such criteria may be historically bound up in philosophical and theological questions of "determinism." The previous chapters already substantiate the usefulness and tractability of such an approach.

One of the most fundamental reasons to consider non-Lipschitz dynamics[1] is that it may be necessary within the context of biological signaling. Consider, for example typical neuronal signals: an analogue signal is first produced through the basic physical processes of channel gating through ions and other transductions (quantum?), the result, however, is a pulse of such relative magnitude that it may be considered a digital signal. And certainly, this fits in with the general theory of coding: analogue signals are easily corruptible, whereas digital codes are easily retrievable from a noisy environment (Yockey 1992). When passing over a relatively long distance, digital codes are preferable; whereas the relatively local environment

[1] For the present, we consider these dynamics under the general term "non-Lipschitz." In the next chapter we identify these dynamics specifically as "terminal," within the context of a more formal theory of their behavior.

does not require the enhancement of digital code. Deterministic dynamics cannot describe this transduction process: it takes an infinite time of approaching a singular point, and if a pulse is created, its length and timing may be arbitrary (see Chap. 7). Only non-Lipschitz dynamics can handle this problem.

All the possibilities of non-Lipschitz dynamics have not yet been exhausted, but what has been noted thus far and confirmed with preliminary experiments is that these dynamics: a) are not a slave to initial conditions; b) can model the singular pauses in biological oscillators; c) can resolve the multiple scales of activity; and d) can be self-organizing and "learn" as part of a larger system. Certainly, it is not inferred that chaos is not possible in biological activity; however, it is suggested that it may not be the only way of interpreting the observed phenomena.

6.2 Biological Time Series

6.2.1 Review of Preliminary Data

In previous work, which attempted to understand heart rate dynamics, it was noted that sudden transitions could occur often in relatively healthy individuals. To reduce the relevant degrees of freedom, the heart rate dimensions of heart transplant recipients were calculated, whose hearts, by virtue of the surgery, were denervated. Unexpectedly, relatively high dimensions were found (~5–7), while, at the same time, stationarity was difficult to obtain (Zbilut et al. 1988). This led to experimental animal models where greater control was deemed possible. Specifically, the spontaneously beating isolated perfused rat heart was employed. Surprisingly, this model was capable of exhibiting a variety of complicated patterns, including a "three state" oscillation which was successfully modeled with the aid of a piece-wise linear map. For the experimental data, an averaged pointwise dimension of ~2.1 and a Kolmogorov entropy value of ~0.1 was calculated. An important feature of this map was the necessity of introducing noise to force new trajectories – otherwise the dynamics would remain in a quiescent state. Although the interpretation at that time was that of "noise induced intermittency," clearly an alternate interpretation is that of single trajectories, which return to a stable node, only to be forced out again by the noise.

The requirement that noise be an important force for the dynamics was not limited to cardiac dynamics, nor simple one-dimensional maps. Similar results were found in the modelling of cat phrenic nerve activity driven by pathologic concentrations of carbon dioxide, but this time using modified Bonhoefer-van der Pol (BVP) equations (Hübler 1992): for a certain set of parameters, trajectories converged on a stable node; by the addition of noise, the dynamics were forced off the node and back to the node by a circuitous route (Zbilut at al. 1991). As with the cardiac data, it was noted that although each of the oscillations was a single trajectory, it was possible to obtain a fractal dimension as well as a positive Liapunov exponent (4.68 and 0.162 respectively).

These results were at once both interesting, yet dismaying: clearly, by definition, dimensions and Liapunov exponents are the properties of systems with multiple trajectories, yet these single trajectory dynamics were capable of generating so-called chaotic measures. Some of these "chaotic" results can be relegated, in part, to the aforementioned problems associated with the calculation of dimensions and exponents of experimental data. Yet a more fundamental question centers on the correct characterization of the dynamics: do they consist of multiple continuous trajectories, or are there "pauses" centered about nodes? And why does noise seem to be so important? Careful consideration of the involved physiology, would suggest that "single orbit" dynamics are an appropriate model for real organisms, in that real organisms must constantly change and respond to environmental factors, whereas continuous dynamical models would limit this ability. These questions led to a reconsider the classical formulation of deterministic dynamics.

Based upon the classical (often Laplacian) perspective of determinism, and the mathematical formalism of Lipschitz conditions (boundedness of derivatives), all motions of a system are uniquely determined by their initial positions, and initial velocities. The Lipschitz condition provides for uniqueness of solutions; however, as suggested by biological data, this leads to a discrepancy between the biological phenomena and the functions by which they are described. This is to say that the formal mathematical conditions do not necessarily follow from the physical phenomena. This determinism counterdefines "nondeterministic" dynamics as systems whose initial conditions do not uniquely determine the state of a system at all later points in time. Such dynamics, however, should not be confused with chaotic dynamics: chaotic dynamics are deterministic in that their motions can be predicted (in principle) if their initial conditions were known with infinite precision. Neither should the term "nondeterministic" imply randomness in the sense of lacking structure. Such dynamics do have structure; however, the structures are unique about singular points, and cannot be predicted for the future even if infinite precision were possible, although by virtue of their combinatorial nature they have a predictable probabilistic structure. The explanation for these "nondeterministic" dynamics is that solutions to the dynamics are not unique for all time, and indeed have multiple solutions (branching).

A simple biological example elucidates this concept: simple, oscillatory arm motion, has been shown to be easily modeled by equations of the van der Pol (VDP) type in that they describe the global symmetry of the limit cycle and the response of the motions to external perturbations. However, for $F = 0$ the dynamics of the VDP oscillators are always periodic, whereas

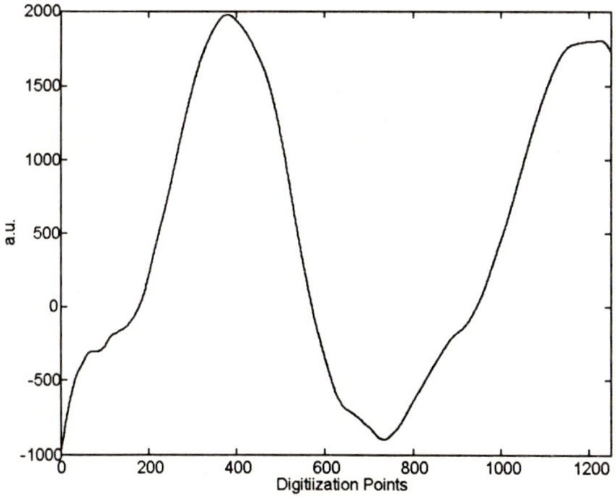

Fig. 6.1. Acceleration of arm movement. Note the brief "stops" near the end points, indicated by a slight change in the slopes and possibly indicative of lack of smoothness of the motion

the natural motions of the arm are not. Instead there is a slight pause of the motion at extreme locations, and after a rather undetermined period of time starts to move again. This means the VDP oscillator does not adequately model the interference of putatively higher neural inputs. A model which uses one VDP for motion in one direction and another for motion in the opposite direction can perfectly model the global geometry as well as the pause near the ends. However, the corresponding motion is no longer deterministic at the endpoints. After a finite period of time, the smallest control force (e.g., noise) will make the arm move (Fig. 6.1).

6.2.2 Examples

Non-Lipschitz (terminal) dynamics is congruent with the observation that biological systems require stability (bounded trajectories) with adaptability (sensitivity to noise and other modulations). That such dynamics are possible in the real physical world include the examples dealing with filaments (Chap. 5). To demonstrate this plausibility in the biological realm is more difficult: biological systems contain large amounts of noise which are amplified and filtered along with the signals of interest (e.g., ECG) (Fig. 6.2). Because the trajectories arrive at equilibrium singular points in finite time, the second derivative diverges. This divergence, however, which appears as a delta function in simulated data, is transformed into a smooth function in processed signals, as is commonly appreciated with the impulse response of filters, and makes detection of these divergences problematic. Other techniques, discussed below, however, are useful for the possible recognition of these dynamics.

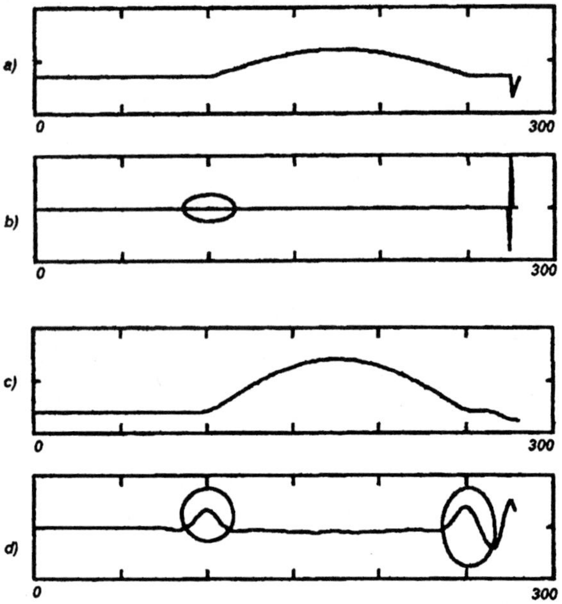

Fig. 6.2. Effect of filtering upon singular points: a simimulated "P" wave of an ECG contains a singular point (a), with a result small 2nd derivative divergence (b). After filtering (c), discontiuities are replaced by small "hillocks" (d)

Respiratory System

Analysis of respiratory signals (pressures, volumes) in both the time and frequency domains have contributed to the understanding of the neural organization and regulation of respiratory control. Mathematical simulations of respiratory patterns have drawn extensively upon continuous variable models. But these models are fully deterministic: they lack nondeterministic components typical of living systems. It has been hypothesized that inspiratory and expiratory phases represent automatic (deterministic) regulation of the lung pressure and volume changes, and that pauses between these phases represents nonautomatic (non-Lipschitz) regulation. If the system is to function properly under varied environmental conditions, this type of regulation strategy is critical since it permits cycle-by-cycle updates. Failure of this update mechanism may be implicated in the variety of pathological patterned breathing processes.

In anesthetized rats, pleural pressure tracings demonstrate pauses in phase plots (see below) (Zbilut et al. 1996a). Their presence may provide a more rational approach to the control of mechanically ventilated patients: further research into the effect of stimuli at points of singularity may indicate that desired control may be obtained with less disruption to physiological processes. In this respect, a benefit is already obtained by the so-called "synchronized" mode of ventilation, which, in fact, initiates a breath near the inception of a spontaneous breath by the patient; i.e., near the point of singularity.

Cardiovascular System

Another way to look for non-Lipschitz dynamics is to examine the probabilities of the dynamics of the motion. In the case of an isolated perfused rat heart, all neurological and hydraulic connections with the organism are severed. The only control of such a beating heart is the spontaneous discharges of the SA node fluctuating about a mean level. If indeed, the dynamics of such a system are governed by singularities, the only control factor would be the random fluctuations themselves, and would be demonstrated by a one dimensional random walk. In fact this is what was found in such an experimental model (Fig. 6.3). It was determined that the beats behaved as a random walk in one dimension. Even as a variable length random walk, the distribution of step sizes was a symmetric normal curve. And, as expected, the power spectrum exhibited a scaling region with an exponent of 2.

To contrast these findings, normal human, severe congestive heart failure (CHF), and non-rejecting transplant hearts were also studied. The CHF and transplant hearts were also chosen, due to reports of decreased variability in heart failure, as well as decreased variability in transplanted hearts due to a lack of nervous system control. The normal human heart exhibited an approximately normal distribution for a variable length random walk, with slight deviations related to baroreceptor and respiratory oscillations, whereas the CHF and transplant hearts demonstrated markedly skewed distributions. The power law, however, for CHF hearts indicated increased correlation in time, whereas the transplant less, approaching a 1/f process, or white noise. These departures from normality suggest varying degrees restriction in dynamical behavior, which may provide a basis for a distinctive feature classification of pathologic cardiac conditions, based upon alteration in the distribution of responses at the singularities.

6.2.3 Conclusions

Phase plane portraits of both respiratory and cardiac signals have in the past suggested the possibility of complex or chaotic dynamics. Careful analysis of results using methods to detect such dynamics, puts doubt upon such approaches: relatively large dimensions and concerns regarding adequacy of data length and stationarity have put such findings into question. Indeed, it is possible to mimic respiratory signals with singular dynamics to obtain positive Liapunov exponents (Zbilut et al. 1991). Similar misgivings have prompted Haken to use the term "quasi-attractors" in the case of physiological oscillators needing choices with respect to their phases, so as to find a compromise between stability and adaptability (Haken 1991). These

TYPE	TIME SERIES	RANDOM WALK (SYMBOLIC DYNAMICS-Pr +/-)	VARIABLE STEP RANDOM WALK	POWER LAW SCALING EXPONENT
ISOLATED RAT HEART		50.1/49.9	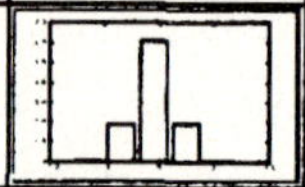	2
NORMAL HUMAN MALE	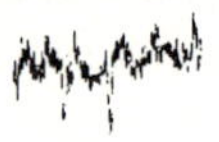	51.7/48.3	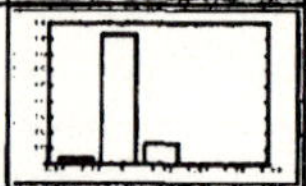	2.13
CONGESTIVE HEART FAILURE		50/50	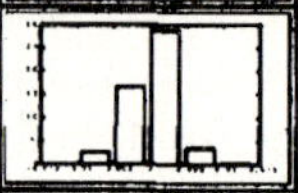	2.4
TRANSPLANT		50.1/49.9	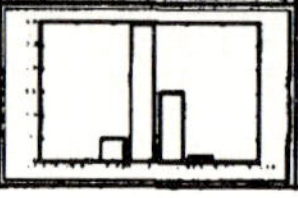	0.54

Fig. 6.3. Random walk of isolated rat heart compared to normal human, human with congestive heart failure, and one with a healthy transplanted heart. Clearly, beats are modified by neural and hydraulic connection

observations require alternative perspectives which recognize the need to explain noise, constant control parameter change, and the need to remain relatively stable against naturally occurring biological perturbations.

Non-Lipschitz dynamics appear to be a natural consequence of physiological complexity. With several levels of control, it appears that multiple branching solutions to basic equations of motion are necessary to give an organism the ability to "fine tune" a given system according to second-by-second changing physiological requirements. By contrast, transformation of the system on some levels toward complete determinism or toward complete randomness may be a marker of pathology.

The traditional approach to physiological regulation has been that of negative feedback loops – frequently multiple and highly redundant. In some cases, however, experimental evidence finds that adjustments are made before the supposed controller has experienced the feedback. Increasingly, evidence suggests that regulation might be based upon learning and experience, and that input is matched to stored traces out of a wide range of output patterns. The non-Lipschitz theory presented here may be a model for such control. Noise is be an important ingredient functioning within the context of the singularities in order to optimize successful choice of trajectories as part of a natural "fuzzy logic," and could be a requisite for primitive control, i.e., prior to learning.

6.3 Importance of Non-Lipschitz Dynamics in Biology

The above has suggested that biological systems may depend upon more than one form of chaos, yet the skeptic might respond with *cui bono*? It might be remarked that extant models are sufficient to explain observed processes, and moreover, are more easily tractable from a computational view.

In the case of non-Lipschitz dynamics, the implications are significant not only from a theoretical viewpoint, but from a practical one as well. First of all, non-Lipschitz dynamics can give rise to a material randomness. This is to say that there is no artificial random generator. That this is important can be simply appreciated by the restrictions, and complications of random number theory, especially as it is applied (and presumed) in various models of natural phenomena: to make models realistic, noise of some sort is required. As a practical matter, this means that choices must be made as to the distribution, strength, and form (additive or multiplicative). Changes in any of these variables can have major consequences in the ultimate correspondence to reality. In living systems, however, the noise is rarely static: it has varying shades of color, correlation, and dynamics. Transients which can be a nuisance in models, can be an important factor – especially when the length of observation is shorter than the transient itself (Crutchfield and Kaneko 1988). There is also the question of the adequacy of numerical simulations, given that random number generators are based upon deterministic rules. Material randomness, in a sense, does not "care" about these considerations, and yet we do not understand the full implications of such a phenomenon.

Secondly, non-Lipschitz dynamics are important from the viewpoint that their control differs markedly from deterministic dynamics: everything centers about singular points. Here exquisite control can be easily accomplished, while at the same time it allows for a relatively robust (against noise) dynamic away from the singularity. Given that biological systems are relatively noisy, this is an important feature – it allows for tremendous flexibility in the expression of dynamical regimes. At the same time, it allows for conservation of energy, especially for repetitive processes, such as neural systems.

6.4 Detecting Non-Lipschitz Dynamics

If non-Lipschitz dynamics is important, can they be detected? The answer to this problem is intimately intertwined with the problem of detecting chaos, from forms of correlated randomness, or forms of quasi-periodic oscillations. The usual methods to make this determination has been to employ fractal dimensions, and positive Liapunov exponents. Whereas these methods have been extraordinarily useful when dealing with simulated systems, their application in the material world, especially the biologic world has been somewhat disappointing. It has become clear that noise, data length, stationarity, and filtering are important variables which may confound the original data so as to give the impression of chaos, when there is none.

Some of the very same problems occur in the attempt to identify non-Lipschitz dynamics, with filtering being an important factor. A main point of filtering is to "smooth out" irregularities or noise (see Fig. 6.2). In so doing, singular points are easily destroyed. So too, in some cases, by inadequate sampling. Nonetheless, some inferences can be made by studying putative methods to detect singularities in theoretical systems.

6.4.1 Divergence of the Second Derivative

Clearly, if motion arrives at zero in a finite time as in non-Lipschitz dynamics, the velocity goes to zero, with a divergence of the second derivative. A physically motivated example of non-Lipschitz equations comes from a simple model describing the dynamics of neutron star magnetic fields. The model envisions two equal but oppositely charged concentric spherical shells which are allowed to rotate differentially. After suitable transformations the equations [see Dixon, et al. (1993) for an extensive discussion; also Dixon (1995)] of interest are

$$\dot{x} = \frac{\lambda x z}{x^2 + z^2} - \epsilon x \tag{6.1}$$

$$\dot{z} = \frac{\lambda z^2}{x^2 + z^2} - \overline{\epsilon} z - (\lambda - \overline{\epsilon}) \tag{6.2}$$

where λ is the scaled Landau damping, and $\overline{\epsilon}$ is the scaled viscous damping parameter tensor. Numerical integration of these equations yields a phase plot as in Fig. 6.4, with apparent intersection of trajectories near the origin, and a time series of oscillations of varying amplitudes. The origin has been shown to be non-Lipschitz, and the solution is multivalued.

Simple inspection of the time series does not suggest this singularity, and indeed it may resemble any number of equations which describe oscillations. Inspection of the second derivative, however clearly demonstrates a divergence (Fig 6.6). The problems previously discussed regarding noise and filtering are easily demonstrated in the following: a very small amount of normally distributed random noise is added to the neutron star equations. After filtering the noise enhanced realization (bidirectional, 10th order Butterworth), the divergence is clearly gone (Fig. 6.7).

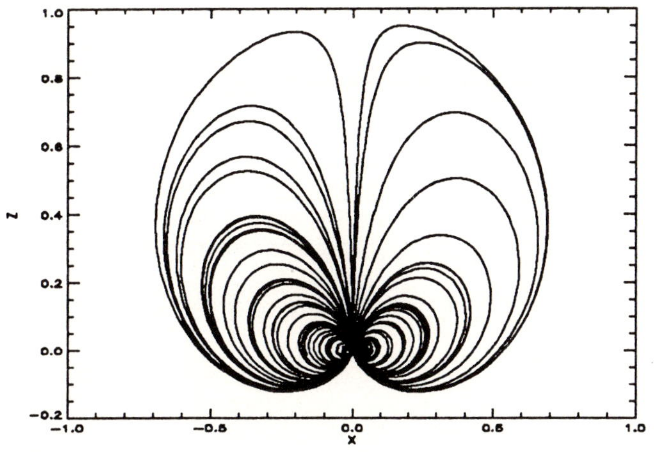

Fig. 6.4. Phase plot of solutions in the neutron star model for $\lambda = 1.0,\ \epsilon = 0.6,\ \bar{\epsilon} = 0.7$ (courtesy of D.D. Dixon)

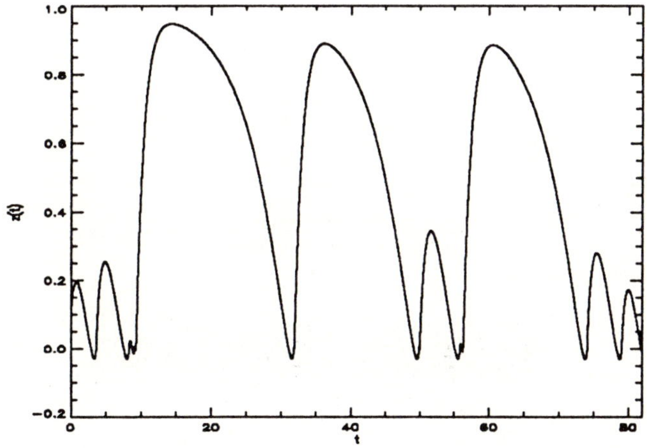

Fig. 6.5. Time series $z(t)$ vs. t for the neutron star model with parameter values $\lambda = 1.0,\ \epsilon = 0.6,\ \bar{\epsilon} = 0.7$ (courtesy of D.D. Dixon)

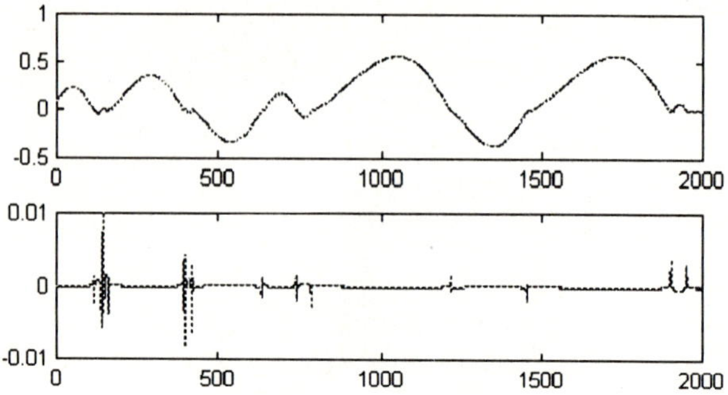

Fig. 6.6. Time series of neutron star equations (top panel), and the divergences of the 2nd derivative at singular points (bottom panel)

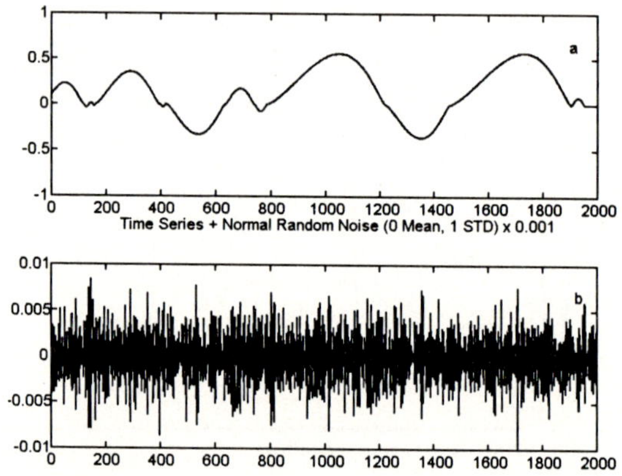

Fig. 6.7. Original neutron star equations with a minimal amount of noise added (top panel). Resulting 2nd derivatives (bottom panel)

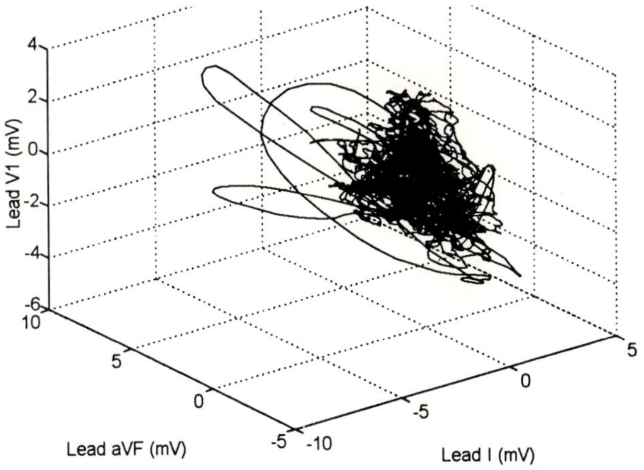

Fig. 6.8. Random walk of rotor as monitored by surface leads

6.4.2 Random Walk

Non-Lipschitz dynamics can be characterized by a random walk (see Chap. 7). Such a characterization has been documented in a freely beating rat heart decoupled corporeally; i.e., with no control modulations via nervous or hydraulic connections to the rest of the body (Giuliani 1996) (see Fig. 6.3).

It is suggested that a similar random walk often exists in the case of rotor waves in the myocardium putatively causative of ventricular fibrillation. Figure 6.8 demonstrates the movement of a rotor wave in two planes, based upon the approximate orthogonal ECG leads I and aVF, and I and V1. These were recorded in a person undergoing routine electrophysiological testing. Movement in both planes exhibited a random walk in 2 dimensions. This was based upon calculations from a spatial Fourier transform, with scaling exponents of 1.8 and 1.9 respectively.

6.4.3 Phase Plane Portraits

Detection of non-Lipschitz dynamics in the respiratory system is demonstrated in Fig. 6.9 for intrapleural pressures which were recorded in the conscious rat using an intrathoracic balloon technique (Zbilut et al. 1996a; Webber and Zbilut 1996). The intrapleural pressure trace was digitized at 500 Hz, low-pass filtered at 4 Hz (5th order Butterworth). Representation of this physiological variable in phase space inscribes counter-clockwise trajectories. Lack of superposition of the individual trajectories indicates they are of the branching type. This means that at the beginning of each and every cycle only one pressure trajectory of many can be opted. Non-

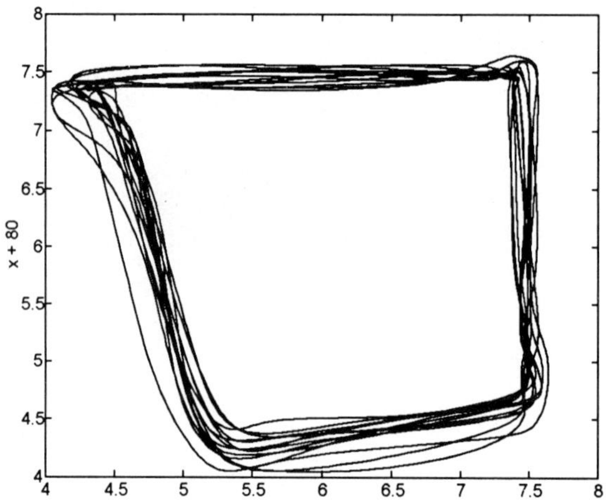

Fig. 6.9. Plural phase plane plot describing orthogonal vectors in upper right corner

Lipschitz dynamics in the plot is seen only during the expiratory pauses (upper right corner) where the trajectory coincides with the horizontal and vertical axes. It is in this region where the vectors are orthogonal, rendering the solutions to the operating differential equations of the non-Lipschitz type. This area actually constitutes a singularity for the system. With inspiration, each trajectory leaves the area from a slightly different position, approaches the inspiratory point (lower left corner), and contracts back toward the orthogonal vectors. Each loop is unique (differing initial conditions at the point of orthogonality).

6.4.4 Wavelet Singularities

Wavelet Decomposition

One of the useful applications of wavelet decomposition has been it application in singularity detection in the form of edges and discontinuities. In this respect Mallat and Hwang (1992) have done pioneering work.

Wavelets are basis functions which rescale a signal at multiple levels. In some ways they are similar to Fourier transforms, except that they are able to localize the frequencies which are divided into octave bands. The type of wavelet used can be important, and certainly one of the most frequently used has been the Daubechies wavelets.

In the present case, it is theorized that non-Lipschitz singularities constitute discontinuities or "edges" (which are often discovered by a discontinuity of the second derivative), and may be a useful way to detect non-Lipschitz processes. To test this idea, the neutron star equations, the chaotic Lorenz system, and a simple sine wave function to wavelet decompo-

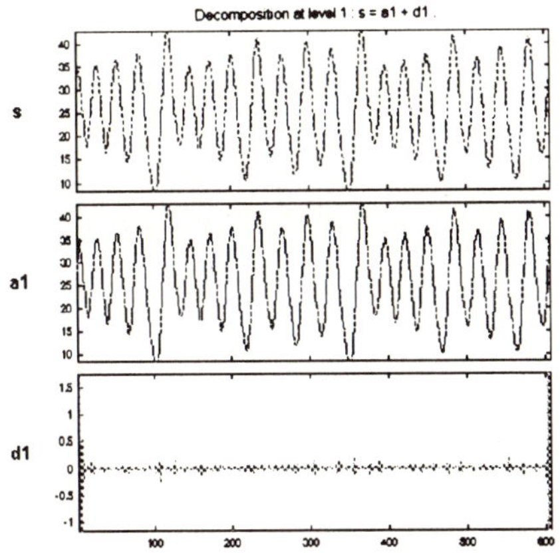

Fig. 6.10. Time series for Lorentz equations (top panel), and wavelet decomposition below. Note the lack of any clear wavelets. Compare with results from the neutron star equations

sition. Because noise is a common problem, it was added to these systems for comparison. Finally, the lessons from these simulations were applied to an experiment with forearm rotation, which putatively should include a non-Lipschitz singularity as suggested by simultaneous measurement of acceleration.

Methods

The above systems were analyzed using the Matlab® Wavelet Toolbox. It has been shown that parabolic signals can be represented by D_6 and D_8 wavelets, and these were chosen for the decomposition of the neutron star, Lorenz, and sine wave equations.

The arm experiment consisted of a subject placing his arm in a manipulandum, which is designed to carefully measure rotation about the elbow joint, and is frequently used in motor control laboratories. The subject was told rotate his forearm comfortably at a rate of about one cycle per second (an auditory cue was provided). The signal was sampled at 40 kHz with 2048 bits of resolution. The subject was allowed to practice the procedure several times prior to recording. Acceleration was measured simultaneously.

The data were decimated to 1 kHz off-line and subjected to the wavelet decomposition.

Results

Inspection of the decompositions clearly indicate the ability of wavelets to detect the singularities of the neutron star equations. Oscillations of the chaotic Lorenz system (Fig. 6.10), and simple sine waves (Fig. 6.11), however revealed no singularities, as was expected. On the other hand, the

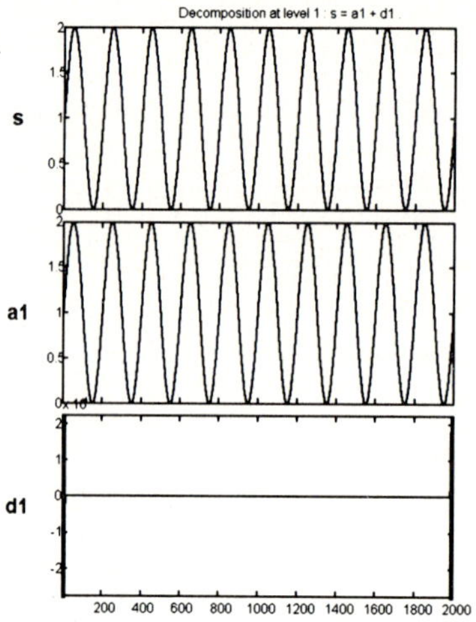

Fig. 6.11. Regular sine wave (top panel) and its wavelet decompositions below

neutron star equation decompositions clearly exhibited singular wavelets (Fig. 6.12). The clarity of the neutron star decomposition, however, was lost with the addition of uniformly distributed random noise (of 0.001 amplitude). The clear wavelets of the d1 level became swamped with the inclusion of noise (Fig 6.13).

Because the arm movement is inherently noisy, a profile approximating the noisy neutron star equations was expected; i.e., no distinguishing wavelets at the low levels of decomposition, but perhaps at d4 and d5. Surprisingly, this was not the case. Figure 6.14 demonstrates relatively large (with respect to signal to noise ratio) wavelets at the d1 level (near points 600 and 950).

6.4.5 Noise Scaling

Non-Lipschitz dynamics may also be discerned on the basis of noise effects on the trajectories: with non-Lipschitz dynamics, the noise has relatively no effect on orbits, whereas with classical dynamics the trajectories scale with the level of noise with respect to first passage times. Hübler (1992) has discussed a physical experiment describing this result.

It is suggested that a similar process governs the ECG: the variability of the heartbeats occurs between the trajectories – at the singular points where noise and control systems exert their effect. Otherwise, the ECG trace could be severely distorted, and timing altered by the noise (Webber and Zbilut 1994).

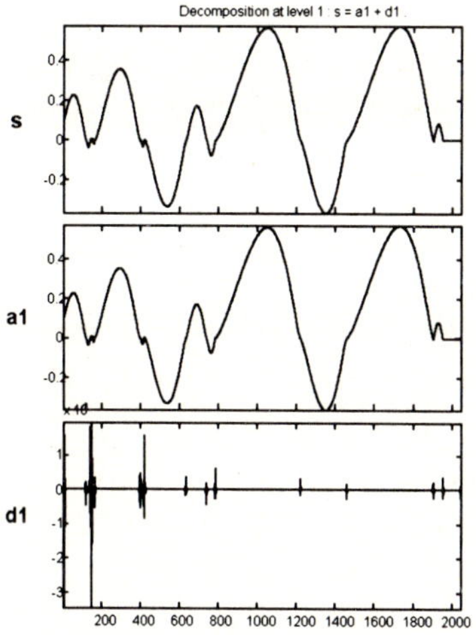

Fig. 6.12. Neutron star equations time series (top panel) and wavelet decompositions below

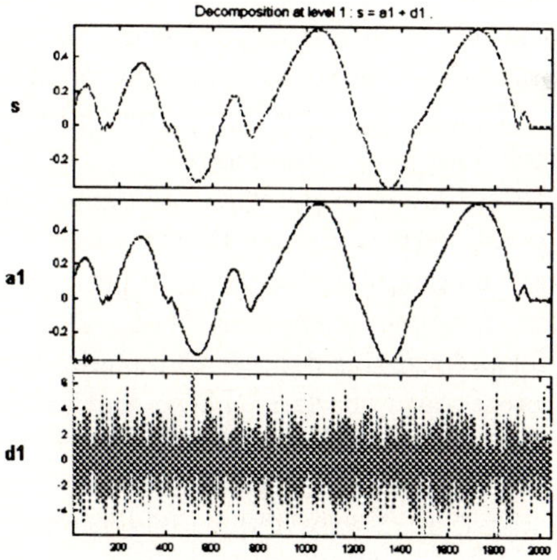

Fig. 6.13.

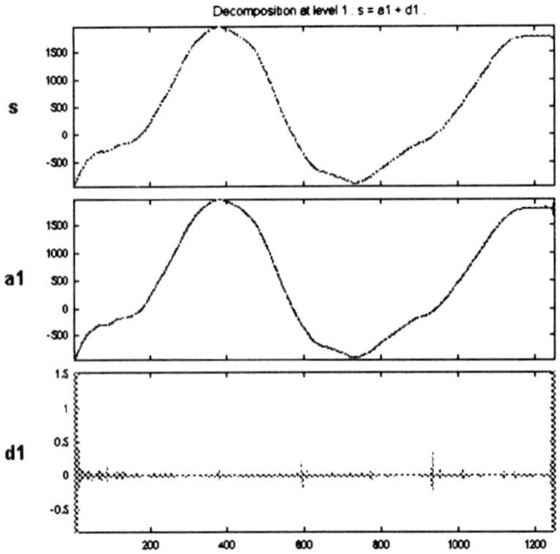

Fig. 6.14.

6.4.6 Recurrence Analysis

Although wavelets hold some promise in the analysis of singular non-Lipschitz dynamics, clearly they are a form of filtering, with the attendant difficulties outlined previously. Still another possible way to evaluate these singular points is that of recurrence quantification (RQ) [For a detailed discussion of the method, see (Zbilut and Webber 1992; Webber and Zbilut 1995; Trulla et al.1996)].

RQ is different from these other measures in that the values of the dynamics are not mathematically transformed, instead it quantifies the number of recurrences as previously defined by Eckmann et al. (1987). These recurrences are further distinguished by their type and entropy, as well as a rough estimate of their Liapunov exponents.

Specifically, the process involves counting the number of recurrences centered on an index point with radius r. Points which form a line on the diagonal on the square matrix of distances have been found to be inversely related to the largest Liapunov exponent ($1/L_{\max}$). Webber and Zbilut (1994b, 1996) have formalized this process to include a running time series of these values to detect state changes. These routines were used to analyze the arm movement data (a different trial). For comparative purposes, a simple sine wave of the same length and cycle was created with a small amount of normally distributed noise added. The results indicate that there was no state change as expected by the Liapunov exponent estimate for the sine wave (Fig. 6.16). The arm motion data, however displayed a divergence at the point where there was an expected pause (Fig. 6.15).

A recurrence analysis was performed for ECG data and showed similar but more complicated results consistent with the known physiology.

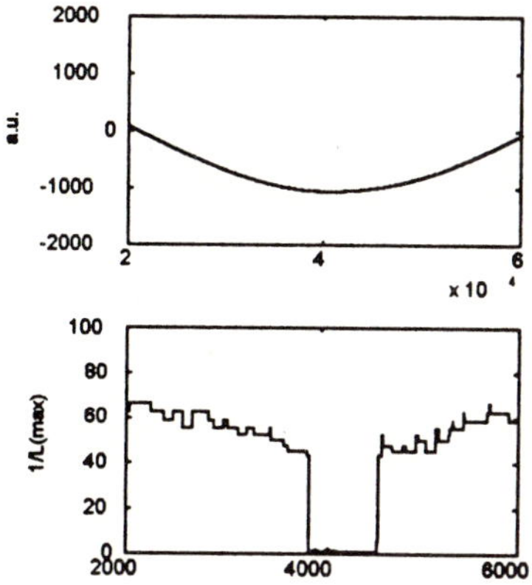

Fig. 6.15. Arm motion (top panel), and estimated local Liapnov exponents [1/L(max)] (bottom panel). Note the divergence at the singular point of sudden deceleration

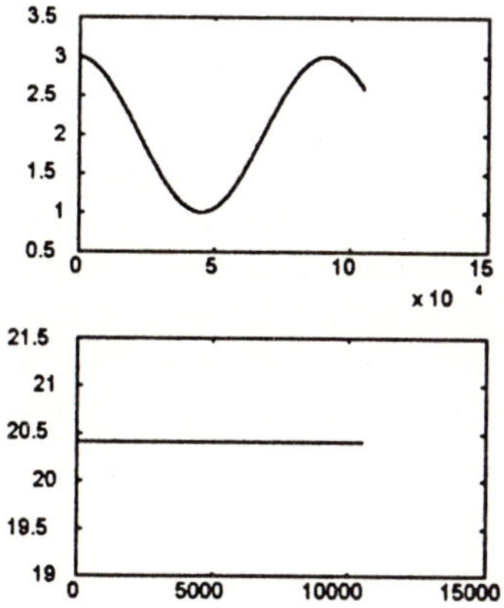

Fig. 6.16. Sine wave with noise (top panel), and estimated Liapunov exponents (bottom panel – 1/L(max)). Note the lack of any change in their value

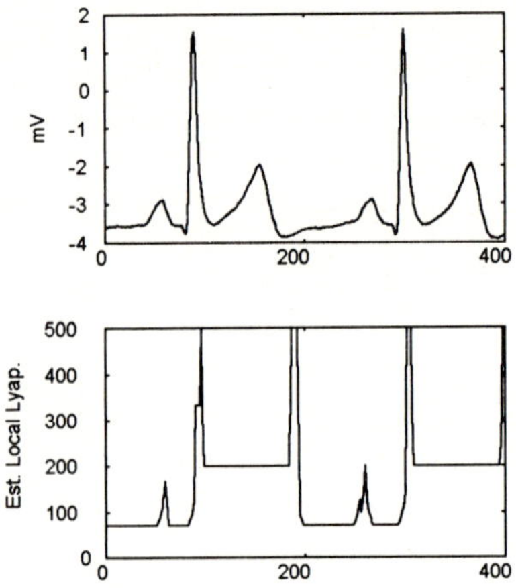

Fig. 6.17. ECG (top panel) and estimated local Liapunov exponents [1/L(max) – bottom]. Note the divergence at the P, QRS, and terminus of the T wave, coincident with the change in singularity status

Specifically, the heart has two separate anatomically regions designed for depolarization: the atria and the ventricles. Although the signal is initiated at the SA node of the right atrium, it is conducted to the ventricles via a delay at the AV node. Functionally this means two areas of depolarization and repolarization (P wave and QRS complex) followed by pauses between these regions. The recurrence analysis demonstrates this as divergences of the estimated local Liapunov exponent (Fig. 6.17).

An important fact to note is that the data were oversampled by many factors above the Nyquist criteria for the main cycle for the arm movement. This may be necessary is some cases, as previously suggested, to prevent loss of the singular point, which, practically speaking, develops in sampled data as a very high frequency. This, of course, also allows for more noise. As a result, the suggested strategy would involve some degree of data oversampling and careful inspection prior to any naive determination of filtering and sampling rates whenever there is a possibility of singular points.

6.5 Discussion

In sum, it is proposed that there are at least two types of chaos operative within biological systems (Table 6.1): deterministic and non-Lipschitz. Recognition of this allows for a resolution to the quandary regarding observed physiological processes which require adaptability as well as many degrees of freedom for their adequate description.

The coexistence of these two forms of dynamics helps also to elucidate the debate between "order and complexity": it has been pointed out that statements regarding biological existence as being "highly ordered and complex" are *non sequiturs* (Yockey 1992). Complex systems are described by high entropic values; whereas ordered systems have low entropy. And yet intuitively, scientists have noted that this does not adequately describe the situation in biological sciences: there are obvious highly ordered structures–yet their function or activity is often impossible to predict precisely because of their complexity (high entropic value). Non-Lipschitz dynamics can accommodate both: the order resides in the piecewise determinism of trajectories, but the complexity resides at the singular points. Here the probabilistic nature of the trajectories preclude deterministic prediction in contradistinction to deterministic chaos which in theory can be predicted given infinite precision. Again, material randomness presents itself as an important entity in the generation of life, and puts into question statements regarding functional order out of chaos (Prigogine et al. 1972a,b).

Table 6.1

Chaotic Dynamics	Non-Lipschitz Dynamics
Dependent upon initial conditions	Dependent on singularity
Attractor present	No attractor (in the usual sense)
Spreads uncertainty over entire attractor; i.e., global instability	Uncertainty at singular points
Behavior varies	Dynamics is well behaved away from singular points
Smooth stretching and folding of attractor	Random spread of points in region of phase space (stochastic attractor)
Information decreases with time	Information infinite at singularity (knowledge of past is zero)
Cantor set	Coutable combinatorial set
Predictability based on stabilization principle	Predictability based on probability evolution
Controlled by external forces	Easily controlled by perturbations
Time is continuum	Near singularity, future time evolutionis decoupled from past

7

Terminal Model of Newtonian Dynamics

Sudden illumination is a manifest sign of long, unconscious work. *–Henri Poincaré*

7.1 Basic Concepts

The governing equations of classical dynamics based upon Newton's laws:

$$\frac{d}{dt}\frac{\partial L}{\partial \dot{q}_i} = \frac{\partial L}{\partial q_i} - \frac{\partial R}{\partial \dot{q}_i}, \; i = 1, 2, ..., n \tag{7.1}$$

where L is the Lagrangian, and $q, \dot{q}_i$ are the generalized coordinates and velocities, include a dissipation function $R(\dot{q}_i \dot{q}_j)$ which is associated with the friction forces:

$$F_i(\dot{q}_1, \dot{q}_2, \cdots \dot{q}_n) = -\frac{\partial R}{\partial \dot{q}_i} \tag{7.2}$$

The structure of the functions (7.2) does not follow from the Newton's laws, and, strictly speaking, some additional assumptions should be made in order to define it. The "natural" assumption (which has been never challenged) is that these functions can be expanded in Taylor series with respect to an equilibrium state

$$\dot{q}_i = 0 \tag{7.3}$$

Obviously this requires the existence of the derivatives:

$$\mid \frac{\partial F_i}{\partial \dot{q}_j} \mid < \infty \text{ at } \dot{q}_j \to 0 \tag{7.4}$$

i.e., F_i must satisfy the Lipschitz condition. This condition allows one to describe the Newtonian dynamics within the mathematical framework of classical theory of differential equations. However, there is a certain price paid for such a mathematical "convenience": the Newtonian dynamics with dissipative forces remains fully reversible in the sense that the time-backward motion can be obtained from the governing equations by time inversion, $t \to -t$. As stressed by Prigogine (1980), in this view, future and past play the same role: nothing can appear in future which could not already exist in past since the trajectories followed by particles can never cross (unless $t \to \pm\infty$). This means that classical dynamics cannot explain the emergence of new dynamical patterns in nature.

In order to trivialize the mathematical part of our argumentation let us consider an one-dimensional motion of a particle decelerated by a friction force:

$$m\dot{v} = F(v) \tag{7.5}$$

in which m is mass, and v is velocity. Invoking the assumption (7.4) one can linearize the force F with respect to the equilibrium $v = 0$:

$$F \to -\alpha v \text{ at } v \to 0, \ \alpha = -(\frac{\partial F}{\partial v})_{v=0} > 0 \tag{7.6}$$

and the solution to (7.5) for $v \to 0$ is:

$$v = v_0 e^{-\frac{\alpha}{m}t} \to 0 \text{ at } t \to \infty, \ v_0 = v(0) \tag{7.7}$$

As follows from (7.7), the equilibrium $v = 0$ cannot be approached in finite time. The usual explanation of such an effect is that, to accuracy of our limited scale of observation, the particle "actually" approaches the equilibrium in finite time. In other words, eventually the trajectories (7.7) and $v = 0$ become so close that we cannot distinguish them. The same type of explanation is used for the emergence of chaos: if two trajectories originally are "very close," and then they diverge exponentially, the same initial conditions can be applied to either of them, and therefore, the motion cannot be traced.

Hence, there are variety of phenomena whose explanations cannot be based directly upon classical dynamics: in addition, they require some "words" about a scale of observation, "very close" trajectories, etc.

In this section we describe a new structure of the dissipation forces which eliminates the paradox discussed above and makes the Newtonian dynamics irreversible (Zak 1989d, 1992b). The main properties of the new structure are based upon a violation of the Lipschitz condition (7.4). Turning to the example (7.5), let us assume that

$$F = -\alpha v - \alpha_1 v^k, \alpha_1 \ll \alpha, \ k = \frac{p}{p+2} < 1, \ p \gg 1 \tag{7.8}$$

in which p is an odd number.

By selecting large p, one can make k close to 1 so that (7.6) and (7.8) will be almost identical everywhere excluding a small neighborhood of the equilibrium point $v = 0$, while, as follows from (7.8), at this point:

$$\left| \frac{\partial F}{\partial v} \right| = (\alpha + k\alpha_1 v^{k-1}) \to \infty \text{ at } v \to 0, \text{ i.e. } F \to -\alpha^1 v^k \text{ at } v \to 0 \quad (7.9)$$

Hence, the condition (7.4) is violated, the friction force grows sharply at the equilibrium point, and then it gradually approaches the straight line (7.6). This effect can be interpreted as a jump from static to kinetic friction.

It appears that this "small" difference between the friction forces (7.6), (7.8) leads to fundamental changes in Newtonian dynamics.

Firstly, the time of approaching the equilibrium $v = 0$ becomes finite. Indeed, as follows from (7.5) and (7.9):

$$t_0 = -\int_{v_0}^{0} -\frac{mdv}{\alpha_1 v^k} = \frac{mv_0^{1-k}}{\alpha_1(1-k)} < \infty \quad (7.10)$$

Obviously this integral diverges in the classical case.

Secondly, the motion described by (7.5), (7.8) has a singular solution $v \equiv 0$ and a regular solution

$$v = [v_0^{1-k} - \frac{\alpha_1}{m}(1-k)t]^{\frac{1}{1-k}} \quad (7.11)$$

In a finite time the motion can reach the equilibrium and switch to the singular solution, and this switch is irreversible. It is interesting to note that the time-backward motion

$$v_- = \{[v_0^{1-k} - \frac{\alpha}{m}(1-k)(-t)]^{p+2}\}^{1/2} \quad (7.12)$$

is imaginary. [One can verify that the classical version of this motion (7.7) is fully reversible if $t < \infty$].

The equilibrium point $v = 0$ of (7.8) represents a "terminal" attractor which is "infinitely" stable and is intersected by all the attracted transients. Therefore, the uniqueness of the solution at $v = 0$ is violated, and the motion for $t < t_0$ [see (7.10)] is totally "forgotten." [This is a mathematical implication of irreversibility of the dynamics (7.8)].

So far we were concerned with stabilizing effects of dissipative forces. However, as well-known from dynamics of non-conservative systems, these forces can destabilize the motion when they feed the external energy into the system [the transmission of energy from laminar to turbulent flow in fluid dynamics (Drazin 1984), or from rotations to oscillations in dynamics of flexible systems]. In order to capture the fundamental properties of these effects in the case of a "terminal" dissipative force (7.8) by using the simplest mathematical model, let us turn to (7.5) and assume that now the friction force feeds energy into the system:

$$m\dot{v} = \alpha_1 v^k, \ k = \frac{p}{p+2} < 1, v \to 0 \quad (7.13)$$

One can verify that for (7.13) the equilibrium point $v = 0$ becomes a terminal repeller, and since

$$\frac{d\dot{v}}{dv} = \frac{k\alpha_1}{m} v^{k-1} \rightarrow \infty \text{ at } v \rightarrow 0 \tag{7.14}$$

it is "infinitely" unstable. If the initial condition is infinitely close to this repeller, the transient solution will escape it during a finite time period:

$$t_0 = \int_{\varepsilon \rightarrow 0}^{v_0} \frac{mdv}{\alpha_1 v^k} = \frac{m v_o^{1-k}}{\alpha_1 (1-k)} < \infty \tag{7.15}$$

while for a regular repeller, the time would be infinite.

As in the case of a terminal attractor, here the motion is also irreversible: the solution

$$v = \pm [\frac{\alpha_1}{m}(1-k)t]^{\frac{1}{1-k}} \tag{7.16}$$

and the solution (7.11) are always separated by the singular solution $v \equiv 0$, and each of them cannot be obtained from another by time inversion: the trajectory of attraction and repulsion never coincide.

But in addition to this, terminal repellers possess even more surprising characteristics: the solution (7.16) becomes totally unpredictable. Indeed, two different motions described by the solution (7.16) are possible for "almost the same" ($v_0 = +\varepsilon \rightarrow 0$, or $v_0 = -\varepsilon \rightarrow 0$ at $t \rightarrow 0$) initial conditions. The most essential property of this result is that the divergence of these two solutions is characterized by an unbounded rate:

$$\sigma = \lim_{t \rightarrow t_0} (\frac{1}{t} \ln \frac{t^{1/1-k}}{| v_0 |}) \rightarrow \infty \text{ at } | v_0 | \rightarrow 0 \tag{7.17}$$

In contrast to the classical case where $t_0 \rightarrow \infty$, here σ can be defined in an arbitrarily small time interval t_0, since during this interval the initial infinitesimal distance between the solutions becomes finite. Thus, a terminal repeller represents a vanishingly short, but infinitely powerful "pulse of unpredictability" which is pumped into the system via terminal dissipative forces. Obviously failure of the uniqueness of the solution here results from the violation of the Lipschitz condition (7.4) at $v = 0$.

As known from classical dynamics, the combination of stabilizing and destabilizing effects can lead to a new phenomenon: chaos. In order to describe similar effects in dynamics with terminal dissipative forces let us slightly modify (7.13):

$$m\dot{v} = \alpha_1 v^k \cos \omega t \tag{7.18}$$

Here stabilization and destabilization effects alternate. With the initial condition $v \rightarrow 0$ at $t \rightarrow 0$ the exact solution to (7.18) consists of a regular solution:

$$v = \pm [\frac{\alpha_1(1-k)}{m\omega} \sin \omega t]^{\frac{1}{1-k}}, \quad v \neq 0 \tag{7.19}$$

and a singular solution:

$$v = 0 \tag{7.20}$$

During the first period $0 < t < \pi/2\omega$ the equilibrium point (7.20) is a terminal repeller. Therefore, within this interval, the motion can follow one of two possible trajectories (7.19) (each with the probability 1/2) which diverge with unbounded Liapunov exponent (7.17) at $v = 0$. During the next period $\pi/2\omega < t < 3\pi/2\omega$ the equilibrium point (7.20) becomes a terminal attractor; the solution approaches it at $t = \pi\omega$ and it remains motionless until $t > 3\pi/2\omega$. After this, the terminal attractor converts into terminal repeller, and the solution escapes again, etc.

It is important to notice that each time the system escapes the terminal repeller, the solution splits into two symmetric branches, so that there is 2^n possible scenarios of the oscillations with respect to the center $v = 0$, while each scenario has the probability 2^{-n} (n is the number of cycles). Hence, the motion (7.19) resembles chaotic oscillations known from classical dynamics: it combines random characteristics with the attraction to a center. However, in the classical case the chaos is caused by a supersensitivity to the initial conditions, while the uniqueness of the solution for fixed initial conditions is guaranteed. In contrast to this, the chaos in the oscillations (7.19) is caused by the failure of the uniqueness of the solution at the equilibrium points, and it has a well organized probabilistic structure. Since the time of approaching the equilibrium point $v = 0$ by the solution (7.19) is finite, this type of chaos can be called "terminal."

Let us turn now to the general case, i.e., to the governing equations (7.1) and (7.2), and introduce the following dissipation function:

$$R = \frac{1}{k+1} \sum_i \alpha_i \mid \sum \frac{\partial \mathbf{r}_i}{\partial q_j} \dot{q}_j \mid^{k+1} \tag{7.21}$$

in which $\mathbf{r}_i$ is the radius-vector of the ith point of the system. One can verify that the classical case corresponds to $k = 1$. As in classical dynamics, this function expresses the dissipation rate of the total energy E:

$$\frac{dE}{dt} = -\sum_i \dot{q}_i \frac{\partial R}{\partial \dot{q}_i} = -(k+1)R \tag{7.22}$$

Within a small neighborhood of an equilibrium state (where the potential energy can be set zero) the energy E and the dissipation function R have the order, respectively:

$$E \sim \dot{q}_i^2, \ R \sim \dot{q}_i^{k+1} \text{ at } E \to 0 \tag{7.23}$$

Hence, the asymptotic form of (7.22) can be presented as:

$$\frac{dE}{dt} = AE^{\frac{k+1}{2}} \text{ at } E \to 0, \ A = \text{Const} \tag{7.24}$$

Obviously, (7.24) is equivalent to (7.5) expressed in terms of energy. This means that all the new properties introduced above are preserved in the general case of Newtonian dynamics with a terminal dissipation function (7.21), i.e., when $k = \frac{p}{p+2} < 1$. Indeed, since

$$|\frac{d\dot{E}}{dt}| \to \infty \text{ at } E \to 0 \text{ for } k < 1 \tag{7.25}$$

the equilibrium states are represented by terminal attractors or repellers, and therefore, the dynamics becomes irreversible. Within the framework of terminal dynamics, formations of new patterns of motion can be understood as chains of terminal attractions and repulsions: as shown above, during each terminal repulsion the solution splits into two symmetric branches, and the motion can follow each of them with equal probability. Such a scenario can be represented by terminal chaos, which has an exact mathematical formulation, and does not depend upon the accuracy to which the initial conditions are known. Driven by non-uniqueness of solutions at terminal repellers, terminal chaos, and consequently, the process of emergence of new patterns of dynamical motions, possess well organized probabilistic structures.

For illustration of the theory let us turn to fluid dynamics. One of its central problems is to explain how a motion which is described by fully deterministic Navier-Stokes equations can be random. Starting with the simplest shear flow

$$\rho\frac{\partial v_x}{\partial t} = \frac{\partial \sigma_{xy}}{\partial y} \tag{7.26}$$

where x, y are Cartesian coordinates, ρ, v_x and σ_{xy} are density, velocity in x-direction, and shear stress, respectively, we will introduce as an analog to (7.9), the following constitutive law

$$\sigma_{xy} = \mu_1(\frac{\partial v_x}{\partial y})^k \text{ at } \frac{\partial v_x}{\partial y} \to 0 \tag{7.27}$$

which coincides with the Newton's formula for $k = 1$.

Combining (7.26) and (7.27) one obtains a terminal analog of the diffusion equation:

$$\frac{\partial v_x}{\partial t} = k\nu_1(\frac{\partial v_x}{\partial y})^{k-1}\frac{\partial^2 v_x}{\partial y^2}, \; \nu_1 = \frac{\mu_1}{\rho}, \; \frac{\partial v_x}{\partial y} \to 0 \tag{7.28}$$

First we will show that (7.28) has a random solution, and then we will discuss its physical interpretation.

Assuming that $v_x(t, y) = v_1(t)v_2(y)$ and separating the variables one arrives at the following equations:

$$\dot{v}_1 = \lambda v_1^k, v_2''(v_2')^{k-1} = \frac{\lambda}{\nu_1 k}v_2, \; \lambda = \text{Const} \tag{7.29}$$

It can be verified (by substitution) that

$$v_2 = \gamma(y+c)^{\frac{k+1}{k-1}}, \ \gamma = [\frac{\lambda(k-1)^{k+1}}{2\nu_1 k(k+1)^k}]^{\frac{1}{k-1}}, \ \lambda > 0, c = (\frac{v_o}{\gamma})^{\frac{k-1}{k+1}}, \ v_0 = v_2(0)\cdots \tag{7.30}$$

while for v_1 the solution is similar to (7.16):

$$v_1 = \pm[\lambda(1-k)t]^{\frac{1}{1-k}} \tag{7.31}$$

and therefore:

$$v_x = \pm\gamma[\lambda(1-k)t]^{\frac{1}{1-k}} \ [y + (\frac{v_0}{\gamma})^{\frac{k-1}{k+1}}]^{\frac{k+1}{k-1}} \tag{7.32}$$

It follows from (7.32), $v = 0$ at $t = 0$. But because of "infinite" instability of the solution (7.31), any infinitesimal disturbance having the same form as (7.30) will become finite during a finite time interval. Moreover, since (7.32) has two symmetric branches, with the equal probability 0.5 the solution (7.30) can be positive or negative. Therefore, (7.32) produces random solutions without any finite random input.

In support to our formal mathematical analysis let us discuss a physical interpretation of the phenomena. The classical version of (7.28) describes the velocity field induced by a sudden move of an infinite plane boundary. But if this boundary has a finite length, it should be replaced by equations of the boundary layer on a flat plate which [with the constitutive law (7.27)] read:

$$\frac{\partial v_y}{\partial t} + v_x\frac{\partial v_x}{\partial x} + v_y\frac{\partial v_x}{\partial y} = k\nu_1(\frac{\partial v_x}{\partial y})^{k-1}\frac{\partial^2 v_x}{\partial y^2}, \ \frac{\partial v_x}{\partial x} + \frac{\partial v_y}{\partial y} = 0 \tag{7.33}$$

Suppose that one considers these equations within an infinitesimal neighborhood of the plate leading edge $y = 0$ where

$$v_x, v_y, \frac{\partial v_x}{\partial x}, \frac{\partial v_x}{\partial y} \to 0 \text{ at } x, y \to 0 \tag{7.34}$$

In the classical case ($k = 1$) the conditions (7.34) would lead to zero acceleration of the fluid at the leading edge:

$$\frac{\partial v_x}{\partial t} = 0 \text{ at } x, y, \to 0 \tag{7.35}$$

which is in contradiction to the sudden relative motion between the plate and the fluid. However, for $k < 1$, (7.33) [with the conditions (7.34)] they reduce to (7.28) and have the solution (7.32). This solution describes the behavior of a fluid characterized by an "infinite" viscosity at the equilibrium state [see (7.14)] which can be associated with a static dry friction. That is why a sudden motion of a plate does not lead to an immediate concentrated jump of velocity gradients: instead it causes a smooth velocity distribution around the leading edge of the plate. Driven by the instability, an infinitesimal velocity distribution (7.30) becomes finite during a finite time interval.

Obviously the behavior of the solution to (7.33) beyond the infinitesimal neighborhood of the leading edge is described by the classical boundary layer equations. For supercritical Reynolds numbers this solution becomes unstable and it amplifies the contributions (7.32) coming from the leading edge of the plate. Thus, the combination of the classical mechanism of instability (due to inertia effects) and terminal instability at the fluid equilibrium leads to the probabilistic solutions describing turbulent motions.

In order to illustrate terminal attraction, let us consider a plane incompressible flow with a stream function ψ and the constitutive law:

$$\sigma_{xy} = \mu_1\left(\frac{\partial^2\psi}{\partial y^2} - \frac{\partial^2\psi}{\partial x^2}\right)^{k+1}, \quad v_x = \frac{\partial\psi}{\partial y}, \quad v_y = -\frac{\partial\psi}{\partial x}, \quad k < 1 \tag{7.36}$$

Based upon the relationship between the rate of change of the kinetic energy and the dissipation function, (Landau and Lifshitz 1959) one obtains:

$$\frac{\rho}{2}\frac{\partial}{\partial t}\int_V \left[\left(\frac{\partial\psi}{\partial x}\right)^2 + \left(\frac{\partial\psi}{\partial y}\right)^2\right] dxdy = -\mu_1 \int_V \left(\frac{\partial^2\psi}{\partial y^2} - \frac{\partial^2\psi}{\partial x^2}\right)^{k+1} dxdy \tag{7.37}$$

where V is the volume occupied by the fluid.

Suppose that $\psi(t, x, y)$ can be represented as a product $\psi = \tilde{\psi}(t)\bar{\psi}(x, y)$. Then (7.37) reduces to the ordinary differential equation with respect to $\varphi(t) = \tilde{\psi}^2(t)$:

$$\dot{\phi} = -\gamma\nu_1\varphi^k, \quad \text{and} \tag{7.38}$$

$$\gamma = \frac{\int_V \left(\frac{\partial^2\bar{\psi}}{\partial y^2} - \frac{\partial^2\bar{\psi}}{\partial x^2}\right)^{k+1} dxdy}{\int_V \left[\left(\frac{\partial\bar{\psi}}{\partial x}\right)^2 + \left(\frac{\partial\bar{\psi}}{\partial y}\right)^2\right] dxdy} = \text{Const}, \quad \nu_1 = \frac{\mu_1}{\rho}$$

Equation (7.38) describes damping of the fluid motion due to viscous stress (7.36). The equilibrium state represents a terminal attractor which is approached in a finite time:

$$t_0 = \frac{\varphi_0^{1-k}}{\gamma\nu_1(1-k)}, \quad \varphi_0 = \varphi(0) \tag{7.39}$$

Equation (7.39) allows one to evaluate k and ν_1 from experimental measurements of t_0.

In conclusion it should be stressed again that all the new effects of terminal dynamics emerge within vanishingly small neighborhoods of equilibrium states which are the only domains where the governing equations are different from classical.

7.2 Terminal Dynamics Limit Sets

7.2.1 General Remarks

The governing equations of classical dynamics may be derived from Lagrangian function, from variational principles, or directly from Newton's laws of motion, and they may be presented in various equivalent forms. However, one mathematical restriction to all of these forms is always applied: the differential equations describing a dynamical system

$$\dot{x}_i = v_i(x_1, x_2 \cdots x_n),\ i = 1, 2, ..., n \tag{7.40}$$

must satisfy the Lipschitz condition which expresses that all the derivatives

$$\left| \frac{\partial v_i}{\partial x_j} \right| < \infty \tag{7.41}$$

must be bounded. Actually this mathematical restriction guarantees the uniqueness of the solution to (7.40) subject to fixed initial conditions. Such a uniqueness seemed to be very important while the dynamical systems have been applied for modelling of energy transformations in mechanics, physics, and chemistry. However, attempts to exploit classical dynamics for information processing with applications to modeling biological and social behaviors have exposed certain limitations of the approach because of determinism and reversibility of solutions to (7.40) with the conditions (7.41). Mathematical and physical aspects of these limitations as well as the consequences of their removal were discussed in the previous section. In this section we will present a general structure of dynamical systems which does not possess a unique solution due to violation of the condition (7.41) at equilibrium points.

7.2.2 Terminal Attractors and Repellers

Terminal dynamics can be introduced as a set of nonlinear ordinary differential equations of the form:

$$\dot{x}_i = v_i^k(x_1, x_2, \ldots, x_n), i = 1, 2, \ldots, n \tag{7.42}$$

in which

$$\left| \frac{\partial v_i}{\partial x_j} \right| < \infty \tag{7.43}$$

and $k < 1$.

Therefore,

$$\left| \frac{\partial \dot{x}_i}{\partial x_j} \right| = k v^{(k-1)}(x_1, \cdots x_n) \left| \frac{\partial v_i}{\partial x_i} \right| \to \infty \text{ if } \dot{x}_i \to 0 \tag{7.44}$$

and the Lipschitz condition (7.41) is violated at all the equilibrium points

$$\dot{x}_i = 0$$

As in the classical case, the equilibrium points are attractors if the real parts of the eigenvalues of the matrix

$$m = \parallel \frac{\partial v_i}{\partial x_j} \parallel \tag{7.45}$$

are negative, that is

$$\operatorname{Re} \lambda_i < 0 \tag{7.46}$$

and are repellers if some of the eigenvalues have positive real parts.

In order to emphasize the difference between classical and terminal equilibrium points we will start with the simplest terminal system:

$$\dot{x} = -x^{1/3} \tag{7.47}$$

This equation has an equilibrium point at $x = 0$ at which the Lipschitz condition (7.41) is violated:

$$\frac{d\dot{x}}{dx} = -\frac{1}{3} x^{-2/3} \to -\infty \text{ at } x \to 0 \tag{7.48}$$

Since here the condition (7.46) is satisfied:

$$\operatorname{Re} \lambda \to -\infty < 0 \tag{7.49}$$

this point is an attractor of "infinite" stability.

The relaxation time for a solution with the initial condition $x = x_0 < 0$ to this attractor is finite:

$$t_0 = -\int_{x_0}^{x \to 0} \frac{dx}{x^{1/3}} = \frac{3}{2} x_0^{2/3} < \infty \tag{7.50}$$

Consequently, this attractor becomes terminal. It represents a singular solution which is intersected by all the attracted transients (Figs. 7.1–7.3).

For the equation:

$$\dot{x} = x^{1/3} \tag{7.51}$$

the equilibrium point $x = 0$ becomes a terminal repeller:

$$\frac{d\dot{x}}{dx} \to \frac{1}{3} x^{-\frac{2}{3}} \to \infty \text{ at } x \to 0, \text{ i.e., } \operatorname{Re} \lambda \to \infty > 0 \tag{7.52}$$

If the initial condition is infinitely close to this repeller, the transient solution will escape the repeller during a finite time period:

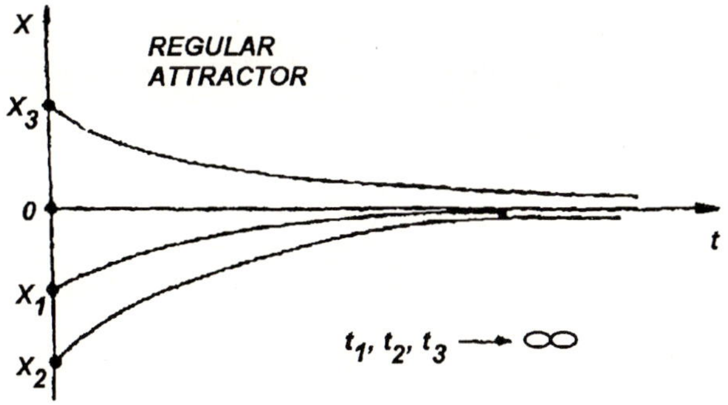

Fig. 7.1. Convergence to a regular attractor

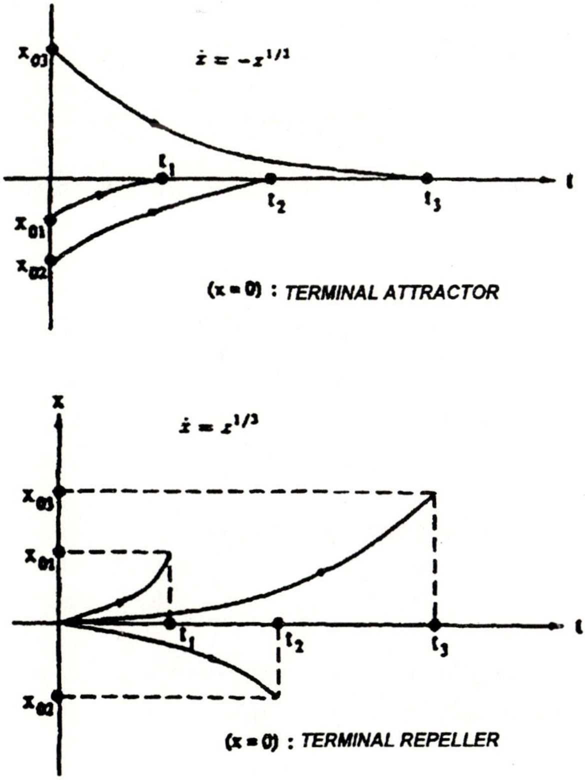

Fig. 7.2.

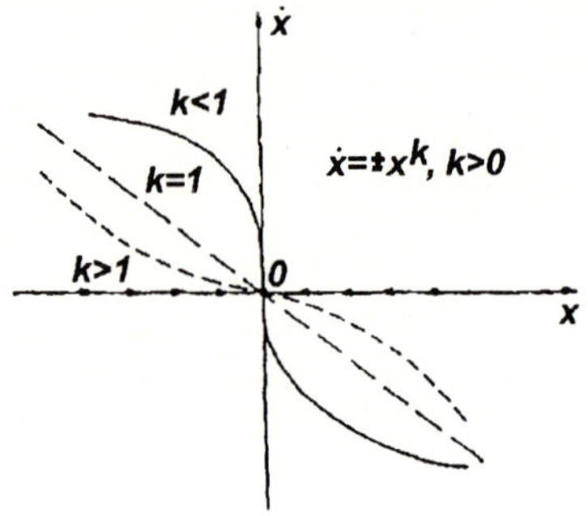

Fig. 7.3.

$$t_0 = \int_{\varepsilon \to 0}^{x_0} \frac{dx}{x^{1/3}} = \frac{3}{2} x_0^{2/3} < \infty \text{ if } x < \infty \tag{7.53}$$

while for a regular repeller, the time would be infinite.

Instead of (7.47) and (7.51), one can consider a more general case:

$$\dot{x} = \pm x^k, k > 0 \tag{7.54}$$

for which the relaxation time (for the attractor) or the escaping time (for the repeller) is:

$$t_0 \begin{cases} \to \infty & \text{if } k \geq 1 \\ = \frac{x_0^{1-k}}{1-k} & \text{if } k < 1 \end{cases} \tag{7.55}$$

As shown in the theory of differential equations, singular solutions in equations:

$$F(x, y, y') = 0 \tag{7.56}$$

are found by eliminating y' from the system:

$$F(x, y, y') = 0, \quad \frac{\partial F}{\partial y'} = 0 \tag{7.57}$$

Hence, static terminal attractors [if they exist in (7.56)] must be among the solutions to the system (7.57).

7.2.3 Static Terminal Attractors and their Trajectories

As shown in nonlinear dynamics, different types of regular attractors (or repellers) can be introduced based upon the second order dynamical system linearized with respect to the origin $x = 0$, $y = 0$:

$$\begin{aligned} \dot{x} &= \quad ax + by \\ \dot{y} &= \quad cx + dy \end{aligned} \tag{7.58}$$

or:

$$\frac{dx}{dy} = \frac{ax + by}{cx + dy} \tag{7.59}$$

Depending upon the eigenvalues of the matrix:

$$M = \begin{pmatrix} a & b \\ c & d \end{pmatrix} \tag{7.60}$$

the attractors (repellers) $x = 0$, $y = 0$ can be a node, a star, a spiral point, or an improper node.

If instead of (7.59) one introduces the following system:

$$\begin{aligned} \dot{x} &= \quad (ax + by)^{1/3} \\ \dot{y} &= \quad (cx + dy)^{1/3} \end{aligned} \tag{7.61}$$

then the equilibrium point $x = 0$, $y = 0$ represents a terminal attractor (or repeller): the Lipschitz condition is violated at this point. Nevertheless, the differential equation of trajectories in configuration space x, y :

$$\frac{dx}{dy} = \left(\frac{ax + by}{cx + dy}\right)^{1/3} \tag{7.62}$$

satisfies the Lipschitz condition and it does not have any singular solutions. This means that both variables x and y are "simultaneously" approaching the terminal attractor, as in the case of a regular attractor [see (7.59)]. Moreover, a similar classification of terminal attractors of the type (7.61) can be performed, based upon the coefficients a, b, c, and d.

This section introduces a more "pathological" situation, when for the differential equations of trajectories in configuration space, the Lipschitz condition is also violated. As will be shown below, such a violation will lead to the loss of the uniqueness of the solutions in the configuration space: the trajectories will merge before approaching the terminal attractor.

Let us start with the following dynamical system:

$$\dot{x} = \quad -(x - x^*)^{1/3} \tag{7.63}$$

$$\dot{y} = \quad [y(x - x^*)]^{1/3} \tag{7.64}$$

It is easily verifiable that the Lipschitz condition here is violated at $x = x^*$, $y = 0$.

The differential equation of trajectories in configuration space x, y can be written as:

$$\frac{dy}{dx} = -y^{1/3} \tag{7.65}$$

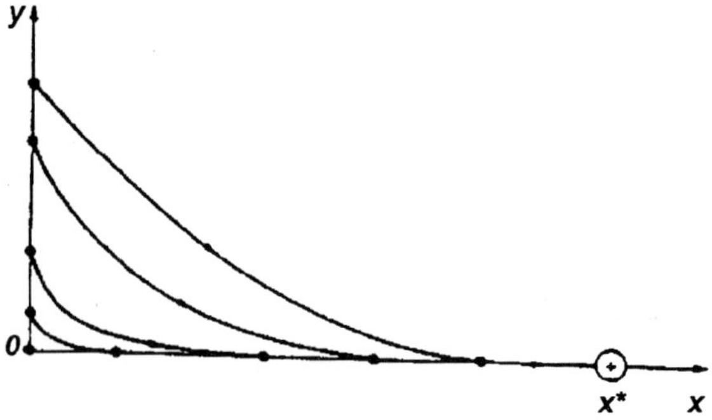

Fig. 7.4. Terminal dynamics trajectory

For this equation, the Lipschitz condition is violated at $y = 0$. This means that $y = 0$ is a singular solution, and all the trajectories in configuration space, x, y first flow to the x axis; i.e., $y = 0$, and then approach the random attractor $x = x^*$, $y = 0$ (Fig. 7.4).

Indeed, as follows from (7.65):

$$y = \begin{cases} (y_0^{2/3} - \frac{2}{3}x)^{3/2} & \text{for } x \leq \frac{3}{2}y_0^{2/3} \\ 0 & \text{for } x \geq \frac{3}{2}y_0^{2/3} \end{cases} \tag{7.66}$$

while $x = x(t)$ follows from (7.63):

$$x = \begin{cases} x^* + [(x_0 - x^*)^{2/3} - \frac{2}{3}t]^{3/2} & \text{for } t \leq \frac{3}{2}(x_0 - x^*)^{2/3} \\ 0 & \text{for } t \geq \frac{3}{2}(x_0 - x^*)^{2/3} \end{cases} \tag{7.67}$$

here x_0 and y_0 are the initial conditions.

The time of approaching the singular solution $y = 0$ by the variable y follows from (7.66) if $x(t)$ is substituted form (7.67):

$$t_1 = \frac{3}{2}(x_0 - x^*)^{2/3} \tag{7.68}$$

The time t_1 of the convergence of the solution to the terminal attractor follows from (7.67):

$$t_2 = \frac{3}{2}[(x_0 - x^*)^{2/3} - \frac{3}{2}(x_0 - x^*)^{2/3}] \tag{7.69}$$

Obviously:

$$t_2 < t_1 \tag{7.70}$$

This means that the trajectory of the motion of the original dynamical system (7.63–7.64) in the configuration space, x, y, first flows into the

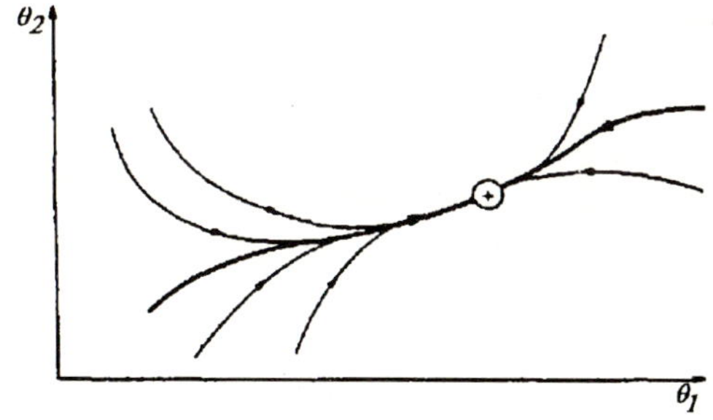

Fig. 7.5.

trajectory $y = 0$, and only then does it approach the terminal attractor $x = x^*$, $y = 0$. Such a trajectory as $y = 0$ we will call a terminal trajectory.

The situation described above can be generalized to the case where a terminal trajectory is a prescribed curve. Indeed, turning again to the system (7.63–7.64) let us transfer to a new system of coordinates

$$x = \vartheta_1, \; y = f(\vartheta_1, \vartheta_2) \tag{7.71}$$

assuming that f is a differentiable function and $\partial f/\partial\vartheta_2 \neq 0$.

Then (7.63–7.64) read:

$$\dot{\vartheta}_1 = -(\vartheta_1 - \vartheta_1^*)^{1/3} \tag{7.72}$$

$$\dot{\vartheta}_2 = \frac{1}{\partial f/\partial\vartheta_2}(\vartheta_1 - \vartheta_1^*)^{1/3}(f^{1/3} + \frac{\partial f}{\partial\vartheta_1}) \tag{7.73}$$

The terminal trajectory $y = 0$ is converted into a curve:

$$f(\vartheta_1, \vartheta_2) = 0 \tag{7.74}$$

Hence, for a desired terminal trajectory (7.74), the corresponding dynamical system is (7.72–7.73) (Fig. 7.5).

7.2.4 Physical Interpretation of Terminal Attractors

As has been pointed out in the introduction of the chapter, the mathematical formalism of terminal dynamics follows from a more general structure of the dissipation function, which allows for the existence of smooth transitions from static to kinetic friction. It should be emphasized that the behavior of the solutions around the equilibrium points in terminal dynamics is more "realistic" than in classical dynamics since actual time of convergence to equilibrium points is finite. However, in order to make it finite, one has to violate the Lipschitz condition (7.2) since all the trajectories must intersect at the equilibrium point. In classical dynamics, the Lipschitz condition

(7.2) is not violated, and the infinite time of convergence is accounted for some "small dissipative forces" which are always present. Actually terminal dynamics incorporates these forces via the parameter k [see (7.3)] which can be found from measurement of the convergence time, [see (7.16)].

It can be shown that the terminal attractor as a mathematical concept has other physical interpretations, and one of them is the energy-cumulation effect, although in this case one deals with the finite time of convergence of a propagating wave rather than a motion of an individual particle. As an example, consider a propagation of an isolated pulse in an elastic continuum along the x axis. In general, the speed of propagation $\dot{x} = \lambda$ depends on x. Suppose there exists such a point x^* where $\lambda(x^*) = 0$. Then the time t^* during which the leading front of the propagating pulse will approach this point is expressed via the following integral:

$$t^* = \int_{x_0}^{x \to x^*} \frac{dx}{\lambda(x)} \tag{7.75}$$

If λ can be presented in the form:

$$\lambda = (x^* - x)^k, \; 0 < k < 1 \tag{7.76}$$

then this integral converges and, therefore, the time t^* is finite. It is easily verifiable that in this case the differential equation:

$$\dot{x} = (x^* - x)^k) \tag{7.77}$$

describing the dynamics of the pulse propagation has a terminal attractor at $x = x^*$. But if the leading and the trailing fronts of the propagating pulse approach the same point x^* during finite time, then eventually the width of the pulse will shrink to zero, and all the energy transported by the pulse will be distributed over a vanishingly small length. Hence, the existence of a terminal attractor in such models leads to an unbounded concentration of energy in the neighborhood of the attractor.

Based upon this model, one can explain and describe the formation of a supersonic snap at a free end of a filament suspended in a gravity field, as well as the cumulation of the shear strain energy at the soil surface as a response to an underground explosion. In these models, the free end of the filament and the free surface of the soil serve as random attractors (see Chap. 3, Sect. 4).

Some terminal dynamics effects in fluid dynamics were introduced and discussed in Chap. 5.

7.2.5 Periodic Terminal Limit Sets

So far, we have concentrated on static random attractors. We will now demonstrate the existence of periodic terminal attractors. For this purpose, let us consider a dynamical system separable in polar coordinates r, θ:

$$\dot{r} = r(R - r)^{1/3}, \ (r \leq R) \tag{7.78}$$

$$\dot{\theta} = \omega \tag{7.79}$$

Here, $d\dot{r}/dr \to -\infty$ at $r \to R$ [compare with (7.9)] and, therefore, the solutions $r = R$, $\theta = \omega t + \theta_R$ is a terminal limit cycle. Its basin is defined by the condition: $r > 0$. For the solution with the initial condition $r_0 < R$ the relaxation time is finite:

$$t_0 = \int_{r_0}^{R} \frac{dr}{r(R-r)^{1/3}} < \int_{r_0}^{R} \frac{dr}{r_0(R-r)^{1/3}} = \frac{2}{3r_0}(R - r_0)^{2/3} < \infty \tag{7.80}$$

It is easily verifiable that a periodic terminal repeller can be obtained by changing the sign in the right-hand side of (7.78).

The terminal analog of chaotic attractor was introduced and discussed in the introduction to this chapter [see (7.18–7.19)].

7.2.6 *Unpredictability in Terminal Dynamics*

The concept of unpredictability in classical dynamics was introduced in connection with the discovery of chaotic motions in nonlinear systems. Such motions are caused by the Liapunov instability, which is characterized by a violation of a continuous dependence of solutions on the initial conditions during an unbounded time interval ($t \to \infty$). That is why the unpredictability in these systems develops gradually. Indeed, if two initially close trajectories diverge exponentially:

$$\varepsilon = \varepsilon_0 \exp \lambda t, \ 0 < \lambda < \infty \tag{7.81}$$

then for an infinitesimal initial distance $\varepsilon_0 \to 0$, the current distance ε becomes finite only at $t \to \infty$. For this reason, the Liapunov exponents (the mean exponential rate of divergence) are defined in an unbounded time interval:

$$\sigma = \lim(\frac{1}{t})\ell n\frac{\varepsilon}{\varepsilon_0}, \ t \to \infty \tag{7.82}$$

In distributed dynamical systems, described by partial differential equations, there exists a stronger instability discovered by Hadamard. In the course of this instability, a continuous dependence of a solution on the initial conditions is violated during an arbitrarily small time period. Such a blow-up instability is caused by a failure of hyperbolicity and transition to ellipticity, and was discussed in Chap. 3, Sect. 3.

This section will show that a similar type of a blow-up instability leading to "discrete pulses" of unpredictability can occur in dynamical systems which contain terminal repellers.

Let us analyze the transient escape from the terminal repeller in the equation

$$\dot{x} = x^{1/3}, \; x_0 = x(0) \tag{7.83}$$

assuming that $|x_0| \to 0$. The solution to (7.83) reduces to the following:

$$x = \pm t^{3/2}, \quad x \neq 0 \tag{7.84}$$

Hence, two different solutions are possible for "almost the same" initial conditions. The most essential property of this result is that the divergence of the solutions (7.84) is characterized by an unbounded parameter:

$$\sigma = \lim_{t \to t_0} \left(\frac{1}{t} \ln \frac{2t^{3/2}}{2|x_0|} \right) = \infty, \; |x_0| \to 0 \tag{7.85}$$

where t_0 is an arbitrarily small (but finite) positive quantity. In contrast to (7.82), here the rate of divergency can be defined in an arbitrarily small time interval, since during this interval the initial infinitesimal distance between the solutions becomes finite. Thus, a terminal repeller represents a vanishingly short but infinitely powerful "pulse of unpredictability" which is "pumped" into the dynamical system.

In order to illustrate the unpredictability in such terminal dynamics, we will turn to the following equation:

$$\dot{x} - yx^{1/3} = 0, \text{ while} \tag{7.86}$$

$$y = \cos \omega t \tag{7.87}$$

Assuming that $x \to 0$ at $t \to 0$, one obtains regular solutions:

$$x = \pm \left(\frac{2}{3\omega} \sin \omega t \right)^{3/2}, \; x \neq 0 \tag{7.88}$$

and a singular solution (an equilibrium point)

$$x = 0 \tag{7.89}$$

During the first time period

$$0 < t < \frac{\pi}{2\omega} \tag{7.90}$$

the equilibrium point (7.89) is a terminal repeller (since $y > 0$). Therefore, within this period, the solutions (7.81) have the same property as the solutions (7.84): their divergence is characterized by an unbounded rate σ.

During the next time period:

$$\frac{\pi}{2\omega} < t < \frac{3\pi}{2\omega}$$

the equilibrium point (7.89) becomes a terminal attractor (since $y < 0$), and the system which approaches this attractor at $t = \pi\omega$ remains motionless until $t > 3\pi/2\omega$. After that, the terminal attractor converts into the terminal repeller, and the system escapes again, etc.

It is important to notice that each time the system escapes the terminal repeller, the solution splits into two symmetric branches, so that the total trajectory can be combined from 2^n pieces, where n is the number of cycles, i.e., it is the integer part of the quantity $(t/2\pi\omega)$. As one can see, here the nature of the unpredictability is significantly different from the unpredictability in chaotic systems.

One can notice that the motion (7.88) resembles chaotic oscillations known from classical dynamics: it combines random characteristics with the attraction to a center. However, in contradistinction to classical chaos, the motion (7.88) is driven by failure of uniqueness of the solution at the equilibrium point, and it has a well-organized probabilistic structure. Since the time of approaching the equilibrium point $x = 0$ by the solution (7.88) is finite, this type of chaos can be called terminal chaos.

Equations (7.86–7.87) can be presented in autonomous form:

$$x = yx^{1/3} \tag{7.91}$$

$$\dot{y} = -\omega z + y(1 - y^2 - z^2) \tag{7.92}$$

$$\dot{z} = \omega y + z(1 - y^2 - z^2) \tag{7.93}$$

if one takes into account that the last two equations have a periodic attractor:

$$y = \cos \omega t, \ \ z = -\sin \omega t$$

Although, in general, the structure of (7.91–7.93) resembles the structure of the Lorenz or the Rössler attractors, the violation of the Lipschitz conditions here is important for appearance of terminal solutions. Indeed, if (7.86) is replaced by the following:

$$\dot{x} = yx \tag{7.94}$$

then the solution to the system (7.91–7.93)

$$x = x_0 e^{-\frac{1}{\omega} \sin \omega t}, x_0 = x(0) \text{ at } t \rightarrow \infty$$

becomes periodic.

7.2.7 Irreversibility of Terminal Dynamics

Classical dynamics describe processes in which time t plays the role of a parameter: it remains fully reversible in the sense that the time-backward motion can be obtained from the governing equation by time inversion, $t \to -t$. This means that classical dynamics cannot explain the emergence of new dynamical patterns in nature.

However, there exists another class of phenomena where past and future play different roles, and time is not invertible: by definition (the second law of thermodynamics) irreversibility is introduced in thermodynamics by postulating the increase of entropy.

As stressed by Prigogine (1980), irreversible processes play a fundamental constructive role in physical world; they are at the basis of important coherent processes that appear with particular clarity on the biological level.

In this connection let us compare dynamical behavior of solutions in a small neighborhoods of classical and terminal repellers:

$$\dot{x} = x \tag{7.95}$$

and

$$\dot{x} = x^{1/3} \tag{7.96}$$

The solution to (7.95)

$$x_+ = x_0 e^t \tag{7.97}$$

describing an escape from a classical repeller is reversible since

$$u_- = x_0 e^t \tag{7.98}$$

is a possible motion describing a convergence to a classical attractor $x = 0$.

The solution to (7.96)

$$x_+ = \sqrt{(\frac{2}{3}t)^3} \tag{7.99}$$

is irreversible since the time-backward motion:

$$x_- = \sqrt{-(\frac{2}{3}t)^3} \tag{7.100}$$

does not exist (x has an imaginary value).

[Strictly speaking, the conventional definition of reversibility requires a stronger condition; namely, that $\overline{v}(t) = -\overline{v}(-t)$, which separates conservative from non-conservative systems in classical dynamics, while the one-to-one mapping between the past and future always exists. In terminal dynamics such mapping may not exist, and this is why the definition of irreversibility given above should include (both conservative and non-conservative) systems whose past can be uniquely reconstructed from the future in contradistinction to those systems for which it cannot be done.]

This mathematical formalism expresses deeper roots of irreversibility of terminal dynamics which can be understood if one turns to the solution of (7.86–7.87). This solution consists of regular (7.88) and singular (7.89) parts. When the regular solution approaches the equilibrium point $x = 0$ (in finite time), it switches to the singular solution $x \equiv 0$, and this switch is irreversible.

7.3 Probabilistic Structure of Terminal Dynamics

As shown above, the terminal version of Newtonian dynamics is different from its classical version only within vanishingly small neighborhoods of equilibrium states, and therefore, it contains classical mechanics as a special case. This means that terminal dynamics is not necessarily always unpredictable and irreversible: in some domains it is identical with the classical dynamics. However, in this section our attention will be concentrated on specific effects of terminal dynamics, and in particular, on its probabilistic structure.

There is a fundamental difference between the probabilistic properties of terminal dynamics and those of stochastic or chaotic differential equations Indeed, the randomness of stochastic differential equations is caused by random initial conditions, random force or random coefficients; in chaotic equations small (but finite!) random changes of initial conditions are amplified by the mechanism of instability. But in the both cases the differential operator itself remains deterministic. In contradistinction to this, in terminal dynamics, randomness results from the violation of the uniqueness of the solution at equilibrium points, and therefore, the differential operator itself generates random solutions.

7.3.1 Terminal Version of the Liouville-Gibbs Theorem

The Liouville-Gibbs theorem in classical dynamics expresses the relationship between the governing differential equations and equations for probability distribution functions. For the dynamical system:

$$\dot{x}_i = v_i(x_1, x_2, ..., x_n), \ i = 1, 2, ..., n \tag{7.101}$$

it has two equivalent forms:

$$\frac{\partial f}{\partial t} + \text{div}\ (f v_i) = 0 \tag{7.102}$$

or

$$f = f_0 \exp(-\int_0^t \text{div}\ v_i dt) \tag{7.103}$$

Here x_i are considered as random variables, while randomness is introduced only through initial conditions x_i^0 possessing a given joint distribution with the joint density f_0, and f is the current joint distribution. Since the operator v_i is deterministic, the system (7.101) can be solved in a deterministic way, and to make the solution vector x a random vector, it suffices to treat the initial conditions as random variables.

It should be recalled that (7.102–7.103) were derived for the case when the Lipschitz conditions are satisfied, i.e.,

$$\left|\frac{\partial \dot{x}_i}{\partial x_j}\right| < \infty, \ i, j = 1, 2, ..., n \tag{7.104}$$

which means that

$$\left|\frac{\partial \dot{v}_i}{\partial x_j}\right| < \infty, \ i, j = 1, 2, ..., n \tag{7.105}$$

For terminal dynamics given by (7.42–7.43), these conditions do not hold [see (7.44)]. Additionally, one can verify that

$$|\text{div } v_i| = k \left|\sum_{i=1}^{n} \frac{\partial v_i}{\partial x_i} v_i^{k-1}\right| \to \infty \text{ if } \dot{x}_j \to 0 \tag{7.106}$$

i.e., $|\text{div } v_i|$ is unbounded at equilibrium points.

This means that (7.102–7.103) are valid everywhere excluding the equilibrium points.

In order to define the distribution of f at $\dot{x}_j$, first of all we will evaluate the functions

$$x_i = x_i(t) \text{ at } x_i \to \overset{*}{x}_i = 0 \tag{7.107}$$

where $\overset{*}{x}_i$ are the coordinates of an equilibrium point.

For simplicity (but without loss of generality) we will assume that

$$\overset{*}{x}_i = 0 \tag{7.108}$$

Then

$$\dot{x}_i \sim (\alpha_i x_i)^k \text{ at } x_i \to 0, \ x_{j \neq i} = 0 \tag{7.109}$$

where

$$\alpha_i = \frac{\partial v_i}{\partial x_i} \text{ at } x_i = 0, \ x_2 = 0, ..., x_n = 0 \tag{7.110}$$

Let us assume first that

$$\alpha_i > 0 \tag{7.111}$$

i.e., the equilibrium point (7.102) is a terminal repeller. Then in a small neighborhood of this point:

$$x_i \sim t^{\frac{1}{1-k}} \text{ at } x_i \to 0 \tag{7.112}$$

and therefore

$$\text{div } v_i \sim k \sum_i (\alpha_i x_i)^{k-1} \sim \frac{1}{t} \text{ at } x_i \to 0 \tag{7.113}$$

Hence

$$-\int_0^t \text{div } v_i dt = \int_0^{\varepsilon \to 0} \text{div } v_i dt \sim \ln \frac{\varepsilon \to 0}{t} \text{ at } x_i \to 0 \tag{7.114}$$

i.e.,

$$\exp(-\int_0^t \text{div } v_i dt) \sim \frac{\varepsilon \to 0}{t} \text{ at } x_i \to 0 \tag{7.115}$$

Therefore, as follows from (7.103):

$$f \to 0 \text{ at } x_i \to 0 \text{ if } f_0 < \infty \tag{7.116}$$

This means that those trajectories which originated outside of the terminal repeller, will never approach it; as following from (7.116), the terminal repeller generates a probability even if the initial conditions are "almost" deterministic. In other words, it represents a "vacuum" of the probability density.

For a terminal attractor, i.e., when

$$\alpha_i < 0 \tag{7.117}$$

after following the similar transformations as those performed in (7.106–7.109), one obtains:

$$f \to \infty \text{ at } x_i \to 0 \tag{7.118}$$

Hence, those trajectories which originated outside of the terminal attractor, will definitely approach it, i.e., the terminal attractor represents a center of concentration of all the probability "mass."

7.3.2 Terminal Dynamics Model of Random Walk

A random walk is a stochastic process where changes occur only at fixed times. In this section we will introduce terminal dynamics which describes this process.

Let us start with the following dynamical system:

$$\dot{x} = \gamma \sin^{1/3} \frac{\sqrt{\omega}}{\alpha} x \sin \omega t, \ \gamma = \text{Const}, \ \omega = \text{Const}, \ \alpha = \text{Const} \tag{7.119}$$

It can be verified that at the equilibrium points:

$$x_m = \frac{\pi m \alpha}{\sqrt{\omega}} \quad m =, \ldots, -2, -1, 0, 1, 2, \ldots, \text{etc.} \tag{7.120}$$

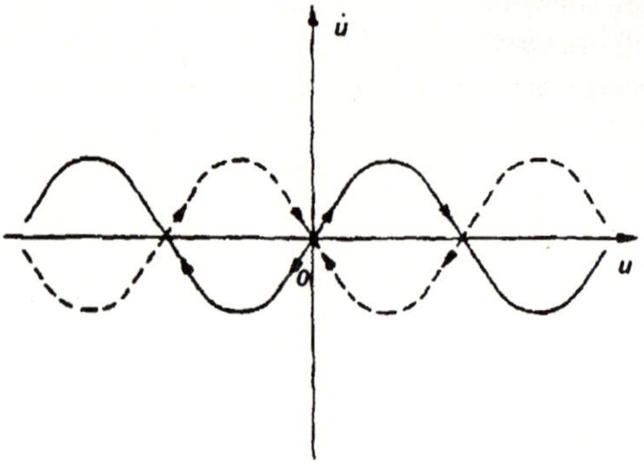

Fig. 7.6.

the Lipschitz condition is violated:

$$\partial \dot{x}/\partial x \to \infty \text{ at } x \to x_m \tag{7.121}$$

If $x = 0$ at $t = 0$, then during the first period

$$0 < t < \frac{\pi}{\omega} \tag{7.122}$$

the point $x_0 = 0$ is a terminal repeller since $\sin \omega t > 0$ and the solution at this point splits into two (positive and negative) branches whose divergence is characterized by unbounded parameter σ [see (7.82)]. Consequently, with an equal probability x can move into the positive or the negative direction. For the sake of concreteness, we will assume that it moves in the positive direction. Then the solution will approach the second equilibrium point $x_1 = \pi\alpha/\sqrt{\omega}$ at

$$t^* = \frac{1}{\omega} \arccos \left[1 - \frac{B(\frac{1}{3}, \frac{1}{3})}{2^{1/3}} \frac{\alpha\sqrt{\omega}}{\gamma} \right] \tag{7.123}$$

in which B is the Beta function.

It can be verified that the point x_1 will be a terminal attractor at $t = t_1$ if

$$t_1 \leq \pi/\omega, \text{ i.e., if } \frac{\gamma}{\alpha} \geq \frac{B(\frac{1}{3}, \frac{1}{3})}{2^{4/3}} \sqrt{\omega} \tag{7.124}$$

Therefore, x will remain at the point x_1 until it becomes a terminal repeller, i.e., until $t > t_1$. Then the solution splits again: one of two possible branches approach the next equilibrium point $x_2 = 2\pi\alpha/\sqrt{\omega}$, while the other returns to the point $x_0 = 0$, etc. The periods of transition from one equilibrium point to another are all the same and are given by (7.123) (Fig. 7.6).

It is important to notice that these periods t^* are bounded only because of the failure of the Lipschitz condition at the equilibrium points. Otherwise they would be unbounded since the time of approaching a regular attractor (as well as the time of escaping a regular repeller) is infinite.

Thus, the evolution of x prescribed by (7.119) is totally unpredictable: it has 2^m different scenarios where $m = E(t/t^*)$, while any prescribed value of x from (7.120) will appear eventually. This evolution is identical to random walk, and the probability $f(x,t)$ is governed by the following difference equation:

$$f(x, t + \frac{\pi}{\omega}) = \frac{1}{2} f(x - \frac{\pi\alpha}{\sqrt{\omega}}, t) + \frac{1}{2} f(x + \frac{\pi\alpha}{\sqrt{\omega}}, t) \tag{7.125}$$

For better physical interpretation we will assume that

$$\frac{\pi\alpha}{\sqrt{\omega}} \ll L, t^* \ll T, \text{ i.e., } \omega \to \infty \tag{7.126}$$

in which L and T are the total length and the total time period of the random walk. Setting

$$\frac{\pi\alpha}{\sqrt{\omega}} \to 0, t^* \to 0 \tag{7.127}$$

one arrives at the Fokker-Planck equation:

$$\frac{\partial f(x,t)}{\partial t} = \frac{1}{2} D^2 \frac{\partial^2 f(x,t)}{\partial x^2}, \quad D^2 = \pi\alpha^2 \tag{7.128}$$

Its unrestricted solution for the initial condition that the random walk starts from the origin $x = 0$ at $t = 0$:

$$f(x,t) = \frac{1}{\sqrt{(2\pi D^2 t)}} \exp(-\frac{x^2}{2D^2 t}) \tag{7.129}$$

qualitatively describes the evolution of the probability distribution for the dynamical equation (7.119). It is worth to notice that for the exact solution one should turn to the difference equation (7.124) since actually $\omega < \infty$.

Equation (7.124) can be presented in the following operator form:

$$[E_t - \frac{1}{2}(E_x + E_x^{-1})]f = 0 \tag{7.130}$$

where E_t and E_x are shift operators:

$$E_t f(x,t) = f(x, t+\tau), E_x f(x,t) = f(x+h, t), \quad h = \frac{\pi\alpha}{\sqrt{\omega}} \tag{7.131}$$

Utilizing the relationships between the shift and the differential operator D:

$$E_t^r = e^{r\tau D_t}, \; E_x^r = e^{rhD_x}, \; D_t = \frac{\partial}{\partial t}, D_x = \frac{\partial}{\partial x} \tag{7.132}$$

one can transfer from (7.124) to (7.128) if $\omega \to \infty$, i.e., $\tau, h \to 0$.

For further analysis it will be more convenient to modify (7.119) as follows:

$$\dot{x} = \gamma \sin^k(\frac{\sqrt{\omega}}{\alpha} x) \sin \omega t \tag{7.133}$$

assuming that

$$k = \frac{1}{2n+1}, \; n \to \infty \tag{7.134}$$

while n is an integer.

This replacement does not change the qualitative behavior of the dynamical system (7.133): it changes only its quantitative behavior between the critical points in such a way that one has explicit control over the period of transition from one critical point to another. Indeed, since

$$\lim \sin_{n\to\infty}^{1/2n+1} X = \operatorname{sgn} \sin X$$

one obtains the solution for x which is valid between critical points $x^{(m)}$ and $x^{(m+1)}$:

$$x = \frac{\gamma}{\omega}(1 - \cos \omega t) \tag{7.135}$$

Obviously the distances between the equilibrium points will not depend upon the step m:

$$h_m = x_m - x_{m-1} = \frac{\pi \alpha m}{\sqrt{\omega}} - \frac{\pi \alpha (m-1)}{\sqrt{\omega}} = \frac{\pi \alpha}{\sqrt{\omega}} \tag{7.136}$$

The period of transition from the $(m-1)^{\text{th}}$ to m^{th} critical point follows from (7.135–7.136):

$$t^* = \frac{1}{\omega} \arccos(1 - \frac{h_m}{\gamma}) \le \frac{\pi}{\omega} \tag{7.137}$$

This means that

$$\delta \ge \omega h_m \tag{7.138}$$

since it should not exceed the period between the conversions of terminal attractors into terminal repellers and vice-versa.

7.3.3 Numerical Implementation

Introduction

The terminal attractor behaves in theory like a dynamical system with a Gaussian distribution. To test this property, a numerical simulation of the attractor was executed on a SUN system and the TPL-CRAY. Programming was done in Mathematica® and C.

The system (compare this implementation with Chap. 11)

$$F = \alpha_1 v^k \cos \omega t \tag{7.139}$$

was chosen for the implementation. It alternates attractors with repellers. A modification of this equation leads to

$$\dot{x} = \gamma \sin \left[\frac{\sqrt{\omega}}{\alpha} \operatorname{erf m} \left(\frac{x}{\sigma \sqrt{2}} \right) \right]^{\frac{k}{k+2}} \sin \omega t \tag{7.140}$$

where

$$\operatorname{erf m}(x) = \begin{cases} -1 \text{ if} & x < -N \\ \operatorname{erf}(x) \text{ if} & |x| \leq N \\ 1 & x > N \end{cases} \tag{7.141}$$

The first sine term reaches zero whenever its argument equals a multiple of π. The number of values of x determines the levels of the system. The distance between two levels is the step size. Due to the special form of the erfm function, these levels are close to each other around the origin and the steps increase leaving from either side of the origin.

This first term takes care of the switching from one level to another level. As soon as the next level is reached, this term will be zero, and the system is in equilibrium. If a noise term is introduced, this equilibrium will still be stable, as long as the sign of the most right sine term does not change.

This sine term regulates the occurrence of changes in direction of the system. Every time this term changes sign – after every half period (semicycle), the slightest bit of noise can cause the state either to increase or decrease. This results in either a switch to the next level or to the previous level. After every semicycle the system switches from one level to a neighboring level. If the noise is uniform, the chances of increase and decrease are equal. This means that the system simulates a discrete random walk, and it behaves like a normally distributed Markov chain.

Simulations

The purpose of the numerical simulations was to show the normal distribution property of the system from (7.140).

Two ways of showing this property are available: a) The average number of occurrences of the state in a certain region of the state space can be

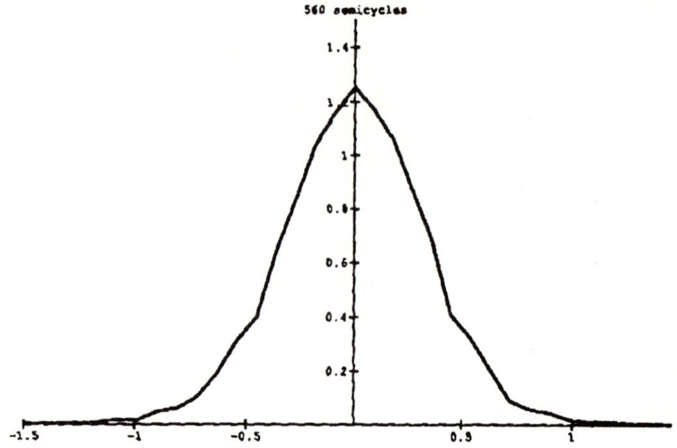

Fig. 7.7.

used. Since this average is merely an average in time instead of the mean, this method is not as strong as the second one. b) The other possibility is to simulate a large number of processes simultaneously, and take the frequency for each state space region at each time point,

Ten thousand processes were simulated simultaneously using the second method.

Results

If the highest (or lowest) level was reached, then the next level could not be reached, simply because it did not exist. To avoid this, a boundary condition was introduced. Every time a simulation crossed this boundary, the noise term was chosen in such a way that the state would return inside the legal area. This introduces a new problem, since now the uniformness of the noise is affected. The expectation was to see small peeks arise near the boundaries, but they would disappear after a sufficient number of semicycles. This can be compared with (unforced) waves in a bounded area; after the waves hit the boundaries, the reflections cause resonance and eventually settle down.

Simulations were carried out for 1000 semicycles, leading to the distribution as shown in Fig. 7.7.

7.3.4 Multidimensional Systems

The results presented in the previous sections can be generalized to multidimensional dynamics. For that purpose consider the following terminal dynamical system:

$$\dot{x}_i = \gamma_i \sin^k\left(\frac{\sqrt{\omega}}{\alpha_i}\sum T_{ij}x_j\right)\sin\omega t, \; T_{ij} = \text{Const} \tag{7.142}$$

assuming that

$$T_{ij} = T_{ji}, T_{11} > 0, \begin{vmatrix} T_{11} & T_{12} \\ T_{12} & T_{22} \end{vmatrix} > 0, ..., \text{etc.} \tag{7.143}$$

i.e. $| T_{ij} |$ is a symmetric positive-definite matrix.

Here k is defined by (7.134). The properties (7.143) provide stability (if $\sin\omega t < 0$), or instability (if $\sin\omega t > 0$) of the system (7.142) at the terminal dynamics equilibrium points $\overset{*}{x}_i$:

$$\overset{*}{x}_i = \lambda_i \sum_{j=1}^{n} m_{ij} \frac{\partial\Delta}{\partial T_{ij}} \tag{7.144}$$

Here

$$\lambda_i = \frac{\pi\alpha_i}{\Delta\sqrt{\omega}}, \; \Delta = \det | T_{ij} | \tag{7.145}$$

m_i is the number of steps made by the variable x_i, and $\frac{\partial\Delta}{\partial T_{ij}}$ is a cofactor of the element T_{ij}.

After one step of a variable x_i, the corresponding value of m_i will change to $m_i + 1$ or $m_i - 1$ with the same probability 0.5. Hence, the length of a step h_i made by the variable x_i will have 2^n equally probable values:

$$h_i = \lambda_i \sum_{j=1}^{n} (\pm\beta_{ij}) \frac{\partial\Delta}{\partial T_{ij}}, \; \beta_{ij} = 1 \tag{7.146}$$

depending on 2^n combinations of the signs of β_{ij} in (7.146).

Denoting each of these combinations by $q (q = 1, 2, \cdots 2^n)$, and introducing a shift operator E_i for each variable x_i:

$$E_i f(t, x_1, ..., x_i, ..., x_n) = f(t, x_1, ..., x_i + 1, ..., x_n) \tag{7.147}$$

one arrives at the following governing equation for the joint probability density of the solution to (7.142):

$$E_t - 2^{-n} \sum_{q=1}^{2^n} \Pi_{i=1}^{n} E_i^{h_{iq}}) f = 0 \tag{7.148}$$

where h_{iq} is a particular value of h_i taken from (7.143) at a particular q.

As follows from (7.148), with increase of n the dynamics of (7.142) becomes less and less predictable.

For $n = 2$, (7.148) reduces to:

$$[E_t - \frac{1}{4}(E_1^{h_{11}} E_2^{h_{21}} + E^{h_{12}} E^{h_{22}} + E^{h_{13}} E^{h_{23}} + E^{h_{14}} E^{h_{24}})]f = 0 \qquad (7.149)$$

where

$$h_{11} = -h_{14} = \frac{\pi}{\Delta\sqrt{\omega}}(\alpha_1 T_{22} - \alpha_2 T_{12}),$$

$$h_{12} = -h_{13} = \frac{\pi}{\Delta\sqrt{\omega}}(\alpha_1 T_{22} + \alpha_2 T_{12}),$$

$$h_{21} = -h_{24} = \frac{\pi}{\Delta\sqrt{\omega}}(\alpha_2 T_{11} - \alpha_1 T_{12}),$$

$$h_{22} = -h_{23} = \frac{\pi}{\Delta\sqrt{\omega}}(\alpha_2 T_{11} + \alpha_1 T_{12}) \qquad (7.150)$$

If $\omega \to \infty$, i.e., $h_{ij}\tau \to 0$, (7.149) transforms into a two-dimensional Fokker-Planck equation:

$$\frac{\partial f}{\partial t} = \frac{1}{2}(D_{11}\frac{\partial^2 f}{\partial x_1^2} + D_{12}\frac{\partial^2 f}{\partial x_1 \partial x_2} + D_{22}\frac{\partial^2 f}{\partial x_2^2}) \qquad (7.151)$$

where

$$D_{11} = \frac{\pi}{\Delta^2}(\alpha_1^2 T_{22}^2 + \alpha_2^2 T_{12}^2)$$

$$D_{12} = \frac{2\pi T_{12}}{\Delta^2}(\alpha_1^2 T_{22} + \alpha_2^2 T_{11})$$

$$D_{22} = \frac{\pi_2}{\Delta^2}(\alpha_2^2 T_{11}^2 + \alpha_1^2 T_{12}^2) \qquad (7.152)$$

It should be noted that all the coefficients D_{ij} in (7.150) which governs the evolution of the probability density f, are uniquely defined by the fully deterministic parameters T_{ij} of the original dynamical system (7.142).

7.4 Stochastic Attractors in Terminal Dynamics

All the dynamical systems considered above exhibited an unrestricted random walk. As a result of that, the joint probability density of their solutions vanishes at $t \to \infty$. In this section we will describe a new phenomena – an attraction of the solution to a stationary stochastic process whose joint density function is uniquely defined by the parameters of the original dynamical system.

7.4.1 One-dimensional Restricted Random Walk

We will start with the following one-dimensional dynamical system:

$$\dot{x} = \gamma \sin^k\left(\frac{\sqrt{\omega}}{\alpha} \sin x\right) \sin \omega t \tag{7.153}$$

It has the following equilibrium points:

$$\overset{*}{x}_m = \arcsin\left(\frac{\pi\alpha}{\sqrt{\omega}} m\right), \quad m = \cdots - 1, -1, 0, 1, 2 \cdots \text{etc.} \tag{7.154}$$

Obviously now the distances between these points depend upon the number of the step m:

$$h_m = \overset{*}{x}_m - \overset{*}{x}_{m-1} = \arcsin\left(\frac{\pi\alpha}{\sqrt{\omega}} m\right) - \arcsin\left[\frac{\pi\alpha}{\sqrt{\omega}}(m-1)\right] \tag{7.155}$$

Let us introduce a new variable:

$$y = \sin x \tag{7.156}$$

Then

$$\overset{*}{y}_m = \frac{\pi\alpha}{\sqrt{\omega}} m, \quad \overset{*}{y}_m - \overset{*}{y}_{m-1} = \frac{\pi\alpha}{\sqrt{\omega}} \tag{7.157}$$

and (7.157) becomes identical to (7.136). This means that the probability as a function of y satisfies the following equation:

$$\left[E_t - \frac{1}{2}(E_y + E_y^{-1})\right] f(t, y) = 0 \tag{7.158}$$

However, in contradistinction to (7.130), here y is bounded:

$$| y | = | \sin x | \leq 1 \tag{7.159}$$

If the solution of (7.158) subject to the boundary condition (7.159) is found:

$$f = \overset{*}{f}(t, y) \tag{7.160}$$

then the solution to the original problem is:

$$f = \overset{*}{f}[t, y(\sin x)] \; |\cos x| \tag{7.161}$$

For better physical interpretation of the solution (7.161) consider a limit case when

$$\sqrt{\omega} \to \infty, \text{ i.e., } \tau, h_m \to 0 \tag{7.162}$$

Then (7.158) transfer to the Fokker-Planck equation:

$$\frac{\partial f}{\partial t} = \frac{1}{2} D^2 \frac{\partial^2 f}{\partial y^2} \tag{7.163}$$

with the boundary conditions:

$$\frac{\partial f}{\partial y} \mid_{y=1} = \frac{\partial f}{\partial y} \mid_{y=-1} = 0 \tag{7.164}$$

Its solution subject to the initial conditions:

$$f(0, y) = \varphi(y), \ \varphi(y) \geq 0 \text{ and } \int_{-1}^{1} \varphi(y) dy = 1 \tag{7.165}$$

is

$$f(t, y) = \frac{1}{2} + \sum_{n=1}^{\infty} a_n e^{-\frac{1}{2}\pi^2 D^2 n^2 t} \cos \frac{n\pi}{2}(y+1), \ \mid y \mid \leq 1 \tag{7.166}$$

$$a_n = 2 \int_{-1}^{1} \varphi(z) \cos \frac{n\pi}{2}(z+1) dz, \ n = 1, 2, ..., \text{etc.} \tag{7.167}$$

and, therefore,

$$f(t, y) \to \frac{1}{2} \text{ at } t \to \infty, \ \mid y \mid \leq 1 \tag{7.168}$$

Returning to the original variable x, one obtains instead of (7.168):

$$f(x) = 0.5 \mid y' \mid = 0.5 \cos x, \ -\frac{\pi}{2} < x < \frac{\pi}{2}, \ f = 0 \tag{7.169}$$

otherwise.

Hence, any solution originated within the interval

$$-\frac{\pi}{2} < x < \frac{\pi}{2} \tag{7.170}$$

always approaches the same stationary stochastic process (7.169) which plays the role of a stochastic attractor.

It should be emphasized that this is a new phenomena which does not exist in classical version of nonlinear dynamics. Unlike chaotic attractors, here the probability density can be uniquely controlled by the parameters of the original dynamical system, while the limit stochastic process does not depend upon the initial conditions if they are within the basin of attraction.

The results presented above were obtained under the assumption (7.161) which allowed us to replace the original difference equation (7.158) by the differential equation (7.163). But if $\sqrt{\omega}$ is finite, and therefore, the steps (7.155) are finite too, in some cases the solution to (7.153) can overcome the barrier (7.169), and, after a slow diffusion, eventually approach the "universal" attractor

$$f = 0 \tag{7.171}$$

Let us investigate such a possibility in details. Turning to the condition (7.138) which synchronizes the conversions of terminal attractors into terminal repellers and vice-versa, we will assume first that this condition is violated:

$$\gamma = \omega h_m - \varepsilon^2, \ \varepsilon \ll 1, \ \text{at } h_m < \overset{*}{h}_m \tag{7.172}$$

Invoking (7.154) one concludes that if

$$\left| \frac{\pi}{2} - \overset{*}{h}_m \right| < \overset{*}{h}_m \tag{7.173}$$

then the solution to (7.153) can jump over the barrier $|x| = \pi/2$ and escape the region (7.170). Conversely, if

$$\left| \frac{\pi}{2} - \overset{*}{h}_m \right| > \overset{*}{h}_m \tag{7.174}$$

then this solution will be trapped within the region:

$$|x| \leq \left| \frac{\pi}{2} - \overset{*}{h}_m \right| \tag{7.175}$$

Qualitatively, the solution to (7.153) under the condition (7.174) behaves as the solution (7.169) representing a stochastic attractor. Obviously, (7.153) has infinite number of such an attractors with the basins:

$$\frac{\pi n}{2} < x < \frac{\pi(n+2)}{2}, \ n = \cdots - 2, -1, 0, 1, 2, \cdots, \text{etc.} \tag{7.176}$$

Under the condition (7.173), this solution will penetrate the barriers and diffuse over all the basins (7.176) approaching the attractor (7.171).

Now one can generalize (7.153) by requiring that its solution should have a stochastic attractor with a prescribed density function $f(x)$ under the only restrictions that

$$f(x) = 0 \text{ for } |x| > N, \ N < \infty, \text{ and } \int_{-N}^{N} f(x)dx = 1 \tag{7.177}$$

Based upon (7.169), one arrives at the following equation instead of (7.153):

$$\dot{x} = \gamma \sin^k \left[\frac{\sqrt{\omega}}{\alpha} p(x)\right] \sin \omega t, \ \ p(x) = 2\int_{-N}^{x} f(\xi)d\xi - 1 \tag{7.178}$$

Indeed, introducing a new variable [compare with (7.156)]:

$$y = p(x), \ y(-N) = -1, \ y(N) = 1$$

one obtains instead of (7.169)

$$f(x) = \frac{1}{2}|y'| = \frac{1}{2}\frac{dp}{dx}$$

In order to illustrate that this constraint does not overdetermined the solution, let us integrate (7.163) over x:

$$\int_{-1}^{1} f(y)dy = 1$$

in addition to the boundary conditions (7.164).

In order to illustrate that this constraint does not overdetermined the solution, let us integrate (7.163) over x:

$$\int_{-1}^{1} \frac{\partial f}{\partial t}dy = \frac{\partial}{\partial t}\int_{-1}^{1} f dy = \int_{-1}^{1} \frac{\partial f}{\partial y}dy = 0, \text{ i.e. } \int_{-1}^{1} f dy = \text{Const}$$

This means that if the initial conditions satisfy this constraint, then the solution will satisfy it automatically.

7.4.2 Multi-dimensional Restricted Random Walk

In order to illustrate the existence of stochastic attractors in multidimensional systems we will consider the following two-dimensional case:

$$\dot{x}_1 = \gamma_1 \sin^k[\sqrt{\omega}\sin(x_1 + x_2)]\sin\omega t \tag{7.179}$$

$$\dot{x}_2 = \gamma_2 \sin^k[\sqrt{\omega}\sin(x_1 - x_2)]\sin\omega t \tag{7.180}$$

By changing variables:

$$x_1 + x_2 = y_1$$

$$x_1 - x_2 = y_2$$

one obtains:

$$\dot{y}_1 = \gamma_1^* \sin^k \sqrt{\omega}\sin y_1 \sin\omega t \tag{7.181}$$

$$\dot{y}_2 = \gamma_2^* \sin^k \sqrt{\omega}\sin y_2 \sin\omega t \tag{7.182}$$

where

$$\gamma_1^* = 2\gamma_1\sqrt{1-y_1^2},\ \gamma_2^* = 2\gamma_2\sqrt{1-y_2^2} \tag{7.183}$$

Equations (7.181–7.182) have the form of (7.153), and therefore, their formal solutions follow from (7.169):

$$f(y_1) = 0.5\ \cos y_1,\ \frac{\pi m_1}{2} < y_1 < \frac{\pi(m_1+2)}{2},\ m_1 = \cdots -1, 0, 1, 2, \cdots$$

$$f(y_1) = 0,\ \text{otherwise} \tag{7.184}$$

$$f(y_2) = 0.5\ \cos y_2,\ \frac{\pi m_2}{2} < y_2 < \frac{\pi(m_2+2)}{2},\ m_2 = \cdots -1, 0, 1, 2, \cdots$$

$$f(y_2) = 0,\ \text{otherwise} \tag{7.185}$$

However, not all of these solutions are stable. Applying the stability conditions (7.143) to linearized versions of (7.179–7.180) yields:

$$\cos y_1 - \cos y_2 < 0,\ \cos y_1 \cos y_2 < 0 \tag{7.186}$$

i.e.,

$$\cos y_1 < 0,\ \cos y_2 > 0$$

and therefore, the solutions are stable if

$$m_1 = \cdots -7, -3, 1, 5, 9,\ \text{etc.};\ m_2 = \cdots -5, -1, 3, 5, 7, \cdots, \text{etc.} \tag{7.187}$$

in (7.184–7.185).

Returning to the original variables one obtains:

$$f(x_1, x_2) = 0.25\ \cos(x_1 + x_2)\cos(x_1 - x_2) \tag{7.188}$$

$$\frac{\pi m_1}{2} < x_1 + x_2 < \frac{\pi(m_1+2)}{2},\ \frac{\pi m_2}{2} < x_1 - x_2 < \frac{\pi(m_2+2)}{2}$$

$$f(x_1, x_2) = 0,\ \text{otherwise} \tag{7.189}$$

The solution (7.188) represents a stationary stochastic process which attracts all the solutions with the initial conditions within the area (7.189). Each pair m_1 and m_2 from the sequences (7.186) defines a corresponding stochastic attractor with the joint density (7.188). Obviously, the solutions for which m_1 and m_2 do not belong to (7.186), are unstable, and eventually, they will be attracted to one of the stochastic attractors (7.189).

Turning to an n-dimensional dynamical system, we will confine ourself by a special form:

$$\dot{x}_i = \gamma_i \sin^k \frac{\sqrt{\omega}}{\alpha} p_i(y_i) \sin \omega t \tag{7.190}$$

where

$$y_i = \sum_{j=1}^{n} T_{ij} x_j, \; T_{ij} = \text{Const} \tag{7.191}$$

It will be assumed that

$$\frac{dp_i}{dy_i} = \begin{cases} > 0 \text{ for } & |y_i| < N_i \\ = 0 \text{ for } & |y_i| > N_i \end{cases} \quad N_i < \infty \tag{7.192}$$

and T_{ij} form a symmetric positive-definite matrix, i.e. the conditions (7.143) are satisfied.

Based upon the conditions (7.143) and (7.192), one concludes that the system (7.190) will be locally stable, or locally unstable, depending upon the sign of $\sin \omega t$, and that synchronizes the conversions of random attractors into random repellers, and vice-versa.

Exploiting the result (7.178), one obtains that the solution to (7.190) has the following density functions in terms of the variables y_i:

$$f(y_1, \cdots y_n) = \Pi_{i=1}^{n} p_i'(y_i), \; p' = \frac{dp}{dy} \tag{7.193}$$

In terms of the variables x_i, the joint density of the solution is:

$$f(x_1, \cdots x_n) = \Pi_{i=1}^{n} p_i'(y_i) \; \det | \, T_{ij} \, | \tag{7.194}$$

where y_i is expressed via x_i by (7.191).

7.4.3 Examples

1. We will start with the following problem: find a dynamical system whose solution is attracted to a stochastic process with the normal density:

$$f(x) = z(\frac{x - \mu}{\sigma}) = \frac{1}{\sigma\sqrt{2\pi}} e^{\frac{(x-\mu)^2}{2\sigma^2}} \tag{7.195}$$

where μ and σ are the mean and the standard deviation, respectively, and $z(y)$ is the standard normal density function. In order to apply (7.178), first of all, (7.195) should be modified, since it does not satisfy the restriction (7.177).

We will introduce a truncated standard normal density function:

$$\tilde{z}(y) = \begin{cases} > z(y) \text{ if } & |y| < N \\ = 0 \text{ if } & |y| > N \end{cases} \quad N < \infty \tag{7.196}$$

Then, with reference to (7.178), one obtains:

$$\dot{x} = \gamma \sin^k [\frac{\sqrt{\omega}}{\alpha} \widetilde{erf}(\frac{x-\mu}{\sqrt{2}\sigma})] \sin \omega t, \ \widetilde{erf}(y) = \frac{2}{\sqrt{\pi}} \int_o^y \tilde{z}(u) du \tag{7.197}$$

Thus, (7.197) represent a dynamical system whose solution is attracted to a stochastic process with the density function (7.196). For a sufficiently large N it will be close to a Gaussian process with μ and σ as the mean and the standard deviation, respectively.

2. Let us assume now that the density $f(x)$ of a sought stochastic process is characterized by $\mu = \mu_0, \sigma = \mu_1$ and higher central moments μ_r. Utilizing the Gram-Charlier series expansion:

$$f(x) = \frac{1}{\sigma} \sum_{r=0}^{\infty} c_r \tilde{z}^{(r)} (\frac{x-\mu}{\sigma}) \tag{7.198}$$

where

$$c_0 = 1, c_1 = c_2 = 0, c_3 = -\frac{1}{3!}\mu_3, \ c_4 = \frac{1}{4!}(\mu_4 - 3)$$

$$c_5 = -\frac{1}{5!}(\mu_5 - 10\mu_6), c_6 = \frac{1}{6!}(\mu_6 - 15\mu_4 + 30), \ \text{etc.} \tag{7.199}$$

$$\tilde{z}^{(r)} = \frac{d^r \tilde{z}(y)}{dy^r} \tag{7.200}$$

and applying (7.178) one obtains:

$$\dot{x} = \gamma \sin^k \{\frac{\sqrt{\omega}}{\alpha} [\widetilde{erf}(\frac{x-\mu}{\sqrt{2}\sigma}) + \sum_{r=3}^{\infty} c_r \tilde{z}^{r-1} (\frac{x-\mu}{\sigma})\} \sin \omega t \tag{7.201}$$

Hence, the solution to the dynamical system (7.201) is attracted to a stochastic process whose density function is characterized by the moments μ_r.

3. In this example we will pose the following problem: find such a dynamical system whose solutions $x_i(t)$ are attracted to a stochastic process characterized by the column of means and the matrix of moments:

$$Mx_i = \mu_i, \ \sigma_{ij} = M(x_i - \mu_i)(x_j - \mu_j), \ i,j = 1,2,\cdots n \tag{7.202}$$

First of all, one can find such an orthogonal transformation:

$$y_i = \eta_i + \sum_{j=1}^{n} T_{ij}(x_j - \mu_j) \tag{7.203}$$

that

$$My_i = \eta_i = 0, \ \sigma'_{ik} = \sum_{j=1}^{n}\sum_{\ell=1}^{n} \sigma_{j\ell} T_{ij} T_{kl} = \delta_{ik} = \begin{cases} 1 \text{ if } & i = k \\ 0 \text{ if } & i \neq k \end{cases} \tag{7.204}$$

where y_i are non-correlated standard normally distributed variables.

Combining (7.190–7.191) and (7.197) one obtains:

$$\dot{x}_i = \gamma_i \sin^k [\frac{\sqrt{\omega}}{\alpha} \widetilde{erf}(\frac{y_i}{\sqrt{2}})] \sin \omega t, \ y_i = \sum_{j=1}^{n} T_{ij}(x_j - \mu_j) \tag{7.205}$$

Some comments concerning the stability of (7.205) should be made. Since T_{ij} is an orthogonal matrix, it does not satisfy the conditions (7.143). However, the real parts of the eigenvalues of T_{ij} are

$$\operatorname{Re} \lambda_i = \cos \varphi_i > 0, \ \text{for } 0 \leq \varphi < \frac{\pi}{2} \tag{7.206}$$

where φ_i are the angles of rotation of the coordinate axes, and since

$$\frac{d}{dy_i} \widetilde{erf}(y_i) > 0 \text{ for } \mid y \mid < N_i \tag{7.207}$$

i.e., the condition (7.192) is satisfied; (7.205) linearized with respect to their equilibrium points, have eigenvalues whose real part are all positive (if $\sin \omega t > 0$) or negative (if $\sin \omega t < 0$), and that synchronizes conversions from random attractors to random repellers, and vice-versa.

Thus, the solution to the dynamical system is attracted to a stochastic process with prescribed probabilistic structure (7.202) if the initial conditions are within the basin of attraction: $\mid y_i \mid < N_i$.

7.5 Self-Organization in Terminal Dynamics

A dynamical system is called self-organizing if it acquires a coherent structure without specific interference from the outside. In this section we will show that terminal dynamics possesses a powerful tool for self-organization coming from a possibility of coupling between the original dynamical system and its own associated probability density dynamics.

We will start with the dynamical system (7.197), and represent it in the form (7.119)

$$\dot{x} = \gamma \sin^k (\frac{\sqrt{\omega}}{\alpha} y) \sin \omega t, \ y = \widetilde{erf}(\frac{x}{\sqrt{2}\sigma}) \tag{7.208}$$

Then the probability density function $f(y, t)$ satisfies (7.128):

$$\frac{\partial f}{\partial t} = \frac{\pi\alpha^2}{2}\frac{\partial^2 f}{\partial y^2}, \quad -N \leq y \leq N \tag{7.209}$$

Its solution subject to the boundary and the initial conditions (7.164–7.165), respectively, is given by (7.166).

In terms of x this solution is:

$$f_*(x,t) = \{\frac{1}{2} + \sum_{n=1}^{\infty} a_n e^{-\frac{1}{2}\pi^3\alpha^2 n^2 t} \cos\frac{n\pi}{2}[\widetilde{erf}(\frac{x}{\sqrt{2}\sigma})]\}\tilde{z}(\frac{x}{\sigma}) \tag{7.210}$$

where $\tilde{z}$ is defined by (7.195–7.196).

In all the previous cases the parameter σ was prescribed in advance denoting the variance of the stationary density:

$$f_*(x) = \tilde{z}(\frac{x}{\sigma}) \quad \text{at } t \to \infty \tag{7.211}$$

Let us assume now that the parameter σ depends upon moments of the current density (7.210), for instance:

$$\sigma^2 = \sigma_0^2 - \text{ var }(x), \; \sigma_0 = \text{Const} \tag{7.212}$$

where

$$\text{var }(x) = \int_{-N}^{N} x^2 f_*(x,t)dx \tag{7.213}$$

First of all, it should be noticed that the time scale t' of changing σ is defined by (7.210) and it has the order:

$$t' \sim \frac{1}{\alpha^2} \tag{7.214}$$

Since the time scale t'' of changing x in (7.208) has the order $t'' \sim \frac{1}{\sqrt{\omega}} \to 0$ [see (7.124)], and therefore

$$t \gg t'' \tag{7.215}$$

the variable σ can be considered as a slow changing parameter in (7.208) [but not in (7.210)!].

Thus, now the dynamical system (7.208) is guided by the probability density via the parameter σ. This parameter is found from (7.212) after the substitution of (7.212) in the integrand. For the final stationary state ($t \to \infty$) one obtains from (7.212–7.213).

$$\text{var}(x) = \sigma^2 = \sigma_0^2 - \text{var}(x) \tag{7.216}$$

whence

$$\sigma^2 = \frac{1}{2}\sigma_0^2 \text{ at } t \to \infty \tag{7.217}$$

Hence, the solution to the dynamical system (7.208) approaches a stochastic attractor with the probability density $\tilde{z}(\frac{x}{\sigma_o/\sqrt{2}})$. It should be stressed that this attractor has not been "stored" in the prescribed coefficients of (7.208): the dynamical system "found" it as a result of coupling with its "own" probability equations.

In general case, parameters of the dynamical system (7.190) can be coupled with the moments of the probability density (7.194), and that will lead to new self-organizing architectures.

7.6 Guided Systems

Turning to (7.119), let us assume that this dynamical system is driven by a vanishingly small input $\varepsilon(t)$:

$$\dot{x} = \gamma \sin^{1/3}(\frac{\sqrt{\omega}}{\alpha}y)\sin\omega t + \varepsilon(t), \quad |\varepsilon(t)| \ll \gamma \tag{7.218}$$

This input can be ignored when $\dot{x} \neq 0$ or when $x = 0$, but the system is stable, i.e., $x = \pi\alpha/\sqrt{\omega}$, $3\pi\alpha/\sqrt{\omega}, ...,$ etc.

However, it becomes significant during the instances of instability when $\dot{x} = 0$ at $x = 0$, $2\pi/\sqrt{\omega}$, etc. Since actually a vanishingly small noise is always present, one can interpret the unpredictability discussed above as a consequence of small random inputs to which the dynamical system (7.218) is extremely sensitive.

However, the function $\varepsilon(t)$ is not necessarily random; it can be associated with a device which controls the behavior of the dynamical system (7.119) through a string of signs. Indeed, actually the only important part in this input is the sign of $\varepsilon(t)$ at the critical points. Consider, for example (7.218), and suppose that

$$\text{sgn } \varepsilon(t_m) = +, +, -, +, -, -, \text{etc. at } t_m = \frac{\pi m}{\omega}, \quad m = 1, 2, ..., \text{etc.} \tag{7.219}$$

The values of $\varepsilon(t)$ in between the critical points are not important since, by our assumption, they are small in comparison to values of the derivative, x, and therefore, can be ignored. Hence, the only part of the input which is significant in determining the solution to (7.218) is the sign of the string (7.219): specification of this string fully determines the dynamics of (7.218). Figure 7.8 demonstrates three different scenarios of motions for different strings, when $\varepsilon(t) = \varepsilon_0$ sgn $\sin\Omega t$. In this section we will discuss more complex dynamical systems when the string (7.219) is undetermined in some critical points. We will assume that

$$\varepsilon(t) = \varepsilon_0 ax, \ \varepsilon_0 \to 0 \tag{7.220}$$

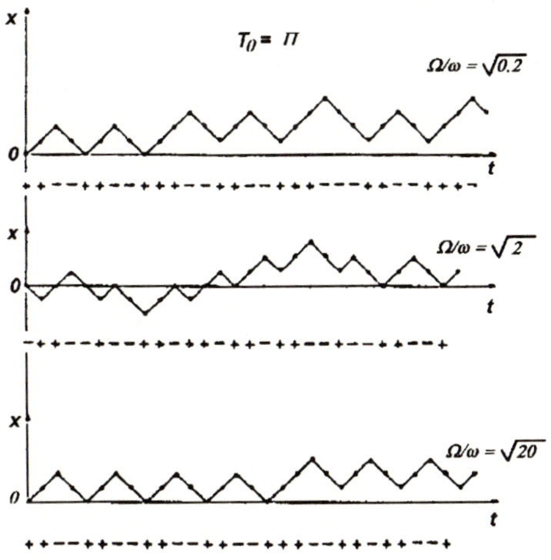

Fig. 7.8. Temporal patterns and their codes

while

$$a \begin{cases} = 0 \text{ if } & 1 < x < -1 \\ < 0 \text{ if } & x = 1 \\ > 0 \text{ if } & x = -1 \end{cases} \tag{7.221}$$

The conditions (7.221) can be implemented via the additional random dynamical system:

$$\dot{a} = a^{1/3}(x-1)(x+1) - \varepsilon_0 a, \ \varepsilon_0 \to 0 \tag{7.222}$$

Indeed, (7.222) has a terminal equilibrium point

$$a = 0 \tag{7.223}$$

which is a terminal attractor if

$$-1 < x < 1 \tag{7.224}$$

and it is a terminal repeller otherwise. One can also verify that the solution escapes the terminal repeller such that

$$a < 0 \text{ if } x = 1, \text{ and } a > 0 \text{ if } x = -1 \tag{7.225}$$

Hence, the dynamical system (7.222) fully implements the conditions (7.221).

Now let us return to (7.218) supplemented by (7.220) and (7.222). Within the domain (7.224) the solution describes a random walk governed by (7.124), or by its continuous approximation (7.128), and it is fully reflected from the boundaries $x = x_1$ and $x = x_2$. Indeed, as follows from (7.225)

$$\text{sgn } \varepsilon(t) = \begin{cases} - \text{ at } & x = 1 \\ + \text{ at } & x = -1 \end{cases} \tag{7.226}$$

Hence, one arrives at a restricted random walk with boundary conditions:

$$\left.\frac{\partial f}{\partial x}\right|_{x=1} = \left.\frac{\partial f}{\partial x}\right|_{x=-1} = 0 \tag{7.227}$$

The solutions to (7.128) subject to the boundary conditions (7.227) at $t \to \infty$

$$f(x) = 0.5, \quad -1 < x < 1 \tag{7.228}$$

i.e., with the same probability the solution will visit all the critical points within the domain (7.224).

Obviously, the solution (7.228) represents a stochastic attractor. But in contradistinction to the stochastic attractors discussed above, here the boundaries (7.224) are explicitly imposed by the guiding dynamics (7.222).

In the general case, the guided version of terminal dynamics can be presented as:

$$\dot{x}_i = \gamma_i \sin^k\left(\frac{\sqrt{\omega}}{\alpha} y_i\right) \sin \omega t + \varepsilon_i(t), \; \xi_i = \sum_j T_{ij} x_j \tag{7.229}$$

while

$$\varepsilon_i(t) = \varepsilon_0 \sum_j a_{ij} x_j, \; \varepsilon_0 \to 0 \tag{7.230}$$

As in the one dimensional case (7.220), the coefficients a_{ij} can be given by inequalities of the type (7.221) by means of additional terminal dynamics of the type (7.222). This will lead to a multidimensional random walk restricted by reflecting boundaries, while the limit form of solution at $t \to \infty$ represents a stationary stochastic attractor of the dynamical system (7.229).

Thus, there are two kinds of coupling between the variables x in the terminal dynamics (7.229): the coefficients T_{ij} carry out a probabilistic coupling via a joint density distribution, while the coefficients a_{ij} perform deterministic, but "qualitative" rather than quantitative coupling (since only the sign of $\varepsilon_i(t)$ at the terminal equilibrium points is important).

7.7 Relevance to Chaos

One of the central problems in Newtonian dynamics is to explain how a motion which is described by fully deterministic governing equations can be random. In order to discuss it, let us consider an exponential growth of a variable α:

$$\alpha = \alpha_0 e^{-\lambda t}, \; 0 < \lambda < \infty \tag{7.231}$$

Obviously, the solution with infinitely close initial condition

$$\tilde{\alpha} = \alpha + \varepsilon, \quad \varepsilon \to 0 \tag{7.232}$$

will remain infinitely close to the original one:

$$| \tilde{\alpha} - \alpha_0 | = \varepsilon e^{-\lambda t} \to 0 \ \text{ if } \varepsilon \to 0, \quad t < \infty \tag{7.233}$$

during all the bounded time intervals. This means that random solutions can result only from random initial conditions when ε in (7.233) is small, but finite rather than infinitesimal. In other words, classical dynamics can explain amplifications of random motions, but cannot explain their origin. According to the terminal model of Newtonian dynamics, random motions are generated by unstable equilibrium states at which dissipation forces do not vanish with velocities, i.e., where the Lipschitz condition is violated. It should be recalled that the evolution of these random motions amplified by the mechanism of instability, can be predicted by using the stabilization principle discussed in Chap. 4.

It is worth noticing that because of a finite precision with which the initial conditions are known, terminal equilibrium points can be incorporated into classical dynamics in the following way. Let us consider a first-order dynamical equation:

$$\dot{v} + \alpha v = 0 \tag{7.234}$$

and assume that the variable v can be observed with a finite error

$$| v_* | << v_0 \tag{7.235}$$

where v_0 is a representative value of v characterizing the scale of motion.

The actual time of approaching the attractor

$$| v | \leq | v_* | \tag{7.236}$$

is finite:

$$t_1 = \frac{1}{\alpha} \ell n \left| \frac{v_0}{v_*} \right| < \infty \tag{7.237}$$

a terminal version of (7.237) describing the same process

$$\dot{v} + \alpha v [1 + (1 - k)(\frac{v}{v_0})^{k-1}] = 0 \tag{7.238}$$

has a solution which at $k \to 1$ is infinitely close to the solution of (7.234) everywhere, except a small neighborhood of the attractor $v = 0$, while the time of approaching this attractor is

$$t_2 = \frac{1}{\alpha(1 - k)^2} \tag{7.239}$$

Equating t_1 from (7.237) and t_2 from (7.239), one finds the order of an "equivalent" value of k:

$$k \sim 1 - \frac{1}{\sqrt{\ell n \mid \frac{v_0}{v_*} \mid}} \tag{7.240}$$

Hence, the fact that dynamical parameters cannot be observed or measured with infinite precision, is mathematically formalized by introducing terminal equilibrium points, while the parameter $k < 1$ is defined by the relative error v_*/v_0. The terminal version of the dynamical system (7.40):

$$\dot{x}_i = v_i[1 + (1-k)(\frac{v_i}{v_0})^{k-1}], \ v_0 \gg v_i \tag{7.241}$$

allows one to explain appearances of random solutions without random inputs within the framework of differentiable dynamics. However, additional terms in (7.241) cannot be interpreted as physical quantities [such as dissipation forces (7.2)], since they are not invariant with respect to coordinate transformations. This emphasizes a computational origin of "classical" chaos in contrast to the physical origin of terminal chaos.

7.8 Relevance to Classical Thermodynamics

When Newton's laws are applied to dissipative dynamical systems, a certain link to thermodynamics must be introduced. Usually such a link is carried out by friction forces which enter the Lagrange equations (7.1) via the dissipation function. Actually the existence of all the types of attractors in classical dynamics (including chaotic attractors) is due to these forces. However, strictly speaking, in order to define the friction forces, one has to introduce additional thermodynamical relationships which couple classical mechanics and classical thermodynamics. But since the effects of irreversibility and unpredictability in classical mechanics come from the friction forces, the same properties must be in classical thermodynamics. In order to demonstrate this, let us turn to the problem of heat conductivity. The time change of temperature T is described by the classical Fourier equation:

$$\frac{\partial T}{\partial t} = \frac{\partial}{\partial x}\left(\varkappa \frac{\partial T}{\partial x}\right)^k \tag{7.242}$$

in which $\varkappa$ is the heat conductivity ($\varkappa > 0$), and k is given by (7.8). Obviously, in the classical case $k = 1$.

Let us introduce a function:

$$\theta = \int \left(\frac{\partial T}{\partial x}\right)^{k+1} dx \tag{7.243}$$

and find its time derivative:

$$\frac{d\theta}{dt} = (k+1)\int\left(\frac{\partial T}{\partial x}\right)^k \frac{\partial^2 T}{\partial x \partial t}dx = \int\left(\frac{\partial T}{\partial x}\right)^k \frac{\partial}{\partial x}\left[\left(\frac{\partial T}{\partial x}\right)^{k-1}\frac{\partial^2 T}{\partial x^2}\right]dx$$

$$= -k(k+1)\int\left[\left(\frac{\partial T}{\partial x}\right)^{k-1}\frac{\partial^2 T}{\partial x^2}\right]^2 dx + \left.\left(\frac{\partial T}{\partial x}\right)^{2k-1}\frac{\partial^2 T}{\partial x^2}\right|_0^L \leqslant 0 \qquad (7.244)$$

The last term in (7.244) vanishes if at the boundaries $x = 0$ and $x = L$:

$$\left.\frac{\partial T}{\partial x}\right|_0 = \left.\frac{\partial T}{\partial x}\right|_L = 0 \qquad (7.245)$$

As follows from (7.244), θ can be considered as a Liapunov function for the problem. Since it decreases indeed to its minimum, the process will approach equilibrium which plays the role of a static attractor. But what is the time period for that process? In order to evaluate it, let us assume that

$$T(x,t) = T_1(t)T_2(x) \qquad (7.246)$$

and substitute (7.246) into (7.244).

Then

$$\frac{dT_1}{dt} = -A^2 T_1^k,\; A^2 = k(k+1)\int\left[\left(\frac{\partial T_2}{\partial x}\right)^{k-1}\frac{\partial^2 T_2}{\partial x^2}\right]^2 dx = \text{Const} \qquad (7.247)$$

Now it is clear that in the classical case ($k = 1$), the time of approaching the equilibrium is infinite. This means that from a formal mathematical viewpoint, the classical thermodynamics is fully reversible and predictable: starting from a nonequilibrium state, a system will move toward an equilibrium forever. Such a paradox is eliminated if k is taken according to (7.8). Indeed, in this case the time of approaching an equilibrium is finite:

$$t_0 = \frac{T_0^{1-k}}{A^2(1-k)},\; T_0 = T|_{t=0} \qquad (7.248)$$

and for $t \geqslant t_0$, the system completely forgets its past (when $t < t_0$). This means that nonequilibrium thermodynamics modified as follows from (7.242), will attain the same properties of irreversibility and unpredictability as classical dynamics with friction forces taken in a terminal form. In both cases, the terminal models are in full agreement with physical reality.

8
Terminal Neurodynamics

The moment of truth, the sudden emergence of new insight, is an act of intuition. –*A. Koestler*

8.1 Introduction

This chapter is motivated by an attempt to link two fundamental concepts of neurodynamics – irreversibility and creativity – in connection with a new architecture of neural networks based upon the terminal dynamics, discussed in Chap. 7.

There is some evidence (Harth 1983, Osovets et al. 1983, Freeman 1987) that primary mode of computation in the brain is associated with a relaxation procedure, i.e., with settling into a solution in the same way in which a dissipative dynamical system converges to an attractor. This is why many attempts were made to exploit the phenomenology of nonlinear dynamical systems for information processing as an alternative to the traditional paradigm of finite-state machines.

Over the past decade, there has been an increasing interest in dynamical aspects of artificial neural networks in the areas of global stability, adaptive pattern recognition, content-addressable memory, and cooperative-competitive decision making by neural nets. The number of publications in the area of neurodynamics is still growing exponentially. But along with this, the number of limitations of current models as well as the number of unanswered questions concerning the relevance of these models to brain performance grow too. Indeed, the biggest promise of artificial neural networks as computational tools lies in the hope that they will resemble the information processing in biological systems. However, the performance of current neural networks is still too "rigid" in comparison with even simplest biological systems. This rigidity follows from the fact that the behavior of a dynamical system is fully prescribed by initial conditions. The system never "forgets" these conditions: it carries their "burden" all the time. In contrast to this, biological systems are much more flexible: they can forget (if necessary) the past, adapting their behavior to environmental changes.

The thrust of this chapter is to discuss the substantially new type of dynamical system for modeling biological behavior introduced in Chap. 7. The approach is motivated by an attempt to remove one of the most fundamental limitations of current models of artificial neural networks – their "rigid" behavior compared to biological systems. As has been already mentioned, the mathematical roots of the rigid behavior of dynamical systems are in the uniqueness of their solutions subject to prescribed initial conditions. Such a uniqueness was very important for modeling energy transformations in mechanical, physical, and chemical systems which have inspired progress in the theory of differential equations. This is why the first concern in the theory of differential equations as well as in dynamical system theory was for the existence of a unique solution provided by so-called Lipschitz conditions. On the contrary, for information processing in a brain-style fashion, the uniqueness of solutions for underlying dynamical models becomes a heavy burden which locks up their performance into a single-choice behavior.

The new architecture for artificial neural networks considered in this chapter exploits a novel paradigm in nonlinear dynamics based upon the concept of terminal attractors, terminal repellers, and terminal chaos. Due to violations of the Lipschitz conditions at certain critical points, the neural network forgets its past as soon as it approaches these points; the solution at these points branches, and the behavior of the dynamical system becomes unpredictable. Since any vanishingly small input applied at critical points causes a finite response, such an unpredictable system can be controlled by a neurodynamical device which operates by sign strings, and as a genetic code, uniquely defines the system behavior by specifying the direction of the motions in the critical points. By changing the combinations of signs in the code strings, the system can reproduce any prescribed behavior to a prescribed accuracy. This is why unpredictable systems driven by sign strings are extremely flexible and are highly adaptable to environmental changes. The supersensitivity of critical points to external inputs appears to be an important tool for creating chains of coupled subsystems of different scales whose range is theoretically unlimited.

Due to existence of the critical points, the neural network becomes a weakly coupled dynamical system: its neurons (or groups of neurons) are uncoupled (and therefore, can perform parallel tasks) within the periods between the critical points, while the coordination between the independent units (i.e., the collective part of the performance) is carried out at the critical points where the neural network is fully coupled. As a part of the new architecture, weakly coupled neural networks acquire the ability to be activated not only by external inputs, but also by internal periodic rhythms. (Such a spontaneous performance resembles brain activity).

One of the most fundamental features of the new architecture is terminal chaos which incorporates an element of "arationality"/irrationality into dynamical behavior, and can be associated with the creativity of neural network performance. Terminal chaos is generated by those critical points at which the sign of the driving "genetic code" is not prescribed. The solution oscillates chaotically about such critical points, and its behavior qualitatively resembles "classical" chaos. However, terminal chaos is characterized by the following additional property: it has a fully predictable probabilistic structure and therefore, it can learn, and be controlled and exploited for information processing. It appears that terminal chaos significantly increases the capability of dynamical systems via a more compact and effective storage of information, and provides an ability to create a hierarchical parallelism in the neurodynamical architecture which is responsible for the tremendous degree of coordination among the individual parts of biological systems. It also implements a performance of high level cognitive processes such as formation of classes of patterns, i.e., formation of new logical forms based upon a generalization procedure.

Special attention will be paid to a new dynamical paradigm – a collective brain which mimics collective purposeful activities of a set of units of intelligence. Global control of the unit activities is replaced by the probabilistic correlations between them. These correlations are learned during a long term period of performing collective tasks, and are stored in the synaptic interconnections. The model is represented by a system of ordinary differential equations with terminal attractors and repellers, and does not contain any "man-made" digital devices.

It is worth emphasizing the phenomenological similarity between brain activity and this new dynamical architecture of neural networks: due to terminal chaos, the dynamical systems can be activated spontaneously driven by a global internal periodic rhythm. In the course of such a spontaneous activity they can move from one attractor to another, change locations of attractors, create new attractors and eliminate the old ones. Another phenomenological similarity follows from the duality of the dynamical performance caused by violation of the Lipschitz condition at equilibrium points: the dynamical motion is continuous, but it is controlled by a string of numbers stored in the microdynamical device: as a genetic code, the combinations of these numbers prescribe the continuous motion of the dynamical system. In this regard one can view the new dynamical architecture as a symbiosis of two fundamental concepts in brain modeling: analog and digital computers.

Although attention will be focused on the basic properties of unpredictable neurodynamics such as irreversibility and coherent (temporal and spatial) structures in terminal chaos and the collective brain, we will start with the discussion of some simple computational advantages of neural nets with terminal attractors.

8.2 Terminal Attractors in Neural Nets

The concept of a terminal attractor was introduced in Chap. 7, Sect. 1 and 2. In this section, we will incorporate it into additive neural nets [see (3.35)], and demonstrate certain advantages in their performance (Zak 1989a, 1986b).

8.2.1 Terminal Attractors: Content Addressable Memory

The idea of storing memory states as static attractors of the neural network dynamics implies that initial configurations of neurons in some neighborhood of a memory state will be attracted to it. This section will show that incorporation of terminal attractors in neural nets allows one to solve the following problem: given n vectors to be stored, find a neural network which has n (and only n) stable equilibrium points representing these stored vectors.

We will start with the simplest case of a single neuron in which a terminal attractor can be incorporated as follows

$$\dot{u} + u = TV - (u - \widetilde{u}^{(1)})^{1/3}, \quad u > 0 \tag{8.1}$$

where $u(t)$ is the mean soma potential of the neuron, T is a constant, $V(u)$ is a sigmoid function (for instance, $V = \tanh u$), $\widetilde{u}^{(1)}$ is a desired static terminal attractor if T is selected as follows

$$T = \widetilde{u}^{(1)}[V(\widetilde{u}^{(1)})]^{-1} \tag{8.2}$$

It is easily verifiable that the last term in the right-hand side of (8.1) does not affect the location of the fixed point (see Fig. 8.1), but it significantly changes the degree of its stability. Indeed, linearizing (8.1) with respect to a point $u^* > 0$ one obtains

$$\dot{u} = a(\widetilde{u}^{(1)} - u) \tag{8.3}$$

in which

$$a = -1 + \frac{T}{\cosh^2 \widetilde{u}^{(1)}} - \frac{1}{3}(u^* - \widetilde{u}^{(1)})^{-2/3} \tag{8.4}$$

Obviously,

$$a \to -\infty \text{ if } u^* \to \widetilde{u}^{(1)} \tag{8.5}$$

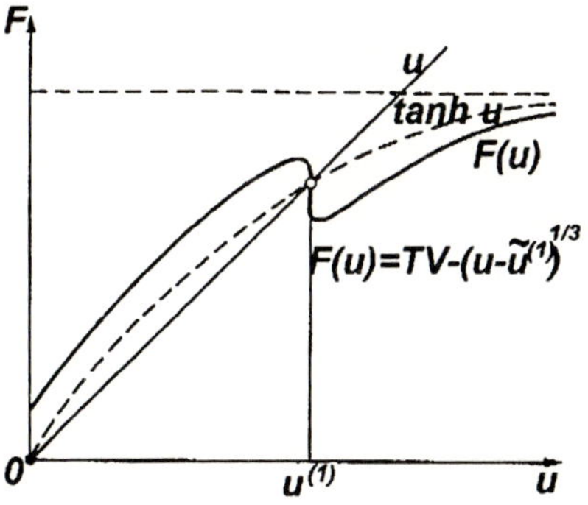

Fig. 8.1. Location and stability of terminal attractor

Thus, incorporation of a terminal attractor into an equilibrium point of (8.1) makes this point "infinitely" stable and suppresses all possible instabilities coming from other terms.

Let us turn now to a n-neuron network

$$\dot{u}_i + u_i = \sum_{j=1}^{n} T_{ij} V_j, \quad i = 1, 2, ..., n \tag{8.6}$$

in which T_{ij} are constants and $V_j = V(u_j)$ are linearly independent and single-valued vectors.

Its equilibrium points are defined by the following algebraic equations

$$\tilde{u}^{(k)} = \sum_{j=1}^{n} T_{ij} V(\tilde{u}_j^{(k)}) \tag{8.7}$$

in which $\tilde{u}_j^{(k)}$ $(i, k = 1, 2, ..., n)$ are the vectors to be stored. If these vectors are linearly independent, one can find the weight matrix T_{ij} from (8.7)

$$\begin{pmatrix} T_{11} & \dots & T_{1n} \\ \dots & \dots & \dots \\ \dots & \dots & \dots \\ T_{n1} & \dots & T_{nn} \end{pmatrix} = \begin{pmatrix} \tilde{u}_1^{(1)} & \dots & \tilde{u}_1^{(n)} \\ \dots & \dots & \dots \\ \dots & \dots & \dots \\ \tilde{u}_n^{(1)} & \dots & \tilde{u}_n^{(n)} \end{pmatrix} \times \begin{pmatrix} V\tilde{u}_1^{(1)} & \dots & V\tilde{u}_1^{(n)} \\ \dots & \dots & \dots \\ V\tilde{u}_n^{(1)} & \dots & V\tilde{u}_n^{(n)} \end{pmatrix}^{-1} \tag{8.8}$$

However, so far, the stability of the equilibrium points $\tilde{u}_i^{(k)}$ is not yet guaranteed. In order to make all the desired equilibrium points stable, we will incorporate a terminal attractor into each fixed point by modifying (8.6) as follows

$$\dot{u}_i + u_i = \sum_{j=1}^{n} T_{ij} V_j - \sum_{k=1}^{n} \alpha_i^{(k)} (u_i - \tilde{u}_i^{(k)})^{1/3} e^{-\gamma^{(k)}(u_i - \tilde{u}_i^{(k)})^2} \tag{8.9}$$

in which

$$\left|\gamma_i^{(k)}(\widetilde{u}_i^{(k)} - \widetilde{u}_i^{(l)})\right| \gg 1, \ \widetilde{u}_i^{(k)} \neq \widetilde{u}_i^{(l)} \text{ if } k \neq l \tag{8.10}$$

while $\gamma_i^{(k)}$ and $\alpha_i^{(k)}$ are positive constants.

The exponential multipliers are introduced into (8.9) in order to localize effects of terminal attractors [see the conditions (8.10)]. It is easily verifiable that all the equilibrium points of the original equation (8.6) are among the equilibrium points of (8.9). Indeed, substituting $u = \widetilde{u}_i^{(k)}$ into (8.9) one arrives at (8.7). But now all these equilibrium points are "infinitely" stable. In order to prove this, let us linearize the system (8.9) with respect to points $\widetilde{u}_i^{*(k)}$ which are sufficiently close to the equilibrium points $\widetilde{u}_i^{(k)}$, respectively, so that

$$\left|u_i^{(k)} - \widetilde{u}_i^{(k)}\right| = \varepsilon \to 0 \tag{8.11}$$

The linearized versions of the system (8.9) in the neighborhoods of the corresponding equilibrium point read

$$\dot{u}_i = \sum_{j=1}^{n} a_{ij}^{(k)}(\widetilde{u}_j^{(k)} - u_j), \ k = 1, 2, ..., n \tag{8.12}$$

in which

$$a_{ij}^{(k)} = \frac{T_{ij}}{\cosh^2 \widetilde{u}_j^{(k)}}, \ i \neq j \tag{8.13}$$

$$a_{ij}^{(k)} = -1 + \frac{T_{ij}}{\cosh^2 \widetilde{u}_j^{(k)}} - \alpha_i^{(k)} \left(\frac{1}{3}\varepsilon^{-2/3} - 2\gamma_i^{(k)}\varepsilon^{4/3}\right) e^{-\gamma_i^{(k)}\varepsilon^2} \tag{8.14}$$

Now the stability of the equilibrium points $u_i^{(k)}$ depends on the eigenvalues λ_k of the matrices a respectively. In order to find these eigenvalues, let us multiply the matrices $a_{ij}^{(k)}$, by $\varepsilon^{-2/3}$ $(\varepsilon \to 0)$

$$a_{ij}^{*(k)} = \varepsilon^{2/3} a_{ij}^{(k)} \tag{8.15}$$

Then, as follows from (8.13) and (8.14), respectively

$$a_{ij}^{*(k)}(i \neq j) \to 0, \ a_{ii}^{*(k)} \to -\frac{1}{3}\alpha_i^{(k)} \text{ at } \varepsilon \to 0 \tag{8.16}$$

that is, the matrix $a_{ii}^{*(k)}$ is a diagonal at $\varepsilon \to 0$, and its eigenvalues are

$$\lambda_k^* = -\frac{1}{3}\alpha_i^{(k)}; \ i, k = 1, 2, ..., n \tag{8.17}$$

Hence,

$$\lambda_k = \varepsilon^{2/3}\lambda_k^* = -\frac{1}{3}\alpha_i^{(k)}\varepsilon^{2/3} \to -\infty \text{ at } \varepsilon \to 0 \tag{8.18}$$

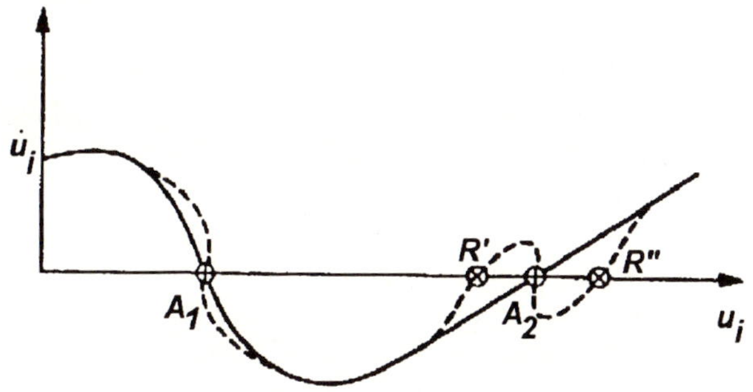

Fig. 8.2. Conversion of repellers into terminal attractors

Thus, all the eigenvalues of the matrices $a_{ij}^{(k)}$ are negative and unbounded at the equilibrium points $\widetilde{u}_i^{(k)}$, which means that all the desired equilibrium points are terminal attractors. Their basins can be controlled by an appropriate selection of the coefficients $\alpha_i^{(k)}$ and $\gamma_i^{(k)}$[see (8.9)].

Obviously, the system (8.9) has additional equilibrium points (repellers) which do not satisfy (8.7). In order to demonstrate this, let us consider (8.9) in a small neighborhood of its fixed points $A_i^{(1)}$ and $A_i^{(2)}$ which were converted into terminal attractors from an attractor and a repeller, respectively (Fig. 8.2)

$$\dot{u}_i = \zeta_i^{(k)}(u_i - \widetilde{u}_i^{(k)}) - \alpha_i^{(k)}(u_i - \widetilde{u}_i^{(k)})^{1/3} e^{-\gamma_i^{(k)}(u_i - \widetilde{u}_i^{(k)})^2}, \quad k = 1, 2$$

in which $\zeta^{(1)} < 0$ and $\zeta^{(2)} > 0$.

In the neighborhood of the point $A_i^{(1)}$, this equation has two additional fixed points (repellers) R' and R'' which coordinates $u_i^{*(2)}$are within the region

$$\frac{\alpha_i^{(2)}}{\zeta_i^{(2)}} > \left| u^{*(2)} - \widetilde{u}_i^{(2)} \right| > \frac{1}{\gamma_i^{(2)} + \zeta_i^{(2)}/\alpha_i^{(2)}}$$

while in the neighborhood of $A_i^{(2)}$ there are no additional equilibrium points.

Thus, all the vectors $\widetilde{u}_i^{(k)}(i, k = 1, 2, ..., n)$ are stored terminal attractors. But, as follows from (8.8), the vectors uniquely define the weight matrix T_{ij} since the vectors $V_j = V(u_j)$ are linearly independent and single-valued. Hence, $\widetilde{u}_i^{(k)}$ are the only vectors which are stored as static attractors. All the additional equilibrium points, that have been demonstrated above, are repellers. It means that this model does not have false static attractors. However, in general, if the weight matrix is not symmetric ($T_{ij} \neq T_{ji}$), the existence of periodic or chaotic attractors cannot be excluded, and only in a particular case of a symmetric weight matrix ($T_{ij} = T_{ji}$) will such a model not have any false attractors, since periodic or chaotic attractors do not exist in gradient systems.

Thus by incorporating terminal attractors in a neural network, one can store desired vectors as stable equilibrium points with no false static attractors. It is worth mentioning that, in general, the corresponding weight matrix T_{ij} is not symmetric.

One of the advantages of the terminal attractor approach is in the fact that the weight matrix T_{ij} corresponding to the desired fixed points is defined by (8.7) and, therefore, can be found by conventional methods (Lapedes and Farber 1986), while the stability of these fixed points is provided by the incorporation of terminal attractors in each equilibrium point. The last procedure can be represented by additional (variable) diagonal terms T^*_{ii} in the weight matrix T_{ij}

$$T^*_{ii} = \frac{1}{V(u_i)} \sum_{k=1}^{n} \left(\tilde{u}_i^{(k)} - u_i\right)^{1/3} e^{-\gamma_i^{(k)}(u_i - \tilde{u}_i^{(k)})^2}$$

It should be emphasized that the dynamical system (8.9) obtained as a result of this approach is characterized by a failure of the Lipschitz conditions at the desired equilibrium points. This leads to a loss of the uniqueness of the solution at these points: a family of regular solutions (representing transient motions of the system) is intersected by a singular solution (representing the equilibrium of the system); and as a result of this, the theoretical time of approaching these equilibrium points becomes finite.

8.2.2 Terminal Attractors: Pattern Recognition

Pattern recognition is based upon a selection of a category (from N given categories) into which a certain pattern must be stored. As shown by Hopfield (1982), this problem can be attacked with a neural network in which patterns are stored in stable equilibrium points, and the decision surface coincides with the basins' boundaries. Hopfield also gave a partial solution to the problem in the case of a symmetric weight matrix. However, one of the most serious limitations of his approach is a very small degree of control over the basins of attraction. In this section, we will demonstrate the usefulness of terminal attractors in Hopfield's approach to pattern recognition by solving the following problem: given n linearly independent vectors $\tilde{u}_i^{(k)}$, find such a neural network which has $n/2$ stable equilibrium point with prescribed basins.

First, let us consider an auxiliary differential equation:

$$\dot{x} = [P(x)]^{1/3} \tag{8.19}$$

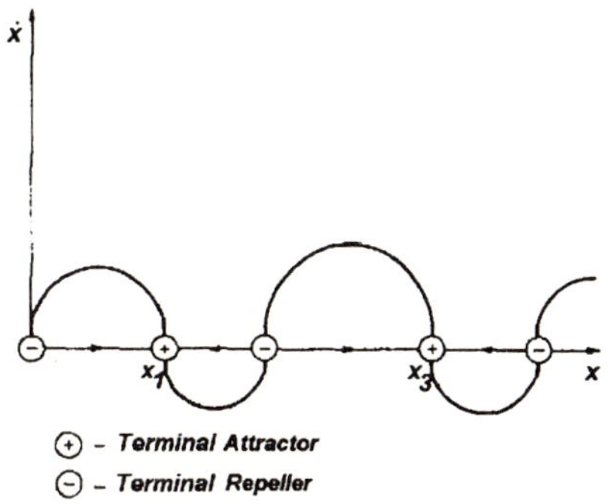

Fig. 8.3. Array of terminal attractors and repellers

with a polynomial

$$P(x) = (x - x_1)(x - x_2)...(x - x_{2n}) = \prod_{k=1}^{2n}(x - x_k), \; x_k < x_{k+1} \qquad (8.20)$$

This equation has $2n$ equilibrium points

$$x = x_k, \; k = 1, 2, ..., 2n \qquad (8.21)$$

It is easily verifiable that

$$x = x_{2k-1}, \; k = 1, 2, ..., n \qquad (8.22)$$

are terminal attractors, and

$$x = x_{2k}, \; k = 1, 2, ..., n \qquad (8.23)$$

are terminal repellers (Fig. 8.3).

Indeed, in small neighborhoods of (8.22–8.23), the polynomial (8.20) can be linearized and represented as

$$P(x) = \alpha_i(x - x_k)$$

in which

$$\alpha = (x_k - x_1)...(x_k - x_{k+1})(x_k - x_{k=1})...(x_k - x_{2n}) = \prod_{e \neq k}(x_k - x_e)$$

and (8.19) is reduced to $x = -x^{1/3}$ and $x = x^{1/3}$ for the neighborhoods (8.22) and (8.23), respectively.

Returning to our original problem, let us modify the neural net dynamical model given by (8.6) as follows

$$\dot{u}_i + u_i = \sum_{j=1}^{n} T_{ij}V_j + [P(u_i)]^{1/3}, \; i = 1, 2, ..., n \qquad (8.24)$$

in which $P(u_i)$ is a polynomial

$$P(u_i) = \prod_{k=0}^{n} (u_i - \widetilde{u}_i^{(k)}) \tag{8.25}$$

If the weight matrix T_{ij} is found from (8.8), then all the vectors $u_i^{(k)}$ represent equilibrium points of (8.15), since

$$P(\widetilde{u}_i^{(k)}) = 0 \tag{8.26}$$

Now we will show that

$$u_i = \widetilde{u}_i^{(2v-1)}, \quad v = 1, 2, ..., n/2 \tag{8.27}$$

are terminal attractors and

$$u_i = \widetilde{u}_i^{(2v)}, \quad v = 1, 2, ..., n/2 \tag{8.28}$$

are terminal repellers, if n is an even number.

First of all, we will recall that similar results were already obtained for a simplified version of (8.24)

$$\dot{u}_i = [P(u_i)]^{1/3} \tag{8.29}$$

[see (8.19–8.23)].

The proof of the same result for the original version of (8.24) is based upon the fact that in a sufficiently small neighborhood of an equilibrium point, a terminal attractor dominates over "regular" terms, and since (8.24) can be obtained from (8.25) by addition of "regular" terms, the stability (or instability) of the equilibrium point will persist.

Indeed, turning to the linearized version of the system (8.24) in the neighborhoods of the corresponding equilibrium points which can be written in the form (8.12), one finds that

$$a_{ij}^{(k)} = \frac{T_{ij}}{\cosh^2 \widetilde{u}_j^{(k)}}, \quad i \neq j \tag{8.30}$$

$$a_{ii} = -1 + \frac{T_{ij}}{\cosh^2 \widetilde{u}_j^{(k)}} - \frac{1}{3}\alpha_i^{(k)} \varepsilon^{-2/3} \tag{8.31}$$

Here, ε is given by (8.11), and $\alpha_i^{(k)}$ is found from the linearization of the polynomial $P(u_i - \widetilde{u}_i^{(k)})$ in the neighborhood of $u_i^{(k)}$

$$P(u_i) \simeq \alpha_i^{(k)} \left(\widetilde{u}_i^{(k)} - u_i\right), \text{ that is}$$

$$\alpha_i^{(k)} = \prod_{e \neq k}^{n} (\widetilde{u}_i^k - \widetilde{u}_i^{(e)}) \begin{cases} > 0 \text{ for } & k = 2v - 1 \\ < 0 \text{ for } & k = 2v \end{cases} \tag{8.32}$$

Following the same procedure as performed above [see (8.15–8.17)], one obtains the expression for the eigenvalues of the matrices $a_{ij}^{(k)}$ [cf. (8.18)]

$$\lambda_k = -\frac{1}{3}\alpha_i^{(k)}\varepsilon^{2/3} \to \begin{cases} -\infty \text{ if } & k = 2v-1 \\ \infty \text{ if } & k = 2v \end{cases} \tag{8.33}$$

which proves that all the equilibrium points (8.27) are "infinitely" stable, that is, they are terminal attractors, and all the equilibrium points (8.28) are "infinitely" unstable, that is, they are terminal repellers.

The basins of attractions can be characterized by the distances between the neighboring repellers; for instance, the basin of the attractor $\widetilde{u}_i^{(2v-1)}$ can be defined by its "projections" $b_i^{(2v-1)}$

$$b_i^{(2v-1)} = \widetilde{u}_i^{(2v)} - \widetilde{u}_i^{(2v-2)} \tag{8.34}$$

Hence, if the desired attractors $\widetilde{u}_i^{(2v-1)}$ as well as their basins $b_i^{(2v-1)}$ are prescribed, then the corresponding repellers can be found from (8.34). This allows one to construct the polynomial (8.25), which when incorporated into the original neural net model [see (8.24)] will solve the problem.

So far it has been assumed that the number of neurons n is an even number. If n is an odd number, then all the equilibrium points $\widetilde{u}_i^{(2v-1)}$ become repellers and $\widetilde{u}_i^{(2v)}$ become attractors.

8.2.3 Models with Hierarchy of Terminal Attractors

This section introduces a sketch of more complex neural nets which consist of a hierarchy of coupled terminal attractors. Such models can perform complex associative memory as well as some useful self-organization properties.

Complex Associative Memory

We will start with the model described by (8.9), which has n terminal attractors $\widetilde{u}_i^{(k)}$, $k = 1, 2, ..., n$ if the weight matrix T_{ij} is defined by (8.11). Let us introduce n "replicas" of this model such that all the replicas are functionally identical to the original model, but each of which is driven by only one of the terminal attractors. The difference in dynamical evolution of the replicas is due only to the difference in their initial conditions, which are assigned to the basins of attraction of the corresponding vectors. Hence, instead of (8.9) we will consider the following $n \times n$ system

$$\dot{u}_i^{(k)} + u_i^{(k)} = \sum_{j=1}^{n} T_{ij} V_j^{(k)} + \left(\widetilde{u}_i^{(k)} - u_i^{(k)}\right)^{1/3}, \quad i, k = 1, 2, ..., n \tag{8.35}$$

in which $u_i^{(k)}$ are the variables of the kth replica, while $V_j^{(k)} = V(u_j^{(k)})$.

So far, the coordinates of the terminal attractors $\tilde{u}_i^{(k)}$ which enter (8.35) not only explicitly, but also implicitly (via the matrix T_{ij}) have been considered as prescribed parameters. Let us now introduce a secondary (a higher level) neural net having the terminal attractor coordinates $\tilde{u}_i^{(k)}$ of the system (8.35) as new variables. Such a model will describe a dynamical interaction between the terminal attractors $\tilde{u}_i^{(k)}$ if one presents it in the form

$$\tilde{u}_i^{(K)} + \tilde{u}_i^{(k)} = (\sum_{l=1}^{n}\sum_{j=1}^{n} T_{ij}^{kl} V_i^{(l)})\chi, \ \tilde{V}_j^{(l)} = \tilde{V}(u_j^{(l)}) \tag{8.36}$$

Here χ is an activation coefficient defined as

$$\chi = \begin{cases} 1 \text{ if } u^{(l)} = \tilde{u}^{(l)}, & i, l = 1, 2, ..., n \\ 0 \text{ if } u^{(l)} \neq \tilde{u}^{(l)}, & i, l = 1, 2, ..., n \end{cases} \tag{8.37}$$

It can be presented in the following analytical form

$$\chi = \lim_{\gamma \to \infty} e^{-\gamma \sum_{i=1}^{n} \sum_{l=1}^{n} (u_i^{(l)} - \tilde{u}_i^{(l)})^2} \tag{8.38}$$

Hence, due to the activation coefficient χ, the secondary neural net (8.36) becomes active only after all the transients $u_i^{(k)}(t)$ of the original neural net approach the corresponding terminal attractors $\tilde{u}_i^{(k)}$. (It should be recalled that the theoretical time for this is finite.) The weight coefficients T_{ij}^{kl} form a tensor of the fourth rank. Since the dimension of a vector $\tilde{u}_i^{(k)}$ is n^2, while the number of the weight coefficient is n^4, generally speaking, (8.36) has n^2 equilibrium points. For the sake of simplicity we will confine our analysis by the case

$$T_{ij}^{kl} = 0 \text{ if } l \neq l, \text{ or } i \neq j \tag{8.39}$$

Then (8.36) reduces to

$$\tilde{u}_i^{(k)} + \tilde{u}_i^{(k)} = T_{ii}^{kk} \tilde{V}_i^{(k)} \chi \tag{8.40}$$

Equation (8.40) has only one equilibrium point $\tilde{\tilde{u}}_i^{(k)}$, while T_{ii}^{kk} is found from an equation similar to (8.2)

$$T_{ii}^{kk} = \tilde{\tilde{u}}_i^{(k)} [V(\tilde{\tilde{u}}_i^{(k)})]^{-1} \tag{8.41}$$

In order to provide stability of this equilibrium point, one should incorporate a terminal attractor into (8.40) [see (8.1)]

$$\dot{\tilde{u}}_i^{(k)} + \tilde{u}_i^{(k)} = [T_{ii}^{kk} \tilde{V}_i^{(k)} + (\tilde{\tilde{u}}_i^{(k)} - \tilde{u}_i^{(k)})]^{1/3} \chi \tag{8.42}$$

Thus, as soon as all the transients $u_i^{(k)}(t)$ in the original neural net approach the corresponding terminal attractors $u_i^{(k)}$, the secondary neural net

(8.42) becomes active and drives the original terminal attractors $u_i^{(k)}$ in a new [prescribed by (8.41)] position $\approx u_i^{(k)}$. In other words, the neural nets (8.35) and (8.42) perform associative memory.

If one eliminates the restriction (8.39), then the original terminal attractors $\tilde{u}_i^{(k)}$ driven by (8.42) with full weight tensor T_{ij}^{kl}, may occupy n^2 different positions, depending upon their initial dislocations.

In the same way one can introduce the third, forth, etc., neural nets and obtain a chain of associations

$$\tilde{u}_i^{(k)} \begin{array}{l} \nearrow \ \approx u_i^{(k)} \rightarrow \approx u_i^{(k)} \\ \searrow \ \approx u_i^{(l)} \rightarrow \approx u_i^{(l)} \end{array} \tag{8.43}$$

Active Terminal Attractors

The concept of an attractor (including a terminal attractor) implies that the attractor passively "waits" for a transient solution which is supposed to approach it. In order to improve the convergence, we will introduce a concept of an active attractor which temporarily "leaves" its position in order to "search" for the corresponding transient solution; after "catching" this solution, it returns to the original position. (Strictly speaking, such an active attractor is not a true static attractor any more.) In order to trigger a mobility of terminal attractors of the model (8.35), we will slightly modify (8.36)

$$\tilde{u}_i^{(k)} = \sum_{l=1}^{n}\sum_{j=1}^{n} T_{ij}^{kl}[u_i^{(l)}(\tilde{V}_i^{(l)} - \tilde{V}_j^{(l)})] + (u_i^{0(k)} - \tilde{u}_i^{(k)})^{1/3} \tag{8.44}$$

in which $u_i^{0(k)}$ describe the original positions of the terminal attractors $\tilde{u}_i^{(k)}$ of the system (8.36).

As follows from (8.44), the activity of the secondary neural net is triggered by inputs to the original neural net

$$u_j^{(l)} \neq 0 \tag{8.45}$$

In the course of the transient process, the coordinates of the original terminal attractors become functions of time

$$\tilde{u}_i^{(k)} = \tilde{u}_i^{(k)}(t) \tag{8.46}$$

until all the attractors "catch" the corresponding inputs $u_j^{(l)}$, that is,

$$u_j^{(l)} = \tilde{u}_j^{(k)}, \text{ that is, } V_j^{(l)} = \tilde{V}_j^{(l)} \tag{8.47}$$

After this, the double sum in (8.44) becomes zero, and the "wandering" attractors $\tilde{u}_j^{(k)}$ return to their original static positions $u_i^{0(k)}$. The "safe" return is guaranteed by the last term in the right-side of (8.44), which makes $u_i^{0(k)}$ infinitely stable. Obviously, such a mobility of terminal attractors may change their actual basins.

From the viewpoint of control theory, the secondary neural net performs a parametric control of the original neural net. The usefulness of such a control is supposed to be provided by an appropriate selection of the weight tensor T_{ij}^{kl}; for instance, the strategy of this control can include a temporary symmetrization of the original weight matrix T_{ij} by minimizing (in dynamics) its antisymmetric terms [see (8.8)]

$$\sum_{i,j=1}^{n}\left[\sum_{l=1}^{n}\tilde{u}_i^{(l)}(V_l^{(i)})^{-1}-\sum_{l=1}^{n}\tilde{u}_j^{(l)}(V_l^{(i)})^{-1}\right]^2 \rightarrow \min \tag{8.48}$$

in which $V_l^{(i)}$ are the elements of the matrix

$$\begin{pmatrix} V_1^{(1)} & \dots & V_1^{(n)} \\ \dots & \dots & \dots \\ V_n^{(1)} & \dots & V_n^{(n)} \end{pmatrix}^{-1} \tag{8.49}$$

Such a symmetrization may destroy possible periodic or chaotic attractors, where the transient solution can be trapped and, thereby, control actual basins of attraction.

Certainly, this example is not a proposal for an actual control strategy, nor any claim for a vigorousness in it. The only purpose of the example is to illustrate the concept of an active terminal attractor.

Terminal Attractors in Neural Nets with Programmed Delays

The previous sections dealt only with neural nets which are described by ordinary differential equations. The solutions to these systems $u_i(t_0, u_i^0, t)$ can be viewed as continuous transformations of the space of initial points $u_0 \in E^n$ defined on $t = t_0$ to points in E^n representing the state variables. In other words, the pattern to be stored or to be recognized is represented by an initial vector u_i^0, while the time of its exposure is not perceived by the neural net. Hence, neural nets based upon ordinary differential equations cannot be utilized for patterns which are characterized not only by the vector of the state variables u_i^0 , but also by durations of the exposure τ_i^0 of its components.

This section introduces neural nets described by delay-differential equations which can store as terminal attractors not only vectors, but also the durations of their exposures. For this purpose, we will consider a neural network with programmed delays τ_i in the response of the outputs of changes in the inputs

$$\dot{u}_i(t) + u_i(t) = \sum_{i=j}^{n} T_{ij}V[u_j(t-\tau_j)] + \alpha_i[\tilde{u}_i - u_i(t-\hat{\tau}_i)]^{1/3}, \quad \alpha_i > 0 \tag{8.50}$$

while the delays τ_i are variables of another dynamical subsystem

$$\dot{\tau}_i + \tau_i = \sum_{j=1}^{n} A_{ij} V(\tau_j) + (\widetilde{\tau}_i - \tau_i)^{1/3} \tag{8.51}$$

in which T_{ij} and A_{ij} are constant weight matrices. The delay $\widetilde{\tau}_i$ will be selected later from stability considerations. The dynamical subsystem (8.51) is not coupled with (8.50) and, therefore, can be considered independently.

If A_{ij} selected such that

$$\widetilde{\tau}_i = \sum_{i=1}^{n} A_{ij} V(\widetilde{\tau}_j) \tag{8.52}$$

then $\widetilde{\tau}_i = \tau_i$ is a terminal attractor, which is approached by transient solutions during a theoretically finite period t^*.

One can conclude that for $t < t^*$, the first group of equations, that is, (8.50), represents a system of nonlinear differential equations with deviating arguments, that is, a system of functional difference equations. However, for

$$t \geq t^* \tag{8.53}$$

all the delays become constant

$$\tau_i = \widetilde{\tau} = \text{Const}$$

and, consequently, (8.50) reduces to the following

$$\dot{u}_i(t) + u_i(t) = \sum_{i=j??}^{n} T_{ij} V[u_j(t - \widetilde{\tau}_j)] + \alpha_i [\widetilde{u}_i(t) - u_i(t - \hat{\tau}_i)]^{1/3} \tag{8.54}$$

which is a system of nonlinear delay-differential equations.

Unlike ordinary differential equations, for (8.54) we require not a constant initial vector u_i^0, but rather a vector function defined over the lag periods

$$\varphi_i = u_i^*(-\tau_i^0 \leq t \leq 0) \tag{8.55}$$

As shown above, all these lags τ_i^0 and, therefore, the corresponding periods of exposures of the input patterns u_i^0 are stored at the terminal attractors $\widetilde{\tau}_i$ of the subsystem (8.51).

Equation (8.54) has an equilibrium point $u_i = \widetilde{u}_i$ if the weight matrix T_{ij} is selected such that

$$\widetilde{u}_i = \sum_{j=1}^{n} T_{ij} V(\widetilde{u}_j), \text{ since } \widetilde{u}_i(t) = \widetilde{u}_i(t - \widetilde{\tau}_j) \text{ if } \dot{u}(t) \equiv 0 \tag{8.56}$$

Now we will prove that this equilibrium point is a terminal attractor.

First of all, one has to notice that the concepts of stability or instability in delay-differential equations are different from those in ordinary differential equations, since here one should analyze the effect of disturbances not at the initial point of a trajectory, but rather of an initial function. Nevertheless, under certain reasonable assumptions, the classical Liapunov theory of stability developed for ordinary differential equations, after some modifications, can also be applied to delay-differential equations. For instance, global stability can be analyzed using a Liapunov functional (instead of a Liapunov function).

For local stability, one can apply a linearization procedure (with respect to an equilibrium point to be considered) which is followed by the Laplace transform. After this, stability analysis is reduced to evaluation of the eigenvalues λ_i of the matrix (3.10) given by the following formulas

$$a_{ij} = \frac{T_{ij}}{\cosh^2 \widetilde{u}_j} e^{-\tau_j \lambda}, \; a_{ij} = -1 + \frac{T_{ij}}{\cosh^2 \widetilde{u}_j} e^{-\tau_j \lambda} - \frac{\alpha_i}{3} \varepsilon^{-2/3} e^{-\tau_j \lambda} \tag{8.57}$$

in which $\varepsilon \to 0$ [see (8.11)]. This yields [cf. (8.17)]

$$\lambda_j = -\gamma_j e^{-\hat{\tau}_j \lambda_j}, \; \gamma_j = \frac{\alpha_j}{3\varepsilon^{2/3}} \to \infty \tag{8.58}$$

while

$$\lambda_j = x_j + i y_j, \; i = \sqrt{-1} \tag{8.59}$$

Separating (8.58) into real and imaginary parts yields

$$x_j + \gamma_j e^{-\hat{\tau}_j \lambda_j} \cos \hat{\tau}_j y_j = 0, \; \hat{\tau}_j y_j - \gamma_j e^{-\hat{\tau}_j \lambda_j} \sin \hat{\tau}_j y_j = 0$$

whence

$$\hat{\tau}_j x_j = - \hat{\tau}_j y_j \cot \hat{\tau}_j y_j, \; e^{\hat{\tau}_j y_j \cot \hat{\tau}_j y_j} \frac{\sin \hat{\tau}_j y_j}{\hat{\tau}_j y_j} = \gamma_j^{-1} \to 0 \tag{8.60}$$

As follows front the second equation in (8.60)

$$\hat{\tau}_j y_j = \pi k - \delta^2, \; \delta^2 \to 0, \; k = 1, 2, ..., \text{etc.} \tag{8.61}$$

and consequently

$$\hat{\tau}_j x_j = -(\pi k - \delta^2) \cot(\pi k - \delta^2) \to \infty \text{ if } \delta^2 \to 0$$

that is

$$x_j \to -\infty \text{ if } \hat{\tau} < 0 \tag{8.62}$$

Hence, all the real parts of the eigenvalues λ_i of the matrix (3.10) are negative, while their moduli are unbounded. This means that the equilibrium point $u_i = \widetilde{u}_i$ of the system (8.54) is a terminal attractor, if $\hat{\tau}_i$ are selected to be negative, for instance $\hat{\tau}_i = -|\tau_i|$.

Thus, the neural net described by (8.50–8.51) allows one to store and recognize patterns which are characterized by a vector $\overline{u}^0$ as well as by durations τ_i^0 of exposure of its components u_i^0.

Generally speaking, this model can be utilized for even more complex problems: to store a vector function $\widetilde{u}_i(t)$ defined for $0 \leq t \leq \widetilde{\tau}_i$. The reason for such a possibility is due to the fact that delay-differential equations require initial conditions in the form of function of time [see (8.55)]. Hence, if this initial condition, that is, the functions $u_i^*(t)$ are "close" to the corresponding stored functions $\widetilde{u}_i(t)$, then they may be attracted to these stored functions. Therefore, the weight matrix T_{ij} must be selected such that $\widetilde{u}_i(t)$ represent a periodic terminal attractor [the example of such an attractor is given by (7.78)]. However, practical realization of this procedure is much more difficult than in the previous case and, therefore, requires a special analysis. We will not go into further details of the problem, confining ourselves to this brief discussion only.

Dynamical Training using Terminal Attractors

The weight coefficients T_{ij} or T_{ij}^{kl} considered in the previous sections were defined by the corresponding analytical formulas (8.8) and (8.41), but we have not yet specified how they will be incorporated into neural nets. Because of a very large number of neurons and their interconnections in future generations of artificial neural nets, it does not seem possible to program the weight coefficients according to the analytical formulas mentioned above. In order to avoid such programming, we will describe a dynamical training procedure by introducing a dynamical equation for the weight matrix which governs its relaxation to the corresponding stable equilibrium point. This point will be represented by a terminal attractor which guarantees its local stability and fast convergence.

Let us introduce a strength energy

$$S = \frac{1}{2} \sum_{i=1}^{n} \sum_{j=1}^{n} T_{ij}^2 (i-j)^2 \tag{8.63}$$

as a measure of interconnections between neurons, requiring that for each prescribed performance of the neural network this energy as a function of T_{ij} must be minimum. Each prescribed performance imposes additional limitations on the weight coefficients T_{ij}. In case of content-addressable memories, these limitations are expressed by the following conditions

$$g_{ik} = \sum_{j=1}^{n} T_{ij} V(\widetilde{u}_j^{(k)}) - \widetilde{u}_j^{(k)} = 0, \; i = 1, ..., n \; k = 1, ..., m \tag{8.64}$$

has to minimize a modified strength energy

$$\widetilde{S}(T_{ij}, \lambda_{ik}) = S + \frac{1}{2} \sum_{i=1}^{n} \sum_{k=1}^{m} \lambda_{ik} g_{ik}^2 \tag{8.65}$$

in which λ_{ik} are constants prescribed to represent the contributions of the constraints (8.64) into the energy (8.65).

Since the energy $\widetilde{S}$ is a quadratic form of the variables T_{ij} [see (8.63–8.64)], its gradient $\nabla\widetilde{S}$ is a linear form of its arguments

$$\frac{\partial\widetilde{S}}{\partial T_{ij}} = (i-j)^2 T_{ij} + \sum_{k=1}^{m} \lambda_{ik} \widetilde{V}_j^{(k)} g_{ik} \tag{8.66}$$

If

$$\det\left(\frac{\partial\widetilde{S}}{\partial T_{ij}}\right) \neq 0 \tag{8.67}$$

and

$$\frac{\partial^2\widetilde{S}}{\partial T_{ij} T_{\alpha\beta}} > 0 \tag{8.68}$$

at

$$\frac{\partial\widetilde{S}}{\partial T_{ij}} = 0 \tag{8.69}$$

the system (8.69) defines a unique stable solution which is the minimum of the energy $\widetilde{S}$.

In order to use (8.68) for dynamical learning, it will be assumed that all the variables T_{ij} are time-dependent. Then

$$\dot{\widetilde{S}} = \nabla\widetilde{S} \cdot \dot{x} \tag{8.70}$$

in which x denotes the vector with components T_{ij}, and ∇S is the vector gradient of the strength energy $\widetilde{S}$ with the components (8.66).

The dynamical system

$$\dot{x} = -\alpha^2 \nabla\widetilde{S} \tag{8.71}$$

represents a gradient system which converges asymptotically to the minimum defined by (8.66), while the strength energy S plays the role of the Liapunov function, since

$$\dot{\widetilde{S}} = \alpha^2 (\nabla S)^2 < 0 \text{ if } \nabla S \neq 0 \tag{8.72}$$

So far, the multiplier α^2 has not been specified. Now we will show that by an appropriate selection of this multiplier the convergence of the dynamical system (8.71) can be significantly improved. Indeed, let us seek α^2 in the form

$$\alpha^2 = \left|\nabla\widetilde{S}\right|^{\gamma}, \; \gamma > 0 \tag{8.73}$$

Since $\nabla\widetilde{S}$ is a linear function of its arguments, it has the order

$$\nabla\widetilde{S} \sim x \text{ at } \nabla\widetilde{S} \to 0 \tag{8.74}$$

and consequently

$$\nabla \dot{\tilde{S}} \sim \dot{x} \text{ at } \nabla\tilde{S} \to 0 \tag{8.75}$$

Hence, with reference to (8.71) and (8.73)

$$\left|\nabla \dot{\tilde{S}}\right| \sim |\nabla S|^{1+\gamma} \text{ at } \nabla\tilde{S} \to 0$$

that is,

$$\frac{d\left|\nabla\tilde{S}\right|}{dt} = \beta^2 \left|\nabla\tilde{S}\right|^{1+\gamma} \text{ at } \nabla\tilde{S} \to 0 \tag{8.76}$$

in which β^2 is a constant.

Now we can evaluate the relaxation time

$$t = \int_{\left|\nabla\tilde{S}\right|_0}^{\left|\nabla\tilde{S}\right| \to 0} \frac{d\left|\nabla\tilde{S}\right|}{\left|\nabla\tilde{S}\right|^{1+\gamma}} = \begin{cases} \infty & \text{if } \gamma \geq 0 \\ \frac{1}{-\gamma}\left|\nabla\tilde{S}\right|_0^{-\gamma} < \infty & \text{if } \gamma < 0 \end{cases} \tag{8.77}$$

Thus, for $\gamma \geq 0$, the relaxation time is infinite, while for $\gamma < 0$ it is finite. The differential equation (8.76) suffers a qualitative change at $\gamma < 0$: it loses the uniqueness of the solution, while the equilibrium point $\left|\nabla\tilde{S}\right| = 0$ becomes a singular solution being intersected by all the transients.

It can be seen that at the equilibrium point $\left|\nabla\tilde{S}\right| = 0$ the Lipschitz condition is violated

$$\frac{\partial}{\partial\left|\left|\nabla\tilde{S}\right|\right|}\left(\frac{d\left|\nabla\tilde{S}\right|}{dt}\right) = \beta^2\left|\nabla\tilde{S}\right|^{\gamma} \to -\infty \text{ at } \left|\nabla\tilde{S}\right| \to 0 \text{ if } \gamma < 0$$

and, consequently, this point is a terminal attractor. In order to be consistent with the results presented in previous sections, we will set

$$\gamma = -\frac{2}{3}$$

providing in (8.77) the following relaxation time

$$t = \frac{3}{2}\left|\nabla\tilde{S}\right|_0^{2/3} < \infty \tag{8.78}$$

Now, the dynamical system governing the learning process can be written in the following final form

$$\dot{T}_{ij} = \alpha^2 \left[(i-j)^2 T_{ij} + \sum_{k=1}^{m} \lambda_{ik}\tilde{V}_j^{(k)} g_{ik}\right], \quad i,j = 1,2,...,n \tag{8.79}$$

and

$$\alpha^2 = \left| \sum_{i=1}^{n} \sum_{j=1}^{n} \left[(i-j)^2 T_{ij} + \sum_{k=1}^{m} \lambda_{ik} \widetilde{V}_j^{(k)} g_{ik} \right] \right|^{-2/3} \tag{8.80}$$

If the conditions (8.67–8.68) are satisfied, this system has a unique stable solution which is approached during a finite time. It defines the weight coefficients T_{ij} by minimizing the strength energy $\widetilde{S}$ while satisfying the conditions (8.64) required for a prescribed content-addressable memory.

Instead of conditions (8.64), other types of limitations can be used, such as continuous mapper conditions

$$\sum_{j=1}^{n} T_{ij} \widetilde{V}_j^{(k)} - \widetilde{u}_i^{(k)} - I_i^{(k)} = 0 \tag{8.81}$$

where $I_i^{(k)}$ is a constant vector representing an external input. It is easy to verify that the finite relaxation time can be provided by using the system (8.79) after replacement of the conditions (8.64) by the conditions (8.81).

Thus, it has been demonstrated that the learning procedure based upon minimization of the strength energy leads to a dynamics in the weight space along a learning "trajectory" which converges to a terminal attractor.

Terminal Attractors in Neural nets with Spatial Self-organization

So far, we have considered the terminal attractor as a tool for improving the convergence of neural nets to their equilibria in phase space. However, terminal attractors can be also useful in neural nets with spatial self-organization (see Chap. 7, Sect. 4) where the main concern is a convergence to an equilibrium which has a certain configuration in actual space. In order to illustrate this, let us turn to convective neurodynamics described by (3.67):

$$\frac{\partial u}{\partial t} + v \frac{\partial u}{\partial x} = 0, \; v = (1-x)^k \tag{8.82}$$

Let us take the initial image in the form of a unit square on the plane, as in (3.72):

$$\dot{u}(x) = \begin{cases} 1 \text{ at } & 0 < x < 1 \\ 0 \text{ at } & x < 0 \text{ and } x > 1 \end{cases} \tag{8.83}$$

Since $v > 0$ for $x < 1$, all the points of the square (8.83) will move to the right and stop at the critical point $x = 1$ where $v = 0$. In other words, the original square will be absorbed by its right side $u = 1$ at $x = 1$ and therefore, it will degenerate into a line.

Let us calculate the time of convergence to the point $x = 1$ of the image (8.83). Referring to (8.82) and taking into account that $v = dx/dt$ one obtains: $dx/dt = 1 - x$ and therefore: $t = -\ln(1-x) \to \infty$ if $x \to 1$, for $k = 1$.

But if, instead of $k = 1$ one takes $k = 1/3$, i.e.,

$$v = \sqrt[3]{1 - x} \tag{8.84}$$

then the time of convergence is finite: $t = 3/2$, in the full analogy with terminal attractors in the phase space discussed above. Hence, in the classical case ($k = 1$) the image (theoretically) will never degenerate into a line, while in the terminal dynamics case, the collapse of dimension can occur in a finite time.

Terminal Attractors in Discrete Models of Neural Nets

In the previous sections, terminal attractors were associated with continuous-time operating neural nets. In this connection, it is interesting to learn what will happen if a continuous-time model with a terminal attractor is converted into a discrete-time model. In this section, we will analyze how a discrete-time model, whose underlying continuous-time model has if a terminal attractor behaves in a small neighborhood of this attractor.

Let us consider a simple differential equation

$$\dot{u} = f(u) - u^{1/3}, \quad f(0) = 0 \tag{8.85}$$

and assume that the derivative df/du exists and is bounded. Then one can easily verify that $u = 0$ is a terminal attractor,

Rewriting (8.85) in the form

$$\dot{u} + u^{-2/3}u = f(u) \tag{8.86}$$

let us introduce the following discrete analog of this differential equation

$$u^{(v+1)} + (u^{(v)})^{-2/3}u^{(v+1)} = f(u^{(v)}) + u^{v} \tag{8.87}$$

whence

$$u^{(v+1)} = \frac{1}{1 + (u^{(v)})^{-2/3}}[f(u^{(v)}) + u^{(v)}] = A(u^{(v)})u^{(v)} \tag{8.88}$$

The algorithm (8.87) converges if A is a contracting operator, that is,

$$||A|| < 1 \tag{8.89}$$

while the less this norm, the faster the convergence process.

We will now emphasize the role of the terminal attractor of the original equation (8.85) at $u = 0$: as one can easily verify:

$$[u^{(v)}]^{-2/3} \to \infty \text{ if } u^{(v)} \to 0 \tag{8.90}$$

and, therefore,

$$||A|| \to 0, \text{ if } u^{(v)} \to 0 \tag{8.91}$$

Thus, in a sufficiently close neighborhood of the terminal attractor, the norm of the contracting operator becomes infinitesimal and, therefore, the convergence process becomes "infinitely" fast.

The same result can be obtained for a general neural net

$$\dot{u}_i = f_i(u_1, u_2 ... u_n) + \sum_{k=1}^{n} \alpha_i^{(k)} (\widetilde{u}_i^k - u_i)^{1/3} \tag{8.92}$$

with terminal attractors $u_i = \widetilde{u}_i$.

Hence, despite the fact that terminal attractors were introduced for a continuous-time operating system, they exhibit similar properties in the corresponding discrete-time analogs.

8.2.4 Spontaneously Activated Neural Nets

In this section we will introduce a new type of dynamical system which can acquire a large number of different structures without a specific interference from the outside: driven by a vanishingly small noise, identical dynamical systems perform significantly different patterns of behavior. In other words, these systems extract information from "nothing" (Zak 1989b,c,d; 1990a,c; 1991a,b,c).

Unpredictable Neural Nets

One of the fundamental limitations of artificial computational systems is that they behave too rigidly when compared with even the simplest biological systems. With regard to neurodynamical systems, this point has a simple phenomenological explanation: all such systems satisfy the Lipschitz condition that guarantees the uniqueness of the solutions subject to prescribed sets of initial conditions. Indeed, a dynamical system

$$\dot{x}_i = f_i(x_1, x_2 ... x_n), \; i = 1, 2, ..., n \tag{8.93}$$

subject to the initial conditions

$$x_i = \overset{0}{x}_i \; \text{ at } t = t_0 \tag{8.94}$$

has a unique solution:

$$x_i = f_i(t, x_0, t_0) \tag{8.95}$$

if all the derivatives exist and are bounded

$$\left| \frac{\partial f_i}{\partial x_j} \right| < \infty \tag{8.96}$$

The uniqueness of the solution (8.95) can be considered as a mathematical interpretation of rigid or predictable behavior of the corresponding dynamical system.

With the limitations (8.96), artificial neural networks can only process information: they can learn by examples, memorize and recognize patterns, and so on. But such abilities are completely predictable, since they are prescribed in advance by the original dynamic structure.

Can a man-made dynamic system create new information? An answer to this question would help us better understand the evolution of intelligence.

We will start with the dynamical paradigm introduced in Chap. 7, and consider the following one-neuron dynamical system:

$$\dot{u} = \gamma \sin^{1/3} \frac{\omega}{\alpha} u \sin \omega t, \ \gamma = \text{Const}, \ \omega = \text{Const}, \ \alpha = \text{Const} \tag{8.97}$$

As shown in Chap. 7 [see (7.119–7.124)] at the equilibrium points:

$$u_k = \frac{\pi k \alpha}{\omega}, \ k = \cdots - 2, -1, 0, 1, 2, \cdots \text{etc.} \tag{8.98}$$

the Lipschitz condition is violated:

$$\partial \dot{u} / \partial u \rightarrow \infty \text{ at } u \rightarrow u_k \tag{8.99}$$

If $u = 0$ at $t = 0$, then during the first period

$$0 < t < \frac{\pi}{\omega} \tag{8.100}$$

the point $u_0 = 0$ is a terminal repeller since $\sin \omega t > 0$ and the solution at this point splits into two (positive and negative) branches whose divergence is characterized by unbounded rate (Zak 1989b,c,d). Consequently, with an equal probability u can move into the positive or the negative direction. For the sake of concreteness, we will assume that it moves in the positive direction. Then the solution will approach the second equilibrium point $u_1 = \pi\alpha/\omega$ at

$$t^* = \frac{1}{\omega} \text{arc cos} \left[1 - \frac{B(\frac{1}{3}, \frac{1}{3})}{2^{1/3}} \frac{\alpha}{\gamma} \right] \tag{8.101}$$

in which B is the Beta function, while the point u_1, will be a terminal attractor at $t = t_1$, if

$$t_1 \leq \pi/\omega, \text{ i.e., if } \frac{\gamma}{\alpha} \leq \frac{B(\frac{1}{3}, \frac{1}{3})}{2^{4/3}} \tag{8.102}$$

Therefore, u will remain at the point u_1 until it becomes a terminal repeller, i.e., until $t > t_1$. Then the solution splits again: one of two possible branches approach the next equilibrium point $u_2 = 2\pi\alpha/\omega$, while the other returns to the point $u_0 = 0$, etc. The periods of transition from one equilibrium point to another are all the same and are given by (8.101).

It is important to emphasize again that these periods t^* are bounded only because of the failure of the Lipschitz condition at the equilibrium points

(Zak 1989b,c,d). Otherwise they would be unbounded since the time of approaching a regular attractor (as well as the time of escaping a regular repeller) is infinite.

Thus, the evolution of u prescribed by (8.97) is totally unpredictable: it has 2^m different scenarios where $m = E(t/t^*)$, while any prescribed value of the neuron potential from (8.98) will appear eventually.

Let us assume that the dynamical system (8.97) is driven by a vanishingly small input $\varepsilon(t)$

$$\dot{u} = \gamma \sin^{1/3} \frac{\omega}{\alpha} u \sin \omega t + \varepsilon(t), \quad |\varepsilon(t) \ll \gamma| \tag{8.103}$$

This input can be ignored when $\dot{u} \neq 0$, or when $\dot{u} = 0$, but the system is stable, i.e. $u = \pi\alpha/\omega,\ 3\pi\alpha/\omega, ...,$ etc. However, it becomes significant during the instants of instability when $\dot{u} = 0$ at $u = 0,\ 2\pi\alpha/\omega, ...,$ etc. Since actually a vanishingly small noise is always present, one can interpret the unpredictability discussed above as a consequence of small random inputs to which the one-neuron dynamical system (8.103) is extremely sensitive.

Thus, in contradistinction to classical dynamical systems, the variable u in (8.103) "forgets" its past motion at each critical point u_k [see (8.98)], while any infinitesimal inputs applied to these points cause finite responses of the neuron. This property appears to be an important tool for creating a chain of coupled subsystems of different scales whose range is theoretically unlimited. As will be demonstrated below such a chain can play the role of a main tool for hierarchical parallelism in neurodynamics.

The function $\varepsilon(t) \ll \gamma$ is not necessarily random: it can be associated with a microsystems which controls the neuron behavior through a string of signs (Zak 1990b). Indeed, actually the only important part in this input is the sign of $\varepsilon(t)$ at the critical points. Consider, for example, (8.103), and suppose that

$$\operatorname{sgn} \varepsilon(t_k) = +, +, -, +, -, -, \text{ etc. at } t_k = \frac{\pi k}{\omega} ..., \quad k = 0, 1, 2, ..., \text{etc.} \tag{8.104}$$

The values of $\varepsilon(t)$ in between the critical points are not important since, by our assumption, they are small in comparison to values of the derivative $\dot{u}$, and therefore, can be ignored. Hence, the only part of the input $\varepsilon(t)$ which is significant in determining the motion of the neuron (8.103) is the sign string (8.104): specification of this string fully determines the dynamics of (8.103).

Does God Play Dice?

Let us return to the simplest unpredictable dynamical system (8.97) and consider a set of such systems:

$$\dot{u}_i = \gamma \sin^{1/3} \frac{\omega}{\alpha} u_i \sin \omega t + \varepsilon_i(t), \quad \varepsilon_i(t) \to 0, \quad i = 1, 2, \ldots, n \tag{8.105}$$

These equations describe dynamics of n identical uncoupled neurons driven by vanishingly small random noise. But since the components $\varepsilon_i(t)$ of this noise are not necessarily identical, the motions of each neuron, in general, will be different from the motions of the others, while these differences will be finite rather than infinitesimal. Such a phenomenon can be interpreted as an "emergence from chaos" of a "society" of neurons with different "personalities" despite the fact that the initial conditions for all of them are almost identical. It is remarkable that the "character" of each "personality" has a random nature coming from the random origin of the noise $\varepsilon_i(t)$.

The next step in the development of the neurons' "society" (8.105) can be based upon the following scenario: let us assume that there exists an objective function to be achieved by the neurons. It can be, for instance, a functional Φ to be minimized:

$$\Phi = \int_0^t [u(t) - f(t)]^2 dt \to \min \tag{8.106}$$

in which $\tilde{t} = \text{Const} > 0$, and $f(t)$ is a prescribed function. Suppose that the performance of each neuron is measured by the value:

$$\Phi_i = \int_0^{\tilde{t}} [u_i(t) - f(t)]^2 dt \tag{8.107}$$

and let

$$\dot{\omega}_i = \left(\omega_o + \Phi_i - \frac{1}{n} \sum_{j=1}^{n} \Phi_j \right) - \omega_i \tag{8.108}$$

in which ω_i is a new frequency of the periodic excitations which will be different for each neuron. As follows from (8.108), those neurons whose performance is above the average:

$$\Phi_i > \frac{1}{n} \sum_{j=1}^{n} \Phi_j \tag{8.109}$$

will receive higher energy input through the frequency ω_i and will act faster. On the contrary, those neurons whose performance is below the average:

$$\Phi_i < \frac{1}{n} \sum_{j=1}^{n} \Phi_j \tag{8.110}$$

will slow down their activity and eventually they will be out of competition.

This primitive scenario of a natural selection is only an illustration of possible development of discrimination in the neuron's "society." In addition to this, the neurons can learn from their experience memorizing useful behavior, and therefore, improving their performances; they can also cooperate and compete with each other, etc. But it is worth emphasizing

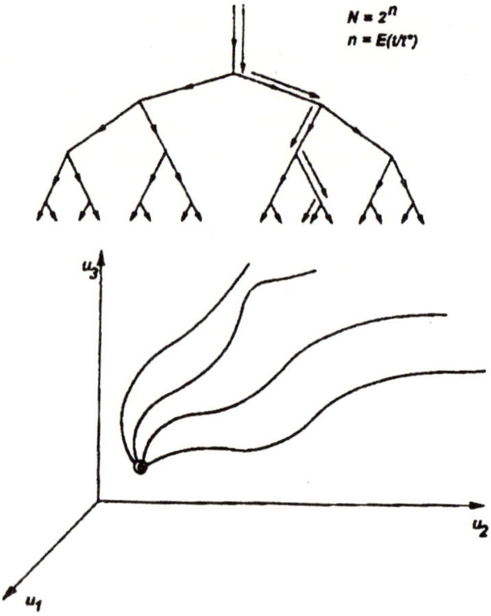

Fig. 8.4. Information born of chaos

that all these activities can be based upon conventional tools of information processing such as learning by examples, memorizing and recognition of patterns, local and global minimization, etc. What is the most important in our scenario is the first step, the "act of creation" when identical neurons in the same environment acquired different patterns of behavior without a special interference from outside. Such a creativity, in principle, cannot be performed by classical (Lipschitz) dynamical systems for which the uniqueness of the solutions subject to prescribed sets of initial conditions is guaranteed. Even the discovery of chaos has not shaken the rigid behavior of the Lipschitz dynamical systems since the chaotic motions are structurally stable. This means that although a particular trajectory of a chaotic motion cannot be predicted, a certain set of average characteristics of the motion as well as the global structure of the limit sets are fully predictable. It is also important to bear in mind that in chaotic systems the unpredictability is caused by a supersensitivity to the initial conditions, while the uniqueness of the solution is still guaranteed by the Lipschitz condition. On the contrary, in terminal dynamics the unpredictability is caused by the failure of the uniqueness of the solution at some of the equilibrium (critical) points. This is why a set of possible trajectories in phase space is not a Cantor set (as in chaotic systems), but rather a countable set of combinatorial nature. Because of this, the global unpredictability in terminal dynamics is associated with an exponential complexity (see Fig. 8.4). Hence, in this respect the unpredictable systems introduced above has some connections with the "digital world." Indeed, as follows from (8.104), they can be encoded by a string of signs: $+, -, -, + + +, \ldots$, etc.

It is worth emphasizing that the most important property of the unpredictable systems introduced above is the ability to change spontaneously their structure as a result of parametric periodic excitations. Such a performance phenomenologically resembles brain activity.

In addition to this, it should be recalled that these unpredictable systems forget their past at each critical point, and therefore, their motion can be easily redirected by a vanishingly small input at the critical points. The last property makes these systems highly adaptable to environmental changes.

Let us now discuss the connection between the unpredictability and creativity of dynamical systems. We have defined earlier the unpredictability as a property of a dynamical system to have a multi-choice response to a periodic parametric excitation. However, the ability to have such a multi-choice is a necessary, but not a sufficient condition for a dynamical system to be creative since any chosen behavior must be useful. In other words, the unpredictable dynamical system, in addition, must have an ability to evaluate the usefulness of different choices and to select the best (or, at least, a good one). As a possible tool for such an evaluation, a functional (8.106) can be utilized. In this respect, we will discuss two extreme cases.

Let us first assume that the unpredictable dynamical system does not "know" the analytical structure of the functional (8.106), and it can only compute the values of (8.106) for each particular behavior. (Actually biological systems face such a problem in real life.) Then the only way the system can minimize the functional (8.106) is to sort out all the possible behaviors by direct computations. Clearly, this strategy inevitably leads to a combinatorial explosion of the number of possible behavior, and therefore, it is unacceptable for practical applications.

As an alternative to this situation, one can come up with another extreme by assuming that the dynamical system has the complete analytical model of the functional (8.106). (This situation is as unrealistic as the previous one.) Then, before acting, the system must "think": it should find the best behavior by minimizing the functional (8.106).

As usual, the real situation falls between these two extremes: the dynamical system does not have a complete analytical model of the functional (8.106), but it has some approximation to this model. Actually, it means that the values of the function $f(t)$ at $t = \pi\omega/k$ are known for most of $k = 1, 2, \ldots,$ etc. but not for all of them. In other words, the code strings will contain some special critical points where the signs of the micro-inputs are undetermined. At these points the dynamical system plays dice: exploiting its ability of multi-choice behavior, it sorts out all the possible solutions by direct computation of the functional (8.106) at the special (undetermined) critical points, and thereby finds the best behavior. However, in contradistinction from the first case, now this strategy does not lead to exponential complexity since, by definition, the number of special critical points is limited by definition.

One Neuron–One Synapse Dynamical System

In this section, based upon the terminal approach to dynamical systems, we will introduce a self-developing dynamical system which spontaneously changes the locations of its attractors. The simplest version of such a system consists of a one-neuron activation dynamics

$$\dot{u} = -(u - Tu^2)^{1/3} \sin \omega t, \ \omega = \text{Const} \tag{8.111}$$

and one-synapsis learning dynamics

$$\dot{T} = (u - Tu^2)^{1/3} \sin \omega t \tag{8.112}$$

while an external input is represented by periodic parametric excitation.

The system (8.111–8.112) possesses two pathological properties. First, it has zero Jacobian:

$$J = \begin{vmatrix} \partial \dot{u}/\partial u & \partial \dot{u}/\partial T \\ \partial \dot{T}/\partial u & \partial \dot{T}/\partial T \end{vmatrix} \equiv 0 \tag{8.113}$$

Because of this, the system has infinite number of equilibrium points which occupy two curves in the configuration space u, T:

$$u_0 = 0 \text{ and } u_0 = \frac{1}{T_0} \tag{8.114}$$

Second, at all the equilibrium points, the Lipschitz condition fails since

$$\left|\frac{\partial \dot{u}}{\partial u}\right| = \left|\frac{(2Tu - 1)\sin \omega t}{3(u - Tu^2)^{2/3}}\right| \to \infty \text{ if } u \to 0 \text{ or } u \to \frac{1}{T} \tag{8.115}$$

$$\left|\frac{\partial \dot{T}}{\partial T}\right| = \left|\frac{2(2Tu - 1)\sin^2 \omega t}{3(u - Tu^2)^{1/3}}\right| \to \infty \text{ if } u \to 0 \text{ or } u \to \frac{1}{T} \tag{8.116}$$

As a result of (8.115–8.116), the characteristic roots λ_1 and λ_2 of the Jacobian (8.113) at the equilibrium points must be

$$\lambda_1 = 0, \ |\lambda_2| \to \infty \tag{8.117}$$

Indeed, linearizing (8.111–8.112) with respect to the points (8.114) one finds

$$\lambda_1 = 0, \ \lambda_2 = \frac{(2Tu - 1)\sin \omega t}{3(u - Tu^2)^{2/3}} - \frac{2u_0 \sin^2 \omega t}{3(T_0 u_0^2 - u_0)^{1/3}} \tag{8.118}$$

It is easy to verify that

$$\lambda_2 \to \begin{cases} \infty \text{ if } & u_0 = 1/T_0 \\ -\infty \text{ if } & u_0 = 0 \end{cases} \quad \text{for } \sin \omega t > 0 \tag{8.119}$$

$$\lambda_2 \to \begin{cases} \infty \text{ if} & u_0 = 1/T_0 \\ -\infty \text{ if} & u_0 = 0 \end{cases} \quad \text{for } \sin \omega t < 0 \tag{8.120}$$

Hence, when the equilibrium points $u_0 = 0$ are stable (they become terminal attractors), the equilibrium points $u_0 = 1/T_0$ are unstable (they become terminal repellers) and vice versa.

One should note that, strictly speaking, the formula for λ_2 in (8.118) can be applied only if the explicit time t in (8.111–8.112) is considered as a slow changing parameter, i.e., if

$$\omega \ll \lambda_2 \tag{8.121}$$

However, since $|\lambda_2| \to \infty$ [see (8.117)] the inequality (8.121) holds for all bounded ω.

The Lipschitz condition fails not only in actual space, but in configuration space u, T. Indeed, as follows from (8.111–8.112), the differential equation for trajectories is

$$\frac{dT}{du} = -(u - Tu^2)^{1/3} \sin \omega t \tag{8.122}$$

and

$$\lambda_3 = \left| \frac{\partial \frac{dT}{du}}{\partial T} \right| = \left| \frac{u^2 \sin \omega t}{3(u - Tu^2)^{2/3}} \right| \to \begin{cases} \infty \text{ if} & u = 1/T \\ 0 \text{ if} & u = 0 \end{cases} \tag{8.123}$$

Thus, the Lipschitz condition fails only at the curve $u_0 = 1/T_0$ of the configuration space u, T. All the equilibrium points of this curve are terminal attractors for $\sin \omega t < 0$ and terminal repellers for $\sin \omega t > 0$.

As follows from (8.123), the Lipschitz condition holds at the curve $u_0 = 0$, while the equilibrium points of this curve possess a neutral stability.

Before analyzing the global behavior of the solutions to equations (8.111–8.112), we will first investigate the local properties of the escape from terminal repellers.

The solutions in an infinitesimal neighborhood of a terminal repeller have the following structure:

$$u = \overset{*}{u} e^{\lambda_2 t}, \; T = \overset{*}{T} e^{\lambda_2 t}, \; \lambda_2 \to \infty \tag{8.124}$$

in which $\overset{*}{u}$ and $\overset{*}{T}$ are initial disturbances.

As follows from (8.124) the transient solution may escape the repeller and approach some values $\widetilde{u}$ and $\widetilde{T}$ during a finite time period t_0 even if the initial disturbances are infinitesimal:

$$t_0 = \frac{1}{\lambda_2} \ln \frac{\widetilde{u}}{\overset{*}{u}} = \frac{1}{\lambda_2} \ln \frac{\widetilde{T}}{\overset{*}{T}}, \; \lambda_2 \to \infty, \; \overset{*}{u}, \; \overset{*}{T} \to 0 \tag{8.125}$$

while $\widetilde{u}$ and $\widetilde{T}$ are sufficiently small, but finite.

One should recall that for bounded λ the instability develops gradually: two initially-close trajectories diverge such:

$$\varepsilon = \varepsilon_0 \exp \lambda t, \quad |\lambda| < \infty$$

that for an infinitesimal initial distance $\varepsilon_0 \to 0$, the current distance becomes finite only at $t \to \infty$.

In this contradistinction, the escape from the terminal repeller (8.124) is similar to Hadamard's instability in continuous systems, where the instability can be defined within a finite time interval. This is why the rate for the instability (8.124) can be also defined in a finite time interval.

$$\sigma = \lim_{t \to t_0} \left(\frac{1}{t} \ln \frac{u}{\overset{*}{u}} \right) \to \infty \text{ if } \overset{*}{u} \to 0 \tag{8.126}$$

Thus, the divergence of the solutions (8.124) describing the escape from the terminal repeller are characterized by the unbounded rate of divergence. This means that here as in the previous section, a terminal repeller represents a vanishingly short, but infinitely powerful "pulse of unpredictability" which is "pumped" into the dynamical system.

The solutions in an infinitesimal neighborhood of a terminal attractor have the following structure:

$$u = \overset{*}{u} e^{-\lambda_2 t}, \quad T = \overset{*}{T} e^{-\lambda_2 t}, \quad \lambda_2 \to \infty \tag{8.127}$$

As follows from (8.127), a solution with finite initial condition $u = \overset{*}{u}$ at $t = 0$ may approach the terminal attractor in a finite time interval t_0:

$$t_0 = \frac{1}{\lambda_2} \ln \frac{\overset{*}{u}}{u} < \infty, \quad \lambda_2 \to \infty, \quad u \to 0 \tag{8.128}$$

while for a regular attractor this time is infinite.

The structure of the solutions around terminal repellers, and attractors, in the configuration space u, T is similar to (8.124) and (8.127) with the only difference being the role of the argument is played by u instead of t:

$$T = \overset{*}{T} e^{\lambda_3 u} \tag{8.129}$$

where

$$\lambda_2 \to \begin{cases} \infty \text{ if} & \sin \omega t > 0 \\ -\infty \text{ if} & \sin \omega t < 0 \end{cases} \tag{8.130}$$

Let us turn now to the global behavior of the solutions to (8.111–8.112) and start with the following initial conditions:

$$u = 0.5, \ T = 2 \text{ at } t = 0 \tag{8.131}$$

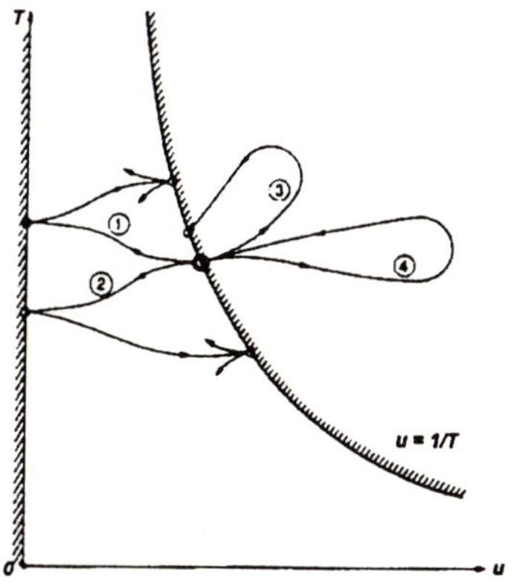

Fig. 8.5. Spontaneous changes of point attractors

According to (8.118) for $0 < t < \pi/\omega$, the point (8.131) is a terminal repeller. In the case of precisely zero disturbances, the system would rest forever at this point. In the presence of infinitesimal disturbances $\overset{*}{u}$ and $\overset{*}{T}$ the system can "choose" an escape scenario from the four combinations:

$$u = \pm \overset{*}{u}\, e^{\lambda_2 t}, \; T = \pm \overset{*}{T}\, e^{\lambda_2 t}, \;\; \overset{*}{u}, \;\; \overset{*}{T} \to 0, \; \lambda_2 \to \infty \tag{8.132}$$

Although initially the differences between the positive and the negative solutions are infinitesimal, their transient divergence is characterized by unbounded rate (8.126) in both actual and configuration spaces. The escaping solutions 1 and 2 (see Fig. 8.5) will approach the corresponding terminal attractors located on the line $u = 0$; they will remain there until $\sin \omega t > 0$, i.e., until all these attractors become repellers. Then, two of the four new branches of each of the solutions will return to the curve $u = 1/T$ giving the rise to another branches of the solutions, and so on.

It is important to note that each time the system escapes the terminal repeller, the solution splits into four possible branches, so that the total trajectory can be combined from 4^n pieces, where n is the integer part of the quantity $t/2\pi\omega$. The number of different structures (i.e., different attractors) which the system can attain is less than the number of different trajectories for two reasons. First, some of the solutions can return to the old attractors. Second, the solutions 1 and 2 do not branch on the curve $u = 0$, because in the configuration space the points of this curve are not terminal – they have neutral stability [see (8.123)]. This is why the number of structural changes in the system (8.111–8.112) has the order of $2^{n/2}$.

The system (8.111–8.112) can be represented in the following autonomous form:

$$\dot{u} = (u - Tu^2)^{1/3} v_2, \; \dot{T} = -\dot{u}^2 \tag{8.133}$$

if the new variable v_2 satisfies the following differential equations:

$$\dot{v}_1 = \omega v_2 + v_1(1 - v_1^2 - v_2^2), \ \dot{v}_2 = -\omega v_1 + v_2(1 - v_1^2 - v_2^2) \tag{8.134}$$

Indeed, equations (8.134) have a stable limit cycle:

$$v_1 = \cos \omega t, \ v_2 = -\sin \omega t, \ \omega = \text{Const} \tag{8.135}$$

and therefore (8.133–8.134) are equivalent to (8.111–8.112).

Multi-Dimensional Systems

Equations (8.111–8.112) can be generalized to the case of n neurons u_i and n^2 synaptic interconnections T_{ij}

$$\dot{u}_i = -\left[u_i + \sum_{j=1}^{n} T_{ij} V(u_j) \right]^{1/3} v_i, \ \dot{T}_{ij} = -\dot{u}_i \dot{u}_j \tag{8.136}$$

$$\dot{v}_i = \omega w_i + v_i(1 - v_i^2 - w_i^2), \ \dot{w}_i = -\omega v_i + w_i(1 - v_1^2 - w_i^2) \tag{8.137}$$

in which $v(u_j)$ is a sigmoid function.

It is easily verifiable that equations (8.136–8.137) possess the same self-developing properties as the original dynamical system. Indeed, in the configuration subspaces,

$$\frac{dT_{ij}}{du_i} = -\dot{u}_j \tag{8.138}$$

which is equivalent to (8.122).

In addition, they can perform some qualitatively new effects: they can spontaneously relocate periodic or chaotic attractors as well as static attractors. To illustrate, we start with the following three-neuron network:

$$\dot{u}_1 = [-u_1 + T_{11} V(u_1)]^{1/3} v, \ \dot{T}_{11} = -\dot{u}_1^2 \tag{8.139}$$

$$\dot{v} = \omega w + v(1 - v^2 - w^2), \ \dot{w} = \omega v + w(1 - v^2 - w^2) \tag{8.140}$$

$$\dot{u}_2 = -u_2 + T_{22} V(u_2) + T_{23} V(u_3) + T_{21} V(u_1) \tag{8.141}$$

$$\dot{u}_3 = -u_3 + T_{33} V(u_3) + T_{32} V(u) + T_{31} V(u) \tag{8.142}$$

Clearly, (8.141–8.142) represent the conventional part of the neural network, while (8.139–8.140) describe its self-developing part. For simplicity, we assumed that (8.139–8.140) are decoupled from (8.141–8.142) (since $T_{12} = T_{13} = 0$), but (8.141– 8.142) are still affected by (8.139–8.140). Let us set up the synaptic interconnections T_{22}, T_{23}, T_{32}, and T_{33} in (8.139–8.140) such that the solution has periodic attractors in the configuration planes u_1 = Const (Fig. 8.6). The spontaneous relocation of the static attractors for (8.139) will cause the corresponding relocations of limit cycles in the configuration planes u_1 = Const for the system (8.141–8.142)

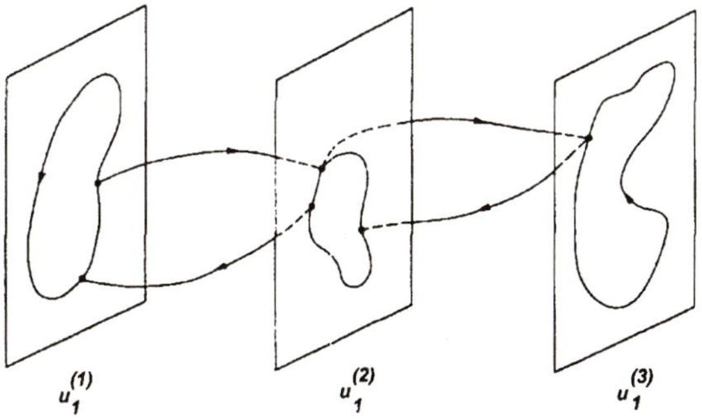

Fig. 8.6. Spontaneous changes of limit cycles

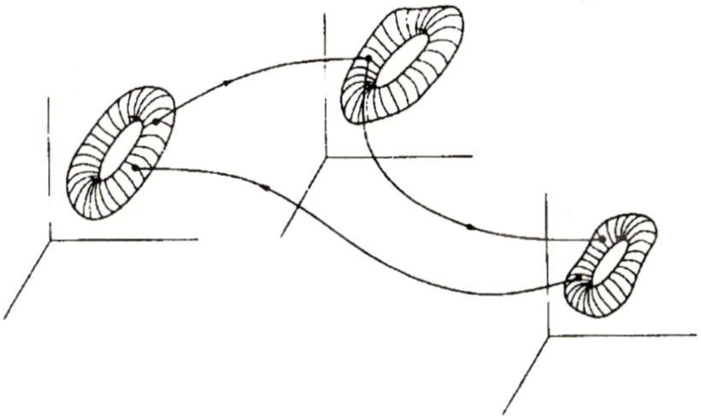

Fig. 8.7. Spontaneous changes of multi-periodic attractors

through the changes of their last terms $T_{21}V(u_1)$ and $T_{31}V(u_1)$, since the locations and the configurations of periodic attractors are parametrically dependent on u_1.

Spontaneous relocations of multiperiodic or chaotic attractors can be organized in the same way. For this purpose, the conventional part of the neural network must consist of at least three neurons, while the coefficients $T_{ij}(i,j=2,3,4)$ should be set up such that the solution has multiperiodic (or chaotic) attractors in the three-dimensional configuration spaces $u_1 =$ Const (Fig. 8.7). The self-developing part of the neural network can be represented by the same system (8.139–8.140). As in the previous case, the spontaneous relocations of the static attractors for (8.139) will cause the corresponding relocations of multiperiodic chaotic attractors in the configuration subspaces $u_1 =$ Const.

Thus, we have introduced self-developing dynamical systems which are able to spontaneously change their structure, i.e., locations and parameters of their attracting sets. Despite the fact that these systems are fully deterministic, their behavior as well as their structure is totally unpredictable.

Although one can argue that maybe the sequence of chaotic attractors spontaneously created by the self-developing systems possesses some hidden order and can be considered as a more complex type of attraction, so far we have no reasons to support such an assumption. It should be recalled that these new effects which are essentially different from the chaotic behavior are due to failure of the Lipschitz condition (8.96) which is not violated in classical dynamics.

What is the usefulness of the self-developing systems if they are totally unpredictable? Let us recall that we introduced such systems as an alternative to the systems with "rigid" behavior in order to develop a mathematical framework for modeling biological systems. So far these systems have not yet been equipped by an internal logic. This is why they do not "know" how to use the freedom they have. In the next item, we will incorporate an internal logic in the form of an objective which should be reached by the system.

Numerical simulation describing unpredictable spontaneous changes of locations of equilibrium points were performed by Zak and Barhen (1989) for the following:

$$\dot{u} = -(u - T \tanh \alpha u + \varepsilon_0 \cos \gamma t)^{1/3} v_2 \tag{8.143}$$

$$\dot{T} = (u - T \tanh \alpha u + \varepsilon_0 \cos \gamma t)^{2/3} v_2^2 \tag{8.144}$$

in which $\varepsilon_0 \to 0$ (but $|\varepsilon_0 \gamma_2| \to \infty$) and

$$\gamma = \omega + \frac{\pi}{2}\zeta, \quad \zeta = 1, 2, ..., \text{etc.}$$

The additional bias can be ignored when the system is stable, but it becomes significant (during the periods of instability different patterns of initial disturbances are simulated by different ζ, while in a real physical model they occur randomly). Figure 8.8a–c demonstrates different scenarios of behavior of the system for different ζ, while in all the cases $\varepsilon_0 = 10^{20}$, $\alpha = 10$. In Fig. 8.8a, with $\omega = 8 \cdot 10^3$ the motion starts with $u = 0$, $T = 5$ and approaches the terminal attractor B. After resting at this point it escapes through the trajectory BC (for $\zeta = 1$) or CD (for $\zeta = 2$). Similar behaviors are demonstrated in Fig. 8.8b,c,d,e with

$$\omega = 5 \cdot 10^3,\ 5 \cdot 10^3,\ 9 \cdot 10^3,\ 5 \cdot 10^3$$

$$u = 1.0,\ 0,\ 0,\ 0$$

$$T = 5.0,\ 0,\ 3.0,\ 4.0$$

Systems with an Objective

Let us return to the one-neuron-one-synapsis dynamical system written in the form (8.133–8.134), and introduce a global objective by requiring that

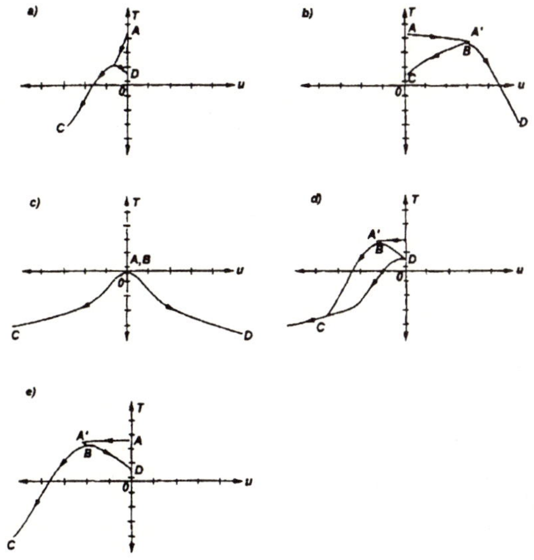

Fig. 8.8. Unpredictable motions

the system settle at a point attractor $u = \widetilde{u}$. It is implied that the system will find itself the corresponding synapsis $\widetilde{T}$ in the course of its spontaneous activity. First, we will modify (8.134) as follows:

$$\dot{v}_1 = [v_2(u - \widetilde{u}) + v_1(1 - v_1^2 - v_2^2)] \tag{8.145}$$

$$\dot{v}_2 = [-v_1(u - \widetilde{u}) + v_2(1 - v_1^2 - v_2^2)] \tag{8.146}$$

Now, instead of (8.43), the stable limit cycle is

$$v_1 = \cos\omega(t), \quad v_2 = \sin\omega(t) \tag{8.147}$$

and

$$\dot{\omega} = -(u - \widetilde{u}) \tag{8.148}$$

Obviously, the spontaneous activity of (8.133) ends when $v_2 = \text{Const} > 0$, because all the points of the curve $Tu = 1$ become terminal attractors. However, $v_2 = \text{Const}$ only if $\dot{\omega} = 0$, i.e., when $u = \widetilde{u}$. Consequently, the system eventually will approach the desirable structure with the prescribed point attractor (Fig. 8.9).

It is important to emphasize that neither the value of $\widetilde{T}$ nor the strategy for defining this value was prescribed in advance.

This approach can be generalized to the case of n neurons and n^2 synaptic interconnections [see (8.136–8.137)] if ω in (8.137) is considered as a prescribed function of $u_1, u_2, ..., u_n$, i.e.,

$$\omega_i = \omega_i(u_1, u_2 ... u_n), \quad i = 1, 2, ..., n$$

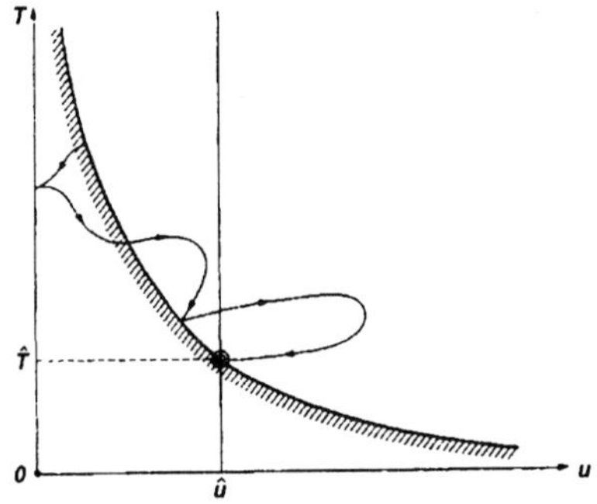

Fig. 8.9. System with an objective

Clearly, the system will stop at such a point attractor whose coordinates satisfy the following equations

$$\omega_i = \omega_i(u_1, u_2, ..., u_n) = 0, \; i = 1, 2, ..., n \tag{8.149}$$

Hence, depending on selections of ω_i, the system can approach a single point attractor, a countable set of possible point attractors, and continuous hypersurfaces of possible point attractors. If (8.149) do not have a solution, then the system will never stop.

If there exists such a function E that

$$\frac{\partial E}{\partial u_i} = \omega_i \tag{8.150}$$

then the point attractor approached by the system corresponds to a minimum of this function and therefore, the dynamical system (8.136–8.137), (8.149) performs optimization of this function.

It should be emphasized that the incorporation of the objective into a self-developing system does not impose any limitations upon the strategy for reaching this objective: the strategy is developed by the system itself. Because of this, however, one does not have any control over the time of convergence of the system to the desirable state. This is why in the next section we will introduce self-developing systems with a microstructure which allows for flexible guidance of their behavior.

Guided Self-Developing Systems

Let us return to the one-neuron-one-synapsis dynamical system [see (8.133–8.134)]. We will slightly modify (8.133) by introducing an infinitesimal bias as follows:

$$\dot{u} = -(u - Tu^2 + \varepsilon_0)^{1/3} v_2, \; \dot{T} = -\dot{u}^2 \tag{8.151}$$

in which

$$|\varepsilon_0| \rightarrow 0, \text{ but } |\varepsilon_0 \lambda_2| \rightarrow \infty \tag{8.152}$$

where λ_2 is given by (8.118) and v_2 is defined by (8.116).

This bias can be ignored when the system is stable, but it becomes significant during the periods of instability. Indeed, in the last case the solution to (8.122) with the bias ε_0 :

$$\frac{dT}{du} = -(u - Tu^2 + \varepsilon_0)^{1/3} v_2 \tag{8.153}$$

in the neighborhood of a terminal repeller has the following structure:

$$T = \frac{\varepsilon_0}{\lambda_2} e^{\lambda_2 u}, \ \lambda_2 \to \infty \tag{8.154}$$

Now the escape from the terminal repeller is controlled by the bias ε_0, and the changes in the structure of the system become predictable.

A compromise between these two extremes can be reached if one sets up

$$\varepsilon = \varepsilon_0 \sin \gamma t \tag{8.155}$$

Then, the unpredictable structural changes will appear only when $\sin \gamma t$ and $\sin \omega t$ vanish simultaneously (which depends on the ratio γ/ω). In other words, here one can control the degree of unpredictability.

More complex situations can occur in a two-neuron dynamical system:

$$\dot{u}_1 = -[u_1 - T_1 V(u_1) + \varepsilon_0 \ \mathrm{sgn}\ f_1(u_2)]^{1/3} v_1, \ \dot{T}_1 = -\dot{u}_1^2 \tag{8.156}$$

$$\dot{u}_2 = -[u_2 - T_2 V(u_2) + \varepsilon_0 \ \mathrm{sgn}\ f_2(u_1)]^{1/3} v_1, \ \dot{T}_2 = -\dot{u}_2^2 \tag{8.157}$$

in which v_1 and v_2 are defined by (8.137) at $i = 1, 2$, while f_1 and f_2 are prescribed functions.

Equations (8.156–8.157) possess a very interesting property: they are coupled only at the moments of escape from a terminal repeller. Indeed, only at that moment the vanishing terms with co-factor cannot be ignored: when the system (8.156) approaches the terminal repeller, the choice of the escape scenario depends upon the sign of its last term, i.e., upon the state of the system (8.157), and vice versa. Such an "impulsive" coupling represents a typical cause-and-effect relationship between two dynamical systems: each of these systems is independent up to a certain "turning point" when it has to choose from several available scenarios. It should be stressed this choice is fully determined by the state of the other system. Therefore, the dynamical systems (8.156–8.157) can be considered as a possible model for "nonrigid" behavior which is typical for biological systems.

Let us assume now that

$$f_1 = 0 \tag{8.158}$$

Then, the system (8.156) becomes totally independent and unpredictable, while the system (8.157) is still dependent on it, i.e., one arrives at a master-slave relationship. This situation can be generalized to the following chain of the master-slave subordination:

$$\dot{u}_1 = -[u_1 - T_1 V(u_1)]^{1/3} v_1, \ \dot{T}_1 = -\dot{u}_1^2 \tag{8.159}$$

$$\dot{u}_2 = -[u_2 - T_2V(u_2) + \varepsilon_0 f_2(u_1)]^{1/3} v_1, \ \dot{T}_2 = -\dot{u}_2^2 \tag{8.160}$$

$$\dot{u}_3 = -[u_3 - T_3V(u_3) + \varepsilon_0 f_3(u_2)]^{1/3} v_3, \ \dot{T}_3 = -\dot{u}_3^2 \tag{8.161}$$

$$\dot{u}_n = -[u_n - T_nV(u_n) + \varepsilon_0 \ f_n(u_{n-1})]^{1/3} v_n, \ \dot{T}_n = -\dot{u}_n^2 \tag{8.162}$$

in which $v_i(i = 1, 2, ..., n)$ are defined by (8.137).

The elements of this chain are not necessarily the one-neuron-one-synapsis dynamical systems. They can be presented in a more general form:

$$u_{i_1} + -[u_{i_1} + \sum_{j_1=1}^{n_1} T_{i_1 j_1} V(u_{j_1})]^{1/3} v_{i_1}, \ \dot{T}_{i_1 j_1} = -\dot{u}_{i_1} \dot{u}_{j_1} \tag{8.163}$$

$$u_{i_2} + -[u_{i_2} + \sum_{j_2=1}^{n_2} T_{i_2 j_2} V(u_{j_2})]^{1/3} + \varepsilon_0 f_{i_2}(u_1, ..., u_{n_1})]^{1/3} v_{i_2} \tag{8.164}$$

$$\dot{T}_{i_2 j_2} = -\dot{u}_{i_2} \dot{u}_{j_2}, \text{ etc.} \tag{8.165}$$

It should be noticed again that the guided self-developing systems introduced above fall between classical (rigid) dynamical systems and totally unpredictable dynamical systems discussed in previous sections. It seems reasonable to assume that such systems may provide a proper mathematical framework for modeling biological systems.

Guided Systems with Objective

In this section, we will simply combine the results of the two previous sections and discuss guided systems with objective. Starting with a one-neuron-synapsis dynamical system, let us write it in the form:

$$\dot{u} = -[u - Tu^2 + \varepsilon_0 \text{ sgn } (u - \widetilde{u})]^{1/3} v_2, \ T = -\dot{u}^2 \tag{8.166}$$

$$\dot{v}_1 = [-v_2(u - \widetilde{u}) + v_1(1 - v_1^2 - v_2^2)] \tag{8.167}$$

$$\dot{v}_2 = [-v_1(u - \widetilde{u}) + v_2(1 - v_1^2 - v_2^2)] \tag{8.168}$$

As follows from (8.167–8.168), the dynamical system has a global objective: to settle at the point attractor $u = \dot{u}$ [compare with (8.145–8.146)]. Besides this, it is guided by the bias in (8.166). Because of this guidance, the system in its critical points selects such a branch among the available solutions which decreases the distance $|u - \widetilde{u}|$ between the current state u and the desirable point attractor $\widetilde{u}$. Indeed. as follows from (8.166), if $u < \widetilde{u}$, then ε_0 sgn $(u - \widetilde{u})$ is negative, and therefore at the critical point $\dot{u} = \varepsilon_0 > 0$, i.e., the selected branch corresponds to the decrease of the difference $\widetilde{u} - u$.

This result can be easily generalized to a dynamical system with n neurons:

$$\dot{u}_i = -[u_i + \sum_{j=1}^{n} T_{ij} V(u_j) + \varepsilon_0 \text{ sgn } (u - \widetilde{u})]^{1/3} v_i, \ \dot{T}_{ij} = -\dot{u}_i \dot{u}_j \tag{8.169}$$

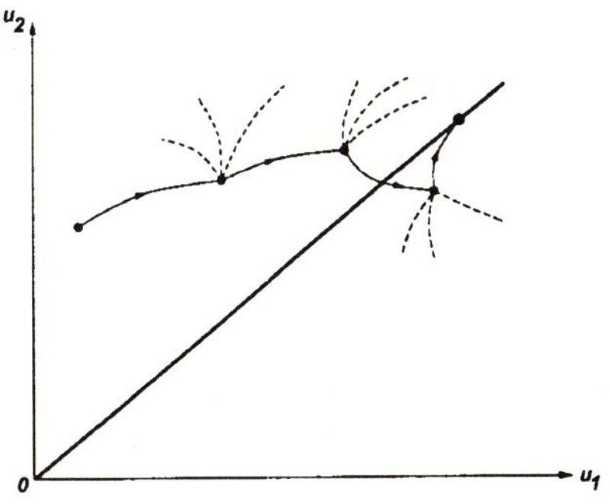

Fig. 8.10. System with an objective that is not fully determined

$$\dot{v}_i = -w_i(u_i - \widetilde{u}_i) + v_i(1 - v_1^2 - w_1^2)$$
$$\dot{w}_i = -v_i(u_i - \widetilde{u}_i) + w_i(1 - v_1^2 - w_1^2) \tag{8.170}$$

The system (8.169–8.170) has an objective: to settle at the point attractor $\widetilde{u}_i$. At critical points where the solution is branching, the system is "pushed" by the bias terms toward its attractor.

One should note that, strictly speaking, both the system (8.166–8.168) and (8.170) are characterized by fully deterministic behavior and objective, although they are extremely sensitive to infinitesimal excitations. In the next example, we will introduce a guided system with an implicit objective which is not fully deterministic.

Suppose that a two-neuron dynamical system has the following form:

$$\dot{u}_1 = -[u_1 - T_1 V(u_1) + \varepsilon_0 \text{ sgn } (u_1 - u_2)]^{1/3} v_1, \ \dot{T}_1 = -\dot{u}_1^2 \tag{8.171}$$

$$\dot{u}_2 = -[u_2 - T_2 V(u_2) + \varepsilon_0 \text{ sgn } (u_2 - u_1)]^{1/3} v_2, \ \dot{T}_2 = -\dot{u}_2^2 \tag{8.172}$$

$$\dot{v}_i = -w_i(u_1 - u_2) + v_i(1 - v_i^2 - w_i^2)$$
$$\dot{w}_i = v_i(u_1 - u_2) + w_i(1 - v_i^2 - w_i^2), \ i = 1, 2 \tag{8.173}$$

The objective of the system is to settle at the point attractor whose position is not fully determined; it can be located at any point of the straight line:

$$u_1 = u_2 \tag{8.174}$$

of the configuration space. At the critical points the system will be "pushed" by the bias terms toward this line. The exact location of the attractor cannot be predicted (Fig. 8.10).

All the previous examples can be generalized by the following model:

$$\dot{u}_1 = -[u_1 + \sum_{j=1}^{n} T_{ij} V(u_j) + \varepsilon_0 \text{ sgn } (u_1 - f_i)]^{1/3} v_i, \ \dot{T}_{ij} = -\dot{u}_i \dot{u}_j \tag{8.175}$$

$$\dot{v}_i = -w_i\omega_i + v_i(1 - v_i^2 - w_i^2), \quad \dot{w}_i = -v_i\omega_i + w_i(1 - v_i^2 - w_i^2) \tag{8.176}$$

in which ω_i are prescribed functions

$$\omega_i = \omega_i(u_1, u_2, ..., u_n) \tag{8.177}$$

and

$$f_i(u_1, ..., u_{i-1}, \ u_{i+1}, ..., u_n) = u_i \tag{8.178}$$

is the explicit expression for u_i from the equation

$$\omega_i(u_1, ..., u_n) = 0 \tag{8.179}$$

As in the case of unguided systems, this system will stop at the point whose coordinates $\widetilde{u}_i$ satisfy (8.179). However, in addition, at each critical point the system will select those branches of the solution which are directed toward the desirable attractor $\widetilde{u}_i$.

If, in particular, the functions (8.177) have a potential E [see (8.150)], then the attractor $\widetilde{u}_i$ will correspond to the minimum of this potential.

8.2.5 Discussion

This section as introduced a substantially new type of dynamical system which spontaneously changes its structure, i.e., locations of its attracting sets. This approach was motivated by an attempt to remove one of the most fundamental limitations of artificial computational systems – their rigid behavior compared with even simplest biological systems. This approach exploits a novel paradigm in nonlinear dynamics based upon the concept of terminal attractors and repellers. Incorporation of these new types of attractors and repellers into dynamical systems requires a revision of some fundamental concepts in the theory of differential equations associated with the failure of Lipschitz condition, such as uniqueness of solutions, infinite time of approaching attractors, bounded Liapunov exponents, and so on. In the course of this revision, it was demonstrated that terminal dynamics exhibits a new qualitative effect: a multichoice response to periodic external excitations. It appears that dynamical systems which possess such a property can serve as an underlying idealized framework for neural nets with "creativity." Based upon this property, a substantially new class of self-developing dynamical systems was introduced and discussed. These systems are represented in the form of coupled activation and learning dynamical equations whose ability to be spontaneously activated are based upon two pathological characteristics. First, such systems have zero Jacobian. As a result, they have an infinite number of equilibrium points which occupy curves, surfaces, or hypersurfaces. Second, at all these equilibrium points, the Lipschitz condition fails, so the equilibrium points become terminal attractors or repellers depending on the sign of the periodic excitation. Both of these pathological characteristics result in self-developing properties of dynamical systems.

Four types of self-developing dynamical systems were introduced and discussed. The first type is represented by totally unpredictable systems which are characterized by unpredictable behavior, unpredictable location of their attracting sets, and unpredictable terminal state. It should be emphasized that, in contradistinction to chaotic systems (which are structurally stable, and therefore, whose averaged properties are predictable), these systems have an unpredictable structure. One should also recall that in the chaotic systems, the unpredictability of a particular trajectory is caused by a supersensitivity to the initial conditions, while the uniqueness of the solution for fixed initial conditions is guaranteed by the Lipschitz condition. In contrast, the unpredictability of self-developing dynamical systems is caused by the failure of the uniqueness of the solution at some of the attracting sets. It is still unclear whether the sequence of attracting sets created by a self-developing system has a hidden order and can be considered as a more complex attracting object.

From the practical viewpoint, self-developing systems of this type can be regarded as a mathematical framework for modeling "nonrigid" dynamical behavior.

The second type of self-developing systems is characterized by a global objective which makes the terminal state of the system fully predictable, although the strategy for approaching this objective is not prescribed: the system must "create" its own strategy. Hence, these systems are self-programmed. However, the price paid for this is an unpredictable time required for approaching the desired terminal state.

The third type of self-developing system (guided systems) has a microstructure: it contains infinitesimal bias terms which control the system behavior at the critical points where the system must make a choice between several different available scenarios of motion. In contrast to the previous case, the behavior of such systems is fully deterministic, although their final state is not prescribed in advance. However, one has to realize that the determinism of the guided systems is as shaky as those in chaotic systems, because they are supersensitive to infinitesimal changes of the bias terms. Obviously, the type of instability in guided self-developing systems is different from chaotic ones: it is characterized by an instantaneous jump from one branch of the solution to another at the critical points, while in chaotic motions the shift from one trajectory to another develops gradually.

The last type of self-developing dynamical system has both a global objective and a microstructure. Its behavior is deterministic, but nonrigid: several subsystems can be uncoupled for most of the time, and they effect each other only during a vanishingly short interval. This is why these systems can model cause-and-effect relationships.

Thus, it has been demonstrated that self-developing dynamical systems which spontaneously change their own structure can be utilized for modeling more complex relationships than those modeled by classical dynamics. From the viewpoint of neural networks, these systems suggest the way of minimizing pre-programming by entrusting this procedure to the dynamical system itself.

8.3 Weakly Connected Neural Nets

In this and the following sections we will turn to a simple model of unpredictable dynamics which is based upon the dynamical paradigm described by (8.97). In this model the unpredictability comes only from the activation dynamics while the synaptic interconnections are prescribed. As will be shown below, such systems are more "manageable" in a sense that their unpredictability can be expressed in terms of probability distributions. Actually, such systems were introduced and described in Chap. 7, (7.142) and (7.229), but in this section we will analyze them from the viewpoint of their neurodynamical performance.

Existing neural net models are represented by nonlinear dynamical systems fully coupled by a set of synaptic interconnections. The expected advantages of such an architecture are supposed to stem from an ability to perform massively parallel, asynchronous and distributed information processing as an alternative to sequential computations. However, fully coupled neural nets actually perform collective, but not truly parallel computations, since any change in a neuron activity instantaneously effects all other neurons. In terms of hardware implementation this means that the firing decision for a particular neuron is allowed only after state information has been received from all other neurons. In order to eliminate such a sequential element in the neural net performance we will introduce a new neural network architecture where the interconnections between the neurons are active only during vanishingly short (critical) periods. Hence, a neuron or a group of neurons can perform independent tasks within a certain period, while the coordination between the independent units is carried out within the critical periods of coupling when the net becomes fully connected. The approach is based upon some effects of terminal dynamics discussed in previous sections.

Let us start with the following differential equation:

$$\dot{u} = \gamma \sin^{1/3} \frac{\omega}{\alpha} u \sin \omega t + \varepsilon, \ \varepsilon \rightarrow 0 \tag{8.180}$$

The bias ε can be ignored when (8.180) is stable, but it becomes significant during the periods of instability since the choice between the splitting branches of the solution is controlled by the sign. Indeed, if $\varepsilon > 0$, then at each terminal repeller $\dot{u} > 0$. Therefore, the positive branch of the solution is always chosen, i.e., u monotonously tends to ∞. Analogously, if $\varepsilon < 0$, then u monotonously tends to -∞.

Let us assume now that

$$\varepsilon = \delta^2 f(u), \ \delta \to 0, \ \left.\frac{df}{du}\right|_{u=0} = \lambda \tag{8.181}$$

Then for positive λ the solution to (8.180) will become unstable:

$$u \to \infty \text{ if } u(0) > 0, \ \text{ and } u \to \infty \text{ if } u(0) < 0 \tag{8.182}$$

The situation is analogous to the instability of the equation associated with (8.180):

$$\dot{u} = \delta^2 f(u) \text{ if } \frac{df}{du} = \lambda > 0 \tag{8.183}$$

For negative λ the solution to (8.180) will tend to $u = 0$ [as the solution to (8.183)], but then it will start oscillating about this point. Indeed, when the point $u = 0$ becomes a terminal repeller, i.e., when $\sin \omega t > 0$, the solution escapes to the neighboring (right or left) equilibrium point. However, as follows from (8.183), $\dot{u} < 0$ at $u_1 = \pi\alpha/\omega$ and $\dot{u} > 0$ at $u_{-1} = -\pi\alpha/\omega$. Therefore, in both cases, the solution returns to the original point $u = 0$. The amplitude and the period of the oscillations about $u = 0$ can be found from (8.98) and (8.101) respectively:

$$\Delta u = \frac{\pi\alpha}{\omega}, \ T = \frac{1}{\omega} \arccos \left[1 - \frac{B(1/3, \ 1/3)}{2^{1/3}} \frac{\alpha}{\gamma}\right] \tag{8.184}$$

Both of these quantities vanish if $\omega \to \infty$. This means that the stability of (8.180) follows from the stability of the associated equation (8.183) if $\omega \to \infty$. Obviously for bounded ω, (8.180) is stable only if the basin of attraction of the equilibrium point $u = 0$ for (8.183) is larger than the amplitude Δu given by (8.184).

Let us now introduce the following neural net

$$\dot{u}_i = \gamma_i \sin^{1/3} \frac{\omega}{\alpha} u_i \sin \omega t + \delta^2 \sum_{j=1}^{n} T_{ij} V(u_j), \ \delta^2 \to 0, \ i = 1, 2, ..., n \tag{8.185}$$

in which u_i is the mean "soma potential" of the neuron, T_{ij} are constant synaptic interconnections, $V(u_j)$ is a sigmoid function, and γ_i, α_i and ω are constant.

Each equation in the system (8.185) is independent of all others and it behaves exactly as (8.183) does, excluding the critical points

$$u_\ell^k = \frac{\pi k_\ell \alpha_\ell}{\omega}, \ k_\ell = \cdots - 2, -1, 0, 1, 2, \cdots \text{ etc.} \tag{8.186}$$

at which the ℓth equation is fully dependent on the rest of the equations:

$$\dot{u}_\ell = \delta^2 \sum_{j=1}^{n} T_{\ell j} V(u_j), \ \delta^2 \to 0, \ \ell = 1, 2, ..., m \leq n \tag{8.187}$$

Here m is the number of equations which simultaneously approach their critical points. As in the one-dimensional case, here the qualitative behavior of (8.185) can be found by analyzing the associated system:

$$\dot{u}_i = \delta^2 \sum_{j=1}^{n} T_{ij} V(u_j), \ \delta^2 \to 0, \ i = 1, 2, ..., n \tag{8.188}$$

which is an n-dimensional analog of (8.183). This means that the synaptic interconnections can be trained by using the same methods as for the Hopfield nets, i.e., backpropagation or the velocity field approach.

However, it should be noted that as in the one-dimensional case, the basins of attraction for (8.185) are embedded into the basins of attraction for (8.188) while the distance between the basins' boundaries in the direction of u_i is equal to the amplitude $\Delta u_i = \pi/\alpha_i \omega_i$. Since for non-convex basin configurations this condition may not be applicable, one should utilize training procedures which provide an explicit control over the basins' configuration (for instance, the velocity field approach, described in Chap. 3. More vigorous stability analyses of (8.185) can be based upon the Baudet contraction theorems which were successfully applied to concurrent asynchronous neurodynamics by Barhen, et al. (1989).

The system (8.185) can be generalized in several ways. Firstly, α_i can be considered as a slow changing parameter controlled by the following differential equation:

$$\dot{\alpha}_i = \lambda_i [\sum_{j=1}^{n} T_{ij} V(u_j) - \alpha_i], \ \lambda_i \ll \omega \tag{8.189}$$

Obviously,

$$\alpha_i \to 0 \text{ when } \sum_{j=1}^{n} T_{ij} V(u_j) \to 0 \tag{8.190}$$

i.e., when the solution approaches a prescribed equilibrium point. But then as a consequence of (8.190) the amplitudes of oscillation about this point vanishes (Fig. 8.11)

$$\Delta u_i = \frac{\pi \alpha_i}{\omega} \to 0 \tag{8.191}$$

and therefore, the prescribed equilibrium point is approached as a regular attractor.

Secondly, instead of the function $\sin^{1/3} \frac{\omega}{\alpha} u$ one can utilize a function $\Phi(u)$ with a more sophisticated distribution of the roots $\Phi(u_k) = 0$ so that each neuron can perform more complex information processing during the period of its independence. The only requirement to the function $\Phi(u)$ is the following:

$$\Phi(u)(u_k - u)^q < \infty \text{ if } u \to u_k, \ q < 1 \tag{8.192}$$

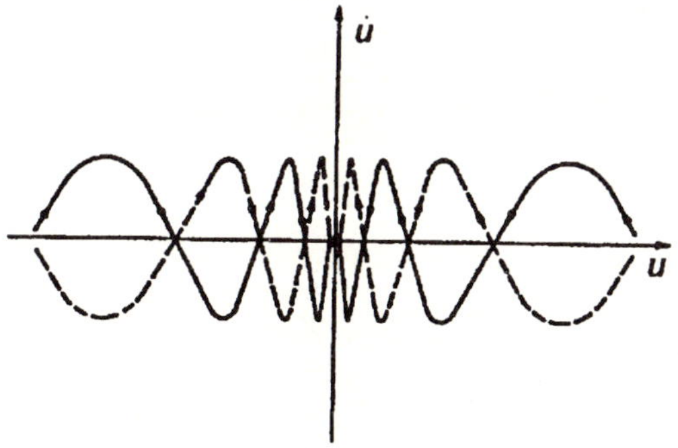

Fig. 8.11. Oscillations with vanishing amplitude

The condition (8.192) guarantees failure of the Lipschitz condition at the critical points, and therefore forces the finite transition periods from one critical point to another.

Thirdly, a weakly coupled unit can consist of several neurons:

$$\dot{u}_{i_1} = \gamma_{i_1}\Phi_{i_1}\left[\frac{\omega}{\alpha_{i_1}}\sum_{j_1} T_{i_1 j_1} V(u_{j_1})\right] \sin\omega t + \delta^2 \sum_{j=1}^{n} T_{ij} V(u_j) \qquad (8.193)$$

$$\dot{u}_{i_2} = \gamma_{i_2}\Phi_{i_2}\left[\frac{\omega}{\alpha_{i_2}}\sum_{j_2} T_{i_2 j_2} V(u_{j_2})\right] \sin\omega t + \delta^2 \sum_{j=1}^{n} T_{ij} V(u_j) \qquad (8.194)$$

etc.

$$i_1,\, j_1 = 1, 2, ..., n_1,\;\; i_2, j_2 = n_1 + 1, ..., n_2,\;\; n = n_1 + n_2 + ..., \text{etc.} \qquad (8.195)$$

Here each of the subsystems (8.193–8.194), etc., is weakly coupled to the others, but all the equations within the subsystem are fully coupled. Hence, the subsystems can perform parallel information processing during the periods between the critical points, while the coordination between them is carried out within vanishingly short critical periods. In this way, the new neural net architecture introduced above can represent a dynamical model of a variety of interconnect topologies for multiprocessor systems including hypercubes.

There are several other advantages of the new architecture. Firstly, the synapse interconnections T_{ij} enter the activation dynamics in the form of vanishingly small quantities [see (8.185)] which should substantially simplify their hardware implementation. Secondly, the neural net performance is robust with respect to external inputs. Indeed, let us include an external input I into (8.180–8.181):

$$\dot{u} = \left(\gamma \sin^{1/3}\frac{\omega}{\alpha}u + I\right)\sin\omega t - \delta^2 f(u),\;\; \frac{\partial f}{\partial u} > 0 \qquad (8.196)$$

Clearly this input shifts the critical points (8.98)

$$u_k = -\frac{\alpha}{\omega} \arcsin \frac{I^3}{\gamma^3} \qquad (8.197)$$

but the attractor $u = 0$ is preserved as long as the associated equations (8.183) are not affected by the input. Thirdly, the new architecture eliminates the synchronization and coordination restrictions on neural computations since most of the time the neurons are uncoupled. The global rhythm for the neural net is generated by a single parameter ω, while the neural net performance is robust with respect to changes of this parameter. Indeed, any changes in ω [see (8.185)] may only shift the critical points (8.186), but they do not affect the associated equations (8.188), and therefore, the original locations of attractors are preserved.

The harmonic excitation $\sin \omega t$ needed can be generated by a dynamical system:

$$\dot{v}_i = \omega v_2 + v_1(1 - v_1^2 - v_2^2), \ \dot{v}_2 = -\omega v_1 + v_2(1 - v_1^2 - v_2^2) \qquad (8.198)$$

which has a stable periodic attractor:

$$v_1 = \cos \omega t, \ v_2 = \sin \omega t \qquad (8.199)$$

It is interesting to note that oscillatory patterns of activity in the brain detected over recent years raise the question about the role of oscillations in the logical structure of neural net performance. Possible answers to this questions were discussed in Chap. 3, Sect. 2. The new neural net architecture introduced above suggests another example of oscillatory activity which is not just a by-product of nonlinear effects, but rather an important element of neural computations.

8.4 Temporal Coherent Structures

The main objective of this section is to analyze the evolution from totally unpredictable to partly organized neural nets which perform elements of "creativity."

8.4.1 Irreversibility and Local Time

In Chap. 7, we came to a fundamental conclusion that the dynamics based upon terminal dynamics architecture is irreversible: a trajectory from past to future never coincides with a trajectory from future to past. Therefore, one can expect the appearance in random neurodynamics of new mechanisms and new coherent structures (which were initially absent in the system) as a result of successive instabilities at the critical points analogously to nonequilibrium thermodynamics (the Brusselator model), chemical kinetics, or ecology (the growth of a population model). In other words, the terminal neurodynamics has a potential for creativity.

However, there is a fundamental difference between terminal neurodynamics and nonequilibrium thermodynamics: the latter is characterized by an infinite time period required for solution to settle down at an attractor or to escape a repeller (if initial disturbance is vanishingly small). For instance, the solution to one-dimensional diffusion has the following structure:

$$u = \frac{1}{\sqrt{t}} e^{-1/t} \to 0 \text{ at } t \to \infty \tag{8.200}$$

Another example of infinite time of approaching an attractor was introduced in Chap. 7, Sect. 8.

In contrast to that, in random neurodynamics, the period of the transition from one equilibrium point to another is finite [see (8.184)]. This period is an important characteristic of the system: it sets up an intrinsic rhythm and can be associated with the scale of a local time $\tilde{t}$. As follows from (8.184), this scale is inversely proportional to the frequency of external excitation ω. In biology such a local time defines the speed of perception of the world, and it is different for different species.

Let us return to the unpredictable dynamical system in the following form:

$$\dot{u}_i = \gamma \sin^{1/3} \frac{\omega}{\alpha} u_i \sin \omega t + \varepsilon_i(t), \ \ \varepsilon_i(t) \to 0, \ \ i = 1, 2, ..., n \tag{8.201}$$

These equations describe dynamics of n identical uncoupled neurons driven by a vanishingly small random input. But since the components $\varepsilon_i(t)$ of this input are not necessarily identical, the motions of each neuron, in general, will be different from the motions of the others, while these differences will be finite rather than infinitesimal. Such a phenomenon can be interpreted as an "emergence from chaos" of a "society" of neurons with different "personalities" despite the fact that the initial conditions for all of them are almost identical. It is remarkable that the "character" of each "personality" has a random nature coming from the random origin of the noise $\varepsilon_i(t)$. Obviously that with equal probability the system (8.201) will eventually approach any prescribed combinations of critical values

$$u_i^{(k)} = \frac{\pi k \alpha}{\omega}, \ \ k = \cdots - 2, -1, 0, 1, 2, \cdots \text{ etc.}, \ i = 1, 2, ..., n \tag{8.202}$$

As in the one-dimensional case, the system (8.201) becomes deterministic if the signs of ε_i are prescribed at each critical point. It can be performed, for instance, through the architecture of weakly connected neural nets, considered in Sect. 4:

$$\dot{u}_i = \gamma_i \sin^{1/3} \frac{\omega}{\alpha_i} u_i \sin \omega t + \delta^2 \sum_{j=1}^{n} T_{ij} V(u_j), \ \delta^2 \to 0, \ \ i = 1, 2, ..., n \tag{8.203}$$

in which u_i is the mean "soma potential" of the neuron, T_{ij} are constant synaptic interconnections, $V(u_j)$ is a sigmoid function, and γ_i, α_i and ω are constant.

Each equation in the system (8.203) is independent of all others and it behaves exactly as (8.180) does, excluding the critical points

$$u_\ell^k = \frac{\pi k_\ell \alpha_\ell}{\omega}, \quad k_\ell = \cdots - 2, -1, 0, 1, 2, \cdots \text{ etc.} \tag{8.204}$$

at which the ℓth equation is fully dependent on the rest of the equations:

$$\dot{u}_\ell = \delta^2 \sum_{j=1}^{n} T_{\ell j} V(u_j), \quad \delta^2 \to 0, \quad \ell = 1, 2, ..., m \leq n \tag{8.205}$$

Here m is the number of equations which simultaneously approach their critical points.

Since from the viewpoint of information processing only the critical values (8.204) of the neuron potential are essential (while the transient trajectories are not important), one can eliminate physical time from (8.201) and replace it by the local time:

$$u_i^{(k+1)} = u_i^{(k)} + \frac{\pi \alpha}{\omega} \text{ sgn } \varepsilon_i(\widetilde{t}_k), \quad \widetilde{t}_k = \frac{\pi k}{\omega}, \quad k = 1, 2, ..., \text{etc} \tag{8.206}$$

in which the local time $\widetilde{t}$ is represented by a discrete variable

$$\widetilde{t}_k = \frac{\pi}{\omega} k, \quad k = 1, 2, ..., \text{etc.} \tag{8.207}$$

One can verify that if the signs of $\varepsilon_i(t_k)$ are prescribed for all i and k, then the system will be fully deterministic and reversible (i.e. it will be invariant with respect to inversion of the local time $\widetilde{t}_k$). As we already mentioned, such a system cannot create new logical structures. On the contrary, if the signs of $\varepsilon_i(\widetilde{t}_k)$ are not prescribed, and therefore, the system has a choice at each critical point (8.202), then it becomes totally unpredictable and irreversible with respect to inversion of local time. Such systems can run into new logical structures which have never existed before. But is this a creativity? Since all these new structures have a random nature, their usefulness is very limited. From this viewpoint, totally unpredictable systems have only a potential for being creative. By creativity we understand the ability to run in new configurations which have a prescribed logical structure, while the "details" may be random. In the next section we will show that creative neural networks fall in between deterministic and unpredictable architectures: they are partially deterministic and partially unpredictable; in particular, they are deterministic with respect to global characteristics of new structures which define their belonging to a certain class of patterns, and they are unpredictable with respect to local, or more individual characteristics.

8.4.2 Terminal Chaos

In this section we will take the next step from unpredictability to creativity of neurodynamics (8.201) by introducing temporal coherence into its performance.

One Dimensional Case

Let us start with a one-dimensional version of (8.201) and assume that

$$\varepsilon(t) = -\delta^2 u, \ \delta^2 \to 0 \tag{8.208}$$

i.e.,

$$\dot{u} = \gamma \sin^{1/3} \frac{\omega}{\alpha} u \sin \omega t - \delta^2 u, \ \delta \to 0 \tag{8.209}$$

It can be verified that the solution to (8.209) will oscillate about the point $u = 0$. Indeed, when the point $u = 0$ becomes a terminal repeller, i.e. when $\sin \omega t > 0$, the solution escapes to the neighboring (right or left) equilibrium point. However, for $\dot{u} < 0$ at $u_1 = \pi\alpha/\omega$ and $\dot{u} > 0$ at $u_1 = -\pi\alpha/\omega$. Therefore, in both cases the solution returns to the original point $u = 0$. The amplitude and the period of the oscillations about $u = 0$ can be found from (8.98) and (8.101) respectively (for simplicity we will normalize them):

$$\Delta u = \frac{\pi\alpha}{\omega} = 1, \ T = \frac{1}{\omega} \arccos \left[1 - \frac{B(1/3,\ 1/3)}{2^{1/3}} \frac{\alpha}{\gamma} \right] = 1 \tag{8.210}$$

However, in contrast to a classical version of (8.209),

$$\dot{u} = -\delta^2 u \tag{8.211}$$

where $u = 0$ is a static attractor, the same point $u = 0$ is not a static, and not even a periodic, but a (terminal) chaotic attractor. Indeed, there are several equally probable patterns of oscillations:

$$0, 1, 0, -1, 0; \ 0, -1, 0, 1, 0; \ 0, 1, 0, 1, 0; \ 0, -1, 0, -1, 0, ... \tag{8.212}$$

which can follow each other in an arbitrary order. In probabilistic terms the oscillations can be characterized as: $u = 0$ at $t = 2\pi n/\omega$, while

$$\Pr \left\{ u \left[t = \frac{(2n+1)\pi}{\omega} \right] = 1 \right\} = 0.5$$

$$\Pr \left\{ u \left[t = \frac{(2n+1)\pi}{\omega} \right] = -1 \right\} = 0.5$$

$$n = -2, -1, 0, 1, 2, ..., \text{etc.} \tag{8.213}$$

so the probability of any combinations of the patterns (8.212) can be found from (8.213).

It can be concluded that the chaotic attractor $u = 0$ of (8.209) is different from chaotic attractors in classical dynamics. Firstly, here the mechanisms of stability and instability act sequentially: during the first period the neuron is attracted to the point $u = 0$, then it is repelled from it (in one of two possible directions). Secondly, the time of approaching the center $u = 0$ is finite (due to failure of the Lipschitz condition at $u = 0$, [see (8.210)]. Clearly, terminal chaos is characterized by a well organized probabilistic structure [see (8.213)] which simplifies its prediction and control.

Two Dimensional Case

The last property provides leading reasonings for utilization of terminal chaos as a compact memory storage. In order to illustrate this, let us consider two independent neurons which have the following microdynamics:

$$\dot{u}_1 = \gamma \sin^{1/3} \frac{\omega}{\alpha} u_1 \sin \omega t - \delta^2 u_1, \ \delta^2 \to 0 \tag{8.214}$$

$$\dot{u}_2 = \gamma \sin^{1/3} \frac{\omega}{\alpha} u_2 \sin \omega t - \delta^2 u_2 (0.5 - u_2)(u_2 - 1) \tag{8.215}$$

As shown above, the first neuron performs chaotic behavior with respect to the center $u_1 = 0$, while the second neuron has two centers of terminal chaos: at $u_2 = 0$ and $u_2 = 1$.

Terminal Chaos as a Tool for Hierarchical Parallelism

Let us introduce now a secondary scale microdynamics with the order of $\delta^4 \to 0$ which couples (8.214) and (8.215):

$$\dot{u}_1 = \gamma \sin^{\frac{1}{3}} \frac{\omega}{\alpha} u_1 \sin \omega t - \delta^2 u_1 + \delta^4 u_2 \tag{8.216}$$

$$\dot{u}_2 = \gamma \sin^{\frac{1}{3}} \frac{\omega}{\alpha} u_2 \sin \omega t + \delta^2 u_2 (0.5 - u_2)(u_2 - 1) + \delta^4 u_1 \tag{8.217}$$

The last terms in these equations are effective only when all the other terms are zero, i.e. at the centers of chaotic attractors $u_1 = 0, u_2 = 0$, and $u_2 = 1$ where the behavior of the neurons is unpredictable. Due to coupling via secondary scale microdynamics, the second neuron makes decisions for the first one at $u_1 = 0$, and the first neuron makes decision for the second one at $u_2 = 0$ and $u_2 = 1$. In other words, the chaotic structure of the neurons with the primary microdynamics (8.214–8.215) reserves room for a match between these two neurons if they work in parallel, without changing their primary microdynamics. In the same way each of these neuron can work in parallel with other neurons while the adjustment between them is carried out by the secondary scale microdynamics due to chaotic structure of their primary microdynamics.

The secondary order microdynamics does not necessarily eliminate chaos. Indeed, if in (8.216–8.217) $u_1 = 0$ and $u_2 = 0$ simultaneously, then the behavior of the first neuron at $u_1 = 0$ is still unpredictable, and the third order scale microdynamics should be incorporated. Hence one arrives at multi-scale microdynamical chains by means of which different neurons are adjusted to each other in their parallel performance on the level of a certain order scale microdynamics. However, room for such an adjustment is provided by the chaotic structure of microdynamics of uncoupled neurons. This gives a possible phenomenological architecture of neurodynamics which is responsible for an enormous degree of coordination among neurons in biological systems.

Two Scale Microdynamics

Now we will make the second step toward the complexity of temporal coherent structures and replace (8.209) by the two-scale micro dynamics:

$$\dot{u}_1 = \gamma \sin^{1/3} \frac{\omega_1}{\alpha_1} u \sin \omega_1 t - \delta^2 u \sin \omega_2 t \tag{8.218}$$

$$\omega_2 \ll \omega_1 \tag{8.219}$$

During the period

$$t < \frac{\pi}{\omega_1} \tag{8.220}$$

the solution to (8.218) behaves exactly as for (8.209): it has a terminal chaotic attractor at $u = 0$ since

$$\text{sgn } u = \text{ sgn } (u \sin \omega_2 t) \text{ at } t < \frac{\pi}{\omega_2} \tag{8.221}$$

But in contrast to the latter, the solution to (8.218) is not locked up in this chaotic attractor: eventually it drifts away from the point $u = 0$ since

$$\text{sgn } u = - \text{ sgn } (u \sin \omega_2 t) \text{ at } t > \frac{\pi}{\omega_2} \tag{8.222}$$

and the scenario of the chaotic oscillations can be the following:

$$0, 1, 0, -1, 0, 1, 2, 1, 2, 3, 4, 5, ..., \text{etc. or}$$

$$0, 1, 0, -1, 0, -1, -2, -1, -2, -3, -2, -3, -4, \text{ etc.} \tag{8.223}$$

This drift can be bounded if one modifies (8.218) as following:

$$\dot{u} = \gamma \sin^{1/3} \frac{\omega_1}{\alpha_1} u \sin \omega_1 t - \delta^2 u \sin \omega_2 t - \delta^4 u \tag{8.224}$$

The last term representing the second-order microdynamics will return the solution to the chaotic attractor $u = 0$ after the period $t > \pi/\omega_2$; and the scenario of the "double-period" chaotic oscillations will be for $\omega_1 = 5\omega_2$:

$$0, 1, 0, -1, 0, 1, 2, 1, 2, 3, 4, 5, 4, 3, 2, 1, 0, -1, 0, -1, -2, -3, -4, \cdots \text{ or}$$

$$0, 1, 0, -1, 0, -1, -2, -1, -2, -3, -2,$$
$$-3, -4, -3, -2, -1, 0, -1, -2, -3, -4, \cdots \tag{8.225}$$

Hence, despite the fact that the solution to (8.224) has a more complex temporal structure and a larger number of unpredictable elements, it is still characterized by global coherence: it oscillates chaotically with respect to $u = 0$, while the amplitudes of this oscillations also changes chaotically from 1 to 5.

Multi Scale Microdynamics

In the same way one can introduce a multi-scale dynamics:

$$\dot{u} = \gamma \sin^{1/3} \frac{\omega_1}{\alpha_1} u \sin \omega_1 t - \delta^2 u \sin \omega_2 t - \delta^4 u \sin \omega_3 t - \delta^4 \cdots \text{ etc., where}$$

$$\omega_1 \gg \omega_2 \gg \omega_3 \text{ etc., } \delta \to 0 \tag{8.226}$$

whose complexity and "creativity" will be proportional to the number of microscales $\delta^2, \delta^4, \ldots$ or the number of local times

$$\tilde{t}_k^{(i)} = \frac{\pi}{\omega_i} k, \ k = 1, 2, \ldots \tag{8.227}$$

This multi-scale model can be associated with a cascade of intrinsic rhythms which characterize the temporal architecture of mental process (Geissler 1987).

Loosely speaking, one can conclude that a neurodynamics – with a limited number of local times, or microscales, has limited creativity in the sense that it is locked up in a set of behaviors which belong to a certain class of patterns, and it can escape this class only if the next micro-level (with the corresponding local time) is added.

In this connection an interesting question can be posed: can a microdynamics "improve" itself by producing additional micro-levels, or the number of such levels is "genetically" prescribed? So far we do not no the answer.

Nonlinear Effects

The structure (8.226) is the simplest way to increase the complexity of coherent temporal behavior. A more sophisticated approach can be based upon nonlinear effects (Zak 1991a,b). In order to illustrate this, let us turn to (8.180) and exploit the following microdynamics:

$$\varepsilon(t) = \delta^2 u(u - 1.5)(u - 4)(4.5 - u)(u - 5) \tag{8.228}$$

Here the solution to (8.180) possesses two different terminal chaotic attractors. The first one has its center at $u = 0$ and is characterized by the probabilities (8.213). The second one has two centers: $u = 4$ and $u = 5$ and is characterized by the probabilities:

$$\Pr(u = 4) = \Pr(u = 5) = 1/3, \quad \Pr(u = 3) = \Pr(u = 6) = 1/6 \tag{8.229}$$

One can verify that their basins of attraction are, respectively:

$$u < 1.5, \quad \text{and } u > 1.5 \tag{8.230}$$

Suppose that the microdynamics (8.228) includes some external input $f(t)$ which can be interpreted as an outside message:

$$\varepsilon(t) = \delta^2[u(u-0.5)(u-4)(4.5-u)(u-5) + f(t)], \quad \delta \to 0 \tag{8.231}$$

Then

$$f(t) > 21 \text{ at } u = 1 \tag{8.232}$$

and the solution which originally was trapped in the first attractor $u = 0$ will escape it and move to the second attractor with two centers: $u = 4$ and $u = 5$. Conversely, if

$$f(t) < -13.5 \text{ at } u = 3 \tag{8.233}$$

the solution will return to the first attractor.

Actually the dynamical systems of this type can be regarded as an implementation of a concept of processing of information which includes semantics. This concept is based upon the idea that a meaning to a message can be attributed only if the response of the receiver (a dynamical attractor) is taken into account (Haken 1985). In this context the inequalities (8.232–8.233) can be utilized for evaluation of the "relative importance" of the messages delivered by the outside input $f(t)$.

Spontaneous Relaxation of Attractors

Let us make the next step toward the complexity of temporal structures and show that under certain conditions the solution can change its attractor; moreover, it can create a new attractor and eliminate the old one. Let us consider the following dynamical system:

$$\dot{u}_1 = \gamma_1 \sin^{1/3} \frac{\omega_1}{\alpha_1} u_1 \sin \omega t - \delta^2 u_1 \tag{8.234}$$

$$\dot{u}_2 = \gamma_2 \sin^{1/3} \frac{\omega_2}{\alpha_2} u_2 \sin \omega_2 t - \delta^2[u_2(u_1 - 0.5)(u_2 - 1)(u_1 + 0.5)] \tag{8.235}$$

in which

$$\Delta u_1 = \frac{\pi \alpha_1}{\omega_1} = \frac{\pi \alpha_2}{\omega_2} = \Delta u_2 = 1, \quad \omega_1 \ll \omega_2 \tag{8.236}$$

Equation (8.234) is identical to (8.209): it has a terminal chaotic attractor at $u_1 = 0$ with patterns of oscillations (8.212). The solution to (8.235) has more complex behavior: it has a terminal chaotic attractor $u_2 = 0$ if u_1 <0.5, but this attractor disappears as soon as $u_1 > 0.5$, and the solution approaches a new terminal chaotic attractor $u_2 = 1$. Obviously such a transition has a random nature since the oscillations of u_1 are chaotic. However the probability of this transition can be found based upon the probability of u_1 given by (8.213).

So far the critical points of the dynamical systems were prescribed in advance. We will introduce now the dynamical system whose critical points change their locations chaotically. For this purpose we introduce the following system:

$$\dot{u}_1 = \gamma_1 \sin^{1/2n+1}\left(\frac{\omega}{\alpha}u_1\right)\sin\omega t - \delta^2\left(u_1 - \frac{2\pi\alpha}{\omega}\right), \; \delta^2 \to 0 \tag{8.237}$$

$$\dot{u}_2 = \gamma_2 \sin^{1/2n+1}\left(\frac{\omega}{\alpha}u_1 u_2\right)\sin\omega t - \delta^2 u_2 u_1, \; n \to \infty \tag{8.238}$$

Here the power 1/3 is replaced by the power $1/(2n+1)$, $n \to \infty$, while n is an integer. This replacement does not change the qualitative behavior of the dynamical system: it changes only its quantitative behavior between the critical points in such a way that one has explicit control over the period of the transition from one critical point to another. Indeed, since

$$\lim \sin^{1/2n+1}_{n\to\infty} X = \text{sgn} \, \sin X \tag{8.239}$$

one obtains the solution for u_1, which is valid between critical points $u_1^{(k)}$ and $u_1^{(k+1)}$

$$u_1 = u_1^{(k)} + \frac{\gamma_1}{\omega}(1 - \cos\omega t) \tag{8.240}$$

Consequently, the periods of transition can be found from the following:

$$u_1^{(k+1)} - u_1^{(k)} = \frac{\gamma_1}{\omega}(1 - \cos\omega T_1) = \frac{\pi\alpha}{\omega}, \text{ i.e.} \tag{8.241}$$

$$T_1 = \frac{1}{\omega}\arccos(1 - \frac{\pi\alpha}{\gamma_1}) \tag{8.242}$$

Analogously,

$$u_2 = u_2^{(k)} + \frac{\gamma_2}{\omega}(1 - \cos\omega t) \tag{8.243}$$

$$u_2^{(k+1)} - u_2^{(k)} = \frac{\gamma_2}{\omega}(1 - \cos\omega T_2) \tag{8.244}$$

while

$$u_1^{(k+1)}u_2^{(k+1)} - u_1^{(k)}u_2^{(k)} = \frac{\pi\alpha}{\omega} \tag{8.245}$$

Consequently, as follows from (8.241), (8.244–8.245):

$$\left(u_1^{(k)} - \frac{\pi\alpha}{\omega}\right)\left[u_2^{(k)} + \frac{\gamma_2}{\omega}(1 - \cos\omega T_2)\right] - u_1^{(k)}u_2^{(k)} = \frac{\pi\alpha}{\omega}$$

whence

$$T_2 = \frac{1}{\omega} \arccos \left[1 - \frac{\pi \alpha}{u_1^{(k)} \gamma_2} \left(1 + u_2^{(k)} \right) \right] \tag{8.246}$$

In contrast to the previous cases, in (8.238) both the values $u_2^{(k+1)} - u_2^{(k)}$ and T_2 depend upon the critical values $u_2^{(k)}$ [see (8.245–8.246)].

Before analyzing the performance of the system (8.237–8.238), we have to select γ_1 and γ_2, such that u_1, and u_2 approach their critical points not later than they become terminal repellers, i.e., when $\sin \omega t = 0$. For (8.237) the restriction imposed upon γ_1, is

$$T_1 \leqslant \frac{\pi}{\omega}, \text{ i.e., } \gamma_1 \geqslant \pi \alpha \tag{8.247}$$

For (8.238) the restriction imposed upon γ_2 [see (8.246)]

$$T_2 \leqslant \frac{\pi}{\omega} \text{ i.e., } \gamma_2 \geqslant \frac{\pi \alpha}{u_1^{(k)}} (1 + u_2^{(k)}) \tag{8.248}$$

requires the values of the critical points $u_1^{(k)}$ and $u_2^{(k)}$ to be bounded from below and from above, respectively. However, such limitations are already incorporated into microdynamics of (8.237–8.238). We will verify this after analyzing the performance of these equations.

The behavior of the solution to (8.237) is similar to that of (8.209), it has a terminal chaotic attractor at $u_1 = 2\pi\alpha/\omega$ and therefore, u_i is bounded:

$$\pi\alpha/\omega < u_1 < 3\pi\alpha/\omega \tag{8.249}$$

Equation (8.238) has a terminal chaotic attractor at $u_2 = 0$, but the behavior of its solution is more complex: the amplitudes of the chaotic oscillations, which is the distance between the critical points $u_2^{(1)} - u_2^{(0)}$, are also changing chaotically. Indeed, according to (8.244) and (8.246):

$$u_2^{(1)} - u_2^{(0)} = \frac{\pi \alpha}{\omega} (1 + u_2^{(1)})$$

and since $u_2^{(0)} = 0$, one obtains:

$$u_2^{(1)} = \frac{1}{\frac{\omega}{\pi \alpha} - 1}$$

Hence, $u_2^{(1)} = u_{2\,\max}$ is bounded if

$$\omega \neq \pi \alpha$$

Now we can return to (8.248) and formulate the restrictions upon the constants ω, α and γ_2, providing the required performance of (8.237–8.238):

$$\gamma_2 \geqslant \omega \left(1 + \frac{1}{\frac{\omega}{\pi \alpha} - 1} \right)$$

Thus, the solution to (8.237) is combined from two independent chaotic motions: the terminal chaotic oscillations with respect to the point $u_2 = 0$ and terminal chaotic oscillations of the amplitudes. But since the probabilistic characteristics of both of these motions are known [they are similar to the characteristics given by (8.213)], one can find all the probabilistic characteristics of the resulting solution.

General Case

Let us consider weakly connected neural nets of the following type:

$$\dot{u}_i = \gamma_i \sin^{1/3}\left(\frac{\omega}{\alpha_i} u_i\right) \sin \omega t - \delta^2 \left[-u_i + \sum_{j=1}^{n} T_{ij}\sigma(u_j)\right],$$

$$\delta^2 \to 0, \; i = 1, 2, ..., n \tag{8.250}$$

in which u_i, is the mean soma potential of the neuron, T_{ij} are constant synaptic interconnections, and γ_i, α_i, ω are constant, and σ is a sigmoid function.

Each equation in the system (8.250) is independent of all others and it behaves exactly as (8.180) does, excluding the critical points

$$u_l^k = \frac{\pi k_l \alpha_l}{\omega} \quad k_l = \cdots - 2, -1, 0, 1, 2, \cdots \text{ etc.} \tag{8.251}$$

at which the lth equation is fully dependent on the rest of the equations:

$$\dot{u}_i = \delta^2 \left[-u_i + \sum_{j=1}^{n} T_{ij}\sigma(u_j)\right], \; \delta^2 \to 0, \; l = 1, 2, ..., m \leqslant n \tag{8.252}$$

Here m is the number of equations which simultaneously approach their critical points.

As follows from previous discussions, the dynamical system (8.250) is driven by the microdynamics

$$\varepsilon_i = \delta^2 \left[-u_i + \sum_{j=1}^{n} T_{ij}\sigma(u_j^{(k)})\right], \; \delta^2 \to 0, \; i = 1, 2, ..., n \tag{8.253}$$

However, actually the only important parts in (8.253) are the signs of ε_i, at the critical points, i.e.

$$\operatorname{sgn} \left[-u_i + \sum_{j=1}^{n} T_{ij}\sigma(u_j^{(k)})\right], \; k = 0, 1, 2, ..., m \tag{8.254}$$

Here we will be interested only in bounded solutions to (8.250). It is easy to conclude that a solution will be bounded by the critical point values $u_i^{(q)}$ and $u_i^{(p)}$:

$$u_i^{(q)} \leqslant u_i \leqslant u_i^{(p)} \tag{8.255}$$

if

$$\mathrm{sgn}\left[-u_i + \sum_{j=1}^{n} T_{ij}\sigma(u_j^{(q)})\right] = +$$

$$\mathrm{sgn}\left[-u_i + \sum_{j=1}^{n} T_{ij}\sigma(u_j^{(p)})\right] = - \tag{8.256}$$

Indeed, as follows from (8.250)

$$\mathrm{sgn}\left.\frac{\partial \dot{u}_i^{(k)}}{\partial u_j}\right|_{u_i^{(k)}=\mathrm{Const}} = \mathrm{sgn}\ T_{ij} = \ \mathrm{Const} \tag{8.257}$$

This means that if the velocities $\dot{u}_i^{(q)}$ or $\dot{u}_i^{(p)}$ at the vertices of the hyperparallelepiped (8.255) are directed inside, then the velocities at all its hypersurfaces are directed also inside since, as follows from (8.257), the direction of these velocities do not change when the points $u_i^{(q)}$ or $u_i^{(p)}$ move along the boundaries of (8.255) in phase space.

Thus, if the solution to (8.255) is bounded in the domain (8.255), the functions ε_i, must change their signs [see (8.256)] and therefore, they must have zero values inside the domain (8.255). Now one can arrive at two different situations: the zeros of ε_i, may or may not coincide with the critical points $u_i^{(k)}$. In the first case the motion will be fully deterministic since each time when the system approaches a critical point its further motion will be predetermined by the corresponding sign of ε_i. In the second case a critical point which coincides with a zero of ε_i, will be a center of terminal chaos described in the previous sections.

How terminal Chaos Can Store Information

Let us consider a dynamical system:

$$\dot{\tilde{u}}_i = f_i(\tilde{u}_1, ..., \tilde{u}_n) \tag{8.258}$$

and find the synaptic interconnection T_{ij} in (8.250) from the following conditions:

$$\mathrm{sgn}\left[-u_i + \sum_{j=1}^{n} T_{ij}\sigma(u_j^{(k)})\right] = \ \mathrm{sgn}\ f_i(u_i^{(k)}, ..., u_n^{(k)}) \tag{8.259}$$

In other words, we require that at the critical points of the system (8.250) the velocities $\dot{u}_i$ and $\dot{\tilde{u}}_i$ have the same directions. Hence, (8.250) stores some "image" of the dynamical system (8.258) if the synaptic interconnections are found from (8.259). However, there are many other dynamical systems:

$$\dot{\tilde{u}}_i = f_i^{(m)}(\tilde{u}_1, ..., \tilde{u}_n), \ m = 1, 2, ..., \text{etc.} \tag{8.260}$$

which satisfy the same condition (8.250) at the critical points $u_i^{(k)}$, and therefore, the synaptic interconnections T_{ij} store a "topological image" of all the set of dynamical systems (8.260) which have only one feature in common: their velocities at the critical points (8.251) have the same directions. Obviously such a topological similarity does not imply similarities in metrical characteristics of the solutions. Loosely speaking, all the stored solutions can be reproduced by elastic deformations of their "topological image."

We will be interested in storing only bounded solutions to (8.258) which means that the functions (8.259) have zero values inside the domain (8.256). Let us assume first that no zero values of (8.259) coincide with a critical point of the system (8.250). Then, as noted above, the solution to (8.250) will be deterministic: it will be uniquely defined by the initial conditions. It means that the synaptic interconnections store a "topological image" of only such particular solutions to (8.259) which have the same initial conditions.

Now we will proceed to the second case when at least some of the zero values of (8.259) coincide with the critical points of the system (8.250). As discussed above, each such zero point will become a center of the corresponding terminal chaos since the velocities $\dot{u}_i$ at these points are not prescribed: they run randomly. From the view point of the theory of differential equations, it means that the dynamical system (8.250) jumps from one particular solution of (8.260) to another exposing different "parts" of the general solution, while the more critical points (8.250) coinciding with zeros of (8.259) the broader "part" of the general solution are exposed. This is why terminal chaos represents a "generalized topological image" of all the possible particular solutions to (8.260), or a class of patterns, and its performance is associated with a higher level cognitive process which leads to the creation of new logical forms such as generalization and abstraction. From this point of view, each particular "run" of (8.250) can be associated with interpretation and specification.

It is easy to conclude that terminal chaos provides the most economical way of storing the information. Indeed, in order to store all the possible scenarios described by the dynamical system (8.260) in static (point) attractors using, for example, additive neural nets, an enormous number of synaptic interconnections would be required. Hence, the procedure of formation of classes of patterns through terminal chaos can be initiated by a tendency of the neural network to minimize the number (or the total strength) of the synaptic interconnections without a significant loss of the quality of prescribed performance; such a tendency can be incorporated into the learning dynamics which controls these interconnections (see Chap. 3, Sect. 3).

Learning Terminal Chaos

So far we did not yet discuss the way in which the synaptic interconnections can be found from (8.259). In this section we will introduce a dynamical learning procedure which solves this problem. The problem is posed as following: find such synaptic interconnection T_{ij} which store the "topological image" of the dynamical system (8.258) in the domain (8.255). In addition we require that the task should be performed with the least number of the synaptic interconnections T_{ij}.

The problem can be reduced to the minimization of the following "energy":

$$\widetilde{E} = \frac{1}{2}\sum_{k=1}^{m}\left\{\text{sgn}\left[-u_i + \sum_{j=1}^{n} T_{ij}\sigma(u_j^{(k)})\right] - \text{sgn}\, f_i(u_i^{(k)}, ..., u_n^{(k)})\right\} \tag{8.261}$$

which is a function of T_{ij} and n.

In order to use the gradient descent method first we will approximate sgn($\cdot$) by the sigmoid function $\sigma(\cdot\ )$ in the first term of (8.261)

$$\widetilde{E} \simeq E = \frac{1}{2}\sum_{k=1}^{m}\left\{\text{sgn}\left[-u_i + \sum_{j=1}^{n} T_{ij}\sigma(u_j^{(k)})\right] - \text{sgn}\, f_i(u_i^{(k)}, ..., u_n^{(k)})\right\} \tag{8.262}$$

$$\sigma(\cdot) = \beta \tanh(\cdot), \ \beta \gg 1 \tag{8.263}$$

which allows us to introduce the following learning dynamics for the sought synaptic interconnections $T_{ij}(i,j\ 1, 2,..., n)$ and the number n:

$$\dot{T}_{ij} = \frac{\partial E}{\partial T_{ij}}, \ i, j = 2, ..., n \tag{8.264}$$

$$n_{i+1} = n_i - \text{sgn}\left[E(n_{i+1}) - E(n_1)\right] \tag{8.265}$$

This is a gradient system of differential and difference equations with respect to T_{ij} and n. Its convergence is provided by the fact that the energy (8.262) plays the role of a Liapunov function. However, because of nonlinearities, the possibility of convergence to a local minimum (min $E \neq 0$) cannot be excluded. In order to find the global minimum (min $E = 0$) one should apply special tools for global optimization (none of them are universal).

In the same way one can store "topological images" of different dynamical systems in the corresponding (non-overlapping) domains.

8.5 Spatial Coherent Structures

Another step from unpredictability to creativity is associated with spatial coherence in neural network performance. As shown in Chap. 3, Sect. 3, organization in actual (physical) space can be achieved by introducing a special type of local interconnection simulating diffusion, dispersion, and convection, while the underlying continuous (in time and space) model is described by field equations in which local interconnections are represented by spatial derivatives of neuron potentials. In this section we will incorporate the diffusion-type local interconnections into the unpredictable neurodynamics (5.155) assuming that the neuron potential discrete distribution u_i over the indices i can be represented by its continuous analog $u(x)$.

Then the system of ordinary differential equations (8.185) reduces to one partial differential equation:

$$u_t = \gamma \sin^{1/3} \frac{\omega}{\alpha} u \sin \omega t + \delta^2 u_{xx} \tag{8.266}$$

in which

$$u_t = \frac{\partial u}{\partial t}, \quad u_{xx} = \frac{\partial^2 u}{\partial x^2}, \quad \alpha, \gamma = \text{Const}$$

while the finite-dimensional version of u_{xx} is:

$$u_{xx} \simeq u_{i+1} - 2u_i + u_{i-1} \tag{8.267}$$

We will select the parameters α and ω such that

$$\frac{|u_{\max}|}{n} \sim \frac{\pi \alpha}{\omega} \tag{8.268}$$

in which n is the number of neurons. In this case the changes of neuron potential per unit length and per unit of (local) time are of the same order.

Without loss of generality one can introduce the following initial and boundary conditions:

$$u(x, 0) = 0, \ \ u(0, t) = u(1, t) = 0 \tag{8.269}$$

Then $u_{xx} = 0$ at $t = 0$, and the solution to (8.266) starts with totally unpredictable behavior: at $t = \pi/\omega$ it approaches the values $\pm\pi\alpha/\omega$ which are randomly distributed over u, and therefore, the function

$$u = u(x, \frac{\pi}{\omega}) \tag{8.270}$$

may have a monstrous configuration. During the next period $\pi/\omega < t < 2\pi/3\omega$ these points at the curve (8.270) where $u_{xx} > 0$ (and therefore $u < 0$) will go up to $u = 0$, and these points where $u_{xx} < 0$ (and therefore, $u > 0$) will go down to $u = 0$. However, as follows from (8.267), these points u_i for which

$$u_{i-1} = u_i = u_{i+1} \tag{8.271}$$

will have

$$u_{xx} = 0 \tag{8.272}$$

The probability of the appearance of such points is:

$$\Pr(u_{i-1} = u_i = u_{i+1}) = (0.5)^3 = 0.125 \tag{8.273}$$

These particular points may return to $u = 0$, but with equal probability they can move away to the value

$$u_i(t = \frac{2\pi}{3\omega}) = \begin{cases} 2\pi\alpha/3\omega \text{ if } & u_i(t = \pi/\omega) > 0 \\ -2\pi\alpha/3\omega \text{ if } & u_i(t = \pi/\omega) < 0 \end{cases} \tag{8.274}$$

Thus, each point of the curve with the probability 0.9375 will return to the initial configuration $u = 0$, and with the probability 0.0625 will move away from it. Actually the curve

$$u = u(x, \frac{2\pi}{3\omega}) \tag{8.275}$$

will be close to $u = 0$ in terms of the mean square distance:

$$\rho = \sqrt{\int_0^1 u^2(x, \frac{2\pi}{3\omega})dx} \to 0 \tag{8.276}$$

It will coincide with the initial configuration almost everywhere excluding some solitary sharp peaks.

The next step of the evolution will be almost the same as the first step: the solution will approach the configuration with the values $\pm\pi\alpha/\omega$ randomly distributed over x, while the probability that this configuration is identical to (8.270) has the order $\sim 2^{-n}$ (n is the number of neurons). The solitary peaks where the magnitude of u_{xx} is large will be pushed back toward $u = 0$, etc.

Thus, the solution to (8.266) with the initial and boundary conditions (8.269) chaotically oscillates about the initial configuration $u = 0$. In other words, it preserves the mean square configuration $u = 0$, while the actual configuration $u(x)$ remains random and unpredictable.

It is worth mentioning that the attraction of the solution to its mean square value is provided by the stability of the solution $u = 0$ to the underlying diffusion equation:

$$u_t = \delta^2 u_x \tag{8.277}$$

subject to the conditions (8.269) which can be obtained by a superposition of terms with exponentially decaying multipliers:

$$u(x,t) = \sum_{p=1}^{\infty} c_n e^{-(\pi p)^2 \delta^4 t} \sin \pi n x \to 0 \text{ at } t \to \infty \tag{8.278}$$

Let us introduce now variable boundary conditions [instead of (8.269)] assuming that they are governed by another dynamical system:

$$u_t(0,t) = \gamma \sin^{1/3}\left[\omega_0 \frac{u(0,t)}{\alpha}\right] \sin \omega t - \delta^2 u(0,t) \tag{8.279}$$

$$u_t(1,t) = \gamma \sin^{1/3}[\omega_0 \frac{u(1,t)}{\alpha}] \sin \omega t - \delta^2 x(1,t) \tag{8.280}$$

where

$$\omega_0 \ll \omega \tag{8.281}$$

Since the general solution to (8.277) is a family of straight lines:

$$x = c_1 s + c_2 \tag{8.282}$$

one concludes that the solution to (8.266) with the boundary conditions (8.279–8.280) will oscillate chaotically with respect to different straight lines of the family (8.282), while the change of these lines is also chaotic [in accord with changes of the boundary conditions following from the dynamics (8.279–8.280)].

Hence, the solution to (8.266) with variable boundary conditions has more unpredictable features, but it still preserves the following property: the mean square solution is always a straight line. In other words, the behavior of the system (8.266), (8.279–8.280) represents the general solution to (8.277) as a mean square of chaotic oscillations, while the behavior of the system (8.266), (8.269) represents one of its particular solution $u = 0$.

If one replaces (8.266) by the following:

$$x_t = \gamma \sin^{1/3} \frac{\omega}{\alpha} x \ \sin \omega t + \varepsilon_o^2 (x_{ss} - \beta^2 x) \tag{8.283}$$

then the solution to the system (8.283), (8.279–8.280) will oscillate chaotically with respect to the curves of the family:

$$u = c_1 \ \text{sgn}\ \beta x + c_2 \cosh \beta x \tag{8.284}$$

The stability of these oscillations follows from the stability of the underlying diffusion equation with chain reactions:

$$u_t = \delta^2 (u_{xx} - \beta^2 u) \tag{8.285}$$

whose exponential decay is defined by the eigenvalues:

$$\lambda_p = -(\beta^2 + \pi p)^2 \delta^2, \ p = 1, 2, ..., \text{etc.} \tag{8.286}$$

As in the previous case, the choice of the curve from the family (8.284) is made by the boundary conditions which are controlled by (8.279–8.280), and therefore, are oscillating chaotically about the values:

$$u(0,t) = 0, \ u(1,t) = 0 \tag{8.287}$$

Again, despite a large number of unpredictable elements, the solution to (8.283) preserves its closeness to curves of the family (8.284).

The system (8.266) can be generalized in many different ways. Firstly, instead of an array of neurons $u(x)$ one can introduce multidimensional distributions $u(x_1, x_2, ..., x_n)$, for instance,

$$u_t = \gamma \sin^{1/3}\left(\frac{\omega}{\alpha}u\right) \sin \omega t - \delta^2 (u_{x_1 x_1} + u_{x_2 x_2}) \tag{8.288}$$

which is a two-dimensional analog of (8.266).

Secondly, more complex structures with diffusion-convection-dispersion effects (see Chap. 3) can be incorporated into the microdynamics:

$$u_t = \gamma \sin^{1/3}\left(\frac{\omega}{\alpha}u\right) \sin \omega t - \delta^2$$

$$\sigma\left(\sum_{i=1}^{n} \alpha_i \frac{\partial u}{\partial x_i} + \sum_{\substack{i=1 \\ j=1}}^{n} \alpha_{ij} \frac{\partial^2 u}{\partial x_1 \partial x_2} + ... + \sum_{\substack{i=1 \\ i_n=1}}^{n} \alpha_{i_1 ... i_m} \frac{\partial^m u}{\partial x_{i_1} ... \partial x_{i_m}}\right) \tag{8.289}$$

where τ is a sigmoid function, and $\alpha_i, \alpha_{ij}, ...$, etc., are time independent functions of $x_1, ..., x_n$.

Finally, spatial coherence can be combined with temporal coherence by exploiting the chain (8.226).

$$u_t = \gamma \sin^{1/3}\left(\frac{\omega_1}{\alpha_1}u\right) \sin \omega_1 t - \delta^2 u_{xx} \sin \omega_2 t - \delta^4 u_{xx} \sin \omega_3 t + \cdots \text{ etc.} \tag{8.290}$$

while the effects described by (8.228), (8.231), (8.234–8.235) can be also included.

8.6 Neurodynamics with a Fuzzy Objective Function

In this section we will discuss another aspect of creative neurodynamics which is associated with the initiation of the performance and a "stop mechanism." So far the dynamical equations (8.185), (8.218), or (8.266) have not had any stable equilibrium states: the neuron potentials could change monotonically or oscillate with respect to a center of attraction, but they cannot acquire constant values. Indeed, the parameter ω which defines the level of the energy input via the harmonic oscillations $\sin \omega t$ into the dynamical systems considered above was constant. This input can be implemented by means of the following autonomous system [see (8.140)]:

$$\dot{v}_i = \omega v_2 + v_1(1 - v_1^2 - v_2^2), \; \dot{v}_2 = -\omega v_1 + v_2(1 - v_1^2 - v_2^2) \qquad (8.291)$$

whose solution is a stable limit cycle:

$$v_1 = \cos \omega t, \; v_2 = -\sin \omega t, \; \omega = \text{Const} \qquad (8.292)$$

Let us assume now that $\omega \neq$ Const, but it is governed by the following dynamical equation:

$$\dot{\omega} = [\omega(1-\omega)]^{1/3} \theta - \delta^2 (\omega - \frac{1}{2}), \; \delta^2 \to 0 \qquad (8.293)$$

One can verify that (8.291) has a static terminal attractor

$$\omega = \begin{cases} 0 \text{ if } & \theta < 0 \\ 1 \text{ if } & \theta > 0 \end{cases} \qquad (8.294)$$

in which θ is a function of the neuron potentials:

$$\theta = \theta(u_1, u_2, ..., u_n) \qquad (8.295)$$

Hence θ plays the role of a "shadow" which opens up ($\theta > 0$) or closes down ($\theta < 0$) the performance of the dynamical system.

Obviously the neurodynamical systems (8.185), (8.218), or (8.266) supplemented by the systems (8.291) and (8.293) can have stable equilibrium states only in the domains where

$$\theta = \theta(u_1, u_2, ..., u_n) < 0 \qquad (8.296)$$

This is why the function θ can be associated with a fuzzy objective function.

Let us now illustrate a conversion of terminal chaos into a static attractor by means of an objective function. For this purpose consider a system of two independent neurons:

$$\dot{u}_1 = \gamma \sin^{1/3}\left(\frac{\omega}{\alpha} u_1\right) \sin \omega t - \delta^2 u_1 (u_1 - 1.5)(u_1 - 4)(4.5 - u_1)(u_1 - 5) \quad (8.297)$$

$$\dot{u}_2 = \gamma \sin^{1/3}\left(\frac{\omega}{\alpha} u_2\right) \sin \omega t - \delta^2 u_2 (0.5 - u_2)(u_2 - 1) \qquad (8.298)$$

$$\Delta u_1 = \Delta u_2 = \frac{\pi \alpha}{\omega} = 1 \qquad (8.299)$$

As noted above [see (8.228) and (8.215)], the first neuron possesses two different terminal chaotic attractors: the first one has its center at $u_1 = 0$ and is characterized by the probabilities (8.213) the second one has two centers: $u_1 = 4$ and $u_1 = 5$ and is characterized by the probabilities (8.229). The second neuron has two centers of terminal chaos: at $u_2 = 0$ and $u_2 = 1$

Without an objective function, i.e., without the additional dynamics (8.293), these neurons are totally independent and their potentials are changing chaotically.

Let us introduce now the following fuzzy objective function:

$$\theta = u_1^2 + u_2^2 - a^2 \tag{8.300}$$

which requires that the equilibrium position of the system (8.297–8.298) must be inside of the circle

$$u_1^2 + u_2^2 = a^2 \tag{8.301}$$

in phase space u_1, u_2.

Involving the dynamical equation (8.293) for ω, one concludes that there are several possibilities for stable equilibria of the system:

$$u_1^{(1)} = 0,\ u_2^{(1)} = 0 \text{ for } 0 < a < 1$$

$$\left.\begin{array}{ll} u_1^{(1)} = 0 & u_2^{(1)} = 0 \\ u_1^{(2)} = 0 & u_2^{(2)} = 1 \end{array}\right\} \text{ for } a \le 1$$

$$\left.\begin{array}{ll} u_1^{(1)} = 0 & u_2^{(1)} = 0 \\ u_1^{(2)} = 0 & u_2^{(2)} = 1 \\ u_1^{(3)} = 4 & u_2^{(3)} = 0 \end{array}\right\} \text{ for } a \le 4$$

$$\left.\begin{array}{ll} u_1^{(1)} = 0 & u_2^{(1)} = 0 \\ u_1^{(2)} = 0 & u_2^{(2)} = 1 \\ u_1^{(3)} = 4 & u_2^{(3)} = 0 \\ u_1^{(4)} = 4 & u_2^{(4)} = 1 \end{array}\right\} \text{ for } a \le \sqrt{17}$$

$$\left.\begin{array}{ll} u_1^{(1)} = 0 & u_2^{(1)} = 0 \\ u_1^{(2)} = 0 & u_2^{(2)} = 0 \\ u_1^{(3)} = 4 & u_2^{(3)} = 0 \\ u_1^{(4)} = 4 & u_2^{(4)} = 1 \\ u_1^{(5)} = 5 & u_2^{(5)} = 0 \end{array}\right\} \text{ for } a \le 5$$

$$\left.\begin{array}{ll} u_1^{(1)} = 0 & u_2^{(1)} = 0 \\ u_1^{(2)} = 0 & u_2^{(2)} = 0 \\ u_1^{(3)} = 4 & u_2^{(3)} = 0 \\ u_1^{(4)} = 4 & u_2^{(4)} = 1 \\ u_1^{(5)} = 5 & u_2^{(5)} = 0 \\ u_1^{(6)} = 5 & u_2^{(6)} = 1 \end{array}\right\} \text{ for } a > 5$$

Hence, as soon as the objective function θ is prescribed, the two neurons become implicitly coupled: their equilibrium positions within the basins of the corresponding chaotic attractors are correlated via the objective function.

Let us assume that first the objective function was:

$$\theta_1 = u_1^2 + u_2^2 - 0.5 \tag{8.302}$$

and consequently, the system (8.291), (8.298) was resting at the only possible equilibrium position:

$$u_1 = 0, \ u_2 = 0 \tag{8.303}$$

At this position $\theta < 0$, and, as follows from (8.293), (8.291),

$$\omega = 0$$

Suppose that the objective function (8.302) changed to

$$\theta_2 = 0.5 - (u_1 + u_2) \tag{8.304}$$

Now at the equilibrium (8.303)

$$\theta_2 > 0$$

and the equilibrium $\omega = 0$ of the system (8.293) switches to the equilibrium $\omega = 1$. This, in turn, activates the dynamical system (8.297–8.298) which leaves the equilibrium (8.303) and starts chaotic oscillations with respect to the values: $u_1 = 0$, $u_2 = 0$, $u_2 = 1$. Eventually the system will settle at the attractors

$$u_1 = 0, \ u_2 = 1 \tag{8.305}$$

which satisfy the new objective function (8.304). Other combinations of equilibrium positions which include values $u_1 = 4$, or $u_1 = 5$, cannot be approached by the systems since they belong to the basin of the different chaotic attractor.

Thus, the system (8.297–8.298), and (8.293) is able to perform tasks which are only partially prescribed: it finds new (not prescribed) equilibrium positions satisfying certain (prescribed) conditions.

In order to emphasize the role of terminal chaos in this type of performance suppose that (8.297–8.298) do not have any microdynamics, i.e., they have a form (8.203) with $\varepsilon = 0$. Then, the system (8.297–8.298) would have totally random motion and it may never find the conditions which satisfy the objective function θ. Due to terminal chaos, the motion of the system is not totally random: it has centers of attractions, and therefore, is bounded. In other words, terminal chaos restricts the domain of stochastic search and eliminates combinatorial explosion in the number of possible trajectories. However, there is a certain price paid for this: the objective function cannot be selected arbitrarily; it should be compatible with the basins of terminal chaotic attractors in the sense that at least one of the centers of these attractors should satisfy the condition $\theta < 0$.

8.7 Discussion and Conclusions

This chapter presents an approach to formalization of the concept of creativity in connection with information processing in neurodynamics. One of the most important and necessary (but not sufficient) conditions for neurodynamics – to be creative, is its irreversibility. In this regard one would recall that classical dynamics is reversible within a finite time interval: trajectories of different particles never intersect, and therefore, they can be traced toward the future or toward the past. In other words, nothing can appear in the future which could not already exist in past.

This means that neural networks based upon classical dynamics are reversible, and therefore, they can perform only prescribed tasks: they cannot invent anything new. However, nature introduced to us another class of phenomena where time is one-sided. These phenomena involve non-equilibrium thermodynamics which are irreversible, and consequently, past and future play different roles. The main tool of transition to fundamentally new effects in irreversible processes is unpredictability caused by instability of equilibriums states with respect to certain types of external excitations. Hence, the unpredictability is the second necessary (but still not sufficient) condition for neurodynamics to be creative. The last condition is the utility of its new (unpredictable) performance. The simplest way to introduce this utility is to require that the new performance preserves certain global characteristics while the details are random. In other words, creative performance is not totally unpredictable: it has certain (prescribed) constraints.

The new (terminal) architecture of neurodynamics discussed in this chapter satisfies all the conditions for creativity. Indeed, it is irreversible due to violations of the Lipschitz conditions at the critical points, it is unpredictable due to terminal repellers at the critical points, and it preserves prescribed global characteristics via temporal and spatial coherence.

As has been demonstrated, the main tool for temporal coherence is terminal chaos which preserves the center of attraction as well as the probabilistic characteristics of the motion with respect to this center; at the same time, the particular values of the neuron potentials remain random until additional information (which arrives from another neuron) eliminates certain degrees of randomness converting the chaotic motion into more (or totally) deterministic motion. Hence, due to redundancy of the degrees of freedom, terminal chaos provide a tool by means of which different neurons are adjusted to each other in their parallel performance. In other words, terminal chaos represent a generalization of certain types of behaviors, or a class of patterns, while each particular member of this class can be identified as soon as additional information is available.

Different levels of complexity of temporal structures with terminal chaos were analyzed. As shown in Sect. 6, each such a level is characterized by the number of microscales with the corresponding local times. As has been demonstrated there, nonlinearities of micro-dynamics on each scale play an important role in ability of the system to move from one terminal chaotic attractor to another, as well as to eliminate one attractor and to create another one.

Spatial coherence of the neurodynamics performance was introduced via local differential interconnections represented by spatial derivatives of neuron potentials and incorporated into microdynamics. Different levels of complexity of spatial and spatial-temporal structures of the neurodynamics performances were analyzed.

Thus, it can be concluded that the new neural network architecture based upon terminal dynamics represents a mathematical model for irreversible neurodynamics capable of creative performance.

9
Physical Models of Cognition

We need to map the tremendous web of connections in the human brain into overlapping classes. *–S. Ulam*

9.1 Introduction

This chapter presents and discusses physical models for simulating some aspects of neural intelligence, and in particular, the process of cognition. The main departure from the classical approach here is in the utilization of a terminal version of classical dynamics introduced in Chap. 7. Based upon violations of the Lipschitz condition at equilibrium points, terminal dynamics attains two new fundamental properties: it is irreversible and nondeterministic. We pay special attention to terminal neurodynamics as a particular architecture of terminal dynamics which is suitable for modeling of information flows. Terminal neurodynamics possesses a well-organized probabilistic structure which can be analytically predicted, prescribed, and controlled, and therefore, which presents a powerful tool for modeling real-life uncertainties. Two basic phenomena associated with random behavior of neurodynamical solutions are exploited. The first one is a stochastic attractor – a stable stationary stochastic process to which random solutions of closed system converge. As a model of cognition, a stochastic attractor can be viewed as a universal tool for generalization and formation of classes of patterns. The concept of stochastic attractor is applied to model a collective brain paradigm explaining coordination between simple units of intelligence which perform a collective task without direct exchange of information. The second fundamental phenomenon (discussed in Chap. 7) is terminal chaos which occurs in open systems. Applications of terminal chaos to information fusion as well as to explanation and modeling of coordination among neurons in biological systems are discussed. It should be emphasized that all the models of terminal neurodynamics are implementable in analog devices, which means that all the cognition processes discussed in the chapter are reducible to the laws of Newtonian mechanics.

9.2 Stochastic Attractor as a Tool for Generalization

As has been remarked, random activity in the human brain has been a subject of discussion in many publications (e.g., Freeman 1987). Interest in the problem was promoted by the discovery of strange attractors. This discovery provided a phenomenological framework for understanding electroencephalogram data in regimes of multiperiodic and random signals generated by the brain. An understanding of the role of such random states in the logical structure of human brain activity would significantly contribute not only to brain science, but also to the theory of advanced computing based upon artificial neural networks. In this section, based upon properties of terminal neurodynamics, we propose a phenomenological approach to the problem: we demonstrate that a stochastic attractor incorporated in neural net models can represent a *class* of patterns, i.e., a collection of all those and only those patterns to which a certain concept applies. Formation of such a class is associated with higher level cognitive processes (generalization). This generalization is based upon a set of unrelated patterns represented by static attractors and associated with the domain of lower level of brain activity (perception, memory). Since a transition from a set of unrelated static attractors to the unique stochastic attractor releases many synaptic interconnections between the neurons, the formation of a class of patterns can be "motivated" by a tendency to minimize the number of such interconnections at the expense of omitting some insignificant features of individual patterns.

Let us first consider a deterministic dissipative nonlinear dynamical systems modeled by coupled sets of first order differential equations of the form:

$$\dot{x}_i = V_i(x_j, T_{ij}), \; i, j = 1, 2, \ldots, n \tag{9.1}$$

in which x_i is an n-dimensional vector function of time representing the neuron activity, and T_{ij} is a constant matrix whose elements represent synaptic interconnections between the neurons.

The most important characteristic of neurodynamical systems (9.1) is that they are dissipative, i.e., their motions, on the average, contract phase space volumes onto attractors of lower dimensionality than the original space.

So far only point attractors have been utilized in the logical structure of neural network performance: they represent stored vectors (patterns, computational objects, rules). The idea of storing patterns as point attractors of neurodynamics implies that initial configurations of neurons in some neighborhood of a memory state will be attracted to it. Hence, a point attractor (or a set of point attractors) is a paradigm for neural net performance based upon the phenomenology of nonlinear dynamical systems. This performance is associated with the domain of lower level brain activity such as perception and memory.

It is easily verifiable that a set of point attractors imposes certain constraints upon the synaptic coefficients T_{ij}. Indeed, for a set of m fixed points $\tilde{x}_i^k (k = 1, 2, \ldots, m)$ one obtains $m \times n$ constraints following from (9.1):

$$0 = V_i(\tilde{x}_j^k, T_{ij}), \; i, j = 1, 2, \ldots, n, \; k = 1, 2, \ldots, m \tag{9.2}$$

In order to provide stability of the fixed points $\tilde{x}_j^k$, the synaptic coefficients must also satisfy the following $m \times n$ inequalities:

$$\operatorname{Re} \lambda_i^k < 0, \; i = 1, 2, \ldots, n, \; k = 1, 2, \ldots, m \tag{9.3}$$

in which λ_i^k are the eigenvalues of the matrices $\|\partial f_i / \partial x_j\|$ at the fixed points $\tilde{x}_i^k$.

How can a neural network minimize the number of interconnections T_{ij} without a significant loss of the quality of a prescribed performance?

Let us assume that the vectors $\tilde{x}_j^k$ have some characteristics in common, for instance, their ends are located on the same circle of a radius r_0, i.e., (after proper choice of coordinates):

$$\sum_{i=1}^{2} (\tilde{x}_i^k)^2 = r_0^2, \; k = 1, 2, \ldots, m \tag{9.4}$$

If for the patterns represented by the vectors $\tilde{x}_i^k$ the property (9.4) is much more important then their angular coordinates $\theta^k (\theta^{k1} \neq \theta^{k2}$ if $k_1 \neq k_2)$, then it is "reasonable" for the neural net to store the circle $r = r_0$ instead of storing m point attractors with at least $2 \times m$ synaptic coefficients T_{ij}. Indeed, in this case the neural net can "afford" to eliminate unnecessary synaptic coefficients by reducing its structure to the simplest form:

$$\dot{r} = r(r - r_0)(r - 2r_0), \quad \dot{\theta} = \omega = \text{Const} \tag{9.5}$$

Equations (9.5) have a periodic attractor

$$r = r_0, \quad \theta = \omega t \tag{9.6}$$

which generates harmonic oscillations with frequency ω. But what is the role of these oscillations in the logical structure of neural net performance? The transition to the form (9.5) can be interpreted as a generalization procedure in the course of which a collection of unrelated vectors $\tilde{x}_i^k$ is united into a class of vectors whose lengths are equal to r_0. Hence, in terms of symbolic logic, the circle $r = r_0$ is a logical form for the class of vectors to which the concept (9.4) applies. In other words, the oscillations (9.6) represent a higher level cognitive process associated with generalization and abstraction. During these processes, the point describing the motion of (9.5) in phase space will visit all those and only those vectors whose lengths are equal to r_0; thereby the neural network "keeps in mind" all the members of the class.

Suppose that a bounded set of isolated point attractors which can be united in a class occupies a more complex subspace of the phase space, i.e., instead of the circle (9.4) the concept defining the class is:

$$\Phi(\tilde{x}_1^k, \tilde{x}_2^k, \ldots \tilde{x}_n^k) = r, \; k = 1, 2, \ldots, m \tag{9.7}$$

Then the formation of the class will be effected by storing a surface:

$$\Phi(x_1, x_2, \ldots, x_n) = r \tag{9.8}$$

as a limit set of the neurodynamics, while all the synaptic coefficients T_{ij} which impose constraints on the velocities along the surface (9.8) will be eliminated.

The character of the motion on the limit set depends upon the properties of the surface (9.8). If (by proper choice of coordinates) this surface can be approximated by a topological product of $(n-1)$ circles [i.e., by an $(n-1)$-dimensional torus] then the motion is quasi-periodic: it generates oscillations with frequencies which are dense in the real. If the surface (9.8) is more complex and is characterized by a fractal dimension, the motion on such a limit set must be chaotic: it generates oscillations with continuous spectrum. In both cases the motion is ergodic: the point describing the motion in the phase space sooner or later will visit all the points of the limit set, i.e., the neural net will "keep in mind" all the members of the class.

Thus, it can be concluded that artificial neural networks are capable of performing high level cognitive processes such as formation of classes of patterns, i.e., formation of new logical forms based upon generalization procedures. In terms of the phenomenology of nonlinear dynamics these new logical forms are represented by limit sets which are more complex than point attractors, i.e., by periodic or chaotic attractors. It is shown that formation of classes is accompanied by elimination of a large number of extra synaptic interconnections. This means that these high level cognitive processes increase the capacity of the neural network. The procedure of formation of classes can be initiated by a tendency of the neural

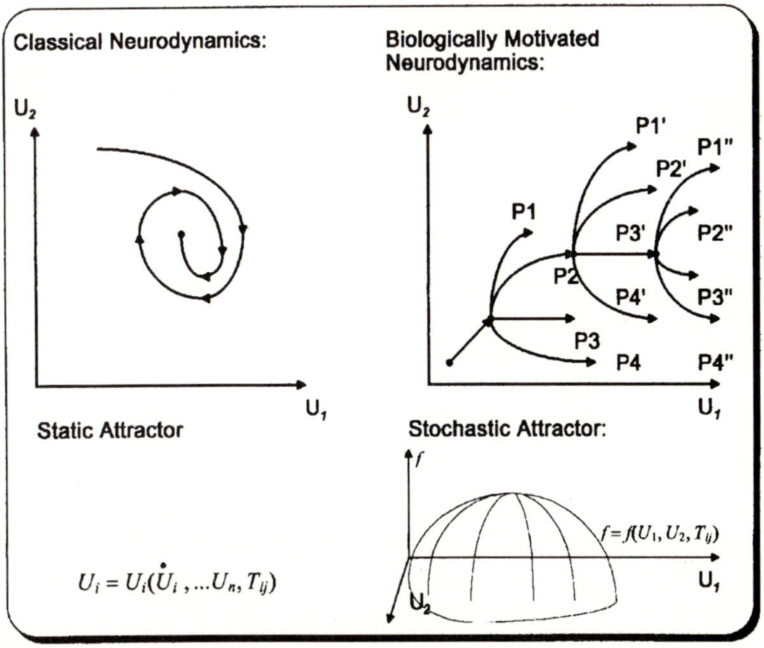

Fig. 9.1.

network to minimize the number (or the total strength) of the synaptic interconnections without a significant loss of the quality of prescribed performance; such a tendency can be incorporated into the learning dynamics which controls these interconnections.

In addition, the phenomenological approach presented above leads to a possible explanation of random activity of the human brain; it suggests that this activity represents the high level cognitive processes such as generalization and abstraction.

Turning to terminal neurodynamics represented by (7.190–7.191) (cf. Chap. 7), one can view a stochastic attractor as a more universal tool for generalization (Fig. 9.1). Indeed, in contradistinction to chaotic attractors of deterministic dynamics, stochastic attractors can provide any arbitrarily prescribed probability distributions (7.194) by an appropriate choice of (fully deterministic!) synaptic weights T_{ij}.

The information stored in a stochastic attractor can be measured by the entropy H via the probabilistic structure of this attractor:

$$H(X_i, X_2, ..., X_n) = -\sum_{X_1} \cdots \sum_{X_n} f(x_1, ..., x_n) \log f(x_1, ..., x_n) \qquad (9.9)$$

where the joint density $f(x_1, ..., x_n)$ is uniquely defined by the synaptic weights T_{ij} by means of (7.194).

For instance, the information stored by the dynamical system (7.197) is measured by the entropy:

$$H = \log_2 \sqrt{2\pi e \sigma^2} \tag{9.10}$$

since this system has a stochastic attractor with normal density distribution (7.195).

As shown by Zak (1993b), random neurodynamical systems can have several, or even, infinite number of stochastic attractors. For instance, a dynamical system

$$\dot{x}_1 = \gamma_1 \sin^k[\sqrt{\omega}\sin(x_1 + x_2)] \sin \omega t \tag{9.11}$$

$$\dot{x}_2 = \gamma_2 \sin^k[\sqrt{\omega}\sin(x_1 + x_2)] \sin \omega t \tag{9.12}$$

has stochastic attractors with the following densities:

$$f(x_1, x_2) = 0.5 \mid \cos(x_1 + x_2)\cos(x_1 - x_2) \mid \tag{9.13}$$

$$\frac{\pi m_1}{2} < x_1 + x_2 < \frac{\pi(m_1 + 2)}{2}, \frac{\pi m_2}{2} < x_1 - x_2 < \frac{\pi(m_2 + 2)}{2} \tag{9.14}$$

$$m_1 = \cdots -7, -3, 1, 5, 9, \text{etc.};\ m_2 = \cdots -5, -1, 3, 5, 7, \cdots \text{ etc.} \tag{9.15}$$

The solution (9.13) represents a stationary stochastic process, which attracts all the solutions with initial conditions within the area (9.14). Each pair m_1 and m_2 from the sequences (9.15) defines a corresponding stochastic attractor with the joint density (9.13).

Hence, the dynamical system (9.11–9.12) is capable of discrimination between different stochastic patterns, and therefore, it performs pattern recognition on the level of classes.

9.3 Collective Brain Paradigm

In this section the usefulness of terminal neurodynamics, and in particular, of the new dynamical phenomenon – stochastic attractor – will be demonstrated by simulating a paradigm of collective brain (Zak and Zak 1994).

9.3.1 General Remarks

The concept of the collective brain has appeared recently as a subject of intensive scientific discussions from theological, biological, ecological, social, and mathematical viewpoints (Huberman 1989). It can be introduced as a set of simple units of intelligence (say, neurons) which can communicate by exchange of information without explicit global control. The objectives of each unit may be partly compatible and partly contradictory, i.e., the units can cooperate or compete. The exchange of information may be at times inconsistent, often imperfect, non-deterministic, and delayed. Nevertheless, observations of working insect colonies, social systems, and scientific communities suggest that such collectives of single units appear to be very successful in achieving global objectives, as well as in learning, memorizing, generalizing and predicting, due to their flexibility, adaptability to environmental changes, and creativity.

In this section collective activities of a set of units of intelligence will be represented by a dynamical system which imposes upon its variables different types of non-rigid constraints such as probabilistic correlations via the joint density. It is reasonable to assume that these probabilistic correlations are learned during a long-term period of performing collective tasks. Due to such correlations, each unit can predict (at least, in terms of expectations) the values of parameters characterizing the activities of its neighbors if the direct exchange of information is not available. Therefore, a set of units of intelligence possessing a "knowledge base" in the form of joint density function, is capable of performing collective purposeful tasks in the course of which the lack of information about current states of units is compensated by the predicted values characterizing these states. This means that actually in the collective brain, global control is replaced by the probabilistic correlations between the units stored in the joint density functions.

Since classical dynamics can offer only fully deterministic constraints between the variables, we will turn to its terminal version discussed in previous sections. Based upon the stochastic attractor phenomenon as a paradigm, we will develop a dynamical system whose solutions are stochastic processes with prescribed joint density. Such a dynamical system introduces more sophisticated relationships between its variables which resemble those in biological or social systems, and it can represent a mathematical model for the knowledge base of the collective brain.

9.3.2 Model of Collective Brain

Let us first turn to an example and consider a basketball team. One of the most significant properties necessary for success in games is the ability of each player to predict actions of his partners even if they are out of his visual field. Obviously, such an ability should be developed in the course of training. Hence, the collective brain can be introduced as a set of simple units of intelligence which achieve the objective of the team without explicit global control; actions of the units are coordinated by ability to predict the values of parameters characterizing the activities of their partners based upon the knowledge acquired and stored during long-time experience of performing similar collective tasks.

We will start with the mathematical formulation of the collective brain for a set of two units considered in the previous section and described by (9.11–9.12). As shown there, this system has a random solution which eventually approaches a stationary stochastic process with the joint probability density (9.13). For further analysis we will take $m_1 = 1$, $m_2 = -1$ from (9.15).

As follows from the solution (9.13), one can find the probability density characterizing the behavior of one unit (say x_1) given the behavior of another, i.e., x_2.

Let us assume now that the unit x_1 does not have information about the behavior of the unit x_2. Then the unit x_1 will turn to the solution (9.13) which is supposed to be stored in its memory. From this solution, the conditional expectation of x_2 given x_1 can be found as:

$$E(x_2 \mid x_1 = x_1) = \int_{-\infty}^{\infty} x_2 f_2(x_2 \mid x_1) dx_2 = \tilde{x}_2 \tag{9.16}$$

in which the conditional density:

$$f_2(x_2 \mid x_1) = \frac{f(x_1, x_2)}{f_1(x_1)} \tag{9.17}$$

where the marginal density

$$f_1(x_1) = \int_{-\infty}^{\infty} f(x_1, x_2) dx_2 \tag{9.18}$$

Actually the integration with respect to x_2 in (7.168) and (7.170) is taken over the square $ABCD$ (Fig. 9.2).

Substituting (9.18) and (9.17) into (9.16) one obtains:

$$\tilde{x}_2 \cong \frac{\pi}{2}, \quad 0 < x_2 < \pi \tag{9.19}$$

similarly,

$$\tilde{x}_1 \cong \frac{\pi}{2}, \quad 0 < x_2 < \pi \tag{9.20}$$

Clearly this result should be expected based upon the symmetry of the square $ABCD$ and the joint probability density (9.13) with respect to the lines (9.19–9.20).

It should be noted that in the general case

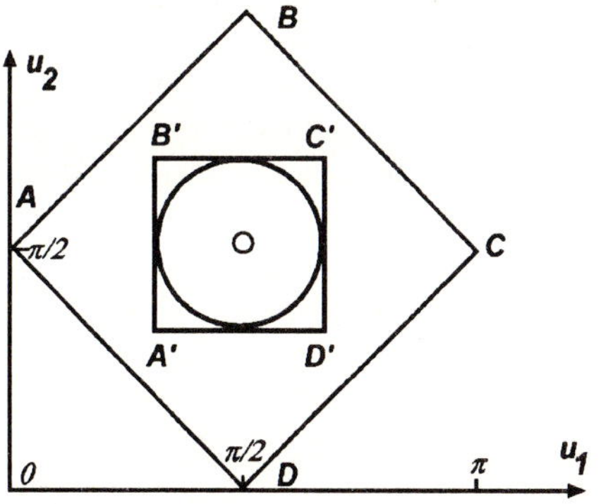

Fig. 9.2.

$$\tilde{x}_2 = \tilde{x}_2(x_1) \text{ and } \tilde{x}_1 = \tilde{x}_1(x_2) \tag{9.21}$$

Substituting (9.19) and (9.20) into (9.11) and (9.12), respectively, one obtains:

$$\dot{x}_1 = \gamma_1 \sin^k[\sqrt{\omega}\sin(x_1 + \frac{\pi}{2})]\sin\omega t \tag{9.22}$$

$$\dot{x}_2 = \gamma_1 \sin^k[\sqrt{\omega}\sin(\frac{\pi}{2} - x_2)]\sin\omega t \tag{9.23}$$

The system (9.22–9.23) represents a collective brain derived from the original system (9.11–9.12).

Both (9.22) and (9.23) are self-contained: they are formally independent since the actual contribution of the other unit is replaced by the "memory" of its typical performance during previous (similar) collective tasks. This memory is extracted from the joint probability density (9.13) in the form of the conditional expectations (9.19–9.20).

The probability densities for the performances of x_1 and x_2 in the collective brain are:

$$\varphi_1(x_1) = \begin{cases} \frac{1}{2}/\cos(x_1 + \frac{\pi}{2}), & 0 \le x_1 \le \pi \\ 0, & \text{otherwise} \end{cases} \tag{9.24}$$

$$\varphi_2(x_2) = \begin{cases} \frac{1}{2}/\cos(\frac{\pi}{2} - x_1), & 0 \le x_2 \le \pi \\ 0, & \text{otherwise} \end{cases} \tag{9.25}$$

and therefore, their joint probability density:

$$\varphi * (x_1, x_2) = \begin{cases} \frac{1}{4}/\cos(x_1 + \frac{\pi}{2})\cos(\frac{\pi}{2} - x_2), & 0 \le x_1, x_2 \le \pi \\ 0, & \text{otherwise} \end{cases} \tag{9.26}$$

Obviously, (9.26) is different from (9.13), and therefore, strictly speaking, the performance (9.24–9.25) of units x_1 and x_2 in the collective brain (9.22–9.23) is different from their original performance (9.13) when all the information about the other unit is available. However, this difference should not be significant if a new task belongs to the same class for which these units were trained.

The dynamical paradigm described above can be easily generalized to an n-dimensional system,

$$\dot{x}_i = \gamma_i \sin^k [\frac{\sqrt{\omega}}{\alpha} p_i(y_i)] \sin \omega t, \; x_i = \sum_{j=1}^{n} T_{ij}, x_j, t_{ij} = \text{ Const} \tag{9.27}$$

while

$$p_i' = \frac{dp_i}{dv_i} \tag{9.28}$$

As shown above, the solution to (9.27) is random, and it eventually approaches a stationary stochastic process with the joint probability density (7.194).

Following the previous example, we will assume that the ith unit x_i has actual input only from the units $x_{k(i)}, k < n$, so that the rest of the inputs should be predicted based upon the joint probability density (7.194) which was learned by each unit in the course of previous collective tasks. Now instead of (9.27) one can introduce the collective brain:

$$\dot{x}_i = \gamma_i \sin^k [\frac{\sqrt{\omega}}{\alpha} p_i(\overset{*}{y}_i)] \sin \omega t \tag{9.29}$$

$$y_i = \sum_{j=1}^{k(i)} T_{ij} y_j + E(\sum_{k(i)+1}^{n} T_{ij} y_i \mid x_i, ..., x_{k(i)}) \tag{9.30}$$

Here the unavailable input from the unit $x_{k(i)1}, ..., x_n$ is replaced by their conditional expectation given $x_1, ..., x_{k(i)}$. As in the two-dimensional case considered above, this expectation is uniquely defined by the joint probability distribution (7.194). In the extreme case $k = n - 1$, i.e., when the actual information from other units is not available, (9.30) reduces to the following:

$$\overset{*}{v}_i = v_i + E(\sum_{j \neq i}^{n} T_{ij} v_j \mid u_i) \tag{9.31}$$

while the conditional expectation in (9.31) depends only on x_i. This means that all the units in the collective brain (9.29–9.30) are formally independent. But as in the example considered above, their performances are coupled by the "memories" of previous collective tasks stored in the form of conditional expectations.

It should be stressed that the main advantage of the collective brain is in its universality: it can perform purposeful activity without global control, i.e. with only partial exchange of information between its units. This is why the collective brain can model many collective tasks which occur in real life. Obviously the new tasks are supposed to belong to the same class for which the units were trained. In other words, too many novelties in a new task may reduce the effectiveness of the collective brain.

9.3.3 Collective Brain with Fuzzy Objective

So far we were concerned with the structure of the model of collective brain regardless of the objective of its performance. In this section we will discuss the collective brain with objectives for its performance. Usually the objective of a performance is reduced to the minimization of a function or a functional subject to some constraints. In this way the problem has, at least, a rigorous mathematical formulation, although its solution may be not simple. However, in the most practical, real life problems (for instance, in operation research) the information about the objective is vague, imprecise, and uncertain. In mathematical terms it means that the analytical structure of the function (or the functional) to be minimized is not available. Clearly, these kinds of problems are the best "match" for the collective brain whose low precision is compensated by a high degree of universality. Actually the main motivation for the development of the mathematical model of a collective brain was its ability to perform in a more "human" way, when rigid rules are replaced by the ordering of multi-choice actions with respect to "preference."

In this section we will discuss fuzzy objectives which are given by a system of inequalities. We will start with a two-dimensional example, and turn to (9.11–9.12). However, now we will assume that $\omega \neq$ Const, and in particular

$$\omega = \begin{cases} 0 \text{ if} & \theta < 0 \\ \omega_0 \text{ if} & \theta > 0 \end{cases}, \quad \omega_0 > 0 \tag{9.32}$$

The inequality (9.32) can be implemented by the following dynamical equation (Zak 1993a):

$$\dot{\omega} = [\omega(\omega_0 - \omega)]^{1/3}\theta - \delta^2(\omega - \frac{1}{2}, \ \delta^2 \to 0 \tag{9.33}$$

(This kind of dynamical system will be discussed in the next section.) Here θ can be an arbitrary function of x_1 and x_2, for instance:

$$\theta = (x_1 - \alpha_1)^2 + (x_2 - \alpha_2)^2 - a^2 \tag{9.34}$$

where α_1, α_2 and a are given constraints.

As follows from (9.34), all the states of the system (9.11–9.12) and (9.32) which are inside of the circle

$$(x_1 - \alpha_1)^2 + (x_2 - \alpha_2)^2 = a^2 \tag{9.35}$$

will correspond to its equilibria, since then $\theta < 0$ and $\omega = 0$. But since the solution to these equations is random, it can approach an equilibrium at different points inside of this circle, i.e., the final equilibrium point will be characterized by some uncertainty.

It is worth emphasizing that this uncertainty is not explicitly imposed by any rigid rule: it is generated by the dynamical system itself as a result of the randomness of its behavior, and the fuzziness of the objective function

$$\theta < 0 \tag{9.36}$$

Obviously the circle (9.35) (or, at least, a part) must be inside of the square $ABCD$ (Fig. 9.2). For this one can take, for instance,

$$\alpha_1 = \alpha_2 = \frac{\pi}{2} \tag{9.37}$$

The objectives of the units u_1 and u_2 are not necessarily identical, but they must be compatible. For instance, instead of (9.11–9.12), (9.32–9.33) one can have:

$$\dot{x}_1 = \gamma_1 \sin^k[\sqrt{\omega_1} \sin(x_1 + x_2)] \sin \omega_1 t \tag{9.38}$$

$$\dot{x}_2 = \gamma_2 \sin^k[\sqrt{\omega_2} \sin(x_1 - x_2)] \sin \omega_2 t \tag{9.39}$$

However, now we will assume that $\omega_i \neq$ Const, and in particular

$$\omega_i = \begin{cases} 0 \text{ if } & \theta_i < 0 \\ \omega_0 \text{ if } & \theta_i > 0 \end{cases}, \quad i = 1, 2 \tag{9.40}$$

while

$$\theta_1 = (x_1 - \frac{\pi}{2})^2 + (x_2 - \frac{\pi}{2})^2 - a^2 \tag{9.41}$$

$$\theta_2 = (a - \ell)^2 - (x_1 - \frac{\pi}{2)}^2 - (x_2 - \frac{\pi}{2})^2 \tag{9.42}$$

In this case the area of the equilibria of the system (9.38–9.42) is within the ring of the width ℓ:

$$(x_1 - \frac{\pi}{2})^2 + (x_2 - \frac{\pi}{2})^2 = a^2, (x_1 - \frac{\pi}{2})^2 + (x_2 - \frac{\pi}{2})^2 = (a - \ell)^2 \tag{9.43}$$

and therefore, the solution can approach any of its points.

We can give the following interpretation of the performance of this system. Let us assume that the units u_1 and u_2 were trained to perform a certain class of collective tasks which led to formation of the soft constraints:

$$T_{11} = 1, \quad T_{22} = -1, \quad T_{12} = T_{21} = 1 \tag{9.44}$$

so that the random behavior of x_1 and x_2 eventually approaches a stationary stochastic process with the joint probability density (9.13). Let us also assume that a new task (which belongs to the same class of tasks for which the system was previously trained) is to optimize some process which depends upon the values x_1 and x_2. Here we will be interested in the cases when this dependence is given in an uncertain, imprecise way, which is typical for the problems in decision making processes. The simplest mathematical formalization of such a dependence is expressed by the "yes-or-no" relationship with respect to certain discrimination surfaces

$$\theta_i(x_1, x_2,) = 0, \quad i = 1, 2, ..., n \tag{9.45}$$

In particular, the values x_1 and x_2 are optimal if

$$\theta_i < 0 \tag{9.46}$$

and non-optimal if

$$\theta_i > 0 \tag{9.47}$$

As follows from (9.38–9.40), in the case (9.46) the dynamical system will remain in equilibrium, while in the case (9.47) this system will evolve until it approaches the area (9.46).

In general, the criterion of optimality may change in time

$$\theta_i(x_1, x_2, t) = 0 \tag{9.48}$$

while θ_i is a slow function of time in a sense that

$$\left| \frac{\partial \theta}{\partial t} \right| \ll \theta \omega \tag{9.49}$$

If the system (9.38–9.40) was in equilibrium, eventually it will be activated again when the inequality (9.46) will change to (9.47), and then it will evolve until a new equilibrium in the area (9.46) will be approached again, etc.

In the most general case the parameters which define the criterion of optimality may themselves be governed by a dynamical process which is controlled by the original dynamical system, for instance:

$$\dot{a}_1 = \lambda_1 \sin^k [\sqrt{\Omega_1} \sin(a_1 + a_2)] \sin \Omega_1 t \tag{9.50}$$

$$\dot{a}_2 = \lambda_2 \sin^k[\sqrt{\Omega_2}\sin(a_1 + a_2)] \sin \Omega_1 t \tag{9.51}$$

$$\omega_i = \begin{cases} 0 \text{ if } & \Gamma_i < 0 \\ \omega_0 \text{ if } & \Gamma_i > 0 \end{cases} \tag{9.52}$$

θ_i in (9.40) and Γ_i in (9.52) are expressed as:

$$\theta_i = \theta_i(x_1, x_2, a_1, a_2), \Gamma_i = \Gamma_i(a_1, a_2, x_1, x_2) \tag{9.53}$$

The relationship between (9.38–9.40) and (9.50–9.53) represents a dynamical game which ends when simultaneously:

$$\theta_i < 0, \quad \text{and} \quad \Gamma_i < 0 \tag{9.54}$$

Hence the equilibrium values of x_1, x_2, a_1, and a_2 are given by the intersections of the inequalities (9.53), and they densely fill up a part of the space x_1, x_2, a_1, a_2. Despite the fact that the dynamical behavior of the parameters x_i, a_i is characterized by uncertainties (coming from the randomness of the solutions and the fuzziness of the optimality criteria), the end of the game, i.e., the stationary values of x_i and a_i can be predicted in probabilistic terms since these values should belong to the stochastic attractor of the system (9.38–9.40), (9.50–9.53) whose joint probability density is uniquely defined by the synaptic interconnections (9.44).

Let us now return to the model of collective brain, and start with the dynamical system (9.37–9.39). In reducing this system to the model of the collective brain, one should assume again that each unit does not have explicit information about another, and therefore, the values of x_2 in (9.38) and the values of x_1 in (9.39) must be replaced by the conditional expectations $\tilde{x}_2$ and $\tilde{x}_1$, respectively [see (9.19–9.20)]. But in addition to this, one has to introduce different rhythms ω_1 and ω_2 which are controlled by different functions θ_1 and θ_2, respectively:

$$\dot{x}_1 = \lambda_1 \sin^k[\sqrt{\omega}\sin(x_1 + \frac{\pi}{2})] \sin \omega_1 t \tag{9.55}$$

$$\dot{x}_2 = \lambda_2 \sin^k[\sqrt{\omega}\sin(\frac{\pi}{2} - x_2)] \sin \omega_2 t \tag{9.56}$$

$$\omega_i = \begin{cases} 0 \text{ if } & \theta_i < 0 \\ \omega_0 \text{ if } & \theta_i > 0 \end{cases}, \quad i = 1, 2 \tag{9.57}$$

while θ_1 and θ_2 are obtained from (9.34) as a result of replacing x_2 and x_1 by their conditional expectations $\tilde{x}_2$ and $\tilde{x}_1$, respectively:

$$\theta_1 = (x_1 - \alpha_1)^2 + (\frac{\pi}{2} - x_2)^2 - a^2 \tag{9.58}$$

$$\theta_2 = (\frac{\pi}{2} + \alpha_1)^2 + (x_2 - \alpha_2)^2 = a^2 \tag{9.59}$$

Equations (9.55–9.56) have random solutions which are attracted to stationary stochastic process with independent probability densities (9.34) and (9.25), respectively. The system will approach an equilibrium state if simultaneously

$$\theta_1 < 0, \quad \text{and} \quad \theta_2 < 0 \tag{9.60}$$

For $\alpha_1 = \alpha_2 = \frac{\pi}{2}$, the area of possible equilibria is the square $A'B'C'D'$

$$x_1 = \frac{\pi}{2} \pm a, \; x_2 = \frac{\pi}{2} \pm a \tag{9.61}$$

Recalling that for the original system (9.38–9.39), the area of possible equilibria is inside of the circle of radius a (Fig. 9.2), one can see that the performance of the collective brain is sufficiently close to the original performance.

It should be stressed that the success of collective brain performance depends upon how close is the new task to the class of tasks for which the system was trained. This means that the collective brain may fail if the new task has too many "novelties." In the selected example, the training area (see the square $ABCD$ in Fig. 9.2) is compatible with the objective (the circle with the center 0 in Fig. 4), and therefore, the performance of the collective brain (the square $A'B'C'D'$ in Fig. 9.2) is satisfactory.

9.4 Open Systems in Terminal Neurodynamics

The neurodynamical models discussed in previous sections are similar to closed thermodynamical systems in a sense that their entropy can only increase until it reaches its maximum at the stochastic attractor. In this section we will introduce open systems which are maintained in their specific states by an influx of energy, so that information stored in the dynamical system, can change, and in particular, can increase.

Let us assume that the dynamical system (7.119) is driven by a vanishingly small input $\varepsilon(t)$:

$$\dot{x} = \gamma \sin^{1/3} \frac{\sqrt{\omega}}{\alpha} x \sin \omega t + \varepsilon(t), \quad |\varepsilon(t)| \ll \gamma \tag{9.62}$$

This input can be ignored when $\dot{x} \neq 0$, or when $\dot{x} = 0$, but the system is stable, i.e. $x = \pi\alpha/\sqrt{\omega}, 3\pi\alpha/\sqrt{\omega}, \ldots$, etc. However, it becomes significant during the instants of instability when $\dot{x} = 0$ at $x = 0, 2\pi\alpha/\sqrt{\omega}, \ldots$, etc.

The function $\varepsilon(t) << \gamma$ can be associated with a microsystem which controls neuron behavior through a string of signs (Zak 1990a,c).

Indeed, actually the only important part in this input is the sign of $\varepsilon(t)$ at the critical points. Consider, for example, (9.62), and suppose that

$$\text{sgn}\ \varepsilon(t_k) = +,+,-,+,-,-,\ \text{etc. at}\ t_k = \frac{\pi k}{\omega}, \ldots\ k = 0,1,2,\ldots, \text{etc.} \quad (9.63)$$

The values of $\varepsilon(t)$ in between the critical points are not important since, by our assumption, they are small in comparison to values of the derivative $\dot{x}$, and therefore, can be ignored. Hence, the only part of the input $\varepsilon(t)$ which is significant in determining the motion of the neuron (9.62) is the sign string (9.63): specification of this string fully determines the dynamics of (9.62).

Figure 7.8 demonstrates three different scenarios of motions for the different strings:

$$\varepsilon = \varepsilon_0 \sin \Omega t,\ \Omega/\omega = \sqrt{0.2}, \sqrt{2} \text{ and } \sqrt{20}, \text{ respectively}$$

It should be emphasized that, although these three solutions are bounded and aperiodic, they are fully deterministic in a sense that each of them is uniquely defined by the corresponding initial conditions. Suppose that

$$\varepsilon(t) = -\varepsilon_0^2 x,\ \varepsilon_0^2 \to 0 \quad (9.64)$$

i.e.,

$$\dot{x} = \gamma \sin^{1/3} \frac{\sqrt{\omega}}{\alpha} x \sin \omega t - \varepsilon_0^2 x, \quad \varepsilon_0 \to 0 \quad (9.65)$$

It can be verified that the solution to (9.65) will oscillate about the point $x = 0$. Indeed, when the point $x = 0$ becomes a terminal repeller, i.e. when $\sin \omega t > 0$, the solution escapes to the neighboring (right or left) equilibrium point. However, $\dot{x} < 0$ at $x_1 = \pi\alpha/\sqrt{\omega} > 0$, and $\dot{x} > 0$ at $x_1 = -\pi\alpha/\sqrt{\omega}$. Therefore, in both cases the solution returns to the original point $x = 0$. The amplitude and the period of the oscillations about $x = 0$ can be found from (7.120) and (7.123), respectively. However, in contrast to a classical version of (9.65),

$$\dot{x} = -\varepsilon_0^2 x \quad (9.66)$$

where $x = 0$ is a static attractor, the same point $x = 0$ is not a static, and not even a periodic, but a stochastic attractor. Indeed, there are several equally probable patterns of oscillations:

$$0,1,0,-1,0; \quad 0,-1,0,1,0; \quad 0,1,0,1,0,; \quad 0,-1,0,-1,0,\ldots \quad (9.67)$$

which can follow each other in an arbitrary order. In probabilistic terms the oscillations can be characterized as: $x = 0$ at $t = \frac{2\pi n}{\omega}$, while

$$\Pr\left\{x\left[t = \frac{(2n+1)\pi}{\omega}\right] = 1\right\} = 0.5$$

$$\Pr\left\{x\left[t=\frac{(2n+1)\pi}{\omega}\right]=-1\right\}=0.5,\ n=-2,-1,0,1,2,...,\text{etc.} \qquad (9.68)$$

so the probability of any combinations of patterns (9.67) can be found from (9.68).

It should be emphasized that the random stationary process (9.68) can be considered a stochastic attractor which is approached by the solution to (9.65) regardless of initial conditions. But in contradistinction to the stochastic attractors in closed system considered in the previous sections, where initial disorder could only increase, here the entropy of the initial distribution of x can be higher than those of the attractor (9.68), i.e., the dynamical system (9.65) may increase the initial information. This fundamental difference between closed and open neurodynamical systems is caused by the fact that evolution of closed systems is driven by pure diffusion [see (7.128)], while evolution of open system is driven by both diffusion and convection. Indeed, the evolution of the probability density of the solution to (9.65) (for $\sqrt{\omega}\to\infty$) is governed by the Fokker-Planck equation with the drift term:

$$\frac{\partial f}{\partial t}=-\alpha\sqrt{\omega}(q-p)\frac{\partial f}{\partial x}+\frac{1}{2}\pi\alpha^2\frac{\partial^2 f}{\partial x^2} \qquad (9.69)$$

where p and q are the probabilities that the process is directed to the right, or to the left, respectively, at each critical point.

Obviously, the last (diffusion) term survives only if

$$(p-1)\sim\frac{1}{\sqrt{\omega}}\to 0 \qquad (9.70)$$

As follows from (9.65),

$$p-q=\begin{cases}\text{sgn } x \text{ if} & x\neq 0\\ 0 \text{ if} & x=0\end{cases},\ p+q=1 \qquad (9.71)$$

and therefore, (9.69) can be rewritten as:

$$\frac{\partial f}{\partial t}=\alpha\sqrt{\omega}\frac{\partial f}{\partial x}\ \text{sgn } x\ +\frac{1}{2}\pi\alpha^2\frac{\partial^2 f}{\partial x^2}\delta[1-(p-q)] \qquad (9.72)$$

where δ is the Dirac function.

Let us assume that the initial density of x is uniform:

$$f_0(x)=\begin{cases}c=1/2 \text{ if} & |x|<\ell\\ 0 & \text{otherwise}\end{cases} \qquad (9.73)$$

Then, the solution to (9.72) is:

$$f(x)=\begin{cases}f_0[x-(\alpha\sqrt{\omega}\ \text{sgn } x)t]+\frac{\alpha\sqrt{\omega}t}{\ell}\delta(0) & \text{if } 0<t\leq\frac{\ell}{\alpha\sqrt{\omega}}\\ \delta(0) & \text{if } t>\frac{\ell}{\alpha\sqrt{\omega}}\end{cases} \qquad (9.74)$$

The Dirac function $\delta(0)$ in the solution (9.74) provides the normalization condition:

$$\int_{-\infty}^{\infty} f(x)dx = 1 \tag{9.75}$$

The process of the probability density evolution can be interpreted as following: the initial uniform density (9.73) "moves" (with the velocity $\pm\alpha\sqrt{\omega}$) toward the point $x = 0$, being absorbed there. At the same time, the Dirac function grows out of the point $x = 0$ in such a way that it compensates the "loss of the area" enveloped by the density $f(x)$. Hence, eventually, the solution approaches the attractor $x = 0$, while the transition period is

$$T = \frac{\ell}{\alpha\sqrt{\omega}} \rightarrow 0 \text{ if } \sqrt{\omega} \rightarrow \infty \tag{9.76}$$

However, for

$$t = \frac{\ell}{\alpha\sqrt{\omega}}$$

the draft term disappears [see(9.71)], and diffusion takes over. For finite $\sqrt{\omega}$ it leads to the random oscillations described by (9.68).

It should be emphasized that stochastic attractors in open systems [see (9.68)] are different from those in closed systems considered in previous sections. Firstly, they may have an entropy which is smaller than the initial entropy. Secondly, the time of approaching these attractors is finite [see (9.76)]. In order to distinguish these two types of stochastic attractors, those in the open systems were called terminal chaotic attractors, or terminal chaos (Zak 1991a), and which we now call terminal chaos. The similarity between the random oscillations (9.68) and chaotic attractors in classical dynamics is in the fact that in both cases the phenomena are based upon combined effects of stability and instability. However, in terminal chaos (9.68) the mechanisms of stability and instability act sequentially. For an extended discussion, see Sect. 8.4.2–8.5.

9.5 Neurodynamical Model of Information Fusion

In this section we will apply open random neurodynamical systems for simulating information fusion.

Let us first consider two uncoupled closed systems of the type (7.197):

$$\dot{x}_1 = \gamma \sin^k [\frac{\sqrt{\omega}}{\alpha} \widetilde{erf}(\frac{x_1}{\sqrt{2}\sigma_1})] \sin \omega t \tag{9.77}$$

$$\dot{x}_2 = \gamma \sin^k [\frac{\sqrt{\omega}}{\alpha} \widetilde{erf}(\frac{x_2}{\sqrt{2}\sigma_2})] \sin \omega t \tag{9.78}$$

Solutions to these equations are random, and they approach stationary stochastic processes with the normal distributions, respectively:

$$f_1(x_1) = \frac{1}{\sigma_1 \sqrt{2\pi}} e^{-\frac{x_1^2}{2\sigma_1^2}}, \; f_2(x_2) = \frac{1}{\sigma_1 \sqrt{2\pi}} e^{-\frac{x_2^2}{2\sigma_2^2}} \tag{9.79}$$

Since $x_1$and $x_2$are independent, their joint density will be:

$$f_{12} = \frac{1}{2\pi\sigma_1\sigma_2} e^{-\frac{x_1^2}{2\sigma_1} - \frac{x_2^2}{2\sigma_2}} \tag{9.80}$$

Suppose now that (9.77–9.78) are coupled via the following microdynamics:

$$\varepsilon_1 = \varepsilon_o^2(x_2 - x_1), \; \varepsilon_2 = \varepsilon_o^2(x_1 - x_2), \; \varepsilon_0^2 \to 0 \tag{9.81}$$

i.e.,

$$\dot{x}_1 = \gamma \sin^k [\frac{\sqrt{\omega}}{\alpha} \widetilde{erf}(\frac{x_1}{\sqrt{2}\sigma_1})] \sin \omega t + \varepsilon_0^2(x_2 - x_1) \tag{9.82}$$

$$\dot{x}_2 = \gamma \sin^k [\frac{\sqrt{\omega}}{\alpha} \widetilde{erf}(\frac{x_2}{\sqrt{2}\sigma_2})] \sin \omega t + \varepsilon_0^2(x_1 - x_2) \tag{9.83}$$

Global behavior of this system is defined by the behavior of an associated dynamical system:

$$\dot{x}_1 = \varepsilon_0(x_2 - x_1), \; \dot{x}_2 = \varepsilon_0(x_1 - x_2) \tag{9.84}$$

The system (9.84) has the following set of equilibrium states:

$$x_1 = x_2 \tag{9.85}$$

All of them are stable since the roots of the corresponding characteristic equation are not positive:

$$\lambda_1 = 0, \; \lambda_2 = -2 \tag{9.86}$$

When the solution to (9.82–9.83) approaches the attractor (9.85), the dynamical system (9.82–9.83) formally reduces to (9.77–9.78) with the joint density (9.80). However, now in (9.80) one must set:

$$x_1 = x_2 = x \tag{9.87}$$

Substituting (9.87) into (9.80), and redefining the constant from the normalization condition, one obtains the following probabilistic property of the solution to (9.82–9.83) at $t \to \infty$:

$$f(x = x_1 = x_2) = \frac{1}{\sigma\sqrt{2\pi}} e^{-\frac{x^2}{2\sigma^2}} \tag{9.88}$$

in which

$$\sigma^2 = \frac{1}{\frac{1}{\sigma_1^2} + \frac{1}{\sigma_2^2}} < \sigma_1^2, \sigma_2^2 \tag{9.89}$$

Hence, the neurodynamical system implements fusion of information coming from two independent sources about the same object. As a result of the fusion, the combined information is characterized by an entropy which is smaller than the entropies of each original components of information, i.e., the information about this object is improved.

This paradigm can be generalized to fusion of n independent sources of information:

$$\dot{x}_1 = \gamma \sin^k [\frac{\sqrt{\omega}}{\alpha} \widetilde{erf}(\frac{x_1}{\sqrt{2}\sigma_1})] \sin \omega t + \varepsilon_0^2 (x_2 - x_1)$$

$$\dot{x}_2 = \gamma \sin^k [\frac{\sqrt{\omega}}{\alpha} \widetilde{erf}(\frac{x_2}{\sqrt{2}\sigma_2})] \sin \omega t + \varepsilon_0^2 (x_3 - x_2)$$

$$\dot{x}_n = \gamma \sin^k [\frac{\sqrt{\omega}}{\alpha} \widetilde{erf}(\frac{x_1}{\sqrt{2}\sigma_1})] \sin \omega t + \varepsilon_0^2 (x_1 - x_n) \tag{9.90}$$

The solution to this system approaches a stationary stochastic process with the probability density (9.88), where

$$\sigma^2 = \frac{1}{\sum_{i=1}^{n} \frac{1}{\sigma_1^2}} \tag{9.91}$$

In the previous examples, the normal distributions were chosen only for the sake of analytical simplicity: any other distributions can be utilized with the same effect of improvement of the combined information.

The next paradigm of information fusion is associated with pattern recognition. In this section, pattern recognition will be considered as a multistep process. In the first step, the pattern is received at a global level when it can be simulated by a multidimensional stochastic attractor. This attractor represents a class to which the pattern belongs. In the second step, when some additional information becomes available, the original stochastic attractor is replaced by a lower dimension stochastic attractor which represent a subclass to which the pattern belongs, etc. A chain of such attractors of lower and lower dimensionalities which identifies the pattern with higher and higher accuracy, can be implemented by terminal neurodynamics as following.

Consider a dynamical system (9.11–9.12) which has a stochastic attractor (9.13), and assume that, as an additional information, the regression of x_2 on x_1 is given in the form:

$$x_2 = \frac{1}{2}x_1 \tag{9.92}$$

Then, modifying (9.11–9.12) as following:

$$\dot{x}_1 = \gamma_1 \sin^k[\sqrt{\omega}\sin(x_1 + x_2)] \sin \omega t + \varepsilon_0^2(x_2 - \frac{1}{2}x_1) \tag{9.93}$$

$$\dot{x}_2 = \gamma_2 \sin^k[\sqrt{\omega}\sin(x_1 + x_2)] \sin \omega t + \varepsilon_0^2(\frac{1}{2}x_1 - x_2) \tag{9.94}$$

and applying the same line of argumentation as those for (9.82–9.83), one concludes that the solution to (9.93–9.94) approaches an one-dimensional attractor which is obtained from (9.13) as a result of replacing x_2 by its expression from (9.92):

$$f(x_1) = \begin{cases} c\left|\cos\frac{3x_1}{2}\cos\frac{x_1}{2}\right|, & \frac{\pi}{3} \le x_1 \le \pi \\ 0 \text{ otherwise} \end{cases} \tag{9.95}$$

and the constant c is found from the normalization condition:

$$c = \frac{1}{\int_{\pi/3}^{\pi} |\cos\frac{3x_1}{2}\cos\frac{x_1}{2}|\, dx_i} = \frac{3\sqrt{3}}{8} \tag{9.96}$$

One can verify that the entropy of the attractor (9.95) is smaller than the entropy of the original attractor (9.13).

In terms of pattern recognition, the original attractor (9.13) can be identified with a class of patterns in which each pattern is characterized by the parameters x_1 and x_2. Any combination of these parameters has a certain probability to appear in a particular pattern. An additional information about dependence between x_1 and x_2 for the pattern to be recognized extracts a subclass of pattern (9.95) to which this pattern belongs.

In order to illustrate a chain of stochastic attractors, start with the dynamical system (7.205) for $i = 1, 2, 3$ and introduce the following two-cascade microdynamics:

$$\dot{x}_1 = \gamma \sin^k[\frac{\sqrt{\omega}}{\alpha}\widetilde{erf}(\frac{y_1}{\sqrt{2}})] \sin \omega t + \varepsilon_0^2(x_2 + x_3 - x_1) \tag{9.97}$$

$$\dot{x}_2 = \gamma \sin^k[\frac{\sqrt{\omega}}{\alpha}\widetilde{erf}(\frac{y_2}{\sqrt{2}})] \sin \omega t + \varepsilon_0^2(x_1 - x_2 - x_3) + \varepsilon_0^4(x_3 - x_2) \tag{9.98}$$

$$\dot{x}_3 = \gamma \sin^k[\frac{\sqrt{\omega}}{\alpha}\widetilde{erf}(\frac{y_3}{\sqrt{2}})] \sin \omega t + \varepsilon_0^2(x_1 - x_2 - x_3) + \varepsilon_0^4(x_2 - x_3) \tag{9.99}$$

If $\varepsilon_0 = 0$, then the solution to this system converges to a three-dimensional stochastic attractor with the joint density given by (7.194) at $n = 3$.

When $\varepsilon_0 \neq 0$ (but $\varepsilon_0 \to 0$), the solution to this system first converges to a two-dimensional attractor on the plane

$$x_1 = x_2 + x_3 \tag{9.100}$$

with the joint density given by (7.194) at $n = 3$ after substitution of (9.100) instead of x_1 and redefining the constant from the normalization condition. Then the solution converges to a one dimensional attractor which dwells on the line

$$x_2 = x_3 \tag{9.101}$$

of the plane (9.100).

9.6 Conclusion

This chapter has presented physical models for simulating some aspects of neural intelligence, and in particular, the process of cognition. The main departure from classical approach here is in utilization of terminal version of classical dynamics introduced in (Zak 1992a,b; 1993a,b). Based upon violations of the Lipschitz condition at equilibrium points, terminal dynamics attains two new fundamental properties: it is irreversible and non-deterministic. Special attention was focused on terminal neurodynamics as a particular architecture of terminal dynamics which is suitable for modeling of information flows. Terminal neurodynamics possesses a well-organized probabilistic structure which can be analytically predicted, prescribed, and controlled, and therefore, which presents a powerful tool for modelling real-life uncertainties. Two basic phenomena associated with random behavior of neurodynamical solutions are exploited. The first one is a stochastic attractor – a stable stationary stochastic process to which random solutions of closed system converge. As a model of cognition, a stochastic attractor can be viewed as a universal tool for generalization and formation of classes of patterns. The concept of stochastic attractor was applied to model a collective brain paradigm explaining coordination between simple units of intelligence which perform a collective task without direct exchange of information. The second fundamental phenomenon discussed was terminal chaos which occurs in open systems. Applications of terminal chaos to information fusion as well as to explanation and modeling of coordination among neurons in biological systems were discussed. It should be emphasized that all the models of terminal neurodynamics are implementable in analog devices, which means that all the cognition processes discussed are reducible to the laws of Newtonian mechanics.

One of the immediate practical application of terminal neurodynamics is the model of the collective brain which mimics collective purposeful activities of a set of units of intelligence without global control. Actually a global control is replaced by the probabilistic correlations between the units. These correlations are learned during a long term period of performing collective tasks, and they are stored in the joint density function. Due to such correlations, each unit can predict the values of parameters characterizing the activities of its neighbors without direct exchange of information. The model of the collective brain describes more sophisticated relationships in dynamics which are typical for biological and social rather than physical systems. In particular, this model can be effective in distributed control and decision analysis. With respect to the last application, the model of collective brain offers a compromise between two polar interpretations of the concept of probability as to whether it is a state of mind or a state of things. Indeed, in the course of training, the units of intelligence in the collective brain learn the objective relationships between them which are represented by physical probability distribution; this distribution is stored in the synaptic interconnections and becomes a property of each unit thereby representing personal or subjective probability.

10

Terminal Dynamics Approach to Discrete Event Systems

I consider it quite possible that physics cannot be based on the field concept, i.e., on continuous structures. –*A. Einstein*

10.1 Introduction

Complexity of dynamical system performance can be significantly enriched by exploiting a terminal model of nonlinear dynamics. As shown in Chap. 7, this model can capture stochastic properties of information processing without utilizing random number generators: a multi-choice response to a deterministic massage is provided by a failure of uniqueness of the solution due to relaxation of Lipschitz conditions at some discrete points. However, since terminal dynamics is differentiable almost everywhere (excluding these discrete points), it preserves most of the analytical structure of mathematical formalism of classical theory of differential equations. A combination of such "contradictory" properties – the analicity and discreetness – gives motivation to simulate discrete event systems using terminal dynamics.

Discrete event dynamics represents a special type of "man-made" system to serve specific purposes of information processing (Cassandras 1989). Models of such systems fall into several broad categories of various levels of abstraction. At the logical level, one is only concerned with the logical order of the events, i.e., with qualitative, or structural behavior of a system; these models are usually deterministic and untimed. At the performance level, the models involve random event lifetimes and other probabilistic entities, and are primary concerned with characterizing the quantitative aspects of the system. The performance models are usually based upon generalized semi-Markov processes and stochastic Petri nets.

This chapter presents and discusses a model for discrete events systems based upon a terminal model of Newtonian dynamics previously introduced in Chap. 7.

10.2 Time-Driven Discrete Systems

A broad class of complex dynamical behaviors can be derived from a simple differential equation as discussed in Chap. 7 :

$$\dot{x} = x^{1/3}\cos\omega t, \quad \omega = \text{ Const} \tag{10.1}$$

Clearly, the Lipschitz condition at the equilibrium point $x = 0$ fails since

$$\left| \frac{d\dot{x}}{dx} \right| = \frac{1}{3}x^{-\frac{2}{3}}\cos\omega t \to \infty \text{ at } x \to 0 \tag{10.2}$$

Nevertheless, the solution to (10.1) can be presented in a closed form. Indeed, assuming that $x \to 0$ at $t = 0$, one obtains a regular solution:

$$x = \pm\left(\frac{2}{3\omega}\sin\,\omega t\right)^{3/2} \text{ if } x \neq 0 \tag{10.3}$$

and a singular solution (an equilibrium point):

$$x = 0 \tag{10.4}$$

As follows from (10.3), two different solutions are possible for "almost the same" initial conditions. The fundamental property of this result is that the divergence of these solutions from $x = 0$ is characterized by an unbounded parameter, σ:

$$\sigma = \lim_{t\to 0}\left[\frac{1}{t}\ell n\frac{\left(\frac{2}{3\omega}\sin\,\omega t\right)^{3/2}}{2\mid x_0\mid}\right] = \infty, \quad \mid x_0 \mid\to 0 \tag{10.5}$$

where t_0 is an arbitrarily small (but finite) positive quantity. The rate of divergence (10.5) can be defined in an arbitrarily small time interval, because the initial infinitesimal distance between the solutions (10.3) becomes finite during the small interval t_0. One should recall that in the classical case when the Lipschitz condition is satisfied, the distance between two diverging solutions can become finite only at $t \to \infty$ if initially this distance was infinitesimal.

Analysis of the solutions (10.3–10.4) shows that during the first time period

$$0 < t < \frac{\pi}{2\omega} \tag{10.6}$$

equilibrium point (10.4) is a terminal repeller. Therefore, within this period, solutions (10.3) have the property that their divergence is characterized by an unbounded rate σ.

During the next time period

$$\frac{\pi}{2\omega} < t < \frac{3\pi}{2\omega} \tag{10.7}$$

equilibrium point (10.3) becomes a terminal attractor, and the system which approaches this attractor at $t = \pi\omega$ remains motionless until $t > 3\pi/2\omega$. After this point, the terminal attractor converts into a terminal repeller, and the system escapes again.

It is important to note that each time the system escapes the terminal repeller, the solution splits into two symmetrical branches; therefore, the total trajectory can be combined from 2^n pieces, where n is the number of cycles; that is, it is the integer part of the quantity $(t/2\pi\omega)$. The nature of this unpredictability is significantly different from the unpredictability in chaotic systems.

The solutions (10.3–10.4) now can be combined in the following form:

$$x = \begin{cases} \pm\left(\frac{2}{3\omega}\sin\omega t\right)^{3/2} \text{ if } & \frac{2\pi n}{\omega} \le t \le \frac{\pi(2n+1)}{\omega} \\ 0 \text{ if } & \frac{\pi(2n+1)}{\omega} \le t \le \frac{2\pi(2n+1)}{\omega} \end{cases}, \quad n = 1, 2, ..., \text{etc.} \tag{10.8}$$

Let us introduce another variable:

$$\dot{y} = x, \quad (y = 0 \text{ at } x = 0) \tag{10.9}$$

After the first time interval $t = \frac{\pi}{\omega}$

$$y = \pm\int_0^{\pi/\omega} \left(\frac{2}{3\omega} \sin \omega t\right)^{3/2} dt = \left(\frac{2}{3\omega}\right)^{3/2} \frac{\Gamma(\frac{1}{2})}{2\Gamma(\frac{7}{4})} = \pm h \tag{10.10}$$

where Γ is the gamma-function.

During the next time interval $\frac{\pi}{\omega} \le t \le \frac{2\pi}{\omega}$

$$y = \pm h = \text{ Const} \tag{10.11}$$

After the third time interval $t = \frac{3\pi}{\omega}$

$$y = \pm h \ \pm h \tag{10.12}$$

etc.

Obviously, the variable y performs an unrestricted symmetric random walk: after each time period $\tau = 2\pi/\omega$ it changes its value on $\pm h$. The probability $f(y, t)$ is governed by the following difference equation:

$$f(y, t + \frac{2\pi}{\omega}) = \frac{1}{2}f(y - h, t) + \frac{1}{2}f(y + h, t), \int_{-\infty}^{\infty} f(y, t)dy = 0 \quad (10.13)$$

where h is expressed by (10.10).

Equation (10.13) defines f as a function of two discrete arguments:

$$y = \pm kh, \quad \text{and} \quad t = l\tau, \quad \tau = \frac{2\pi}{\omega}, \quad k, l = \ 0, 1, 2, ..., \text{etc.} \quad (10.14)$$

For convenience, we will keep for discrete variable y and t the same notations as for their continuous versions.

By change of variables:

$$z = \varphi(y), y = \varphi^{-1}(z) \quad (10.15)$$

one can obtain a stochastic process with a prescribed probability distribution:

$$\psi(z, t) = f[\varphi^{-1}(z), t] \mid \frac{d\varphi^{-1}}{dz} \mid \quad (10.16)$$

implemented by the dynamical system (10.1), (10.9), and (10.15).

Here z is also considered as a discrete variable changing at each time-step τ according to (10.14–10.15).

It is important to emphasize that the system (10.1), (10.9), (10.15) does not have any random input: the randomness is "generated" by the differential operator in (10.1) due to violation of the Lipschitz condition.

Along with the function $y(t)$ defined by (10.9), consider a time-delay function:

$$y_1(t) = y(t - \frac{\pi}{\omega}) \quad (10.17)$$

As follows from (10.8–10.9)

$$y_1(t) = \ \text{Const if} \ \frac{2\pi n}{\omega} \leq t \leq \frac{\pi(2n + 1)}{\omega} \quad (10.18)$$

and therefore,

$$y_1(t)x(t) = \ \text{Const} \ x(t) \ \text{if} \ \frac{2\pi n}{\omega} \leq t \leq \frac{2\pi(n + 1)}{\omega} \quad (10.19)$$

Obviously, an arbitrary function of $\varphi(y_1)$ possesses a similar property:

$$\varphi(y_1)x = \ \text{Const} \ x \ \text{if} \ \frac{2\pi n}{\omega} \leq t \leq \frac{2\pi(n + 1)}{\omega} \quad (10.20)$$

Based upon (10.20), let us modify (10.1) as following:

$$\dot{u} = \varphi(y_1)u^{1/3}\cos\omega t, \ \varphi(y_1) = z_1 > 0 \tag{10.21}$$

where y_1 is defined by (10.1), (10.9) and (10.17).

Taking into account the property (10.20), the solution to (10.20) can be written in the form similar to (10.8):

$$u = \begin{cases} \pm\left(\frac{2(h)}{3\omega}\sin\omega t\right)^{3/2} \text{ if } & \frac{2\pi n}{\omega} \leq t \leq \frac{\pi(2n+1)}{\omega} \\ 0 \text{ if } & \frac{\pi(2n+1)}{\omega} \leq t \leq \frac{2\pi(2n+1)}{\omega} \end{cases} \tag{10.22}$$

where h is given by (10.10).

One should recall that y_1, and therefore, $\varphi(y_1)$ are random variables. Indeed as follows from (10.17), the probability density $f(y_1, t)$ is described by (10.13). Therefore, the probability density of $\varphi(y_1, t)$ is expressed by (10.16):

$$\psi(z_1, t) = f[\varphi^{-1}(z_1), t] \mid \frac{d\varphi^{-1}}{dz_1} \mid, \ z_1 = \varphi(y_1) \tag{10.23}$$

Here y_1 and t are the discrete variables obtained in the same way as in (10.14). This means that $\varphi(y_1), f(y_1, t_1)$ and $\psi(y_1, t)$ are functions of discrete arguments.

Now we are ready to introduce a dynamical model for the generalized random walk:

$$\dot{v} = \alpha u, \quad \alpha = \frac{2\Gamma(7/4)\omega}{\Gamma(1/2)\sqrt{\varphi(h)}}(\frac{3\omega}{2})^{3/2} = \text{ Const} \tag{10.24}$$

Indeed, after the first time interval $t = \frac{\pi}{\omega}$

$$v = \pm\alpha\int_0^{\pi/\omega}[\frac{2\varphi(h)}{3\omega}\sin\,\omega t]^{3/2}dt = \pm\varphi(\tilde{h}), \ \tilde{h} = \omega h \tag{10.25}$$

During the next time interval $\frac{\pi}{\omega} \leq t \leq \frac{2\pi}{\omega}$

$$v = \pm\varphi(\tilde{h}) = \text{ Const} \tag{10.26}$$

After the third time interval $t = \frac{3\pi}{\omega}$

$$v = \pm\varphi(\tilde{h}) \pm \varphi(\tilde{h}), \text{ etc.} \tag{10.27}$$

Thus, the variable v performs a symmetric generalized random walk: after each time period $\tau = 2\pi/\omega$, it changes its value on $\pm\varphi(\tilde{h})$. But $\varphi(\tilde{h})$, in turn, is also a random variable: its probability density follows from (10.25) and (10.23). Hence, at each step the variable v have a probability p_k to move from any point v_0 to an arbitrarily selected point v_k.

It should be emphasized that this generalized random walk is implemented by the dynamical system (10.1), (10.9), (10.17), (10.21), (10.1) and (10.24). Considering v as a discrete variable at $t = n\tau$

$$v = \pm n\varphi(\tilde{h}) \tag{10.28}$$

one obtains the governing difference equation for the probability distribution of v:

$$\Phi(v, t + \frac{2\pi}{\omega}) = \frac{1}{2}\sum_{k=1}^{n} p_k \; \{\Phi[v - \varphi(k\tilde{h})] + \Phi[v + \varphi(k\tilde{h})]\} \tag{10.29}$$

$$\int_{-\infty}^{\infty} \Phi(v, t)dv = 1 \tag{10.30}$$

where

$$p_k = f[\varphi^{-1}(k\tilde{h}, t)] \; | \frac{d\varphi^{-1}(k\tilde{h}, t)}{dz} | \tag{10.31}$$

and n is the number of the discrete values in (10.14).

Applying the terminology of discrete event systems, both simple and generalized dynamical models of random walks can be characterized as time-driven systems, since here with every "clock tick" the state is expected to change with a prescribed probability.

We will stress that although the dynamical system (10.1), (10.9), (10.17), (10.21), (10.23–10.24) has a random solution, this randomness is well-organized: it can be uniquely described in probability terms with the aid of (10.29). Indeed, if the initial probability distribution

$$\Phi(v, 0) = \Phi_0(v) \tag{10.32}$$

$$f(y, 0) = f_0(y) \tag{10.33}$$

$$y = \varphi(z) \tag{10.34}$$

is given, then all the variables in the right-hand part of (10.29) can be computed at each time step τ, and this uniquely defines the left-hand part of (10.29), and therefore, the evolution of the probability distribution $\Phi(v, t)$.

Based upon the system (10.1), (10.9), (10.17), (10.21), (10.23–10.24), one can find functions $\Phi_0(v)$, $f_0(y)$ and $\varphi(z)$ in (10.32), (10.33) and (10.34), respectively, such that they provide a required evolution of the probability distribution $\Phi(v, t)$.

So far we were concerned with the performance aspects of the time-driven dynamics. On the logical level, the only function $\varphi(z)$ in (10.32) contributes to the logical structure of the system: it prescribes the probabilities p_k in (10.29) that the solution move from any point v_0 to an arbitrarily selected point v_k.

For a physical interpretation of the solution to (10.29), let us consider the simplest case:

$$k = 1, \ \omega \rightarrow \infty, \quad z = y \tag{10.35}$$

Then, as follows from (10.10), (10.14) and (10.25):

$$\tilde{h} \rightarrow 0, \quad \tau \rightarrow 0 \ \text{ and } \ \frac{\tilde{h}^2}{\tau} = \frac{2}{27\pi}[\frac{\Gamma(\frac{1}{2})}{\Gamma(7/4)}]^2 = D \tag{10.36}$$

It is easily verifiable that (10.29) reduces to the Fokker-Planck equation:

$$\frac{\partial \Phi(v,t)}{\partial t} = \frac{1}{2} D \frac{\partial^2 \Phi(v,t)}{\partial v^2} \tag{10.37}$$

and therefore, its solution is:

$$\Phi(v,t) = \frac{1}{\sqrt{2\pi D^2 t}} \exp\,(-\frac{v^2}{2D^2 t}) \tag{10.38}$$

It is easy to verify that (10.37) satisfies the constraint (10.30).

10.3 Event-Driven Discrete Systems

Let us return to (10.1) and assume that it is driven by a vanishingly small input ε:

$$\dot{x} = x^{1/3} \cos \omega t + \varepsilon, \quad \varepsilon \rightarrow 0 \tag{10.39}$$

From the viewpoint of information processing, this input can be considered as a massage, or an event. This massage can be ignored when $\dot{x} \neq 0$, or when $\dot{x} = 0$, but the system is stable, i.e., $x = \pi\omega, 2\pi\omega, ...$, etc. However, it becomes significant during the instants of instability when $\dot{x} = 0$ at $t = 0, \pi/2\omega, ...$, etc. Indeed, at these instants, the solution to (10.39) would have a choice to be positive or negative if $\varepsilon = 0$, [see (10.3)]. However, with $\varepsilon \neq 0$

$$\text{sgn } x = \ \text{sgn } \varepsilon \quad \text{at } t = 0, \pi/2\omega, ..., \text{etc.} \tag{10.40}$$

i.e. the sign of ε at the critical instances of time (10.40) uniquely defines the evolution of the dynamical system (10.39). Actually the event ε may represent an output of a microsystem which uniquely controls the behavior of the original dynamical system (10.39).

The solution to (10.9) now becomes deterministic if $\text{sgn } \varepsilon \neq 0$ at critical points (10.40), and instead of (10.10–10.12), one obtains, respectively:

$$y = \text{ sgn } \varepsilon \int_0^{\pi/2} (\frac{2}{3\omega} \sin \omega t)^{3/2} dt = h \text{ sgn } \varepsilon \tag{10.41}$$

$$y = h \text{ sgn } \varepsilon = \text{ Const} \tag{10.42}$$

$$y = h \text{ sgn } \varepsilon_1 + h \text{ sgn } \varepsilon_2 + \text{ etc.} \tag{10.43}$$

where $\varepsilon_1, \varepsilon_2, ...$ are the values of ε at $t = 0,\ \pi/2\omega$ etc., respectively.

The probability $f(y, t)$, instead of (10.13), is governed by the following difference equation:

$$f(y, t + \frac{2\pi}{\omega}) = pf(y - h, t) + (1 - p)f(y + h, t) \tag{10.44}$$

where

$$p = \begin{cases} 1 \text{ if} & \text{sgn } \varepsilon = 1 \\ 0 \text{ if} & \text{sgn } \varepsilon = -1 \\ \frac{1}{2} \text{ if} & \varepsilon = 0 \end{cases} \tag{10.45}$$

Actually the evolution of the probability distribution in (10.44) is represented by rigid shifts of the initial probability distribution $f(y, 0)$, unless sgn $\varepsilon = 0$.

Let us modify now (10.21) in the same way as (10.1):

$$\dot{u} = \varphi(y_1)u^{1/3} \cos \omega t + \varepsilon, \quad \varepsilon \to 0 \tag{10.46}$$

assuming that y_1 is still defined by (10.1), (10.9) and (10.17). Then the solution to (10.24), instead of (10.25–10.27), is written as:

$$\upsilon = \varphi(\tilde{h}) \text{ sgn } \varepsilon \tag{10.47}$$

$$\upsilon = \varphi(\tilde{h}) \text{ sgn } \varepsilon = \text{ Const} \tag{10.48}$$

$$\upsilon = \varphi(\tilde{h}) \text{ sgn } \varepsilon_1 + \varphi(\tilde{h}) \text{ sgn } \varepsilon_2 + \cdots \text{ etc.} \tag{10.49}$$

where $\varepsilon_1, \varepsilon_2, ...$ are the values of ε at $t = 0, \pi/2\omega$, etc., respectively.

Unlike the previous case [see (10.25–10.27)], now the variable υ defined by (10.1), (10.9), (10.17), (10.24), and (10.46) perform a non-symmetric generalized random walk: after each time period $\tau = 2\pi/\omega$, it changes its value on $\varphi(\tilde{h})$ sgn ε, i.e., in a certain direction defined by the event ε, (unless sgn $\varepsilon = 0$). But $\varphi(\tilde{h})$ is still a random variable whose probability density follows from (10.25) and (10.23). Hence, at each step the variable υ has a probability to move from any point υ_0 to an arbitrary selected point υ_k in the direction defined by sgn ε.

Introducing v by (10.18), one obtains the governing difference equations for the probability distribution of v:

$$\Phi(v, t + \frac{2\pi}{\omega}) = \sum_{k=1}^{n} p_k \ \{p\Phi[v - \varphi(k\tilde{h})] + (1-p)\Phi[v + \varphi(k\tilde{h})]\} \quad (10.50)$$

$$\int_{-\infty}^{\infty} \varphi(v, t) dv = 1 \quad (10.51)$$

where p_k and p are expressed by (10.31) and (10.45), respectively.

Thus, although the dynamical system (10.1), (10.9), (10.17), (10.24) and (10.46) has a random solution, its probabilistic properties are uniquely defined by (10.50) and the constraint (10.51).

It should be noted that the solution to (10.50) or (10.29) does always automatically satisfy the constraint (10.51) [as it was in the case of solution (10.38)]. This is why, in general, it is more convenient to consider Φ in (10.50) or (10.29), as an auxiliary (not normalized) function while introducing the normalized probability distribution by the following formula:

$$\hat{\Phi}(v, t) = \frac{\Phi(v, t)}{\int_{-\infty}^{\infty} \Phi(v, t) dv} \quad (10.52)$$

Then

$$\int_{-\infty}^{\infty} \hat{\Phi}(v, t) dv \equiv 1 \quad (10.53)$$

Comparing the previously considered dynamical system (10.1), (10.9), (10.17), (10.21), (10.23–10.24) with the dynamical system (10.1), (10.9), (10.17), (10.24) and (10.46), one can see a fundamental difference: the latter is driven by a massage ε whose changes are independent upon the "clock tick" of the dynamical system itself. This massage, in general, can be an output of another dynamical system with its own time scale, or it can depend upon the variables (or their probabilities) of the original dynamical system. In all these cases, the massage ε can be treated as an independent event, and therefore, the dynamical system (10.1), (10.9), (10.17), (10.24) and (10.46) is driven by events.

In conclusion, we will review the structure of the solution to (10.50) which describes the evolution of the probability distribution $\Phi(v, t)$.

Introducing the displacement operators

$$E_t \theta(t) = \theta(t + \frac{2\pi}{\omega}), \ E_v \theta(v) = \theta(v + \tilde{h}) \quad (10.54)$$

one can rewrite (10.50) in the following form:

$$\{E_t - \sum_{k=1}^{n} p_k\ [pE^{-k} + (1-p)E^k]\}\Phi(v,t) = 0 \tag{10.55}$$

Applying Bool's symbolic method, i.e. replacing the operator E_v by a constant λ, one arrives at an ordinary difference equation:

$$(E_t - \tilde{\lambda})\Phi = 0, \quad \tilde{\lambda} = \sum_{k=1}^{n} p_k\ [p\lambda^{-k} + (1-p)\lambda^k] = \text{ Const} \tag{10.56}$$

The solution to this equation in a symbolic form is:

$$\Phi(v,t) = \tilde{\lambda}^{\ell}\varphi(v) \tag{10.57}$$

where $\varphi(v) = \Phi(v,0)$ is the initial probability distribution of v, and $\ell = 0, 1, 2, \cdots$ etc. [see (10.14)].

Obviously, for $\ell = 0$:

$$\Phi(v,t) = \varphi(v) \tag{10.58}$$

Then, for $\ell = 1$:

$$\Phi(v,t_1) = \sum_{k=1}^{n} p_k\ [p\varphi(v - \frac{2\pi k}{\omega}) + (1-p)\varphi(v + \frac{2\pi k}{\omega})] \tag{10.59}$$

Continuing this process for $\ell = 2, 3, \cdots$ etc., one arrives at the following recurrent relationships:

$$\Phi(v,t_{\ell+1}) = \sum_{k=1}^{n} p_k\ [p\Phi(v - \frac{2\pi k}{\omega}, t_\ell) + (1-p)\Phi(v + \frac{2\pi k}{\omega}, t_\ell)] \tag{10.60}$$

Hence, based upon the initial condition (10.58), the evolution of the probability distribution $\Phi(v,t)$ can be uniquely defined by (10.60).

The solution to (10.29) can be obtained from (10.60) if one sets $p = 1/2$.

It should be emphasized that (10.50) is piecewise linear: indeed, it is linear as long as the massage ε does not change its sign. However, since this sign may depend upon time t, or upon the state variable v, or even upon its probability $f(v)$, globally (10.50) can be linear with variable coefficients, or even nonlinear.

10.4 Systems Driven by Temporal Events

In this section we will assume that the message ε is given as a function of time :

$$\varepsilon = \varepsilon(t) \tag{10.61}$$

We will start with a simplified version of the system (10.1), (10.9), (10.17), (10.24), (10.46), assuming that in (10.15)

$$y = \begin{cases} z \text{ if } & |z| \leq nh \\ 0 & \text{otherwise} \end{cases} \tag{10.62}$$

This means that at each state the variable v has equal probability to have steps $\pm\tilde{h}, \pm 2\tilde{h}, \cdots \pm n\tilde{h}$.

One can rewrite the system in the following form [see (10.24)]:

$$\dot{u} = y_1 u^{1/3} \cos \omega t + \varepsilon(t), \quad |\, \varepsilon(t) \,| \ll |\, u \,| \tag{10.63}$$

$$\dot{v} = \alpha u, \quad \alpha = \text{ Const} \tag{10.64}$$

here y_1 is a random variable whose probability distribution is governed by (10.13) while

$$y_1 = y(t - \frac{\pi}{\omega}) \tag{10.65}$$

As shown above, (10.13) can be implemented by the dynamical system (10.1), (10.9) which is not coupled with (10.63–10.64).

Now (10.50) describing the evolution of the probability distribution of the variable v reduces to:

$$\Phi(v, t + \frac{2\pi}{\omega}) = \frac{1}{n} \sum_{k=1}^{n} p\Phi(v - k\tilde{h}) + (1-p)\Phi(v + k\tilde{h}) \tag{10.66}$$

while p is given (10.50).

We will illustrate the solution to the dynamical system (10.63) as well as to the associated probability equation (10.66) by assuming that in (10.63):

$$\varepsilon(t) = \varepsilon_0 \sin \Omega t, \quad \varepsilon_0 = \text{ Const } \ll 1 \tag{10.67}$$

First consider the case when ω/Ω is irrational, i.e.,

$$\frac{\omega}{\Omega} \neq \frac{m}{n} \tag{10.68}$$

where m and n are integers.

Then at the critical points:

$$t = \frac{2\pi}{\omega}\ell, \ \ell = 1, 2 \cdots \tag{10.69}$$

$$\varepsilon(t) \neq 0 \tag{10.70}$$

and therefore, in (10.62):

$$\tilde{\lambda} = \begin{cases} \frac{1}{n}\sum_{k=1}^{n} \lambda^k \text{ if } & 1 - \frac{1}{2\ell} \leq \frac{\Omega}{\omega} \leq 1 + \frac{1}{2\ell} \\ \frac{1}{n}\sum_{k=1}^{n} \lambda^{-k} \text{ if } & 1 \leq \frac{\Omega}{\omega} \leq 1 + \frac{1}{2\ell} \end{cases} \tag{10.71}$$

For $n = 1$ the solution to (10.66) can be presented in the form of a propagating wave:

$$\Phi = \lambda^t \varphi(v) = E_v^t \varphi(v) = \varphi(v \pm t) \tag{10.72}$$

where the signs + and – correspond to (10.71).

Actually in this particular case the solution remains fully deterministic if the initial conditions to the dynamical system (10.63–10.64) are deterministic. In terms of the solution (10.72), this would mean that $\varphi(v)$ as well as $\varphi(v \pm t)$ are the δ functions.

Turning to the general case of (10.71), let us apply the solution (10.60). Then, for the first time-step:

$$\Phi(v, t_1) = \frac{1}{n}\sum_{k=1}^{n} \varphi(v - \frac{2\pi k}{\omega}), \ \varphi = \Phi(v, 0) \tag{10.73}$$

Hence, the solution starts with n waves propagating in the same direction, but with different speeds. Application of the solution (10.60) to the next time-steps shows that these waves start interacting and the structure of the solution becomes as complex as the linear wave interference.

For the case (10.71), instead of (10.73) one obtains

$$\Phi(v, t_1) = \frac{1}{n}\sum_{k=1}^{n} \varphi(v + \frac{2\pi k}{\omega}), \ \varphi = \Phi(v, 0) \tag{10.74}$$

i.e., a similar wave train propagates in opposite direction.

Let us replace the condition (10.68) by the following:

$$\frac{\omega}{\Omega} = \frac{m}{n} \tag{10.75}$$

Then at some of the critical points (10.69):

$$\operatorname{sgn} \varepsilon(t) = 0 \tag{10.76}$$

and therefore, all the three cases in (10.45) can occur.

Then, instead of (10.73–10.74), the following solution can be obtained for the first time-step:

$$\Phi(v, t_1) = \frac{1}{2n}\sum_{k=1}^{n} [\varphi(v - \frac{2\pi k}{\omega}), +\varphi(v + \frac{2\pi k}{\omega})], \ \varphi = \Phi(v, 0) \tag{10.77}$$

i.e., the solution starts with two trains of waves propagating in opposite directions.

In terms of the Fokker-Plank equation [which can be considered as a continuous approximation to (10.50) when $\tilde{h} \to 0, \tau \to 0$], the solution (10.73–10.74) can be associated with drift effects, while (10.77) includes diffusion effects.

In all cases when the message ε depends on time, (10.50) remains linear (but with variable coefficients).

10.5 Systems Driven by State-Dependent Events

In this section we will discuss more complex structures of the input message ε assuming that it depends upon the state variable v:

$$\varepsilon = \varepsilon(v) \tag{10.78}$$

We will start with the simplest case when

$$\varepsilon = -\gamma^2 v, \quad \gamma^2 \ll 1 \tag{10.79}$$

It can be verified by qualitative analysis of the system [compare with (10.63–10.64)]:

$$\dot{u} = y_1 u^{1/3} \cos \omega t - \gamma^2 v \tag{10.80}$$

$$\dot{v} = \alpha u \tag{10.81}$$

that its solution will randomly oscillate about the point $v = 0$.

Indeed, when $v > 0$, then $\operatorname{sgn} \varepsilon < 0$, and v decreases; when $v < 0$, then $\operatorname{sgn} \varepsilon > 0$, and v increases. But when $v = 0$, then $\operatorname{sgn} \varepsilon = 0$, and the solution can escape the point $v = 0$ with the same probability 1/2 to the right or to the left.

The same result can be obtained by a formal analysis of the solutions (10.73–10.74), and (10.77). Let us assume that the initial condition $v(t = 0)$ was set up randomly with the probability distribution:

$$\Phi = \begin{cases} a \text{ if } & |v| \leq \frac{1}{2a} \\ 0 & \text{otherwise} \end{cases} \tag{10.82}$$

in which $\tilde{h}$ is given by (10.10) and (10.25), and ℓ is given by (10.14).

Since the area enveloped by the function $\varphi(v, t)$ in (10.82) is shrinking, one has to turn to the normalized distribution [see (10.52)]:

$$\hat{\Phi}(v, t) = \begin{cases} \frac{a}{\frac{1}{2a} - \tilde{h}\ell} \text{ if } & |v| \leq \frac{1}{2a} - \tilde{h}\ell \\ 0 & \text{otherwise} \end{cases} \tag{10.83}$$

As follows from (10.83), in a finite time

$$T = \frac{\pi}{ah\omega^2} \tag{10.84}$$

the solution degenerates into a δ-function.

This means that the solution will arrive at the point $v = 0$ with the probability 1. However, at this point

$$\text{sgn } \varepsilon = 0 \tag{10.85}$$

and (10.72) must be replaced by (10.77) at $n = 1$.

During the next time step the solution will be:

$$\hat{\Phi} = \begin{cases} \frac{1}{2}h\tilde{\omega} \text{ if } & |v| \leq \frac{1}{2a} - h\tilde{\omega} \\ 0 & \text{otherwise} \end{cases} \quad \text{at } \overset{*}{t} = T + \frac{2\pi}{\omega} \tag{10.86}$$

This describes the onset of diffusion of the δ-function in both directions. However, for $t \geq \overset{*}{t}$, (10.77) must be replaced by (10.72) since now $\text{sgn } \varepsilon \neq 0$, and the solution approaches the attractor $v = 0$ again as a δ-function. This periodic (in terms of probability) process corresponds to random oscillations of the dynamical system about the point $v = 0$ which qualitatively was described above.

In general, when $n > 1$, the behavior of the solution following from (10.73) and (10.79) remains qualitatively the same, with the only difference that the periodic behavior of the solution around the point $v = 0$ is replaced by a multi-periodic one, while the largest amplitude of the oscillations:

$$A = nh\tilde{\omega} \tag{10.87}$$

Let us assume that instead of (10.79), the function

$$\varepsilon = \varepsilon(v) \tag{10.88}$$

has several zeros

$$\varepsilon(v_i) = 0, \quad \text{and } \varepsilon(v_j) = 0, \ i, \ j = 1, 2, ..., \text{etc.} \tag{10.89}$$

where

$$\frac{d\varepsilon}{dv}\,|_{v=v_i} < 0, \ \text{ and } \ \frac{d\varepsilon}{dv}\,|_{v=v_j} > 0 \tag{10.90}$$

Start with the case when the largest step in the generalized random walk is larger than the largest distance between the neighboring zeros v_j:

$$nh\omega > \max_{j=1,2\cdots} \mid v_j - v_{j-1} \mid \tag{10.91}$$

Then the solution will eventually visit all the zeros v_i with the probability proportional to the distance $\mid v_j - v_{j-1} \mid$ where

$$v_j < v_i < v_{j+1} \tag{10.92}$$

In the case when the largest step $nh\omega$ is less than some of the distances between the neighboring zeros $\overset{*}{v}_j$:

$$nh\omega < \left|\overset{*}{v}_j - \overset{*}{v}_{j-1}\right| \tag{10.93}$$

the solution will be trapped in the basins of these zeros v_i for which:

$$\overset{*}{v}_j < v_i < \overset{*}{v}_{j+1} \tag{10.94}$$

The zeros v_i can be considered as minima of the function:

$$\varepsilon = \int \varepsilon(v)dv \tag{10.95}$$

This means that the dynamical system (10.80–10.81) can be exploited for finding local minima of an arbitrary function (10.95) such that these minima will be visited by the solution with probabilities proportional to the sizes of their basins.

The "rule" can be rearranged if one introduces in (10.15) the following change of variables:

$$y = z(1 + \beta^2\varepsilon^2), \ \beta = \text{ Const} \tag{10.96}$$

Then, with reference to (10.17), (10.21), and (10.25), one concludes, that in the dynamical system (10.80–10.81), the largest step h of the generalized random walk will depend upon the depth of the minimum of the function (10.95):

$$H = nh\omega(1 + \beta^2\varepsilon^2), \ \beta = \text{ Const} \tag{10.97}$$

Hence, the solution of the dynamical system (10.80–10.81) equipped with the additional condition (10.96), will now visit the local minima of the function (10.95) with probabilities which are proportional to their depths as well as the sizes of their basins. In other words, this dynamical system, with sufficiently large β, will find the global minimum of the function (10.95).

More sophisticated "rules" of performance of the dynamical system (10.80–10.81) can be implemented by changing the function φ in (10.15).

10.6 Events Depending upon State Variable Properties

The complexity of the dynamical systems considered above will be significantly enriched if events depend upon probabilities of the state variable

v, and in particular, upon its statistical invariants such as expectation, variance, etc.

Starting with the dynamical system

$$\dot{u} = y_1 u^{1/3} \cos \omega t + \varepsilon \tag{10.98}$$

$$\dot{v} = \alpha u \tag{10.99}$$

assume that

$$\varepsilon = -\gamma^2 \bar{v}, \ \gamma = \text{ Const } \ll 1 \tag{10.100}$$

in which $\bar{v}$ is the mathematical expectation:

$$\bar{v} = \int_{-\infty}^{\infty} v\phi(v,t)dv \tag{10.101}$$

Two fundamental properties make the system (10.98–10.99) different from all the previous cases.

Firstly, this system is coupled with the associated probability equation (10.50) since it contains the probability distribution ϕ as a new unknown [see (10.101)].

Secondly, the probability equation (10.50) becomes nonlinear since now the variable p explicitly depends upon the unknown distribution $\phi(v,t)$ [see (10.45), (10.100–10.101)].

Nevertheless, (10.98–10.100) are simple enough to be treated analytically. Indeed, consider the problem with the initial condition (10.82). As follows from the symmetry of (10.82) with respect to $v = 0$:

$$\bar{v} = 0 \quad \text{at} \quad t = 0 \tag{10.102}$$

and therefore

$$\text{sgn } \varepsilon = 0 \quad \text{at} \quad t = 0 \tag{10.103}$$

The governing equation for the probability distribution for this case is obtained from (10.50) by setting

$$p = \frac{1}{2} \tag{10.104}$$

Hence, the solution to (10.50) starts from a symmetric diffusion, and therefore, the conditions (10.102–10.104) will persist. Eventually the solution will approach zero:

$$\phi(v,t) \to 0, \quad t \to \infty \tag{10.105}$$

Thus, despite an apparent similarity between the dynamical systems (10.1), (10.81) and (10.98–10.100), their behaviors at the same initial conditions are significantly different.

As a second example, replace (10.100) by the following:

$$\varepsilon = -\gamma^2[L\bar{v} - \overline{(v-\bar{v})^2}], \gamma \ll 1, L = \text{ Const} \tag{10.106}$$

where

$$\overline{(v-\bar{v})^2} = \text{ Var } (\bar{v}) = \sigma^2 = \int_{-\infty}^{\infty} (v-\bar{v})^2 \phi(v,t) dv \tag{10.107}$$

and analyze the solution at the following initial conditions:

$$\phi(v,t) = \begin{cases} \frac{1}{L} \text{ if } & 0 \leq v \leq L \\ 0 & \text{otherwise} \end{cases} \tag{10.108}$$

Since at $t = 0$

$$\bar{v} = \frac{L}{2}, \quad \text{Var } (\bar{v}) = \frac{L^2}{12} \tag{10.109}$$

and therefore

$$\text{sgn } \varepsilon = -1 \tag{10.110}$$

the initial probability distribution (10.108), will shift to the left until

$$\text{sgn } \varepsilon = 0 \tag{10.111}$$

For a simple random walk, i.e., when $n = 1$, the solution to (10.50) for $\phi(v,t)$ has the form of a single wave propagating without deformation [see (10.72)]. In this case the state (10.111) can be found analytically.

Indeed, since the solution $\phi(v,t)$ can be represented by a moving rectangle, one obtains:

$$\bar{v} = \frac{a_1 + a_2}{2}, \quad \text{Var } (\bar{v}) = \frac{L^2}{12}, \quad L = a_1 - a_2 \tag{10.112}$$

where a_1 and a_2 are the coordinates at the right and the left ends of the rectangle, respectively.

Then the condition (10.111) occurs when

$$a_1 = \frac{7}{12} L \tag{10.113}$$

i.e., when

$$\phi(v,t) = \begin{cases} \frac{1}{L} \text{ if } & -\frac{5}{12}L \leq v \leq \frac{7}{12}L \\ 0 & \text{otherwise} \end{cases} \quad \text{at } \overset{*}{t} = \frac{5\pi}{6} \frac{L}{h\omega^2} \tag{10.114}$$

The solution (10.114) is stable since

$$\frac{\partial^2 \varepsilon}{\partial a_1^2} < 0 \quad \text{and} \quad \frac{\partial \varepsilon}{\partial a_1} = 0 \quad \text{at} \quad \varepsilon = 0 \tag{10.115}$$

and therefore, it will be valid for $t \geq \overset{*}{t}$.

Hence, the dynamical system (10.98–10.99), (10.106) subject to the initial conditions (10.108) eventually approaches a stationary stochastic process with the probability distribution (10.114). However, this distribution depends upon the initial conditions (10.108) and therefore, it does not represent an attractor.

Such a simple analytical result could be obtained only for $n = 1$. If $n > 1$, the initial probability distribution changes its original shape.

10.7 Multi-Scale Chains of Events

In many problems of operation research, and especially, in decision analysis, one class of events can be much more important than another, so that in the presence of the first class of events ε_1, the second class ε_2 become decisive. In terms of the dynamical system (10.1), (10.9), (10.17), (10.24) and (10.46), (10.61), this condition can be incorporated by modifying (10.61) as following:

$$\varepsilon = \delta\varepsilon_1 + \delta^2\varepsilon_2, \quad 0 < \delta \ll 1 \tag{10.116}$$

Here ε_1 and ε_2 can be considered as functions of time, state variable and their probabilities, i.e.,

$$\varepsilon_1 = \varepsilon_1(t, v, \bar{v}, \cdots), \; \varepsilon_2 = \varepsilon_2(t, v, \bar{v}, \cdots) \tag{10.117}$$

Obviously, the second term in (10.116) can be ignored if $\varepsilon_1 \neq 0$ at the critical points (10.14). However, if $\varepsilon_1 = 0$, but $\varepsilon_2 \neq 0$ at these points, then the dynamical system is driven only by the event ε_2.

The evaluation of the probability distribution at v is described by the same equations (10.50–10.51), but (10.45) defining the probability p should be modified as following:

$$p = \begin{cases} 1 \text{ if} & \text{sgn } \varepsilon_1 = 1, \text{ or} \quad \text{sgn } \varepsilon_1 = 0, \text{ but} \quad \text{sgn } \varepsilon_2 = 1, \\ 0 \text{ if} & \text{sgn } \varepsilon_1 = -1, \text{ or} \quad \text{sgn } \varepsilon_1 = 0, \text{ but} \quad \text{sgn } \varepsilon_2 = -1, \\ \frac{1}{2} \text{ if} & \text{sgn } \varepsilon_1 = 0, \text{ and} \quad \text{sgn } \varepsilon_2 = 0 \end{cases} \tag{10.118}$$

In the same way one can introduce a multi-scale chain of events by modifying (10.116) as following:

$$\varepsilon = \delta\varepsilon_1 + \delta^2\varepsilon_2 + \cdots + \delta^n\varepsilon_n, \quad 0 < \delta \ll 1 \tag{10.119}$$

with the corresponding modification of (10.118).

10.8 Multi-Dimensional Systems

So far we were discussing the dynamical systems with only one state variable v (while u, x and y played the role of auxiliary variables). However, all the results obtained above can be generalized to dynamical systems which are characterized by the state variables $v_i, i = 1, 2, ..., n$.

Let us start with (10.1) and (10.9) and rewrite them in the following form:

$$\dot{x}_i = x_i^{1/3} \cos \omega t, \quad \dot{y}_i = x_i \tag{10.120}$$

As follows from (10.13), the probabilities $f_i = f_i(y_i, t)$ are governed by the difference equations:

$$f_i(y_i, t + \frac{2\pi}{\omega}) = \frac{1}{2} f_i(y_i - h, t) + \frac{1}{2} f_i(y_i + h, t), \int_{-\infty}^{\infty} f_i(y_i, t) dy_i = 1 \tag{10.121}$$

where h is expressed by (10.10).

By changing variables

$$z_i = \varphi_i(y_1, \cdots y_n), \quad y_i = \varphi_i^{-1}(z_1, \cdots z_n) \tag{10.122}$$

one can obtain a stochastic process with a prescribed probability distribution. Indeed, since y_i in (10.120) are statistically independent, the joint probability

$$f(y_1, y_2 \cdots y_n, t) = \Pi_{i-1}^n f_i(y_i, t) \tag{10.123}$$

and therefore, the joint probability for z_i

$$\psi(z_1, \cdots z_n, t) = \Pi_{i=1}^n f_i[\varphi_i^{-1}(z_1, \cdots z_n)] \mid \det \frac{\partial y_i}{\partial z_j} \mid \tag{10.124}$$

Hence, the dynamical system (10.120), (10.122) characterized by the state variables $z_1, \cdots z_n$ performs a random motion with the joint probability function (10.124) found with the aid of the difference equations (10.121).

For better interpretation of (10.124), reduce (10.122) to the following parameterized form

$$z_i = \sigma(\sum_{i=1}^{n} T_{ij} y_j), \ \sigma(\cdot) = \ \tanh(\cdot), T_{ij} = \ \text{Const} \tag{10.125}$$

Here $\sigma(\cdot)$ is a sigmoid function, while the representation (10.125) is "borrowed" from neural network architecture. Then

$$y_i = \sum_{i=1}^{n} T'_{ij} \sigma^{-1}(z_j), \quad \sigma^{-1}(\cdot) = \ \text{arctanh}\ (\cdot) \tag{10.126}$$

and T'_{ij} are elements of the inverse matrix $||T'_{ij}||$

$$||T'_{ij}|| = ||T_{ij}||^{-1} \tag{10.127}$$

Since

$$\det \frac{\partial y_i}{\partial z_j} = \ \det ||\frac{T'_{ij}}{1 - z_j^2}||, \ -1 < z_j < 1 \tag{10.128}$$

one obtains instead of (10.124) :

$$\psi(z_1, \cdots z_n, t) = \Pi_{i=1}^{n} f_i[\sum_{i+1}^{n} T'_{ij} \sigma^{-1}(z_j)] \ \det ||\frac{T'_{ij}}{1 - z_j^2}|| \tag{10.129}$$

As follows from (10.121), each variable y_i performs a simple symmetric unrestricted random work, and therefore,

$$f_i(y_i) \to 0 \ \text{ at } \ t \to \infty, \ -\infty < y_i < \infty$$

However, as follows from (10.125),

$$-1 < z_i < 1 \tag{10.130}$$

and consequently,

$$\varphi(z_1, \cdots z_n, t) \to 0 \ \text{ at } \ t \to \infty \tag{10.131}$$

(Otherwise the condition $\int_{-\infty}^{\infty} \varphi(z_1, \cdots z_n, t) dz_1 \cdots dz_n \equiv 1$ cannot be enforced).

Thus, the solution to the dynamical system (10.120), (10.122) approaches a steady stochastic process (i.e., a stochastic attractor) with the joint probability expressed by (10.118) at $t \to \infty$:

$$\varphi_0(z_1 \cdots z_n) = \varphi(z_1 \cdots z_n, t) \ \text{ at } \ t \to \infty \tag{10.132}$$

Obviously φ_0 is uniquely defined by the constant T_{ij} via (10.126) and (10.128), and therefore, one can prescribe the joint probability φ_0 by an appropriate choice of these constants.

Applications of stochastic attractors to information processing were discussed in Chap. 8.

Let us turn now to (10.46) and (10.24) and generalize them to the following system:

$$\dot{u}_i = \varphi_i(y_1^{(i)})u_i^{1/3}\cos\omega t + \varepsilon_i(t, v_i), 0 < \varepsilon_i \ll 1, \quad \varepsilon_i = \varepsilon_i(t, v_i, \bar{v}_i \cdots) \tag{10.133}$$

$$\dot{v}_i = \alpha u_i \tag{10.134}$$

where α is expressed by (10.24), $y_1^{(i)}$ are defined as

$$y_1^{(i)}(t) = y_i(t - \frac{\pi}{\omega}) \tag{10.135}$$

[compare with (10.17) and (10.125)], and y_i are defined by (10.120), while φ_1 are some prescribed functions.

Keeping in mind (10.18–10.20), one arrives at the difference equations for the probabilities $\phi_i(v_i t)$ similar to (10.50–10.51):

$$\phi_i(v_i, t+\frac{2\pi}{\omega}) = \sum_{k=1}^{n} p_k^{(i)}\{p_i\phi_i[v_i - \varphi_i(k\tilde{h})] + (1-p_i)\phi_i[v_i - \varphi(k\tilde{h})]\} \tag{10.136}$$

where $\tilde{h}$ is expressed by (10.25) defined by (10.125).

Here, with reference to (10.45):

$$p_i = \begin{cases} 1 \text{ if } & \text{sgn } \varepsilon_i = 1 \\ 0 \text{ if } & \text{sgn } \varepsilon_i = -1 \\ \frac{1}{2} \text{ if } & \varepsilon = 0 \end{cases} \tag{10.137}$$

and with reference to (10.31) and (10.23), respectively:

$$p_k^{(i)} = f_i[\varphi_i^{-1}(k\tilde{h}, t)] \mid \frac{d\varphi^{-1}}{dz_i} \mid, \quad z_i = \varphi_i(y_1^{(i)}) \tag{10.138}$$

where f_i is defined by (10.121).

So far the variables v_i in (10.133–10.134), as well as the probabilities ϕ_i in (10.136), are independent. This is why the joint probability $\phi(v_1, \cdots v_n, t)$ can be found as:

$$\phi(v_1, \cdots v_n, t) = \Pi_{i-1}^{n}\phi_i(v_i, t) \tag{10.139}$$

Let us change variables v_i as follows:

$$w_i = \sigma(\sum_{i=1}^{n} T_{ij}v_j), \quad \text{where } v_i = \sum_{i=1}^{n} T'_{ij}\sigma^{-1}(w_j) \tag{10.140}$$

Here $\sigma, \sigma^{-1}, T_{ij}$ and T'_{ij} are defined by (10.125–10.127).

Then the dynamical system (10.133–10.134) is expressed via the new variables w_i:

$$\dot{u}_i = \varphi_i(y_1^{(i)})u_i^{1/3} \cos \omega t + \varepsilon_i[t, v_i(w_1, \cdots w_n)] \tag{10.141}$$

$$\dot{w}_i = \alpha(1 - w_i^2) \sum_{i=1}^{2} T_{ij} u_j \tag{10.142}$$

The joint probability $\theta(w_1, \cdots w_n, t)$ is found from (10.139) by formal change of variables:

$$\theta(w_1, \cdots w_n, t) = \phi[v_1(w_1, \cdots w_n,), \cdots v_n(w_1, \cdots w_n)] \det \|\frac{T'_{ij}}{1 - w_j^2}\| \tag{10.143}$$

where $v_i(w_1, \cdots w_n,)$ are given by (10.140), while $\phi_i(v_i, t)$ are defined by (10.136).

Thus, the dynamical system (10.141–10.142) represents an n-dimensional generalized non-symmetric random walk.

After each time period $\tau = 2\pi/w$, all the variables w_i change their values on

$$\Delta w_i = (\text{ sgn } \varepsilon_i)\varphi_i(\tilde{h}) \text{ if } \varepsilon_i \neq 0 \tag{10.144}$$

where $\varphi_i(\tilde{h})$ are random variables: their probability densities follows from (10.138) and (10.121).

The sizes of the steps (10.144) can be correlated if one introduces the following constraints:

$$\sum_{i=1}^{m} \Omega_{ij}\varphi_j(y_1^{(j)}) = 0, \ \Omega_{ij} = \text{ Const }, \ m < n \tag{10.145}$$

where, for concreteness, $\varphi_i(\cdot)$ can be chosen as:

$$\varphi_i(\cdot) = \tanh(\cdot) \tag{10.146}$$

By appropriate choice of coefficients Ω_{ij}, one can create more or less preferable transitions of the dynamical system from one state to another.

The directions of the steps (10.144) are governed by the signs of the events ε_i. Therefore, according to (10.141), they depend upon the time t and the state variables w_i. In general they can also depend upon the statistical invariants $\bar{w}_i$, $\bar{w}_i^2$, etc.

Special attention should be paid to the branching points at which

$$\varepsilon_i = 0 \tag{10.147}$$

At these points the direction of the next step is not defined:

$$\triangle w_i = \pm\varphi_i(\tilde{h}) \tag{10.148}$$

so with an equal probability 1/2, the variable w_i can move in positive or negative direction.

From the viewpoint of information processing, the branching points are very important: they incorporate the probability of "sudden" changes in the behavior of the dynamical system. Obviously, the location of these points as well as the domains of positive and negative ε_i can be uniquely prescribed by the constants T_{ij}.

Thus, the behavior of the dynamical system (10.141–10.142) is uniquely defined by the joint probability evolution (10.143), and it can be prescribed by the appropriate choice of the constants T_{ij} and Ω_{ij}.

It should be emphasized that the variables w_i in the system (10.141), (10.142) are coupled in two different ways: the coefficients Ω_{ij} provide statistical correlation between the sizes of steps $\triangle w_i$, while the coefficients T_{ij} correlate the state-dependent events ε_i which are responsible for directions in which the dynamical system moves.

Hence, even a brief analysis of the performance of the dynamical system (10.141–10.142) demonstrates that the complexity of its behavior matches the complexity of behavior of typical discrete-events systems which occur in information processing structures, in social dynamics, in decision making processes, etc. At the same time, this dynamical system possesses a relatively simple and fully tractable analytical structure which allows one to analyze it not only numerically, but qualitatively as well.

10.9 Synthesis of Discrete-Event Systems

So far our attention was focused on analysis of terminal dynamics models for discrete events dynamics. In this section we will draft possible approaches to synthesis of these systems. Since the discrete-event dynamical systems discussed above, are uniquely defined by the constant parameters T_{ij} and Ω_{ij} [in the sense that these parameters uniquely define the evolution of joint probability of the state variables, given by (10.143)], the problem of synthesis can be reduced to finding the parameters T_{ij} and Ω_{ij} in such a way that the objective of the performance is achieved.

We will consider four problems of synthesis associated with systems identification, optimization based upon global objective, optimization based upon local rules, and systems with a collective brain.

10.9.1 System Identification

The problem of system identification arises when the analytical structure of the dynamical process performed by the system is unknown. Then, based upon experimental data, a phenomenological version of the dynamical system which has identical input-output characteristics is developed. For deterministic systems, the process of parameter identification reduces to a nonlinear optimization problem. The same approach can be formally applied to a discrete-event system if the objective is to reproduce its behavior in terms of state variable probability evolution. Indeed, in this case one can turn to (10.143) which uniquely defines this evolution in terms of the parameters T_{ij} and Ω_{ij} and solve the inverse problem of finding these parameters from given input-output data (Ljung 1988).

Recently, along with the formal mathematical approach to system identification, several biologically inspired methods borrowed from neural networks theory were developed. In connection with the discrete event systems, the strategy for application of these methods may be the following.

Let us assume that experimentally observed behavior of the system can be statistically approximated by a histogram which describes the distribution frequency with which certain states are visited by the dynamical system. Such an approximation is biologically meaningful since the frequency mentioned above is proportional to the strength of the memory trace for the corresponding pattern of behavior.

If the experimental histogram is presented as

$$\hat{\phi} = \hat{\phi}(w_1, \cdots w_n) \tag{10.149}$$

then, introducing the "energy" function:

$$E = \int_{w_1, \cdots w_n} [\hat{\phi}(w_1, \cdots w_n) - \phi(w_1, \cdots w_n, T_{ij}, \Omega_{ij})]^2 dw_1, \cdots dw_n \to \min \tag{10.150}$$

one can derive the following learning dynamics:

$$\dot{T}_{ij} = -\frac{\partial E}{\partial T_{ij}}, \quad \dot{\Omega}_{ij} = -\frac{\partial E}{\partial \Omega_{ij}} \tag{10.151}$$

where $\phi(w_1, \cdots w_n, T_{ij}, \Omega_{ij})$ can be found as a solution to (10.143), or it can be reproduced by the dynamical system (10.141–10.142) for each particular $\overset{*}{T}_{ij}$ and $\overset{*}{\Omega}_{ij}$.

The system (10.151) will converge to a minimum (which will be a global minimum if ϕ is a quadratic form of T_{ij}, and Ω_{ij}), since E plays the role of the Liapunov function.

10.9.2 Optimization Based upon Global Objective

In many problems of operation research, the objective of the performance of a discrete event system is to minimize expectations of a certain combination of state variables with possible constraints imposed upon other statistical invariants, for instance:

$$E = \theta(\bar{w}_1, \cdots \bar{w}_n) \to \min \tag{10.152}$$

or

$$E = \int_{t_1}^{t_2} \theta(\bar{w}_1, \cdots \bar{w}_n, t)dt \to \min \tag{10.153}$$

while

$$\bar{w}_i^2 \leq \overset{*}{w}{}_i^2 = \text{ Const} \tag{10.154}$$

Clearly the solution to the problem of finding the optimal values of T_{ij} and Ω_{ij} for the dynamical system (10.141–10.142) minimizing the performance indices (10.152), or (10.153) can be reduced to the case (10.150–10.151) considered above.

10.9.3 Optimization Based upon Local Rules

In many real-life situations, a member of a biological, or a social system does not have an explicitly formulated global objective for the whole system. Instead, it has its own local objective which can be partly compatible with, and partly contradictory to the local objectives of other members. In addition to this, each member may try to copy the behavior of a "successful" neighbor, or a leader, based upon local rules, and these rules couple the evolution of all the members of the system. Eventually such a system may approach a state which can be interpreted as the global objective of the performance.

Let us turn to analytical formulations of the local rules.

The local objectives of each member (or, a dynamical unit) can be introduced by reducing (10.152) to the following:

$$E_i = E_i(w_i, T_{ii}, \Omega_{ii}) \to \min \tag{10.155}$$

where

$$\dot{T}_{ii} = -\frac{\partial E_i}{\partial T_{ii}}, \dot{\Omega}_{ii} = -\frac{\partial E_i}{\partial \Omega_{ii}}, \quad i = 1, 2, ..., n \tag{10.156}$$

Clearly, (10.156) defines only the diagonal elements of the matrices T_{ij} and Ω_{ij}.

In order to define the non-diagonal elements of these matrices, first we will assume that the indices i and j of the elements T_{ij} and Ω_{ij} are related to spatial locations (or coordinates) of the corresponding elements so that the positive integer $| i - j |$ is proportional to the spatial distance between the dynamical units characterized by the variables w_i and w_j. Then, it can be assumed that T_{ij} is inversely proportional to the distance $| i - j |$, $(i \neq j)$. Indeed, it would mean that the close neighbors effect each other's behavior more strongly than the more distant ones.

However, the spatial distance between the dynamical units is not the only measure of the degree of interaction between them: the distance in functional space may be even more important. Such a distance between the units i and j can be introduced as the scalar $| w_i - w_j |$. Then the local rule for the non-diagonal elements can be formulated, for instance, as following:

$$T_{ij} = \frac{\beta}{| i - j |} e^{-\alpha(w_i - w_j)^2} = T_{ji}, \ (i \neq j) \tag{10.157}$$

where α is a constant of the same dimensionality as $1 / w^2$, and β is the constant. The coefficients $\Omega_{ij} (i \neq j)$ can be defined in a similar way.

As follows from (10.157), the interaction between two dynamical units increases with the decrease of both the spatial distance $| i - j |$ and a functional distance $| w_i - w_j |$. In simple words, it means that the strongest interaction occurs between close neighbors who are in the same "income" bracket.

As a result of the local rules (10.156–10.157), the dynamical system (10.141–10.142) will eventually approach some stochastic process which can be associated with a certain optimization problem defined implicitly via these rules. However, in general it is a very difficult (if not an impossible) task to reconstruct the global objective of system performance based only upon local rules, but without an actual run of the system.

11

Modeling Heartbeats Using Terminal Dynamics

In biology, Occam's razor cuts your throat.[1]
–Anonymous

An early and frequent subject for investigation of chaotic dynamics has been the heartbeat. Although long-standing evidence indicated that there were clear rhythmic patterns embedded in the heartbeat time series (Kitney and Rompelman 1980), there were also sufficient divergences from regularity to support a consideration of deterministic chaos (Babloyantz and Destexhe 1988; Garfinkel et al. 1992). Even from the earliest investigations (Zbilut et al. 1988; Mayer-Kress et al. 1988), however, there has been disagreement with this characterization, given some of the stochastic aspects of involved control processes (DeFelice and Isaac 1992; Ruelle 1994; Kanters et al. 1994; Guevara and Lewis 1995).

Additionally, fundamental concerns regarding physiological adaptability, i.e., the ability to respond to continuous changes in perfusion requirements, have added doubts as to whether a fully deterministic system, such as chaotic dynamics can express this (Zbilut et al. 1995, Zbilut et al. 1996b; see also Chap. 6). It has been also noted that he heartbeat signal is composed of both deterministic, and random parts (Sayers 1973; Wilders and Jongsma 1993).

Experiments with isolated rat hearts have suggested that heartbeats might be better described by terminal dynamics (Zbilut et al. 1988; Giuliani

---

[1]Cited by Horgan, J. (1996) at a Santa Fe workshop on the "Limits of Scientific Knowledge." The end of science. Addison-Wesley, New York, p. 234

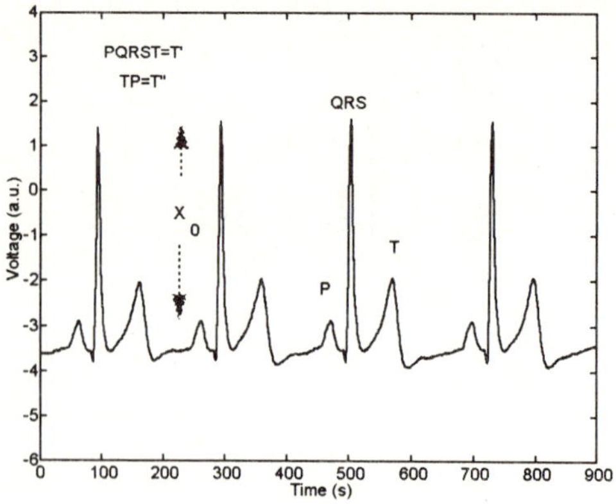

Fig. 11.1. The normal human ECG consists of a relatively fixed PQRST (T') (excluding variances of AV nodal delay and rate dependency of the QT interval), and varying length pauses (T"). Neurohormonal signals as well as ion channels are responsible for their control

et al. 1996). Following strategies outlined in Chap. 7, a terminal dynamics model of the heartbeat was developed.

11.1 Theory

Careful inspection of typical electrocardiograms (ECG) demonstrates that the signal is not really continuous, but rather consists of a relatively constant portion (the PQRST complex) followed by a varying length pause (Fig. 11.1) (Webber and Zbilut, 1994). These pauses, moreover, cannot be simply modeled by adding noise, since the noise spreads over the entire system, and destroys the constant portion (i.e., the entire PQRST) (Hübler 1992). To overcome this obstacle, the pauses may be considered as singular points of terminal dynamics. By doing so, several extraordinary things happen: a) solutions are no longer unique; b) the equilibria are approached in a finite time; c) the points become conditioned to the effects of infinitesimal noise with the possibility of relating micorscopic to macroscopic events; and d) the process is irreversible (no memory) thus freeing the model from the tyranny of initial conditions, and allowing for adaptability

As a concrete example, the heartbeat intervals of an isolated, perfused rat heart, which has been demonstrated to exhibit a random walk in time were modeled using terminal dynamics.[2] This preparation was chosen in

[2] Our work has indicated that the rat heartbeat has a power law scaling of ~2, and correlation dimension of ~2 as would be expected of a random walk in time (Zbilut et al. 1989). Transplanted human hearts, which do not have nervous system connections, also

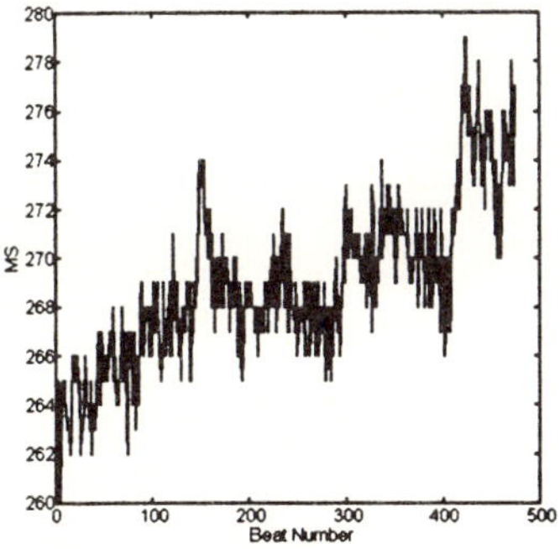

Fig. 11.2. Heartbeat tachogram of an isolated, perfused rat heart describing a random walk. The power law scaling exponent (based on a power spectrum of 1024 intervals) was ˜2. The distribution was Gaussian, with a supposed correlation dimension of ˜2 (Zbilut et al. 1989). It should be emphaised that although the rat ECG is qualitatively similar to to human ECG's, there are some minor differences (Fraser et al. 19671 Kuwahara et al. 1994)

order to reduce the degrees of freedom associated with the heart (i.e., no higher level modulations from the neurohormonal or fluid systems) (Fig. 11.2). The intact, isolated heart does, however, maintain its automaticity, controlled by the SA node. It should be noted, however, that it gradually "runs down" over a time course of minutes to hours from loss of natural substrate and edema formation (Rubboli 1994).

11.2 Theoretical Model

For the model, (Fig. 11.3) the voltages of the PQRST complex were considered to be of the same shape with a fundamental period, and the time between the complexes normally distributed random variables. The dynamics are then implemented by the following three first-order differential equations:

$$\dot{x} = x[(x_0 - x)]^{1/3}\theta - \varepsilon(x - \frac{x_0}{2}) \quad (11.1)$$

$$\dot{\omega} = [(\omega - \overset{*}{\omega})(\omega_0 - \omega)]^{1/3}\theta - \varepsilon(\omega - \frac{\omega_0 - \overset{*}{\omega}}{2}) \quad (11.2)$$

$$\dot{T} = a_1 \sin^{1/3}(a_2 \operatorname{erf} \frac{T'' - \mu}{\sqrt{2\tau}}) \sin \Omega t \quad (11.3)$$

have power law scaling of ˜2 and dimensions of ˜2 (Zbilut et al. 1988). Other species exhibit similar characteristics; e.g., the isolated SA node of a rabbit exhibits a Gaussian distribution of intervals (Wilders, private communication, 1996).

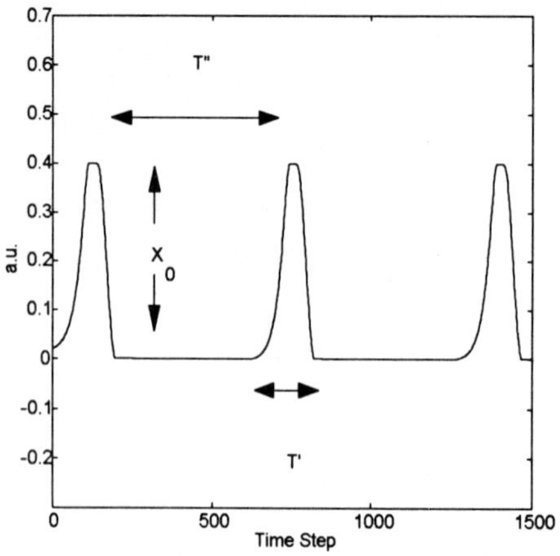

Fig. 11.3. Example of simulated heartbeat intervals from a Euler integration. Statistics were consonant with experiments

where

$$\theta = \sin\left[\int_0^t \omega(t')dt'\right] \tag{11.4}$$

and

$$\overset{*}{\omega} = 2\pi/(a_3 + a_4 T') \tag{11.5}$$

Here x_0 is the height of the pulse, and $T' = 2\pi/(a_3 + a_4 T)$ is the duration of the pulse; μ, and τ are the expectation and variance of the pauses' durations, T'', respectively, so as to maintain the shape of a Gaussian distribution consonant with experimental intervals. The positive constants (a_{1-4}): a) control the rate of change of amplitude of $T(a_1)$ and may relate to ion channel kinetics; b) control the mapping of the periodicity into $2\pi(a_2)$; and c) fix the mean and variance of the frequency such that decreasing them stops automaticity, whereas increasing them approaches fibrillation (a_{3-4}). The number ε is positive, such that $\varepsilon \ll \omega_0$, and functions as a control parameter.[3]

If $\varepsilon = 0$, (11.1) has two equilibrium points,

$$x = 0, \text{ and } x = x_0 \tag{11.6}$$

at $\theta = 0$, the second point is stable, and the first point is unstable. Therefore, during the period,

[3]The attempts here to relate the variables to specific physiological variables are heuristic, in the same sense that the original Hodgkin-Huxley equations were. Considerably more work needs to be performed to verify the correspondences.

$$x = x_0 \tag{11.7}$$

when $\theta < 0$, this point becomes unstable, and the solution switches to the first point, $x = 0$. An infinitesimal noise drives the motion to the regular solutions, and the control parameters prevents the solution from escaping to infinity. A side effect of this, is that the solution will have small oscillations with respect to the points (11.6), since actually $\varepsilon \neq 0$. If θ changes its sign periodically, the pulses and pauses alternate.

11.3 Discussion and Conclusions

Two fundamental properties of (11.1) should be emphasized: since the Lipschitz condition at the equilibrium points is violated; i.e.,

$$\left|\frac{d\,\dot{x}}{dx}\right|_{x=0} \to \infty, \quad \left|\frac{d\,\dot{x}}{dx}\right|_{x=x_0} \to \infty \tag{11.8}$$

a) both points represent terminal attractors or terminal repellers; i.e., they are infinitely stable or infinitely unstable, and b) the transition time from one point to another is theoretically finite. Without these properties, which are different from those of classical mechanics, the alternations of pulses and pauses could not be simulated at all. If θ were presented by a simple periodic function:

$$\theta = \sin \omega_0 t \tag{11.9}$$

then the durations of the pulses T' and pauses in T'' would be equal

$$T' = T'' = T/\omega_0 \tag{11.10}$$

In order to keep the durations of the pulses constant, but the durations of the pauses variable (and in particular normally distributed) (11.9) is replaced by (11.4) where $\omega(t)$ is a solution to (11.2). This equation has the same terminalproperties as (11.1); i.e., its solution is $\omega_0 =$ Const, if $\theta < 0$ and $\omega = \overset{*}{\omega}$ if $\theta > 0$. The values of $\overset{*}{\omega}$ are found from (11.5) providing the normal distribution of the pause durations T''. The normal distribution of T'' is simulated by (11.3) which also represents terminal dynamics. This equation models a random walk whose final probability distribution is normal.

It is important to emphasize that (11.3) does not have any random input, instead, it is driven by the mechanism of terminal dynamics instability (Zak 1993). Thus a terminal dynamics origin of the model (11.1–11.3) allows one to simulate the beating of an isolated heart as a piece-wise stochastic process without utilizing any man-made devices such as random number generators. This model can be further modified to account for more complicated physiological factors, e.g., respiratory modulations, which can be coupled to (11.1–11.3).

A main result of this theory is that very minimal amounts of energy (and/or noise) are necessary to perform control maneuvers at the singular points (Dixon 1995). One obvious source of microscopic noise are the open-close kinetics of ion channels. There have been very few attempts to model the effect of specific numbers of such channels. One recent attempt employed the isomorphism between open-close configurations of a set of channels, and spins on a lattice. Of course very small amounts of noise are necessary, then, to drive the configurations on the lattice [see (Shirokova et al. 1996)].

Simulations with small to moderate levels of noise (on the order of 10^{-4} to 10^{-8} rms) have confirmed the absence of any significant deterioration of the dynamics (i.e., on the constant portion). This consideration is important from the viewpoint of cardiac dynamics: the heartbeat dynamics are relatively robust against noise once a beat is started, but are easily manipulated between beats. Such dynamics maintain adaptability, while preserving stability (Haken 1991), as well as removing undue complexity (Crutchfield 1993).

12
Irreversibility in Thermodynamics

Mathematics has a restricted range; it has not changed since Archimedes. There are axioms, proofs, lemmas, theorems. In physics it is not clear what one really does and at what point one becomes satisfied that the formulation is correct. –*S. Ulam*

12.1 Introduction

Transport phenomena such as thermal conductivity and diffusion represent nonequilibrium thermodynamical processes which are described by parabolic partial differential equations of the following type:

$$\frac{\partial u}{\partial t} = D_{ij}\frac{\partial^2 u}{\partial x_i \partial x_j}, \quad D_{ij} = \mathrm{Const} \tag{12.1}$$

It is known that (12.1) subject to the initial condition

$$u \mid_{t=0} = u_0(x) \tag{12.2}$$

has a unique bounded solution for $t > 0$.

However, for $t < 0$ the same problem is ill-posed, and expresses the fundamental property of irreversibility of thermal conductivity and diffusion. Actually this property directly follows from the second law of thermodynamics.

Although solutions to (12.1) are in sufficiently good agreement with experiments, there are still some logical difficulties in reconciliation of this macroscopic phenomenological model with the fully reversible Hamiltonian dynamics on the microscopic level, since, actually the irreversible processes described by (12.1), are completely composed of reversible events, and is known as the irreversibility paradox. However, strictly speaking, the formal derivation of (12.1) from the microscopic Hamiltonian mechanics requires some additional arguments of a probabilistic nature. But can these arguments be represented in terms of classical mechanics? Or, more precisely, can they be replaced by some equivalent mechanical forces on the microscopic level? Let us turn to stochastic processes which connect microscopic mechanics and thermodynamics. These processes are based upon some probabilistic arguments which can not be formally derived from Newtonian mechanics. But maybe they can be derived from a non-Lipschitz version of Newtonian mechanics? In Chap. 7 we introduced the non-Lipschitz random walk. In the next section, based upon non-Lipschitz forces, we will introduce a pure mechanical model of random walk – the simplest stochastic process – whose macroscopic interpretation leads to (12.1).

12.2 Mechanical Model of Random Walk

In this section we will revisit the random walk phenomenon with emphasis on its physical meaning in terms of Newtonian mechanics.

A random walk is a stochastic process where changes occur only at fixed times; it represents the position at time t_m of a particle taking a random "step" x_m independently of its previous ones.

In order to implement this process based only upon the Newton's laws, consider a rectilinear motion of a particle of unit mass driven by a non-Lipschitz force:

$$\dot{v} = \nu v^{1/3} \sin \omega t, \quad \nu = \text{Const}, \quad [\nu] = \frac{m^{1-k}}{\sec^{2-k}} \tag{12.3}$$

$$\dot{x} = v \tag{12.4}$$

where v and x are the particle velocity and position, respectively.

Subject to zero initial condition:

$$v = 0 \quad \text{at} \quad t = 0 \tag{12.5}$$

Equation (12.3) has a singular solution:

$$v \equiv 0 \tag{12.6}$$

and a regular solution:

$$v = \pm\left(\frac{4\nu}{3\omega} \sin^2 \frac{\omega}{2} t\right)^{3/2} \tag{12.7}$$

These two solutions coexist at $t = 0$, and this is possible because at this point the Lipschitz condition fails:

$$|\frac{\partial \dot{v}}{\partial v}|_{t\to 0} = k\nu v^{k-1} \sin \omega t|_{t\to 0} \quad \to \quad \infty \tag{12.8}$$

Since

$$\frac{\partial \dot{v}}{\partial v} > 0 \text{ at } |v| \neq 0, \quad t > 0 \tag{12.9}$$

the singular solution (12.6) is unstable, and the particle departs from rest following the solution (12.7). This solution has two (positive and negative) branches [since the power in (12.7) includes the square root], and each branch can be chosen with the same probability 1/2. It should be noted that as a result of (12.8), the motion of the particle can be initiated by infinitesimal disturbances (this never can occur when the Lipschitz condition is in place: an infinitesimal initial disturbance cannot become finite in finite time).

Strictly speaking, the solution (12.7) is valid only in the time interval

$$0 \leq t \leq \frac{2\pi}{\omega} \tag{12.10}$$

and at $t = \frac{2\pi}{\omega}$ it coincides with the singular solution (12.6).

For $t > 2\pi/\omega$, (12.6) becomes unstable, and the motion repeats itself to the accuracy of the sign in (12.7).

Hence, the particle velocity v performs oscillations with respect to its zero value in such a way that the positive and negative branches of the solution (12.7) alternate randomly after each period equal to $2\pi/\omega$.

Turning to (12.4), one obtains the distance between two adjacent equilibrium positions of the particle:

$$x_i - x_{i-1} = \pm \int_0^{2\pi/\omega} (\frac{4\nu}{3\omega} \sin \frac{\omega}{2} t)^{3/2} dt = 64(3\omega)^{-5/2} \nu^{3/2} = \pm h \tag{12.11}$$

Thus, the equilibrium positions of the particle are:

$$x_0 = 0, x_1 = \pm h, x_2 = \pm h \pm h, \text{etc.} \tag{12.12}$$

while the signs are randomly alternated with equal probability, 1/2.

Obviously, the particle performs an unrestricted symmetric random walk: after each time period

$$\tau = \frac{2\pi}{\omega} \tag{12.13}$$

it changes its value on $\pm h$ [see (12.12)].

The probability density $u(x,t)$ is governed by the following difference equation:

$$u(x,t+\tau)=\frac{1}{2}u(x-h,t)+\frac{1}{2}u(x+h,t) \tag{12.14}$$

while

$$\int_{-\infty}^{\infty} u(x,t)dx=1 \tag{12.15}$$

12.3 Phenomenological Force

As demonstrated in Chap. 7, a non-Lipschitz force

$$F=m\nu v^{1/3}\sin\omega t=\pm\gamma\sqrt{\frac{4\nu}{3\omega}}\sin\frac{\omega}{2}t\sin\omega t \tag{12.16}$$

applied to a particle of the mass m, leads to a classical random walk.

It should be stressed that the governing equations (12.3–12.4) are fully deterministic: they are based upon the Newton's laws. The stochasticity here is generated by the alternating stability and instability effects due to failure of the Lipschitz conditions at equilibria.

Let us analyze the properties of the force (12.18).

First of all, the time average of this force is zero:

$$\tilde{F}=0 \tag{12.17}$$

since, as follows from (12.16), the signs + and – have equal probability.

For the same reason, the ensemble average of F is also zero:

$$<F>=0 \tag{12.18}$$

The work done by the force (12.16) during one step is zero:

$$A=\int_0^{2\pi/\omega} Fvdt=\pm\nu\left(\frac{4\nu}{3\omega}\right)^2\int_0^{2\pi/\omega}\sin^4\frac{\omega}{2}t\sin\omega tdt=0 \tag{12.19}$$

Since the time average of the particle's kinetic energy can be expressed via the temperature, one obtains:

$$\tilde{v}^2=\left(\frac{4\nu}{3\omega}\right)^3\int_0^{2\pi/\omega}\sin^6\frac{\omega}{2}tdt=\frac{5\pi}{8\omega}\left(\frac{4\nu}{3\omega}\right)^3=\frac{KT}{m} \tag{12.20}$$

Then the only unspecified parameter ν in (12.16) is expressed via the temperature:

$$\nu = \frac{3\omega}{4}\sqrt[3]{\frac{8\omega KT}{5\pi m}} \tag{12.21}$$

Here T is the absolute temperature, and K is the Boltzmann's constant.

The parameter ω^{-1} has the order of the time period between collisions of the particle:

$$\omega \sim \frac{1}{\tau} \sim 10^{14}\frac{1}{\text{sec}} \tag{12.22}$$

On the macro-scale this is a very large number, and one can consider a continuous approximation assuming that

$$\omega \to \infty \tag{12.23}$$

Then, as follows from (12.11), (12.13), and (12.21):

$$\tau \to 0, \quad h \to 0, \quad \text{and} \quad \frac{h^2}{\tau} \to 0.19\,\frac{KT}{m} = 2D \tag{12.24}$$

and therefore, (12.14) can be replaced by the Fokker-Planck equation, i.e., by an one-dimensional version of (12.1). It is interesting to emphasize that the diffusion coefficient D is defined by the amplitude ν of the non-Lipschitz force (12.16).

Now the following question can be asked: does the force (12.16) exist in a sense that it can be detected by direct measurements on the microscopic level? Probably, not. Indeed, on that level, this force is a resultant of a large number of collisions with other particles. However, on the stochastic level as an intermediate between the micro- and macro-levels, the phenomenological force (12.16) represents a part of the mathematical formalism, and can be accepted.

As follows from (12.16), on a micro-scale of time

$$t \sim \tau \tag{12.25}$$

the system (12.3–12.4) is not conservative, and the motion is irreversible. Moreover, each time the particle arrives at equilibrium point, it totally "forgets" its past.

On the contrary, on a macro-scale of time when

$$t \gg \tau \tag{12.26}$$

the system (12.3–12.4) can be treated as conservative based upon (12.18–12.19), and therefore, it is fully reversible. This means that the particle whose motion is described by (12.3–12.4), can return to its original position passing through all of its previous steps backward; however, the probability of such an event will be vanishingly small (but not zero!), or, in other words, the period of time t_0 during which this event can occur is very large (but finite!):

$$\tau \ll t_0 < \infty \tag{12.27}$$

12.4 Non-Lipschitz Macroscopic Effects

Turning back to the macroscopic equation (12.1), one can notice its inconsistency with the results discussed in the last section, and in particular, with the condition (12.27). Indeed, (12.1) does not have any time scale which would allow the implementation of the condition (12.27): the time of approaching a thermodynamical equilibrium is unbounded, and therefore, (12.1) excludes any reversible solutions even if $t_0 \to \infty$. The only logical way out of this situation is to introduce a time-scale into (12.16) so that the time of approaching an equilibrium would be finite. Then one can argue that this time is not large enough to include reversible solutions. In order to do this, let us turn to (12.1), and, for the sake of concreteness, treat it as an equation for thermal conductivity. Then, the relationship between the heat flow q and the temperature u can be sought in the following form:

$$q = q(\nabla u) \tag{12.28}$$

It should be emphasized that the function (12.28) is not prescribed by any macroscopic laws, and therefore, it must be found from experiments. The basic mathematical assumption about (12.28) is its expendability in a Taylor series. Then, for small gradients:

$$q = -\chi \nabla u + \cdots \text{etc.} \tag{12.29}$$

where χ is the thermal conductivity, this leads to (12.1). But even if higher order gradients of u are taken into account, the time of approaching equilibria would still remain unbounded.

However, there is another possibility in representing (12.28) if one relaxes the Lipschitz condition at $\nabla u = 0$. Indeed, instead of (12.29) one can write:

$$q = -\chi\left(\frac{\nabla u}{\varepsilon_0}\right)^{k-1} \nabla u + \cdots \ \text{etc.} \tag{12.30}$$

where k has the form (7.13), and ε_0 has the dimensionality of ∇u, i.e.

$$[\varepsilon_0] = [\nabla u] \tag{12.31}$$

Equation (12.30) is different from (12.29) only within an infinitesimally small neighborhood of the equilibria states where

$$\nabla u \to 0 \tag{12.32}$$

Otherwise

$$(\frac{\nabla u}{\varepsilon_0})^{k-1} \simeq 1 \tag{12.33}$$

One can verify that the Lipschitz condition for the function (12.30) at $\nabla u \to 0$ is violated:

$$\mid \frac{\partial q}{\partial \nabla u} \mid \to \infty \quad \text{at} \quad \nabla u \to 0 \tag{12.34}$$

Mathematical consequences of this property will be discussed below.

Turning to (12.30), one can write the following equation instead of (12.1) :

$$\frac{\partial u}{\partial t} = D\frac{\partial}{\partial x}[(\frac{\partial u}{\partial x})^k], \quad D = \frac{\chi \varepsilon_0^{1-k}}{\rho c} = \text{Const} > 0 \tag{12.35}$$

where χ, c, and ρ are the coefficient of thermal conductivity, specific heat, and density, respectively. Equation (12.35) reduces to the classical diffusion equation:

$$\frac{\partial u}{\partial t} = D\frac{\partial^2 u}{\partial x^2} \tag{12.36}$$

if $k = 1$.

Let us compare the solutions to (12.35–12.36) subject to the same initial and boundary conditions. Introducing the function:

$$\theta = \int (\frac{\partial u}{\partial x})^{k-1} dx \tag{12.37}$$

one obtains:

$$\frac{d\theta}{dt} = (k+1) \int (\frac{\partial u}{\partial x})^k \frac{\partial^2 u}{\partial x \partial t} dx = (k+1)D \int (\frac{\partial u}{\partial x})^k \frac{\partial^2}{\partial x^2}[(\frac{\partial u}{\partial x})]^k dx$$

assuming separation of variables:

$$u(x,t) = u_1(t)u_2(x) \tag{12.38}$$

one arrives at the following ordinary differential equation:

$$\dot{u}_1 = -Au_1^k \tag{12.39}$$

where

$$A = D \int (\frac{\partial u_2}{\partial x})^k \frac{\partial^2}{\partial x^2}[(\frac{\partial u}{\partial x})^k] dx = \text{Const} \tag{12.40}$$

For $k = 1$ [see (12.36)]:

$$u_1 = \overset{0}{u}_1 e^{-At}, \quad u_1^2 \to 0 \text{ at } t \to \infty \tag{12.41}$$

For $k < 1$ [see (12.35)]:

$$u_1 = [(\overset{0}{u}_1)^{1-k} - A(1-k)t]^{1/1-k} \tag{12.42}$$

Here

$$\overset{0}{u}_1 = u_1 \text{ at } t = 0 \tag{12.43}$$

Thus, as follows from (12.41), the solution to the classical equation (12.36) approaches the equilibrium state in infinite time, while the solution to the "scaled" equation (12.35) approaches the same equilibrium state in a finite time [see (12.42)]:

$$t_0 = \frac{(\overset{0}{u}_1)^{1-k}}{(1-k)A} < \infty \text{ if } k < 1 \tag{12.44}$$

One can notice that although (12.36) can be obtained from (12.35) as a limit at $k \to 1$, the solution (12.41) cannot be obtained as the same limit of (12.42). However, the quantitative difference between these solutions can be detected only within a very small neighborhood of the equilibrium when

$$\left| \frac{\partial u}{\partial x} \right| \sim \varepsilon_0 \tag{12.45}$$

The period t_0 in (12.44) represents the macroscopic time scale

$$0 < \tau \ll t_0 \ll t_{00} < \infty \tag{12.46}$$

where

$$t_{00} \sim \tau n! \tag{12.47}$$

and

$$n \sim \frac{u_0}{\varepsilon_0 \sqrt{D\tau}} \tag{12.48}$$

Here $\tau \sim 10^{-10}$ is the relaxation time, u_0/ε_0 is a macroscopic representative length, and $\sqrt{D\tau}$ is the order of the distance between two adjacent collisions.

Two new constants, k and ε_0, in (12.35) can be found from a simple experiment: turning to (12.44) and recording the time t_0 of approaching the state of equilibrium for different initial conditions, one can calculate k and A, and therefore ε_0 [see (12.35) and (12.40)].

Equation (12.35) possesses many other interesting properties such as non-uniqueness and non-Lipschitz instability with respect to infinitesimal disturbances at equilibria. These properties and their relevance to onset of turbulence [in the case of fluid dynamics interpretation of (12.35), see (Zak 1992b)].

12.5 Microscopic View

In previous sections, the problem of irreversibility in thermodynamics was discussed on the stochastic and macroscopic levels of description. This and all the next sections will be devoted to the same problem, but from the viewpoint of the microscopic level of description. On this level, the microscopic state of a system may be specified in terms of positions and moments of a constituent set of particles: atoms and molecules. Within the Born-Oppenheimer approximation, it is possible to express the Hamiltonian (or the Lagrangian) of a system as functions of nuclear variables, the (rapid) motions of electrons having been averaged out. Making the additional approximation that a classical description is adequate, one can write the Lagrange equations which govern the microscopic motions of the system:

$$\frac{d}{dt}\frac{\partial L}{\partial \dot{q}_i} - \frac{\partial L}{\partial q_i} = 0, \quad i = 1, 2, ..., n, \quad L = W + \Pi \tag{12.49}$$

Here q_i and $\dot{q}_i$ are the generalized coordinates and velocities characterizing the system, W is the kinetic energy including translational components (as well as rotational components if polyatomic molecules are considered), Π is the potential energy representing the effects of an external field (including, for example, the container walls), the particle interactions and elastic collisions.

All the solutions to (12.49) are fully deterministic and reversible if the initial conditions are known exactly. But since the last requirement is physically unrealistic, small errors in initial conditions will grow exponentially in case of instability of (12.49). (Such an instability may have the same origin as the instability in the famous three-body problem). As a result of this, the solution to (12.49) attains stochastic features, i.e., becomes chaotic, and therefore, it looses its determinism and reversibility. The connection between the chaotic instability and the problem of irreversibility in thermodynamics was stressed by Prigogine (1980): "The structure of the equations of motion with 'randomness' on the microscopic level then emerges as irreversibility on the macroscopic level." Based upon the same ideas as those introduced by Prigogine, we propose a different mathematical framework for their implementation. This framework exploits the stabilization principle introduced in Chap. 5. As shown there, this principle imposes some additional constraints upon the motion, and this makes the solutions to (12.49) irreversible [see (8.180–5.473)].

As shown in this example, the original unstable (chaotic) motion is decomposed into the mean motion along the trajectory $q_2 =$ Const with the constant velocity (5.460), and transverse fluctuations whose kinetic energy is proportional to the original error q_2^0 and to the degree of instability $| G_0 |$. It is important to emphasize that both components of the motion are stable in a sense that initial error in q_2 at $t = 0$ does not grow, and initial error in q_1 at $t = 0$ grows linearly with time.

Obviously the mean, or averaged motion represents a macroscopic view on the particle behavior extracted from the microscopic world, while the irreversibility of this motion is manifested by the loss of initial kinetic energy to microscopic fluctuations.

It should be emphasized that the decomposition of the motion into regular and fluctuation components was enforced by the stabilization principle as a supplement to Newtonian mechanics (see Chap. 5), while without this principle any theory where dynamical instability can occur is incomplete.

As another illustration, we will consider the motion of a charged particle (charge $-e$, mass m) in a uniform magnetic field, B in the vicinity of a metallic sphere (radius a) biased to a potential $\mathbf{V}_0 > 0$:

$$m\dot{\mathbf{v}} = -e\ \mathbf{v}\ \times \mathbf{B} + e\nabla\mathbf{v} \tag{12.50}$$

where $\mathbf{v} = \frac{d\hat{r}}{dt}$ is the velocity of the particle, and $v = v_0(a/p)$ is the electrical potential due to the sphere [see (5.523)].

Equation (12.50) can be written in a dimensionless form:

$$\dot{v}_x = -\frac{x}{r^3}v_y, \quad \dot{v}_y = -\frac{y}{r^3} + v_x, \quad \dot{v}_z = -\frac{z}{r^3} \tag{12.51}$$

$$v_x = \dot{x}, \quad v_y = \dot{y}, \quad v_z = \dot{z} \tag{12.52}$$

where

$$\dot{x} = \frac{dx}{d\tau}, \cdots, \text{etc.}, \; r^2 = x^2 + y^2 + z^2$$

$$r = \frac{\hat{r}}{\lambda}, \quad \tau = \omega_e t, \quad \lambda^3 = ev_0 a/m\omega_e^2, \quad \omega = eB/m$$

As reported by Barone (1993), there are certain domains of initial conditions which lead to chaotic trajectories. The system is chaotic, for instance at $x = 1.5$; $y = 0$, $z = 4.0$, $x_x = v_y = v_z = 0$ at $t = 0$. We have reproduced these results [see Fig. 5.23] by solving (12.51–12.52) numerically.

The implementation of the stabilization principle, i.e. simultaneous solution of (12.51–12.52) [after their Reynolds decomposition into the form (5.471)] and the constraint (5.479) were performed numerically. The numerical strategy was very simple: along with the basic solution, a perturbed solution were calculated and compared with the basic one after certain time steps; if the perturbed solution diverged faster than the prescribed time–polynomial, then an appropriate Reynolds force was applied to suppress it; otherwise no actions were taken. The resulting trajectories in the same x, y, z phase space are plotted in Fig. 5.24. These trajectories represent an averaged, or expected motion which is not chaotic any more. It is important to emphasize that this motion is stable in the sense that small changes of the initial conditions will cause small changes in the motion.

Actually this example elucidates the mechanism of transition from Hamiltonian mechanics describing fully reversible mechanical processes on the microscopic level, to irreversible macroscopic motions describing thermodynamical processes. On the same line of argumentation, the stabilization principle implements the preference to more probable states of the system over the less probable states.

12.6 Discussion and Conclusion

The problem of irreversibility in thermodynamics was revisited and analyzed on the microscopic, stochastic, and macroscopic levels of description (Zak 1996). It was demonstrated that Newtonian dynamics (as well as any dynamical theory where chaotic solutions are possible) can be represented in the Reynolds form when each dynamical variable is decomposed into the mean and fluctuation components. Additional equations coupling fluctuations and the mean values follow from the stabilization principle formulated by Zak (1994a) and briefly described in the previous sections. The main idea of this principle is that the fluctuations must be selected from the condition that they suppress the original instability down to a neutral stability. Supplemented by the stabilization principle, the Hamiltonian, or Lagrangian formalisms can describe the transition from fully reversible to irreversible motions as a result of the decomposition of chaotic motions (which are very likely to occur in many-body problems) into regular (macroscopic) motions and fluctuations. Actually the stabilization principle implements the preference to more probable states of the system over the less probable states, and from this viewpoint it can be associated with the averaging procedure exploited in statistical mechanics. However, the averaging procedure was always considered as an "alien intrusion" into classical mechanics, and this caused many discussions around the problem of irreversibility on the macroscopic level. On the contrary, the stabilization principle is a part of Newtonian mechanics (as well as a part of any dynamical theory where chaotic motions can occur), and therefore, it provides a formal mathematical explanation for the transition from fully reversible to irreversible processes.

On the stochastic level of description, a new phenomenological force with non-Lipschitz properties is introduced. This force as a resultant of a large number of collisions of a selected particle with other particles, has characteristics which are uniquely defined by the thermodynamical parameters of the process under consideration, and it represents a part of the mathematical formalism describing random-walk-like processes without invoking any probabilistic arguments.

Additional non-Lipschitz thermodynamical forces were incorporated into macroscopic models of transport phenomena in order to introduce a time scale. These forces are effective only within a small domain around equilibria. Without causing any changes in other domains, they are responsible for finite time of approaching equilibria. Such a property is very important for interpretation of irreversibility on the macroscopic scale. Indeed, there is always an extremely small (but non-zero) probability that a particle performing a random walk can return to its original position passing through all of its previous steps backward, and therefore, this effect should not be excluded from the solutions to the macroscopic equations if they are observed during an infinitely large period of time. However, these practically unrealistic situations may be excluded from consideration in the case of modified macroscopic equations since they are characterized by a limited time scale.

13

Terminal Dynamics Effects in Viscous Flows

If you have ever seen water run smoothly over a dam, and then turn into a large number of blobs and drops as it falls, you will understand what I mean by unstable. –*R. Feynman*

13.1 Introduction

One of the central problems in fluid dynamics is to explain how motion which is described by fully deterministic governing equations can be random. Indeed, let us consider exponential growth of a vorticity component ω:

$$\omega = \omega_0 e^{\lambda t}, \quad 0 < \lambda < \infty \tag{13.1}$$

Obviously a solution with an infinitesimally close initial condition

$$\hat{\omega} = \omega_1 e^{\lambda t}, \ \omega_1 = \omega_0 + \varepsilon, \ \varepsilon \to 0 \tag{13.2}$$

will remain infinitesimally close to the original one:

$$| \omega - \hat{\omega} | = \varepsilon e^{\lambda t} \to 0 \ \text{ at } \ t < \infty \ \text{ if } \ \varepsilon \to 0 \tag{13.3}$$

during all bounded time intervals.

This means that random solutions can result only from random initial conditions when ε in (13.2) is small, but finite rather than infinitesimal. In other words, classical fluid dynamics can explain amplifications of random motions by the mechanism of instability, but it cannot represent their origin using a mathematical formalism.

The discovery of chaotic motions in nonlinear dynamics demonstrates that the same kind of problems exists in the general formalism of Newtonian mechanics when motions described by fully deterministic models appear to be random. A revision of this formalism was presented in Chap. 7. Based upon that introduction, we can formulate the following basic physical assumptions: the dynamics of a fluid, on the macroscopic level, is described by the Navier-Stokes equations which are based upon Newton's laws. However, in addition to this, other physical assumptions are needed to introduce the dissipation function which defines the rheology of the stress-strain relationships in a fluid. On the macroscopic level, these assumptions are based upon the two laws of thermodynamics, as well as upon the principles of kinetics. The rest of the "details" must be found from experiments. However, there is another set of assumptions (which are of a mathematical nature) used in the formulation of Navier-Stokes equations. The most powerful of them is the requirement of differentiability (as many times as necessary) of all the macroscopic parameters with respect to time and space coordinates. Such a requirement is fully compatible with the principles of the macroscopic level of description. However, another mathematical assumption about the expandability of the dissipation function in a Taylor series with respect to the state of equilibrium (which is used for deriving the simplest version of the constitutive law) is not so "innocent" as it may look on first sight. Indeed, from the physical viewpoint, it eliminates the possibility of static friction or plasticity effects which may exist within the infinitely small neighborhood of equilibrium states. The models which describe such effects are well known (Ziegler 1963), and they are fully compatible with the laws of mechanics and thermodynamics. From the mathematical viewpoint, the assumption about the expandability of a Taylor series of the dissipation function enforces the Lipschitz condition at the equilibrium states, and this, in turn, leads to infinite time of approaching these states. The main objective of this chapter is to show that by relaxing the Lipschitz condition in the constitutive law of viscous liquids one will have a much more realistic scenario of behavior of liquids in the domains approaching to and departing from the equilibrium states (Zak and Meyers 1996).

13.2 Constitutive Equations

Following the ideas described in Chap. 7, we will introduce and discuss here the non-Lipschitz version of the dissipation function for a liquid in the same way as it was done in Sect. 7.21.

As follows from extremum principles in irreversible thermodynamics (see Chap. 12), the simplest form of the dissipation function for an isotropic liquid which may incorporate non-Lipschitz properties, is the following:

$$R = D(I_2) \tag{13.4}$$

where D is a positive-definite differentiable function of the second invariant I_2 of the rate-of-strain, tensor ε:

$$\varepsilon = \ \mathrm{def}\ \mathbf{v} = \frac{1}{2}(\nabla\mathbf{v} + \nabla\mathbf{v}^T) \tag{13.5}$$

i.e.,

$$I_2 = \frac{1}{2}\varepsilon_{jk}\varepsilon_{kj} \tag{13.6}$$

Here ε_{jk} are the components of the tensor ε:

$$\varepsilon_{jk} = \frac{1}{2}(\frac{\partial v_j}{\partial x_k} + \frac{\partial v_k}{\partial x_j}) \tag{13.7}$$

while v_j are the components of the velocity vector $\mathbf{v}$.

The dissipation function (13.4) defines the deviatoric stress tensor:

$$\sigma_{jk} = \frac{1}{2}\frac{\rho D}{I_2}\varepsilon_{jk} \tag{13.8}$$

The isotropic part of the stress tensor can be presented in the simplest form (since $\nabla \cdot \mathbf{v} = 0$):

$$\overset{*}{\sigma}_{ii} = -p \tag{13.9}$$

where p is the pressure.

Turning back to (13.4) let us specify the dissipation function as following:

$$\rho D = 4\mu'(\frac{\sqrt{I_2}}{\varepsilon_0})^{k-1}\ I_2 \tag{13.10}$$

where μ' and ε_0 are positive constants with the dimensions of viscosity μ and the rate-of-strain ε, respectively, while $k < 1$ is given by (3.99).

Then the deviatoric stress tensor follows from (13.8) and (13.10):

$$\sigma_{jk} = 2\mu'(\frac{\sqrt{I_2}}{\varepsilon_0})^{k-1}\ \varepsilon_{jk} \tag{13.11}$$

Equation (13.11) is different from a Newtonian liquid only within an infinitely small neighborhood of the equilibria states where

$$I_2 \to 0, \qquad \text{i.e.,} \varepsilon_{jk}, \sigma_{jk} \to 0 \tag{13.12}$$

Otherwise

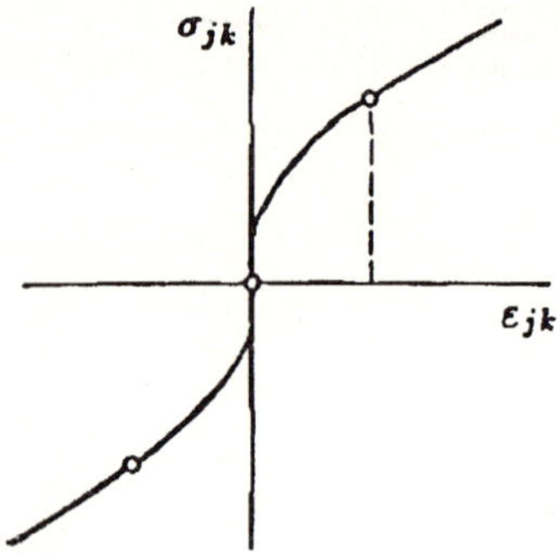

Fig. 13.1. Limit case of stress-strain relationship for viscoplastic body

$$\left(\frac{\sqrt{I_2}}{\varepsilon_0}\right)^{k-1} \simeq 1 \tag{13.13}$$

as follows from (13.11) and (13.13), $\mu' = 2\mu$ where μ is the classical viscosity.

One can verify that the Lipschitz condition for the function (13.11) at $I_2 \to 0$ is violated since

$$\left| \frac{\partial \sigma_{jk}}{\partial \varepsilon_{jk}} \right| \to \infty \qquad \text{at} \qquad I_2 \to 0 \tag{13.14}$$

Mathematical consequences of this property (which are similar to those described in Chap. 3) will be discussed in the next section.

The physical meaning of the property (13.8) is clear: it describes a limit case of a viscoplastic body when the domain of plasticity is vanishingly small (Fig. 13.1).

Let us summarize now all the arguments for selecting k based upon (13.13).

Firstly, k must be close to 1

$$| \; k - 1 \; | \ll 1$$

to preserve classical results in domains which exclude only small neighborhoods around equilibria [see (13.13)].

Secondly, k must be less then 1

$$0 < k < 1$$

to introduce the plasticity effects around equilibria via the relaxation of the Lipschitz condition [see (13.14)].

Thirdly, k must be represented by a fraction with an odd numerator and an odd denominator in order to preserve the stress-strain relationships in the form given in Fig. 13.1. Indeed, in case of an even numerator, the left branch in Fig. 13.1 will be positive, while in case of an even denominator, it will be imaginary. Obviously both cases are physically unrealistic.

Hence, actually (3.99) (cf. Chap. 3 and 7) minimizes the degree of arbitrariness to which the constant k is defined. It should be noted that a similar model was discussed by Ziegler (1963) where he introduced a limit case of a viscoplastic model.

In the case of a two-dimensional flow where the velocity can be expressed via the stream function ψ:

$$v_1 = \frac{\partial \psi}{\partial x_2}, \qquad v_2 = -\frac{\partial \psi}{\partial x_1} \tag{13.15}$$

and therefore,

$$I_2 = \frac{1}{2}(\varepsilon_{11}^2 + \varepsilon_{22}^2 + 2\varepsilon_{12}^2) = (\frac{\partial^2 \psi}{\partial x_1 \partial x_2})^2 + (\frac{\partial^2 \psi}{\partial x_2^2} - \frac{\partial^2 \psi}{\partial x_1^2})^2 \tag{13.16}$$

Equation (13.11) reads:

$$\sigma_{11} = \mu' \varepsilon^{1-k} [(\frac{\partial^2 \psi}{\partial x_1 \partial x_2})^2 + (\frac{\partial^2 \psi}{\partial x_2^2} - \frac{\partial^2 \psi}{\partial x_1^2})^2]^{\frac{k-1}{2}} \frac{\partial^2 \psi}{\partial x_1 \partial x_2} \tag{13.17}$$

$$\sigma_{22} = -\mu' \varepsilon_o^{1-k} [(\frac{\partial^2 \psi}{\partial x_1 \partial x_2})^2 + (\frac{\partial^2 \psi}{\partial x_2^2} - \frac{\partial^2 \psi}{\partial x_1^2})^2]^{\frac{k-1}{2}} \frac{\partial^2 \psi}{\partial x_1 \partial x_2} \tag{13.18}$$

$$\sigma_{12} = -\mu' \varepsilon_o^{1-k} [(\frac{\partial^2 \psi}{\partial x_1 \partial x_2})^2 + (\frac{\partial^2 \psi}{\partial x_2^2} - \frac{\partial^2 \psi}{\partial x_1^2})^2]^{\frac{k-1}{2}} (\frac{\partial^2 \psi}{\partial x_2^2} - \frac{\partial^2 \psi}{\partial x_1^2}) \tag{13.19}$$

In the simplest case of a two-dimensional unidirectional flow:

$$v_1 = u, \ v_2 = v_3 = 0, \ \ \sigma_{11} = \sigma_{22} = \sigma_{33} = \sigma_{13} = \sigma_{23} = 0 \tag{13.20}$$

the only non-zero component of the stress tensor is:

$$\sigma_{12} = \mu' \varepsilon_o^{1-k} \, (\frac{\partial u}{\partial x_2})^k, \ \ \mu' = 2\mu \tag{13.21}$$

Equations (13.16–13.21) will be exploited in our further discussions.

13.3 Governing Equations

Substituting the constitutive equations (13.11) into the momentum equations:

$$\rho(\frac{\partial \mathbf{v}}{\partial t} + \mathbf{v}\nabla\mathbf{v}) = -\nabla p + \nabla \cdot \sigma \tag{13.22}$$

and taking into account the identity

$$\nabla.(\alpha A) = \alpha \nabla \cdot A + A \cdot \nabla \alpha \tag{13.23}$$

holding for arbitrary tensor A and scalar α, one obtains the terminal dynamics version of the Navier-Stokes equation:

$$\rho(\frac{\partial \mathbf{v}}{\partial t} + \mathbf{v}\nabla\mathbf{v}) = -\nabla p + \mu^*[I_2^{\frac{k-1}{2}} \nabla \cdot (\nabla\mathbf{v} + \nabla\mathbf{v}^T) + (\nabla\mathbf{v} + \nabla\mathbf{v}^T) \cdot \nabla(I_2^{\frac{k-1}{2}})] \tag{13.24}$$

where

$$\mu^* = \frac{1}{2}\mu'\varepsilon_0^{1-k} = \text{ Const}$$

This equation must be complemented by the condition of incompressibility:

$$\nabla \cdot \mathbf{v} = 0 \tag{13.25}$$

Equation (13.24) is different from the Navier-Stokes equation only within vanishingly small neighborhoods of equilibria where

$$I_2 \to 0, \quad \text{i.e.} \quad \sigma \to 0, \;\; \varepsilon \to 0 \tag{13.26}$$

Otherwise

$$I_2^{\frac{k-1}{2}} \simeq 1, \;\; \nabla(I_2^{\frac{k-1}{2}}) \simeq 0, \;\; \varepsilon^{1-k} \simeq 1 \tag{13.27}$$

which reduces (13.24) to its classical form:

$$\rho(\frac{\partial \mathbf{v}}{\partial t} + \mathbf{v}\nabla\mathbf{v}) = -\nabla p + \mu\nabla \cdot (\nabla\mathbf{v} + \nabla\mathbf{v}^T) \tag{13.28}$$

In the particular case (13.20–13.21) of a two-dimensional unidirectional flow, (13.24–13.25) reduce to one equation:

$$\frac{\partial u}{\partial t} = \nu^*(\frac{\partial u}{\partial x_2})^{k-1}\frac{\partial^2 u}{\partial x_2^2}, \;\; \nu = \frac{\mu}{\rho} \tag{13.29}$$

which is different from the classical diffusion equation

$$\frac{\partial u}{\partial t} = \nu\frac{\partial^2 u}{\partial x_2^2} \tag{13.30}$$

only if

$$\mid \frac{\partial u}{\partial x} \mid \to 0 \tag{13.31}$$

13.4 Large-Scale Effects

Let us evaluate the range of motion scales where the proposed model describes special effects missed in the classical description.

Turning to the constitutive equation (13.4) and expanding it in a Taylor series:

$$R = a_1\, I_2 + a_2 I_2^2 + \cdots \quad \text{etc.} \tag{13.32}$$

one can verify that the Newtonian liquid described by the Navier-Stokes equation corresponds to the case when only the first term in (13.32) is kept. This is why this simplest model is valid only for velocity gradients which are relatively small in comparison to those on the molecular scale.

The same conclusion can be made based upon statistical mechanical concepts when the non-equilibrium component of the Maxwell distribution function is expanded in a Taylor series.

However, there is another possibility in representing (13.32), for instance:

$$R = a_{-1}(I_2)^{-1} + a_{-2}(I_2)^{-2} + \cdots \quad \text{etc.} \tag{13.33}$$

It has never been exploited because of the mathematical "inconvenience" caused by the singularity at equilibria where $I_2 \to 0$.

The proposed model defined by (13.10) belongs to the same type as (13.33), although it has a weaker singularity:

$$R \to 0, \quad \text{but} \quad \frac{dR}{dI_2} \to \infty \quad \text{at} \quad I_2 \to 0 \tag{13.34}$$

i.e., at equilibria the Lipschitz condition is violated.

It should be expected that contrary to the case (13.32), the constitutive laws of the type (13.33), including the proposed model, are taking into account the large scale motion effects. Indeed, as pointed out above, the proposed model describes new effects when the velocity gradients are small in the sense that

$$\sqrt{I_2} \le \varepsilon_0 \tag{13.35}$$

Here ε_0 is the physical constant of the liquid introduced by the constitutive equation (13.10). Since its dimensionality is:

$$[\varepsilon_0] = \frac{1}{\text{sec}} \tag{13.36}$$

one can introduce the time scale T_0 of the motions described by the proposed model. Indeed, based upon (13.35–13.36), one obtains:

$$\sqrt{I_2} \sim \frac{v_0}{L_0} \sim \frac{1}{T_0} \le \varepsilon_0 \tag{13.37}$$

whence

$$T_0 \geq \frac{1}{\varepsilon_0} \tag{13.38}$$

The length scale L can be found from the condition:

$$L_0 = \sqrt{\nu T_0} \geq \sqrt{\frac{\nu}{\varepsilon_0}} \tag{13.39}$$

The evaluations (13.38–13.39) demonstrate that the proposed model describes large scale motion effects, i.e., motions close to equilibria where the velocities and their gradients are relatively small.

Turning to the governing equations (13.24), one can simplify them by ignoring the convection terms of the acceleration which are small in comparison to similar viscous terms in the domain of large scale motions (13.38–13.39):

$$\rho\frac{\partial \mathbf{v}}{\partial t} = -\nabla p + \mu^*[I_2^{\frac{k-1}{2}}\nabla\cdot(\nabla\mathbf{v}+\nabla\mathbf{v}^T) + (\nabla\mathbf{v}+\nabla\mathbf{v}^T)\cdot\nabla(I_2^{\frac{k-1}{2}})] \tag{13.40}$$

The expression for the energy dissipation:

$$\dot{E}_D = -\frac{\nu^*}{2}\int_v (\frac{\partial v_i}{\partial x_j} + \frac{\partial v_j}{\partial x_i})^{k+1} dv \tag{13.41}$$

does not differ much from the classical case ($k = 1$), which means that it decreases with the growth of the length scale:

$$\dot{E}_D \sim O(L_0^{-\frac{1}{2}}\nu_0^{\frac{1}{2}}R_e^{\frac{1}{2}}) \tag{13.42}$$

The same can be said about the dissipation stresses [see 13.11].

However, the dissipation forces $\nabla \cdot \sigma$, i.e., the contribution of the dissipation stresses to the momentum equation, differ significantly from the classical case $k = 1$: they grow sharply with the decrease of the velocity gradients, becoming unbounded at the equilibrium. As will be shown below, the last property is responsible for a finite time of approaching equilibria. From a physical viewpoint this means that at equilibria the dissipation is carried out by static friction.

Thus, the modification of the constitutive law which relaxes the Lipschitz condition at equilibria by introducing a vanishingly small static friction, eliminates one of the least "damaging" inconsistencies in fluid dynamics (as well as in classical dynamics): theoretically infinite time of approaching equilibria. However, as a "side-effect," it eliminates a more "damaging" inconsistency: the occurrence of stochastic motions in flows which are described by fully deterministic hydrodynamical models. As will be shown below, the relaxation of the Lipschitz condition at equilibria in combination with instability may cause the failure of the uniqueness of the solution to

(13.24–13.25), and this can be represented by additional stochastic components in the solution. The instability mentioned above is a supersensitivity to infinitesimal changes of initial condition. At first sight it seems unlikely that at equilibria where the actual viscosity is very large (strictly speaking, it is a static friction rather than viscosity), any instability can occur at all. However, as well-known from the theory of hydrodynamic stability, viscosity can be a destabilizing factor, for instance, in parallel flows where the conditions (13.26) are well satisfied.

13.5 Behavior Around Equilibria

In this section we will analyze the behavior of a terminal dynamics liquid within vanishingly small neighborhoods of equilibrium states where the condition (13.26) holds.

Our analysis will be based upon the energy balance for the liquid in a volume v with the boundary s which for any isotropic liquid which can be presented in the following form:

$$\frac{\partial}{\partial t}\int_{\mathrm{v}}\frac{\rho v^2}{2}d\mathrm{v} = -\oint_s[\rho\mathbf{v}(\frac{v^2}{2}+\frac{p}{\rho})-(\mathbf{v}\sigma)]d\mathbf{n}-\int_{\mathrm{v}}\sigma_{ik}\frac{\partial v_i}{\partial x_k}d\mathrm{v} \qquad (13.43)$$

Here $\mathbf{v}\sigma$ denotes a vector with the components $v_i\sigma_{ik}$, and $\mathbf{n}$ is the unit normal to the surface s.

Confining our discussion to a two-dimensional flow and utilizing the expressions given by (13.15–13.19), one rewrites (13.44) in terms of the stream function ψ:

$$\frac{1}{2}\frac{\partial}{\partial t}\int_{\mathrm{v}}[(\frac{\partial\psi}{\partial x_1})^2+(\frac{\partial\psi}{\partial x_2})^2]dxdy =$$

$$-\frac{\nu^*}{2}\int_{\mathrm{v}}(\frac{\partial^2\psi}{\partial x_2^2}-\frac{\partial^2\psi}{\partial x_1^2})^{k+1}dxdy+2\nu^*\oint\{[(\frac{\partial^2\psi}{\partial x_1\partial x_2})^2$$

$$+(\frac{\partial^2\psi}{\partial x_2^2}-\frac{\partial^2\psi}{\partial x_1^2})^2]^{\frac{k-1}{2}}[\frac{\partial^2\psi}{\partial x_1\partial x_2}\frac{\partial\psi}{\partial x_2}+(\frac{\partial^2\psi}{\partial x_1^2}-\frac{\partial^2\psi}{\partial x_2^2})\frac{\partial\psi}{\partial x_1}]\cos\alpha_1$$

$$+[(\frac{\partial^2\psi}{\partial x_1\partial x_2})^2+(\frac{\partial^2\psi}{\partial x_2^2}-\frac{\partial^2\psi}{\partial x_1^2})^2]^{\frac{k-1}{2}}$$

$$[(\frac{\partial^2\psi}{\partial x_2^2}-\frac{\partial^2\psi}{\partial x_1^2})\frac{\partial\psi}{\partial x_2}+\frac{\partial^2\psi}{\partial x_1\partial x_2}\frac{\partial\psi}{\partial x_1}]\cos\alpha_2\}ds$$

$$-\oint\{\frac{p}{\rho}+\frac{1}{2}[(\frac{\partial\psi}{\partial x})^2+(\frac{\partial\psi}{\partial y})^2]\}\sqrt{(\frac{\partial\psi}{\partial x_1})^2+(\frac{\partial\psi}{\partial x_2})^2}.\cos\varphi\}ds \qquad (13.44)$$

where

$$\nu^* = \frac{\mu^*}{\rho} = \gamma \varepsilon_0^{1-k} \tag{13.45}$$

$\alpha_1(x_1 x_2)$, and $\varphi(x_1, x_2)$ are angles between the unit normal **n** and the coordinate axes x_1, x_2, and the velocity vector **v**, respectively. It is understood that these angles are known from the boundary conditions.

Let us assume that

$$\mathbf{v} \cdot \mathbf{n} = 0, \text{ i.e., } \cos\varphi = 0, \text{ but } (\mathbf{v}\sigma) \cdot \mathbf{n} \neq 0 \tag{13.46}$$

which means that the external flow does not penetrate the volume boundary S, and therefore, the exchange of energy between the volume v and the external flow is carried out by the viscous term $(\mathbf{v}\sigma)$.

Then the last term in (13.44) vanishes.

Suppose that

$$\psi = \psi_1(t)\psi_2(x_1, x_2) \tag{13.47}$$

Then the (13.44) can be reduced to an ordinary differential equation for $\psi_1(t)$:

$$\dot{\psi}_1 = \gamma^*(-A_1 + A_2)\psi_1^k, \quad A_1 > 0 \tag{13.48}$$

where

$$A_1 = \frac{1}{2} \frac{\int_v \left(\frac{\partial^2 \psi_2}{\partial x_2} - \frac{\partial^2 \psi_2}{\partial x_1}\right)^{k+1} dxdy}{\int_v [(\frac{\partial \psi_2}{\partial x_1})^2 + (\frac{\partial \psi_2}{\partial x_2})^2] dxdy} = \text{Const} \tag{13.49}$$

$$A_2 = \frac{2\int_v a\{[\frac{\partial^2 \psi_2}{\partial x_1 \partial x_2} - \frac{\partial \psi_2}{\partial x_2} + (\frac{\partial^2 \psi_2}{\partial x_1^2} - \frac{\partial^2 \psi_2}{\partial x_2^2})\frac{\partial \psi_2}{\partial x_1}] \cos\alpha_1}{\int_v [(\frac{\partial \psi_2}{\partial x_1})^2 + (\frac{\partial \psi_2}{\partial x_2})^2] dxdy}$$

$$+ \frac{\int_v [\frac{\partial^2 \psi_2}{\partial x_1 \partial x_2} - \frac{\partial \psi_2}{\partial x_1} + (\frac{\partial^2 \psi_2}{\partial x_2^2} - \frac{\partial^2 \psi_2}{\partial x_1^2})\frac{\partial \psi_2}{\partial x_2}] \cos\alpha_2\} ds}{\int_v [(\frac{\partial \psi_2}{\partial x_1})^2 + (\frac{\partial \psi_2}{\partial x_2})^2] dxdy} = \text{Const} \tag{13.50}$$

and

$$a = [(\frac{\partial^2 \psi_2}{\partial x_1 \partial x_2})^2 + (\frac{\partial^2 \psi_2}{\partial x_2^2} - \frac{\partial^2 \psi_2}{\partial x_1^2})^2]^{\frac{k-1}{2}} \tag{13.51}$$

We will analyze (13.37) for two cases when

$$A_2 - A_1 = -B^2 < 0 \tag{13.52}$$

and

$$A_2 - A_1 = B^2 > 0 \tag{13.53}$$

In the case (13.52) assume that

$$\psi_1(t=0) = \psi_1^0 > 0 \tag{13.54}$$

which corresponds to the initial kinetic energy of the flow:

$$E_0 = \frac{\rho}{2}(\psi_1^0)^2 \int_V [(\frac{\partial \psi_2}{\partial x_1})^2 + (\frac{\partial \psi_2}{\partial x_2})^2] dx dy > 0 \tag{13.55}$$

Then (13.48) [under the condition (13.52)] describes the damping of the fluid motion due to viscous stress. It has regular solution

$$\psi_1 = [(\psi_1^0)^{1-k} - \nu^* B^2(1-k)t]^{\frac{1}{1-k}} \tag{13.56}$$

and singular solution

$$\psi_1 \equiv 0 \tag{13.57}$$

In a finite time

$$t_0 = \frac{(\psi_1^0)^{1-k}}{\varepsilon_0^{1-k} \nu B^2 (1-k)} < \infty \text{ if } k < 1 \tag{13.58}$$

the regular solution (13.56) approaches equilibrium, i.e. the singular solution (13.57). This time depends upon the constants k and ε_0 which can be found from experimental measurements of t_0.

It should be stressed that in the classical case, ($k = 1$), the solution to (13.48) approaches the equilibrium (13.57) asymptotically, i.e. $t_0 \to \infty$. This is why the parameter k found from (13.58) must be less than one.

In the case (13.53) assume that

$$\psi_1(t=0) = \psi_1^0 \to 0 \tag{13.59}$$

i.e., the liquid is in equilibrium,

$$E_0 = 0 \tag{13.60}$$

Under the condition (13.4), this equilibrium is unstable. Indeed, (13.53) subject to the initial condition (13.59), has the form:

$$\psi_1 = \pm[\nu^* B^2(1-k)t]^{\frac{1}{1-k}} \tag{13.61}$$

The solution (13.61) possesses a remarkable property: it departs the equilibrium so fast that the velocity becomes finite despite vanishingly small disturbances (13.59) [compare with the classical case (13.3)]. At the same time, with equal probability 1/2, this solution can become positive or negative which means that the solution attains stochastic properties. It should be emphasized that this stochasticity results from the relaxation of the Lipschitz condition at equilibria, and this, in turn, leads to failure of the uniqueness of the solution. Thus, formal incorporation of an infinitesimal static friction in the constitutive equation of liquid allows one to explain the statistical nature of turbulence: in domains of supercritical Reynolds numbers, infinitesimal random components of the solution caused by the failure of the Lipschitz condition, are amplified by the mechanism of instability and lead to fully developed stochastic motions.

13.6 Attraction to Equilibrium After Sudden Move of Boundaries

In the previous section we discussed two fundamentally new properties of the terminal dynamics model of liquid: a finite time of approaching equilibria, and occurrence of stochastic solutions to the modified Navier-Stokes equations. Both of these effects are in full agreement with experiments.

In this section we will illustrate the modified model by example of an unsteady unidirectional flow induced by a sudden simultaneous move of both lower and upper boundaries.

Utilizing the constitutive law (13.21), one can write the following governing equation:

$$\frac{\partial u}{\partial t} = \nu^* (\frac{\partial u}{\partial y})^{k-1} \frac{\partial^2 u}{\partial y^2} \tag{13.62}$$

subject to the following boundary and initial conditions, respectively:

$$u(0, t) = u_0, \frac{\partial u}{\partial y}(\ell, t) = 0 \quad 0 < t < +\infty \tag{13.63}$$

$$u(y, 0) = 0, \quad 0 < y < \ell \tag{13.64}$$

Here u is the flow velocity parallel to the horizontal axis x, y is the axis normal to the flow, 2ℓ is the distance between the lower and upper boundaries, and ν^* is the modified viscosity expressed by (13.45), u_0 is the initial velocity of the boundaries, and k is expressed by (3.99).

The second boundary condition in (13.62) is formulated for the middle line between the boundaries in virtue of the symmetry of the problem.

For $k = 1$ one arrives at the classical diffusion equation:

$$\frac{\partial u}{\partial t} = \nu \frac{\partial^2 u}{\partial y^2} \tag{13.65}$$

The solution to this equation subject to boundary and initial conditions (13.63–13.64) is well known:

$$u(y,t) = u_0 - \frac{4u_0}{\pi}\sum_{n=0}^{+\infty}\frac{1}{2n+1}e^{-\frac{(2n^2+1)^2\pi^2\nu}{4\ell^2}t}\sin\frac{(2n+1)\pi}{2\ell}y \tag{13.66}$$

$$0 < y < \ell, \quad 0 < t < +\infty \tag{13.67}$$

where ν is the kinematic viscosity.

Obviously, this solution is valid for (13.62) in the domains where the condition (13.13) is satisfied, i.e., where

$$|\frac{\partial u}{\partial y}| \sim O(\varepsilon_0) \tag{13.68}$$

As follows from the solution (13.66), the condition (13.68) holds if

$$0 < t < t_* \tag{13.69}$$

where

$$t_* \sim O(\frac{1}{\varepsilon_0}) \tag{13.70}$$

Turning to (13.62), let us introduce a new variable $\tilde{u}$:

$$u = \varepsilon_0 \tilde{u} \tag{13.71}$$

whose dimensionality is :

$$[\tilde{u}] = L \ \text{ since } \ [\varepsilon_0] = T^{-1} \tag{13.72}$$

Therefore,

$$[\frac{\partial \tilde{u}}{\partial y}] = [(\frac{\partial \tilde{u}}{\partial y})^{k-1}] = 1 \tag{13.73}$$

Then (13.62) can be rewritten in the form:

$$\frac{\partial \tilde{u}}{\partial t} = \nu(\frac{\partial \tilde{u}}{\partial y})^{k-1}\frac{\partial^2 \tilde{u}}{\partial y^2} \tag{13.74}$$

Since we are looking for the solution to (13.74) in the domain $t > t_*$, the boundary and initial conditions now are formulated as:

$$\tilde{u}(0,t) = \tilde{u}_0, \quad \frac{\partial \tilde{u}}{\partial y}(\ell,t) = 0, \quad t_* < t < +\infty \tag{13.75}$$

$$\tilde{u}(y,t_*) = \tilde{u}_*, \ 0 < y < \ell \tag{13.76}$$

Here

$$\tilde{u}_0 = \varepsilon_0 \; u_0, \quad \tilde{u}_* = \varepsilon_0 \; u_* \tag{13.77}$$

where u_* is velocity at $t = t_*$ obtained from the classical solution (13.66) which is valid for

$$0 < t \leq t_* \tag{13.78}$$

Seeking the solution to (13.74) in the domain $t > t_*$ for $k < 1$ [see (3.99)] in the form

$$\tilde{u} = \tilde{u}_0 + u_1(t)u_2(y) \tag{13.79}$$

one obtains:

$$\dot{u}_1 + \lambda \nu u_1^k = 0, \;\; (\dot{u}_1 = du_1/dt), \lambda = \;\; \text{Const} \tag{13.80}$$

$$u_2''(u_2')^{k-1} + \lambda \tilde{u}_2 = 0, (u_2' = du_2/dy) \tag{13.81}$$

the general solution to (13.81) has the form:

$$y = \frac{1}{k+1} \int (\frac{\lambda}{2} u_2^2 + C_1)^{-\frac{1}{k+1}} du_2 + C_2 \tag{13.82}$$

where C_1 and C_2 are arbitrary constants.

As follows from (13.82), y is a continuous function of k, so that

$$y(k) \longrightarrow y(1) \text{ if } k \longrightarrow 1 \tag{13.83}$$

Hence, (13.82) can be approximated by the classical solution:

$$y \simeq \frac{1}{2} \int (\frac{\lambda}{2} u_2^2 + C_1)^{-\frac{1}{2}} du_2 + C_2 \tag{13.84}$$

and therefore,

$$u_2 = -\frac{4}{\pi} \sum_{n=0}^{+\infty} \frac{1}{2n+1} \sin \frac{2n+1}{2\ell} \pi y, \quad \lambda_n = [\frac{(2n+1)\pi}{2\ell}]^2 \tag{13.85}$$

For each λ_n in (13.77), one can obtain a particular solution to (13.80):

$$u_1^{(n)} = [C_n - (1-k)\nu \lambda_n t]^{\frac{1}{1-k}}, \quad C_n = \text{ Const} \tag{13.86}$$

However, since (13.4) is essentially nonlinear, the superposition principle is not applicable here. In order to circumvent this difficulty, we will confine ourselves to the solution for sufficiently large time [see (13.70)].

$$t > t_* \tag{13.87}$$

where the lowest mode corresponding to

$$\lambda_0 = \frac{\pi^2}{4\ell^2} \tag{13.88}$$

dominates over the others.

Then the solution to (13.74) reduces to:

$$\tilde{u} = \tilde{u}_0\{1 - \frac{4}{\pi}[C_0 - (1-k)\frac{\pi^2\nu}{4\ell^2}t]^{\frac{1}{1-k}}\}\sin\frac{\pi y}{2\ell}, \quad t \geq t_* \tag{13.89}$$

The constant C_0 can be found by matching the solutions (13.66) and (13.89) at $t = t_*$, $y = \ell$:

$$C_0 = e^{-\frac{\pi^2(1-k)\nu}{4\ell^2\varepsilon_0}} + \frac{\pi^2(1-k)\nu}{4\ell^2\varepsilon_0} \tag{13.90}$$

while

$$\tilde{u}_* = \tilde{u}_0(1 - \frac{4}{\pi}e^{-\frac{\pi^2\nu}{4\ell^2\varepsilon_0}})\sin\frac{\pi y}{2\ell} \tag{13.91}$$

Finally, the solution to (13.62) is:

$$u = u_0\{1 - \frac{4}{\pi}[e^{-\frac{\pi^2(1-k)\nu}{4\ell^2\varepsilon_0}} - (1-k)\frac{\pi^2(1-k)\nu}{4\ell^2}t]^{\frac{1}{1-k}}\sin\frac{\pi y}{2\ell}, \quad t \geq t_* \tag{13.92}$$

while for $0 < t \leq t_*$ the solution can be presented in the classical form (13.66).

Although (13.92) represents an approximate solution to (13.62), it still preserves its fundamental property: the finite time t_0 of approaching the equilibrium:

$$t_* \ll t_0 = \frac{1}{\varepsilon_0} + \frac{4\ell^2}{(1-k)\pi^2\nu}e^{-\frac{\pi^2(1-k)\nu}{4\ell^2\varepsilon_0}} < +\infty \tag{13.93}$$

As could be expected, this time depends upon two new physical constants of the liquid: k and ε_0.

13.7 Sudden Start from Rest

Continuing the analysis of the proposed model of a fluid, in this section we will pose the following problem: find the velocity field and the drag forces induced by a particle of a vanishingly small size suddenly starting from rest. This problem is very important in a variety of physical contexts, such as the settling of sediment in a liquid, and the fall of mist droplets in air. Nevertheless, from a formal mathematical viewpoint, for a Newtonian liquid such a problem does not make much sense: all the hydrodynamical effects vanish when the size of the particle becomes infinitesimal.

Indeed, invoking the Stokes solution for a moving sphere,

$$\psi = ur^2 \sin^2\theta(\frac{3}{4}\frac{a}{r} - \frac{1}{4}\frac{a^3}{r^3}) \tag{13.94}$$

$$F = 6\pi a\mu u \tag{13.95}$$

one obtains:

$$\psi \to 0, \quad F \to 0 \text{ if } a \to 0 \tag{13.96}$$

where a is the radius of the sphere.

But if this sphere moves in an unbounded volume, any finite size is "vanishingly small." This is why the smallness of the size of a particle is actually understood as the smallness of the Reynolds number, R. However, expressing the drag force in (13.95) via the Reynolds number, one arrives at a singularity for the drag coefficient:

$$C_D \cong \frac{24}{Re} \to \infty \text{ if } R \to 0 \tag{13.97}$$

Thus, the classical approach to the problem posed above gives only qualitative rather that quantitative results.

We will start with the plane flow in the domain of small velocity gradients where

$$\frac{\sqrt{I_2}}{\varepsilon_0} \ll 1 \tag{13.98}$$

Therefore, the momentum equations for this case can be reduced to the form (13.40), i.e.,

$$\frac{\partial}{\partial t}(\frac{\partial^2 \psi}{\partial x_1^2} + \frac{\partial^2 \psi}{\partial x_2^2}) = \frac{\partial^2 \sigma_{11}}{\partial x_1 \partial x_2} + \frac{\partial^2 \sigma_{12}}{\partial x_2^2} - \frac{\partial^2 \sigma_{12}}{\partial x_1^2} - \frac{\partial^2 \sigma_{22}}{\partial x_1 \partial x_2} \tag{13.99}$$

in which σ_{11}, σ_{12}, and σ_{22} are expressed by (13.17–13.19), respectively.

We will show that this equation has a class of solutions which is fundamentally different from those in the classical case. For this purpose, let us seek the solution in the following form:

$$\psi = \alpha t^q (x_1^m + x_2^m) \tag{13.100}$$

Substituting (13.100) into (13.99), one obtains:

$$q = \frac{1}{1-k}, \quad m = -\frac{2k}{1-k}, \quad \alpha = \pm(1-k)^{\frac{k}{k-1}}(k-1)^{\frac{2k}{k-1}}\nu_* \tag{13.101}$$

The two signs for α can be expected if one recalls that, as follows from (3.99):

$$k = \frac{2n-1}{2n+1} \tag{13.102}$$

where n is the one of the natural numbers

$$n = 1, 2, \cdots \text{ etc.} \tag{13.103}$$

Then

$$\frac{1}{1-k} = \frac{2n+1}{2} \tag{13.104}$$

and the power $1/(1-k)$ includes the square root operation. [The expression in the square brackets in (13.101) is positive for k given by (13.102)].

The solution (13.100) in terms of velocities can be presented in the form:

$$v_1 = \pm x_2 \varepsilon_0 (\beta \frac{\gamma t}{x_2^2})^{\frac{1}{1-k}}, \; v_2 = \pm x_1 \varepsilon (\beta \frac{\gamma t}{x_1^2})^{\frac{1}{1-k}} \tag{13.105}$$

here

$$\beta = \pm(1+k)^k \; (k-1)^{k+1} \; = \; \text{Const} \tag{13.106}$$

while the physical constant ε_0 is introduced by the constitutive equation (13.10).

One can verify that the expressions in the first brackets in (13.72) have the dimension of velocity, and the expressions in the second brackets are dimensionless.

Substituting (13.72) into the momentum equations:

$$\frac{\partial v_1}{\partial t} = -\frac{1}{\rho}\frac{\partial p}{\partial x_1} + \frac{\partial \sigma_{11}}{\partial x_1} + \frac{\partial \sigma_{12}}{\partial x_2}, \frac{\partial v_2}{\partial t} = -\frac{1}{\rho}\frac{\partial p}{\partial x_2} + \frac{\partial \sigma_{12}}{\partial x_1} + \frac{\partial \sigma_{22}}{\partial x_2} \tag{13.107}$$

one concludes that

$$\sigma_{11} = 0, \quad \sigma_{22} = 0, \quad \nabla p = 0, \quad \text{i.e., } p = \; \text{Const} \tag{13.108}$$

This means that the velocity field represents a shear flow.

We will start with the formal analysis of the solution (13.105). First it should be noted that

$$v_1 = v_1^0 \pm \; x_2 \varepsilon_0 (\beta \frac{\gamma t}{x_2^2})^{\frac{1}{1-k}}, v_2 = v_2^0 \pm \; x_1 \varepsilon_0 (\beta \frac{\gamma t}{x_1^2})^{\frac{1}{1-k}} \tag{13.109}$$

are also the solutions to (13.99), or (13.107), and therefore, (13.105) represent particular solutions to (13.107) subject to the initial conditions:

$$v_1 = 0, \quad v_2 = 0, \quad \text{at} \quad t = 0 \tag{13.110}$$

However, in addition to this, (13.107) have a singular solution for the same initial conditions (13.110):

$$v_1 \equiv 0, \quad v_2 \equiv 0 \tag{13.111}$$

which is not included in the family of the solutions (13.109). Obviously such a non-uniqueness of solution is a result of violation of the Lipschitz condition at the equilibrium. As in cases analyzed in the previous sections [see (13.56)], the solution (13.111) is unstable: infinitesimal initial velocities

$$\pm v_1^0 \to 0, \quad \pm v_2^0 \to 0 \quad \text{at} \quad t = 0 \tag{13.112}$$

transfer it into one of the solutions (13.105) which will rapidly escape from the equilibrium. It is important to emphasize that the signs of the solutions (13.105) are defined by the signs of the initial conditions (13.112) which are random. Actually this is the origin of stochasticity of the solutions to the Navier-Stokes equations modified to the form (13.24).

However, one should recall that the solution (13.105) is valid only for those domains where the condition (13.13) is still true, i.e., when

$$\left| \frac{\partial v_1}{\partial x_2} \right|, \left| \frac{\partial v_2}{\partial x_1} \right| \sim O(\varepsilon_0) \tag{13.113}$$

in which ε_0 is defined by (13.10).

As follows from (13.105):

$$\left| \frac{\partial v_1}{\partial x_2} \right| = \frac{k}{1-k} \beta \varepsilon_0 \left(\frac{\gamma t}{x_2^2} \right)^{\frac{1}{1-k}}, \quad \left| \frac{\partial v_2}{\partial x_1} \right| = \frac{k}{1-k} \beta \varepsilon_0 \left(\frac{\gamma t}{x_1^2} \right)^{\frac{1}{1-k}} \tag{13.114}$$

Hence, the solution (13.105) is not valid for the domain

$$x_1^2, \, x_2^2 \; > \; r_0^2 \tag{13.115}$$

while

$$r_0^2 \cong O\left(\frac{\nu t}{\gamma}\right), \; \gamma = \left(\frac{1-k}{k\beta}\right)^{1-k} \; = \; \text{Const} \tag{13.116}$$

For this domain one has to apply the original version of the momentum equations (13.24) which include the convective components of the acceleration [being dropped in (13.129)].

Let us now concentrate on some physical effects described by the solution (13.105). Consider a rigid particle at rest. Then it must be that:

$$v_1 = v_2 \equiv 0, \quad x_1^2, x_2^2 \; < \; r_0^2 \tag{13.117}$$

This condition can be satisfied if one combines the positive and the negative branches of the solution (13.105) as following:

$$v_1 = v_1' + v_1'', \quad v_2 = v_2' + v_2'' \tag{13.118}$$

where

$$v_1' = \begin{cases} +\beta(x_2\varepsilon_0)(\frac{\gamma t}{x_2^2})^{\frac{1}{1-k}} & \text{for } x \geq -r_0 \\ 0 & \text{otherwise} \end{cases} \tag{13.119}$$

$$v_1'' = \begin{cases} -\beta(x_2\varepsilon_0)(\frac{\gamma t}{x_2^2})^{\frac{1}{1-k}} & \text{for } x_2 \geq -r_0 \\ 0 & \text{otherwise} \end{cases} \tag{13.120}$$

$$v_2' = \begin{cases} +\beta(x_1\varepsilon_0)(\frac{\gamma t}{x_1^2})^{\frac{1}{1-k}} & \text{for } x_1 \geq -r_0 \\ 0 & \text{otherwise} \end{cases} \tag{13.121}$$

$$v_3' = \begin{cases} -\beta(x_1\varepsilon_0)(\frac{\gamma t}{x_1^2})^{\frac{1}{1-k}} & \text{for } x_1 \geq -r_0 \\ 0 & \text{otherwise} \end{cases} \tag{13.122}$$

Thus, the solution (13.119–13.122) describes the flow around a rigid particle of the radius $r_* = r_0$ at rest. But before discussing the cause of this flow, let us find the force of interaction between the flow and the particle.

Obviously, this force can be found as:

$$F_f = \int_0^{2\pi} \sigma_{12}\, r^* d\varphi, \quad \text{i.e.,} \tag{13.123}$$

$$F_f = 4\pi r_* \rho\nu\varepsilon_0 (\frac{k\beta}{1-k})^k \, (\frac{\nu t}{r_*^2})^{\frac{k}{1-k}} \tag{13.124}$$

This equation is valid only until the condition (13.113) is satisfied, i.e. [with reference to (13.116)] until

$$F_f \leq F_{f\max} \cong O\,(4\pi\rho\nu\varepsilon_0) \tag{13.125}$$

Let us assume that some external force F is applied to the particle at rest. In contradistinction to the classical case, the flow starts moving, first raising the reaction force (13.124), and only after this, the particle starts moving. This situation resembles the behavior of a rigid body on a rough surface which can start moving only after the applied force exceeds the maximum static friction.

The maximum force due to infinitesimal static friction of the liquid is evaluated by (13.125). It depends upon ε_0 which is the physical constant characterizing this liquid.

The time-delay for the motion of the particle is found from the condition (13.116)

$$\Delta t \sim O\,(\frac{\gamma r_*^2}{\nu}) \tag{13.126}$$

where ν is given by (13.116).

As follows from (13.126), this delay depends upon k which is another physical constant characterizing the liquid.

For

$$t > \Delta t, \text{ or } F > F_{f\max} \tag{13.127}$$

the particle starts moving, and one has to apply the original version of the momentum equations (13.24) which in this domain will coincide with the Navier-Stokes equations. This means that in the domain (13.127) the velocity field and the drag force can be found from the Stokes formulas (13.94) and (13.97).

13.8 Phenomenological Approach

As emphasized in the previous sections, both constants k and ε_0 [see (3.99) and (13.11)] represent additional physical properties of liquids, and therefore, they must be found from experiments. However, in this section we will find both k and ε_0 based upon phenomenological concepts. For this purpose let us compare the solutions to (13.62) and (13.65) for

$$t \geq \frac{1}{\varepsilon_0} = t_* \tag{13.128}$$

The solution to (13.62) expressed by (13.92) approaches the equilibrium at $t = t_0$ where t_0 is defined by (13.93). The solution to (13.65) expressed by (13.66), theoretically never approaches the equilibrium; however, in finite time it approaches a domain of insensitivity where the velocity u_{00} is so small that it cannot be detected by sensors. Hence, the actual time t_0' of approaching the equilibrium by the solution (13.66) can be found as:

$$t_0' = \frac{1}{\varepsilon_0} - \frac{4\ell}{\pi^2\nu}\ell n[\frac{\pi}{4}(1-\xi)] \tag{13.129}$$

where

$$\xi = \frac{u_{00}}{u_*} \ll 1 \tag{13.130}$$

in which u_{00} is the value of the smallest detectable velocity, and

$$u_* = u \;\text{ at }\; t = t_* \tag{13.131}$$

Based upon up-to-date level of measurement techniques,

$$\xi \cong 0.01 \tag{13.132}$$

Recalling that the solutions (13.66) and (13.92) are different only within a small neighborhood of the equilibrium, let us find "equivalent" values of $k = k^*$ and $\varepsilon_0 = \varepsilon_0^*$ from the condition that the time t_0 of approaching the equilibrium by the solution (13.92), and the time t_0' of approaching the domain of insensitivity are the same. Equating t_0 and t_0' from (13.93), and (13.129), respectively, one arrives at a phenomenological relationship between k^* and ε_0^*:

$$\varepsilon_0^* = -\frac{\pi^2(1-k^*)\nu}{4\ell^2 \ell n\{(k^*-1)\ell n[\frac{\pi}{4}(1-\xi)]\}}, \quad \xi \cong 0.01 \tag{13.133}$$

The second relationship between these parameters can be derived from the following phenomenological concept. Turning to the constitutive law (13.11), one has to provide the property that this law is sufficiently close to the linear one for large velocity gradients (which, however, are smaller than the molecular velocity gradients). Since the molecular velocity gradients are of the order of $1/\tau$ where τ is the Maxwell relaxation time, one can write:

$$(\frac{1}{\varepsilon_0^*\tau})^{k-1} \cong 1-\xi \tag{13.134}$$

This condition guarantees that the difference between the constitutive law (13.11) and the linear law are within the bounds of accuracy of the velocity measurements for the entire domain where the equations for a Newtonian liquid are applicable.

As follows from (13.134):

$$\varepsilon_0^* = \frac{1}{\tau}(1-\xi)^{\frac{1}{1-k_*}} \tag{13.135}$$

Equations (13.133) and (13.135) express the phenomenological versions of the physical parameters k and ε_0 via the physical constant τ and the accuracy of measurements ξ. In contrast to k and ε_0, the constants k^* and ε_0^* are problem-dependent. Indeed, (13.133) was derived from the solution to a particular problem discussed in the Sect. 6, and this is why it contains the representative length ℓ. However, despite this limitation, (13.133) and (13.135) provide a good evaluation of the order of these parameters.

Equations (13.133) and (13.135) solved for water at 20^oC become:

$$\tau = 10^{-9}\text{sec}, \ \xi = 0.01, \quad \nu = 10^{-2}\frac{\text{cm}^2}{\text{sec}}, \quad \ell = 10 \text{ cm} \tag{13.136}$$

and lead to the following values of k^* and ε_0^* :

$$k = \frac{3109}{3111}, \quad \varepsilon_0 = 2.66.10^{-5}\,\text{sec} \tag{13.137}$$

Now we can evaluate the parameters β, γ, r_0, $F_{\max}$ and Δt introduced in the previous section [see (13.106), (13.116), (13.125–13.126)], respectively:

$$\beta \cong 2.10^{-7}, \quad \gamma \cong 1, \quad r_0 \cong \gamma t, \quad F_{\max} \cong 4.266\,\pi 10^{-5} \rho \nu r_*$$

$$\Delta t \simeq \frac{r_*^2}{\nu} \tag{13.138}$$

13.9 Application to Acoustics

So far we have considered only incompressible fluids. However, all the modifications of the constitutive law can be applied to gases too. Indeed, turning to (13.10), one can write:

$$\rho D = 2\mu'(\frac{\sqrt{I_2}}{\varepsilon_0})^{k-1} + (\mu' + \frac{1}{2}\mu'')(\frac{I_1}{\varepsilon_0})^{k-1} \tag{13.139}$$

where μ'' is the second viscosity. Then, instead of (13.10) one obtains:

$$\sigma_{jk} = 2\mu'(\frac{\sqrt{I_2}}{\varepsilon_0})^{k-1}\varepsilon_{jk} + [\mu''(\frac{I_1}{\varepsilon_0})^{k-1}\varepsilon_{\ell\ell} - p]\delta_{jk} \tag{13.140}$$

Here

$$I_1 = \varepsilon_{ii} \tag{13.141}$$

is the first invariant of the rate-of-strain tensor, and δ_{jk} is the Kronecker's delta symbol.

The momentum and mass conservation equations instead of (13.24), and (13.25), take the following form, respectively:

$$\rho(\frac{\partial \mathbf{v}}{\partial t} + \mathbf{v}\nabla\mathbf{v}) = -\nabla p + \mu'\{(\frac{\sqrt{I_2}}{\varepsilon_0})^{k-1}[\nabla.(\nabla\mathbf{v} + \nabla\mathbf{v}^T)]$$

$$+ (\nabla\mathbf{v} + \nabla\mathbf{v}^T).\nabla(\frac{\sqrt{I_2}}{\varepsilon_0})^{k-1}\} + (\mu' + \frac{1}{3}\mu'')[(\frac{I_2}{\varepsilon_0})^{k-1}\nabla\nabla.\mathbf{v} + \nabla.\mathbf{v}\nabla(\frac{I_2}{\varepsilon_0})^{k-1}] \tag{13.142}$$

$$\frac{\partial \rho}{\partial t} + \nabla\cdot(\rho\mathbf{v}) = 0 \tag{13.143}$$

For a one-dimensional compressible viscous flow, the normal stress is given by the following expression:

$$\sigma_{xx} = -p + (\frac{4}{3}\mu'' + \mu')\varepsilon_0^{1-k}(\frac{\partial u}{\partial x})^k \tag{13.144}$$

while the momentum and mass conservation equations describing acoustic waves reduce to:

$$\rho_0 \frac{\partial u}{\partial t} = -\frac{\partial p}{\partial x} + \tilde{\mu}\frac{\partial}{\partial x}(\frac{\partial u}{\partial x})^k \tag{13.145}$$

$$\frac{\partial \rho}{\partial t} = -\rho_0 \frac{\partial u}{\partial x} \tag{13.146}$$

in which

$$\tilde{\mu} = (\frac{4}{3}\mu'' + \mu')\varepsilon_0^{1-k} \tag{13.147}$$

and ρ_0 is the unperturbed value of the density.

After elimination of the pressure p, one arrives at the governing equation for acoustic waves:

$$\frac{\partial^2 u}{\partial t^2} = c^2 \frac{\partial^2 u}{\partial x^2} + \tilde{\nu}\frac{\partial^2}{\partial t \partial x}(\frac{\partial u}{\partial x})^k, \; c^2 = \frac{dp}{d\rho}, \; \tilde{\nu} = \frac{\tilde{\mu}}{\rho} \tag{13.148}$$

Usually the last term describing viscous effects is ignored since it is much smaller than the elastic term. However, in our case the viscous term can be as sizable as the elastic term if the velocity gradients are small, i.e., if

$$\frac{\partial u}{\partial x} \; < \; \varepsilon_0 \tag{13.149}$$

In order to simplify (13.148) let us introduce a new variable τ instead of t:

$$\tau = t - \frac{x}{c} \tag{13.150}$$

Then (13.148) reduces to:

$$\frac{\partial u}{\partial x} = a^2 \frac{\partial}{\partial \tau}(\frac{\partial u}{\partial \tau})^k, \; a^2 = \frac{\nu}{c^3}(\varepsilon_0 c)^{1-k} \tag{13.151}$$

Equation (13.151) is identical to (13.62) if the variables x and τ are replaced by t and y, respectively.

Let us find an approximate solution to (13.151) subject to the following initial condition:

$$u(0, \tau) = \varphi(\tau), \; -\infty \; < \; x \; < \; +\infty \tag{13.152}$$

Seeking the solution in the form:

$$u = \int u_\lambda(x) f_\lambda(\tau) d\lambda \tag{13.153}$$

and substituting (13.153) into (13.151) one obtains:

$$f_\lambda = [(f'_\lambda)^k \,]' \tag{13.154}$$

These equations are similar to (13.81), and therefore, their solutions (13.82) can be simplified to the approximations (13.84) which lead to the following form:

$$f''_\lambda + \lambda^2 f_\lambda = 0, \quad f_\lambda = e^{i\lambda\tau} \tag{13.155}$$

Then the functions u_λ can be found from the following equations:

$$\frac{du_\lambda}{dx} + a^2\lambda^2 x^k = 0 \tag{13.156}$$

whence

$$u_\lambda = u_\lambda^0[1 - \frac{(1-k)a^2\lambda^2}{u_\lambda^0}x]^{\frac{1}{1-k}} \tag{13.157}$$

Substituting (13.155) and (13.157) into (13.153), one arrives at a Fourier integral:

$$u = \int u_\lambda^0[1 - \frac{(1-k)a^2\lambda^2}{u_\lambda^0}x]^{\frac{1}{1-k}}\, e^{i\lambda\tau} d\lambda \tag{13.158}$$

Since

$$\varphi(\tau) = \int u_\lambda^0 \, e^{i\lambda\tau} d\lambda \tag{13.159}$$

the values of u_λ^0 can be found as a Fourier coefficients for $\varphi(\tau)$:

$$u_\lambda^0 = \frac{1}{2\pi}\int \varphi(\xi)\, e^{-i\lambda\tau}\, d\xi \tag{13.160}$$

Substituting (13.160) into (13.153) one obtains the solution in the following integral form:

$$u = \int_\infty^\infty \{[\frac{1}{2\pi}\int_\xi \varphi(\xi)\, e^{-i\lambda\tau}\, d\xi]^{1-k} - (1-k)a^2\lambda^2 x\}^{\frac{1}{1-k}}\, e^{i\lambda\tau}\, d\lambda \tag{13.161}$$

Let us assume that

$$u(0,\tau) = \varphi(\tau) = u_0 \sin\omega\tau \tag{13.162}$$

Then

$$\lambda^2 = \omega^{k+1}, \; u_\lambda^0 = u_0 \tag{13.163}$$

and, as follows from (13.161)

$$u = [u_0^{1-k} - (1-k)a^2\omega^{k+1}x]^{\frac{1}{1-k}} \sin\omega(t - \frac{x}{c}) \tag{13.164}$$

Equation (13.164) describes a travelling acoustic wave generated by a sinusoidal source of sound located at $x = 0$. The wave propagates with the classical acoustic speed $c^2 = dp/d\rho$, but the amplitude of this wave gradually decreases with increase of the distance x from the source.

Eventually, at the distance

$$| x | \geq \ell \frac{u_0^{1-k}\omega^{(1-k)}}{(1-k)a^2\omega^2} = \frac{(1-k)c^3}{\nu\omega^2}\left(\frac{u_0\omega}{c\varepsilon_0}\right)^{1-k} \tag{13.165}$$

the sound vanishes being absorbed due to the viscosity.

As follows from (13.165), the critical distance ℓ depends upon two new physical constants: k and ε_0. As can be expected, this distance decreases with the increase of viscosity ν and the frequency ω.

It should be recalled that in the classical case the solution to the same problem, instead of (13.164), would be:

$$u = u_0 \, e^{-a^2\omega^2 x} \sin\omega\left(t - \frac{x}{c}\right) \tag{13.166}$$

This means that an acoustic wave is never fully absorbed; the distance over which the source of sound can be detected, is infinitely large. Equation (13.165) gives a correction to this idealized result stating that this distance is finite.

13.10 Application to Elastic Bodies

Effects, similar to those described in viscous fluids, occur in elastic bodies if dissipation processes are taken into account. Indeed, in this case the total stress tensor can be combined from the elastic and viscous components:

$$\sigma = \sigma_e + \sigma_v \tag{13.167}$$

while usually

$$\sigma_e \gg \sigma_v \tag{13.168}$$

However, in domains close to equilibria, the elastic stresses vanish, and

$$\sigma_v \gg \sigma_e \tag{13.169}$$

But in these domains the viscous stress can be represented by (13.11), and therefore, the governing equations take the form (13.40). This means that all the effects described above including finite time of approaching equilibria, non-uniqueness of solutions starting from equilibria, as well as finite distance of absorption of an acoustic wave, occur in elastic bodies in the domains close to equilibria.

13.11 Discussion and Conclusions

Plasticity effects are well-pronounced in heavy viscous liquids such as lubricators and dyes, but they have never been studied as classical Newtonian fluids like water, or air, presumably because they were expected to be vanishingly small. Our analysis demonstrates that although the quantitative contribution of these effects is small indeed, qualitatively even infinitesimal static friction leads to two new fundamental properties which are: the theoretically finite times of approaching equilibria, and the non-uniqueness of solutions which start at equilibria.

The first property can be associated with the paradox of irreversibility – one of the most fundamental and yet not fully understood problems in physics. Indeed, the concept of viscosity can be derived, on the microscopic level of description, from the fully reversible equations of Hamiltonian dynamics. This means that the irreversible processes in viscous flows are completely composed of reversible events. One of the most convincing and well accepted explanations of this paradox is that the change from an ordered arrangement to a disordered arrangement on the microscopic level as a source of irreversibility is much more probable than a change in the opposite direction.

In other words, any macroscopic system, in principle, can return to its initial state passing through all of its previous states; however, the probability of such an event is so small (but not zero!), that the period of time during which this event can occur is extremely large in comparison to the time scale of the macroscopic motions. However, the Navier-Stokes equations, or their simplified version – the diffusion equation – do not have any time scale: the time of approaching an equilibrium is unbounded, and therefore, these equations exclude any reversible solutions even if $t \to \infty$. The only logical way out of this situation is to introduce a time-scale into the Navier-Stokes equations so that the time of approaching an equilibrium will be finite. Then one can argue that this time is not large enough to include reversible solutions. Actually this time scale was introduced by relaxing the Lipschitz condition at the equilibrium states [see (13.37)]. Simple experiments which allow one to find the constants defining this scale were also described.

The second property is linked to another fundamental, but still unsolved problem of theoretical physics – the problem of turbulence. From a formal mathematical viewpoint, turbulence results from the dynamical instability of the Navier-Stokes equations when the Reynolds number exceeds certain critical values, and it is described by stochastic solutions. But how can such solutions occur from a fully deterministic model? A physical explanation is very simple: possible uncertainties and small errors (which always can be interpreted as random components of initial conditions) are amplified by the mechanism of instability, and this leads to the stochasticity of the solutions for supercritical Reynolds numbers. In other words, turbulence is caused by a random input into the fully deterministic Navier-Stokes equations. However, a mathematician can argue that, in principle, there is always a possibility that there are no uncertainties or errors in initial conditions at all, and then the solutions will never become stochastic. The modified version of the Navier-Stokes equations attains a very fundamental new property: it generates stochasticity as a result of the non-uniqueness of the solution which, in turn, follows from relaxation of the Lipschitz condition at equilibrium states. In cases of dynamical instability the random components of the solution are amplified and this leads to stochastic solutions simulating turbulence. Otherwise these random components decay and vanish.

It should be stressed that although the qualitative differences between the classical and modified Navier-Stokes equations are fundamental, all the new effects emerge within vanishingly small neighborhoods of equilibrium states which are the only domains where the modified governing equations are different from classical. This means that the formal differences between the solutions to classical and modified models may not be detectable in domains which do not include equilibrium states.

14
Quantum Intelligence

Quantum particles: the dreams that stuff is made of. –*D. Moser*

14.1 Introduction

During the last fifty years, a theory of computations has been based upon classical physics implemented by the deterministic Turing machine. However, along with the many successes of digital computers, the existence of so-called hard problems has placed some limitations on their capabilities, since the computational time for such problems grows exponentially as a function of the dimensionality. It was well understood that the only way to fight the "curse" of the combinatorial explosion is to enrich digital computers with analog devices. In contradistinction to a digital computer which performs operations on numbers symbolizing an underlying physical process, an analog computer processes information by exploiting physical phenomena directly, and thereby, it significantly reduces the complexity of the computations. This idea was stressed by Feynman (1982) who demonstrated that the problem of exponential complexity in terms of calculated probabilities can be reduced to a problem of polynomial complexity in terms of simulated probabilities. However, the main disadvantage of analog computers is a lack of universality. This is why the concept of a quantum computer became so attractive: its analog nature is based upon physical simulations of quantum probabilities, and, at the same time, it is universal (at least, for modeling the physical world).

Although the development of the quantum-mechanical device is still in progress (Turchette et al. 1995) a new quantum theory of computations has been founded (Deutsch 1989; Shor 1994). This theory suggests that there is a second fundamental advantage of the hypothetical quantum computer which is based upon the wave properties of quantum probabilities: a single quantum computer can follow many distinct computational paths all at the same time and produce a final output depending on the interference of all of them. This particular property opened up a new chain of algorithms which solve in polynomial time such hard problems as factorization and discrete log i.e., the problems which are believed to be intractable on any classical computer.

Thus, there are at least two areas where the quantum computer is expected to be superior over the classical one: quantum mechanics (due to simulation of quantum probabilities), and some specific combinatorial problems linked to operations research (due to interference of quantum probabilities).

However, in addition to quantum mechanics, there are other computational "worlds" (biological, psychological, and social dynamics, informatics, artificial intelligence) (although we note there has been some intensive activity in the biological area – see the Epilogue) where a quantum mechanical microstructure has not been observed. Will the quantum computer be superior in these areas too? There are some doubts about this. Indeed, quantum mechanics is fully reversible, while the dynamics associated with biology, psychology and other intellectual activities is fundamentally irreversible: their time evolution describes transformations to higher levels of complexity. Consequently, a quantum computer can provide only calculations rather than simulations in these areas.

A similar disadvantage may occur even in computations in classical physics if the models involve transport phenomena (heat transfer, diffusion, etc.) since then the problems also become irreversible. It should be noted that despite the quantum-mechanical micro-structure of the transport phenomena, the transformation from the fully reversible Hamiltonian models to their macroscopically irreversible versions is very sophisticated (in particular, it includes transition through chaotic instability, averaging, etc.). This is why direct simulations of classical physics by a quantum computer is in question. In addition to this, computations in classical physics are expected to produce deterministic answers while a quantum computer provides the answers in terms of probabilities. It should be noted, however, that the last property of a quantum computer can be useful for classical computing in the case of a Monte-Carlo approach, if the quantum computer is exploited as a random number generator.

This chapter has been motivated by an attempt to incorporate into classical computing the basic ideas of quantum computing: the simulations of probabilities, and the interference between different branches of probabilistic scenarios (Zak 1997a,b).

In contradistinction to the previous chapters, where the origin of the bias term ε_0 [see (7.218), (8.105), etc.] was not specified, here we represent ε_0 as a result of a chaotic time series, and, in particular, of a logistic map, where the source of stochasticity is in finite precision of initial conditions. The combination of chaos and terminal dynamics systems (Zak 1994a,b; 1996) can generate finite-state Markov chains. The corresponding stochastic process is described by two types of equations: the first one implements simulations in the form of random solutions, while the second (which is of the Fokker-Planck type) describes the evolution of the probabilities. Coupling between these two types of equations implements interference of probabilities similar (but not identical) to those in quantum mechanics.

The third, fourth and fifth sections describe a mathematical formalism behind the dynamical simulations of stochastic processes including a probabilistic Turing machine, the concept of stochastic attractors, as well as a new interpretation of simulated conditional probabilities.

The sixth section addresses applications to the modeling of intelligent systems.

In the seventh section, the effect of reducing computational complexity in combinatorial problems due to classical imitations of probability interference is discussed. Based upon this effect, the concept of quantum intelligence is introduced.

The last section introduces a hypothetical deterministic dynamical system in a pseudo-Euclidean space which, in principle, can simulate the Schrödinger equation.

14.2 Proof-of-Concept

Classical dynamics is fully deterministic if initial conditions are known exactly. Otherwise in some nonlinear systems, small initial errors may grow exponentially so that the system behavior attains stochastic-like features, and such a behavior is called chaotic. The discovery of chaos contributed to a better understanding of irreversibility in dynamics, of evolution in nature, and in the interpretation and modeling of complex phenomena in physics and biology. However, there is a class of phenomena which cannot be represented by chaos directly. This class includes so called discrete events dynamics where randomness appears as point events, i.e., there is a sequence of random occurrences at fixed or random times, but there is no additional component of uncertainty between these times. The simplest example of such a phenomenon is a heartbeat dynamics which, in the first approximation, can be modeled by a sequence of pulses of equal heights and durations, but the durations of the pauses between these pulses are randomly distributed (see Chap. 11). Most processes of this type are associated with intellectual activities such as optimal behavior, decision making process, games, etc. In general, discrete events dynamics is characterized by a well-defined probabilistic structure of piecewise deterministic Markov chains, and it can be represented by probabilistic Turing machine. To the contrary, a probabilistic structure of chaos, and even the appearance of chaos at all, cannot be predicted based only upon the underlying model without actual numerical runs. (The last statement can be linked to Richardson's (1968) proof that the theory of elementary functions in classical analysis is undecidable). But is there a "missing link" between chaos and discrete events dynamics? And if it is, can this link be simulated based only upon physical laws without exploiting any man-made devices such as random number generators? A positive answer to this question would make a fundamental contribution to the reductionists view on intrinsic unity of science that all natural phenomena are reducible to physical laws. However, in addition to this philosophical aspect, there is a computational advantage in exploiting simulated probabilities instead of calculated ones in the probabilistic Turing machine: as shown by Feynman (1982), the exponential complexity of algorithms in terms of calculated probabilities can be reduced to polynomial complexity in terms of simulated probabilities.

In this section we demonstrate that the missing link between chaos and a discrete event process can be represented by terminal dynamics, (Zak 1994a, 1996). In order to illustrate the basic concepts of terminal dynamics, consider a rectilinear motion of a particle of unit mass driven by a terminal dynamics force:

$$\dot{v} = \nu v^{1/3} \sin \omega t, \ \nu = \text{Const}, \ |v| = \frac{\text{m}^{2/3}}{\text{sec}^{5/3}} \tag{14.1}$$

$$\dot{x} = v \tag{14.2}$$

where v and x are the particle velocity and position, respectively.

Subject to the zero initial condition

$$v = 0 \text{ at } t = 0 \tag{14.3}$$

(14.1) has a singular solution

$$v = 0 \tag{14.4}$$

and a regular solution

$$v = \pm \left(\frac{4\nu}{3\omega} \sin^2 \frac{\omega}{2} t \right)^{3/2} \tag{14.5}$$

These two solutions coexist at $t = 0$, and this is possible because at this point the Lipschitz condition fails:

$$\left| \frac{\partial \dot{v}}{\partial v} \right|_{t \to 0} = \frac{1}{3} \nu v^{-2/3} \sin \omega t|_{t \to 0} \to \infty \tag{14.6}$$

Since

$$\frac{\partial \dot{v}}{\partial v} > 0 \text{ at } |v| \neq 0, \; t > 0 \tag{14.7}$$

the singular solution (14.4) is unstable, and the particle departs from rest following the solution (14.5). This solution has two (positive and negative) branches [since the power in (14.5) includes the square root], and each branch can be chosen with the probability p and (1-p) respectively. It should be noted that as a result of (14.5), the motion of the particle can be initiated by infinitesimal disturbances (such motion never can occur when the Lipschitz condition holds: an infinitesimal initial disturbance cannot become finite in finite time).

Strictly speaking, the solution (14.5) is valid only in the time interval

$$0 \leq t \leq \frac{2\pi}{\omega} \tag{14.8}$$

and at $t = 2\pi/\omega$ it coincides with the singular solution (14.4). For $t > 2\pi/\omega$ (14.4) becomes unstable, and the motion repeats itself to the accuracy of the sign in (14.5).

Hence, the particle velocity v performs oscillations with respect to its zero value in such a way that the positive and negative branches of the solution (14.5) alternate randomly after each period equal to $2\pi/\omega$.

Turning to (14.2), one obtains the distance between two adjacent equilibrium positions of the particle:

$$x_i - x_{i-1} = \pm \int_0^{2\pi/\omega} \left(\frac{4\nu}{3\omega} \sin^2 \frac{\omega}{2} t \right)^{3/2} dt = 64(3\omega)^{-5/2}\nu^{3/2} = \pm h \tag{14.9}$$

Thus, the equilibrium positions of the particle are

$$x_0 = 0, \; x_1 = \pm h, \; x_2 = \pm h \pm h, \ldots \tag{14.10}$$

while the positive and negative signs randomly alternate with probabilities p and (1-p), respectively.

Obviously, the particle performs an unrestricted random walk: after each time period

$$\pi = \frac{2\pi}{\omega} \tag{14.11}$$

it changes its value on $\pm h$ [see (14.10)]. The probability density $f(x, t)$ is governed by the following difference equation:

$$f(x, t + \tau) = pf(x - h, t) + (1 - p)f(x + h, t) \tag{14.12}$$

which represents a discrete version of the Fokker-Planck equation, while

$$\int_{-\infty}^{\infty} f(x, t) dx = 1 \tag{14.13}$$

Several comments to the model (14.1) and its solution have to be made. Firstly, the "viscous" force

$$F = -\nu v^{1/3} \tag{14.14}$$

includes static friction [see (14.6)] which actually causes failure of the Lipschitz condition. These type of forces are well-known in theory of viscoplasticity (Ziegler 1963). It should be noted that the power, 1/3, can be replaced by any power of the type:

$$k = \frac{2n-1}{2n+1}, \quad n = 1, 2, ..., \text{etc.} \tag{14.15}$$

with the same final result (14.12). In particular, by selecting large n, one can make k close to 1, so that the force (14.13) will be almost identical to its classical counterpart

$$F_c = -\nu v \tag{14.16}$$

everywhere excluding a small neighborhood of the equilibrium point $v = 0$, while at this point

$$\frac{dF}{dv} \to \infty, \text{ but } \left|\frac{\partial F_c}{\partial v}\right| \to 0 \text{ at } v \to 0 \tag{14.17}$$

Secondly, without the failure of the Lipschitz condition (14.6), the solution to (14.1) could not approach its equilibrium $v = 0$ in finite time, and therefore, the paradigm leading to a random walk (14.12) would not be possible.

Finally, we have to discuss the infinitesimal disturbances mentioned in connection with the instability of the solutions (14.5) at $v = 0$. Actually the original equation should be written in the form:

$$\dot{v} = \nu v^{1/3} \sin \omega t + \varepsilon(t), \ \varepsilon \to 0 \tag{14.18}$$

where $\varepsilon(t)$ represents a time series sampled from an underlying stochastic process representing infinitesimal disturbances. It should be emphasized that this process is not driving the solution of (14.18): it only triggers the mechanism of instability which controls the energy supply via the harmonic oscillations $\sin \omega t$. As follows from (14.18), the function $\varepsilon(t)$ can be ignored when $\dot{v} = 0$ or when $\dot{v} \neq 0$, but the equation is stable, i.e. $v = \pi\omega$, $2\pi\omega, ...$, etc. However, it becomes significant during the instants of instability when $\dot{v} = 0$ at $t = 0, \ \pi/2\omega, ...$, etc. Indeed, at these instants, the solution to (14.1) has a choice to be positive or negative if $\varepsilon = 0$, [see (14.5)]. However, with $\varepsilon \neq 0$,

$$\text{sgn } x = \text{sgn } \varepsilon \text{ at } t = 0, \ \pi/2\omega, ..., \text{etc.} \tag{14.19}$$

i.e., the sign of ε at the critical instances of time (14.19) uniquely defines the evolution of the dynamical system (14.18). Thus, the dynamical system (14.18) transforms a stochastic process [via its sample $\varepsilon(t)$] into a binary time series which, in turn, generates a random-walk-paradigm (14.18). Actually the solution to (14.18) represents a statistical signature of the stochastic process ε .

Within the framework of the dynamical formalism, the time series $\varepsilon(t)$ can be generated by a fully deterministic (but chaotic) dynamical system. The simplest of such systems is the logistic map which plays a central role in population dynamics, chemical kinetics and many other fields. In its chaotic domain

$$y_{n+1} = 4y_n(1 - y_n), \ y_0 = 0.2 \tag{14.20}$$

the power spectrum for the solution is indistinguishable from a white noise. However, for a better match with (14.18), we will start with a continuous version of (14.20) represented by the following time-delay equation.

$$y(t + \tau) = 4y(t)[1 - y(t)], \ \tau = \frac{\pi}{2\omega} \tag{14.21}$$

$$y(t^*) = 0.2, \quad -\frac{\pi}{4\omega} < t^* < \frac{\pi}{4\omega} \tag{14.22}$$

The solution to (14.21) at $t = 0, \pi/2\omega$,..., etc., coincides with the solution to (14.20), but due to the specially selected initial condition (14.22), the solution to (14.20) changes its values at $t = -\frac{\pi}{4\omega}, \frac{\pi}{4\omega}$,..., etc., so that at the points $t = 0, \pi/2\omega, ...$, the sign of this solution is well-defined.

Now assume that

$$\varepsilon(t) = \varepsilon_0[y(t) - 0.51], \ \varepsilon_0 \ll 1 \tag{14.23}$$

The subtraction from $y(t)$ of its mean value provides the condition

$$p = 1 - p = \frac{1}{2} \tag{14.24}$$

Indeed, for the first hundred points in (14.23),

$$\operatorname{sgn} \varepsilon = \begin{array}{l} -++-+++--+--+------++++--+ \\ --++-+-+--++----+-+---++-+ \\ +-++-+++--++++++-++++++--+ \\ +----++---+-+---+--- \end{array} \tag{14.25}$$

has an equal number of positive and negative values which are practically not correlated. Therefore, the statistical signature of the chaotic time series (14.23) is expressed by the solution to (14.12–14.13) at $p = \frac{1}{2}$ with the initial conditions

$$f(0,0) = 1, \ f(x,0) = 0 \text{ if } x \neq 0 \tag{14.26}$$

which is a symmetric unrestricted random walk:

$$f(x,t) = C_n^m 2^{-n}, \ m = \frac{1}{2}(n + x), \ n = \text{integer} \left(\frac{2\omega t}{\pi}\right) \tag{14.27}$$

Here the binomial coefficient should be interpreted as 0 whenever m is not an integer in the interval $[0, n]$, and n is the total number of steps.

The connection between the solution (14.26) and the solutions to the system (14.18), (14.21), (14.2) should be understood as follows. Suppose we solve the system (14.18), (14.21), (14.2) subject to the initial condition (14.22) with $v = 0$ and $x = 0$ at $t = 0$.

Since (14.21) is supersensitive to inevitable errors in (14.22), the solution will form an ensemble of chaotic time series, and for any fixed instant of time this ensemble will have the corresponding probability distribution which coincides with (14.26). In other words, the probabilities described by (14.12), are simulated by the dynamical system (14.18), (14.21) and (14.2) without an explicit source of stochasticity [while the "hidden" source of stochasticity is in finite precision of the initial condition (14.22)].

Combining several dynamical systems of the type (14.18), (14.21), (14.2) and applying an appropriate change of variables, one can simulate a probabilistic Turing machine which transfers one state to another with prescribed transitional probabilities. Non-Markovian properties of such a machine can be incorporated by introducing time-delay terms in (14.2):

$$\dot{x} = v(t) + \alpha_1 v(t - \tau_0) + \alpha_2 v(t - 2\tau_0) + \cdots \tag{14.28}$$

However, there is a more interesting way to enhance the dynamical complexity of the system (14.18), (14.21), (14.2). Indeed, let us turn to (14.23) and introduce a feedback from (14.2) to (14.18) as following:

$$\varepsilon = \varepsilon_0(\tilde{y} - x), \ \varepsilon_0 \ll 1, \ \tilde{y} = y - 0.51 \tag{14.29}$$

Then the number of negative (positive) signs in the string (14.25) will prevail if $x > 0$ ($x < 0$) since the effective zero-crossing line moves down (up) away from the middle. Thus, when $x = 0$ at $t = 0$, the system starts with an unrestricted random walk as described above, and $|x|$ grows. However, this growth changes signs in (14.23) such that $\dot{x} < 0$ if $x > 0$, and $\dot{x} > 0$ if $x < 0$. As a result of this

$$x_{\max} \le y_{\max}, \ x_{\min} \ge y_{\min} \tag{14.30}$$

where $y_{\max}$ and $y_{\min}$ are the largest and the smallest values in the time series $y(t)$, respectively. Hence, the dynamical system (14.18), (14.23), (14.2) simulates a restricted random walk with the boundaries (14.30) implemented by the dynamical feedback (14.29), while the probability

$$p(\operatorname{sgn} \varepsilon > 0) = \begin{cases} 0 \text{ if } & x \ge y_{\max} \\ 1 \text{ if } & x \le y_{\min} \end{cases} \tag{14.31}$$

For the sake of qualitative discussion, assume that p changes linearly between $x = y_{\min}$ and $x = y_{\max}$, i.e.,

$$p = \begin{cases} 0 \text{ if } & x > y_{\max} \\ \frac{y_{\max} - x}{y_{\max} - y_{\min}} \text{ if } & y_{\min} \le x \le y_{\max} \\ 1 \text{ if } & x < y_{\min} \end{cases} \tag{14.32}$$

[The actual function $p(x)$ depends upon statistical properties of the underlying chaotic time series $y(t)$. In particular, for the logistic map (14.20), small deviations from (14.32) take place only around the ends (i.e., when $x \cong y_{\max}$ or $x \cong y_{\min}$]. Then the simulated restricted random walk is a solution to (14.12) and (14.32).

Let us modify the feedback (14.29) as

$$\varepsilon = \varepsilon_0 \left[\tilde{y} - (x^2 - x)\right] \tag{14.33}$$

Now when $x = 0$ at $t = 0$, the system is unstable since

$$\operatorname{sgn} x = \operatorname{sgn} \dot{x}, \ -\infty < x < \frac{1}{2} \tag{14.34}$$

and the process is divided into two branches. The negative branch (with the probability 1/2) represents an unrestricted random walk ($x \to \infty$), while the positive branch (with the same probability, 1/2) is eventually trapped within the basin of the attractor, $x = 1$, since

$$\operatorname{sgn} x = -\operatorname{sgn} \dot{x}, \ \frac{1}{2} < x < \infty \tag{14.35}$$

simulating a restricted random walk as that described above with the only difference that its center is shifted from $x = 0$ to $x = 1$.

As a next step in complexity, let us introduce the information H associated with the random walk process described by (14.12–14.13):

$$H = -\int_{-\infty}^{\infty} f \log_2 f dx \tag{14.36}$$

and modify the feedback (14.29) as following:

$$\varepsilon = \varepsilon_0[\tilde{y} - x(1 + H)] \tag{14.37}$$

Following the same line of argumentation as that performed for feedback (14.29), one concludes that the feedback (14.38) becomes active only if the process is out of the domain of maximum information, and therefore, it is always attracted to this domain.

Since (14.31) is still valid, we will apply the approximation similar to (14.32):

$$p = \begin{cases} 0 \text{ if} & x(1 + H) \geq y_{\max} \\ \frac{y_{\max} - x(1+H)}{y_{\max} - y_{\min}} \text{ if} & y_{\min} \leq x(1 + H) \leq y_{\max} \\ 1 \text{ if} & x(1 + H) \leq y_{\min} \end{cases} \tag{14.38}$$

in order to continue our qualitative analysis. It should be noted that now p depends not only on x, but also on f, and that makes (14.12) nonlinear. In addition to this, the system (14.18), (14.2) and (14.37), which simulates probabilities, is coupled with the system (14.12–14.13) and (14.38) describing the evolution of calculated probabilities. Actually due to this coupling, the entire dynamical system attains a self-organizing property so as to maximize information generated by the random walk.

The self-organizing properties of the system (14.18), (14.2), (14.12–14.13), (14.37–14.38), mentioned above have a very interesting computational interpretation: they provide a mutual influence between different branches of probabilistic scenarios. Such an influence or interference, is exploited by the hypothetical quantum computer (Shor 1994) as a more powerful tool, in a complexity theoretic sense, than classical probabilistic computations. However, in the quantum computer, the interference is restricted to a linear unitary matrix transformation of probabilities (which is the only one allowed by quantum mechanics laws), while in the classical system (14.18), (14.2), (14.37) there is no such restriction; by choosing an appropriate probabilistic term in the feedback (14.37), we can provide an optimal interference. The price paid for such a property is the necessity to exploit the calculated probabilities (14.12–14.13) and (14.38), the consequences of which will be discussed in the last section.

14.3 Attractors and Nonlinear Waves of Probability

Let us return to the simplest case of a restricted random walk described by (14.18), (14.29) and (14.2), and analyze the corresponding probability equation, following from (14.12) and (14.32)

$$f(x, t+\tau) = \frac{y_{\max} - x}{y_{\max} - y_{\min}} f(x-h, t) + \frac{x - y_{\min}}{y_{\max} - y_{\min}} f(x+h, t) \quad (14.39)$$

$$y_{\min} \leq x \leq y_{\max} \quad (14.40)$$

For $x = y_{\max}$

$$f(y_{\max} t + \tau) = f(y_{\max} + h, t) \quad (14.41)$$

Hence, if

$$f(y_{\max} + h, t) = 0 \quad (14.42)$$

then

$$f(y_{\max} t + \tau) = 0 \quad (14.43)$$

This means that with the initial conditions (14.26),

$$f(y_{\max}, \tau) \equiv 0 \quad (14.44)$$

For the same reason:

$$f(y_{\min}, \tau) \equiv 0 \quad (14.45)$$

Thus, if the solution starts within the interval (14.40), it is trapped there.

Let us assume now that instead of (14.26), the initial conditions are:

$$f(x_0, 0) = 1, \ f(x, 0) = 0 \text{ if } x \neq 0, \ x_0 > y_{\max} \tag{14.46}$$

Then (14.39) reduces to

$$f(x, t + \tau) = f(x + h, t), \ x \geq y_{\max} \tag{14.47}$$

The solution to this equation can be written in the form of a traveling wave of the delta-function:

$$f = \delta(x - x_0 + t), \ x \geq y_{\max} \tag{14.48}$$

moving toward the interval (14.40).

When x approaches $y_{\max}$, (14.47) must be replaced by (14.39), and the solution is trapped again within the interval (14.40). The same effect occurs if the solution starts with $x_0 < y_{\min}$.

Thus, the interval (14.40) represents a stochastic attractor (or a closed set of states) i.e. a stochastic process to which all the solutions of (14.12), (14.32) converge irrespective of their initial conditions. Obviously then the intervals:

$$\infty > x \leq y_{\max} \text{ and } y_{\min} > x > -\infty \tag{14.49}$$

represent the basin of this attractor.

The feedback (14.29) can be modified as:

$$\varepsilon = \varepsilon_0(\tilde{y} - \alpha x), \ \alpha > 0 \tag{14.50}$$

without any qualitative changes in the results described above if the interval (14.40) is replaced by

$$\frac{y_{\min}}{\alpha} \leq x \leq \frac{y_{\max}}{\alpha} \tag{14.51}$$

However, there are significant qualitative changes if $\alpha < 0$: the interval (14.51) becomes a stochastic repeller, i.e., any solution originated within this interval will eventually escape it.

In the case of nonlinear feedback [instead of (14.50)]:

$$\varepsilon = \varepsilon_0 \left[\tilde{y} - \varphi(x)\right] \tag{14.52}$$

the conditions

$$\frac{d\varphi}{dx} > 0 \text{ or } \frac{d\varphi}{dx} < 0 \tag{14.53}$$

within the interval

$$x' \leq x \leq x'', \text{ where } \varphi(x') = y_{\max}, \ \varphi(x'') = y_{\min} \tag{14.54}$$

lead to stochastic attractor, or repeller, respectively. Several stochastic attractors and repellers within the interval (14.54) are possible if the sign of the derivative $\partial\varphi/\partial x$ is changing.

Let us turn now to non-Markov stochastic processes when the probability of transition to another state depends not only upon the present, but also upon the past states. This properly can be easily simulated by modifying the feedback (14.54) as follows:

$$\varepsilon = \varepsilon_0 \left\{ \tilde{y} - \varphi \left[x(t), x(t-\tau), x(t-2\tau), ..., \text{etc.} \right] \right\} \tag{14.55}$$

In order to apply (14.12), the approximation (14.32) should be replaced by:

$$p = \begin{cases} 0 \text{ if} & \varphi > y_{\max} \\ \frac{y_{\max} - \varphi}{y_{\max} - y_{\min}} \text{ if} & y_{\min} \leq x \leq y_{\max} \\ 1 \text{ if} & \varphi < y_{\min} \end{cases} \tag{14.56}$$

and substituted in (14.12).

Now (14.12) represents a more correlated (non-Markov) stochastic process, while by appropriate selection of the function φ in (14.55), a prescribed probability distributions as well as correlation's functions can be incorporated into the simulated stochastic process.

For an illustration of the dynamical simulation of a non-Markov stochastic process assume that

$$\tau = 1, \ \varphi = \bar{x}(t) + \bar{x}(t-1) - \frac{1}{2}, \ \bar{x}(0) = \frac{1}{4}, \ |y_{\max}| = |y_{\min}|, \ \bar{x} = \frac{x}{|y_{\max}|} \tag{14.57}$$

The last assumption is natural for the mean-zero chaotic process $\bar{y}$ [see (14.29)]. Then

$$p = \begin{cases} 0 \text{ if} & \bar{x}(t) + \bar{x}(t-1) = \frac{1}{2} \\ 1 \text{ if} & \bar{x}(t) + \bar{x}(t-1) = -\frac{1}{2} \end{cases} \tag{14.58}$$

i.e.

$$p = \begin{cases} 0 \text{ if} & \bar{x} = \frac{1}{2} - \frac{1}{4}\cos \pi t \\ 1 \text{ if} & \bar{x} = -\frac{1}{2} + \frac{3}{4}\cos \pi t \end{cases} \tag{14.59}$$

Thus, the non-Markov process characterized by the feedback (14.55), (14.57) has a stochastic attractor with variable length:

$$0 \leq \ell = \frac{1}{2} - \frac{1}{4}\cos \pi t - (\frac{1}{2} + \frac{3}{4}\cos \pi t) = 1 - \cos \pi t \leq 2 \tag{14.60}$$

At the instants

$$t = 0, 2\pi, ..., \text{etc.} \tag{14.61}$$

the attractor disappears, and the process becomes fully deterministic. One can verify that if φ in (14.57) is changed to the following:

$$\varphi = \bar{x}(t) + \bar{x}(t-1) + \frac{1}{2} \tag{14.62}$$

then the domain (14.60) represents a stochastic repeller.

An important step toward a higher complexity of simulated stochastic processes is the dependence of feedback (14.55) upon probability functionals [see, for instance, (14.37)]. There are two fundamentally new effects here. Firstly, the equation (14.12) becomes nonlinear, and secondly, the dynamical simulations become explicitly dependent upon the probabilities.

In order to illustrate this, consider the following feedback instead of (14.50):

$$\varepsilon = \varepsilon_0 \left[\tilde{y} - \frac{\operatorname{sgn} x}{\alpha\, y_{\max}} \int_{-\infty}^{\infty} x^2 f(x) dx \right], \quad \alpha > 0 \tag{14.63}$$

where the variance

$$\sigma^2(t) = \int_{-\infty}^{\infty} x^2 f(x) dx \tag{14.64}$$

is a function of t and a functional of $f(x)$. Following the same line of argumentation as that applied for the derivation of (14.40) and (14.51), one obtains

$$\sigma^2 \leq \alpha y_{\max}^2 \text{ if } |y_{\max}| = |y_{\min}| \tag{14.65}$$

In other words, the feedback (14.63) produces a stochastic attractor whose variance is limited by the condition (14.65). However, the lengths of this attractor can be found only from the solution to (14.12) with p expressed as:

$$p = \begin{cases} 0 \text{ if} & \sigma^2 \geq \alpha y_{\max}^2, \ x > 0 \\ 1 \text{ if} & \sigma^2 \geq \alpha y_{\max}^2, \ x < 0 \\ \frac{1}{2}\left(1 - \frac{\operatorname{sgn} x}{\alpha y_{\max}^2} \sigma^2\right), & \sigma^2 \leq \alpha y_{\max}^2 \end{cases} \tag{14.66}$$

In the same way one can implement other limitations imposed upon probability densities of the simulated stochastic process.

In order to clarify the aspect of interference of probabilities provided by dependence of dynamical simulations upon probability densities, let us turn to (14.2), (14.18) and (14.63). The solution to this system is random; however, each probabilistic scenario is controlled by the feedback (14.63) which includes the probability density. This probability density is governed by (14.12) and (14.66) coupled with (14.2), (14.18) and (14.63). Hence, the mechanism of interference of probabilities here is different from that of quantum mechanics: it is "artificially" organized via a special architecture of the dynamical system.

14.4 Simulation of Conditional Probabilities

In previous sections we dealt with only one state variable x (while v and y played the role of auxiliary variables). Now we will consider multidimensional systems. Such systems could be simply obtained by starting with n identical systems of the type (14.18) and (14.2):

$$\dot{v}_i = \nu_i v_i^{1/3} \sin \omega t + \varepsilon_i(x_i), \ \dot{x}_i = v_i, \ i = 1, 2, \ldots, \text{etc.} \tag{14.67}$$

and then changing variables

$$w_i = \varphi_i(v_i, ..., v_n) \tag{14.68}$$

However, as will be shown below, there are multi-dimensional systems which are more complex than those represented by (14.67–14.68), and therefore, they should be discussed separately. In order to demonstrate this, consider the following two dimensional system:

$$\dot{v}_1 = \nu v_1^{1/3} \sin \omega t + \varepsilon_0(\bar{y} - x_2), \ \dot{x}_1 = v_1 \tag{14.69}$$

$$\dot{v}_2 = \nu v_2^{1/3} \sin \omega t + \varepsilon_0(\bar{y} - x_1), \ \dot{x}_2 = v_2 \tag{14.70}$$

with respect to variables x_1 and x_2.

Equations (14.69–14.70) are coupled via the feedbacks, but their associated probability equations are not coupled:

$$f_1(x_1, x_2, t + \tau) = p_1(x_2) f_1(x_1 - h, x_2, t) + [1 - p_1(x_2)] f_1(x_1 + h, x_2, t) \tag{14.71}$$

$$f_2(x_1, x_2, t + \tau) = p_2(x_1) f_2(x_1, x_2 - h, t) + [1 - p_2(x_1)] f_1(x_1, x_2 + h, t) \tag{14.72}$$

where

$$p_1 = \begin{cases} 0 \text{ if} & \bar{x}_2 > 1 \\ \frac{1}{2}(1 - \bar{x}_2) \text{ if} & |\bar{x}_2| \leq 1 \\ 1 \text{ if} & \bar{x}_2 < -1 \end{cases}, \quad p_2 = \begin{cases} 0 \text{ if} & \bar{x}_1 \geq 1 \\ \frac{1}{2}(1 - \bar{x}_1) \text{ if} & |\bar{x}_1| \leq 1 \\ 1 \text{ if} & \bar{x}_1 < -1 \end{cases} \tag{14.73}$$

$$|y_{\max}| = |y_{\min}|, \ \bar{x}_1 = \frac{x_1}{|y_{\max}|}, \ \bar{x}_2 = \frac{x_2}{|y_{\max}|} \tag{14.74}$$

It should be noted that x_2 and x_1 enter as parameters into (14.71–14.72) respectively. This is why $f_1(x_1 \mid x_2)$ and $f_2(x_2 \mid x_1)$ represent conditional probability densities: f_1 describes the density of x_1 given x_2 and f_2 describes the density of x_2 given x_1.

The solution to (14.71–14.72) subject to the initial conditions (14.26), and the condition (14.13) for a sufficiently small initial time interval t are:

$$f_1 = C_n^{m_1} \left(\frac{1 - \bar{x}_2}{2}\right)^{m_1} \left(\frac{\bar{x}_2 - 1}{2}\right)^{n - m_1}, \ m_1 = \frac{1}{2}(n + \bar{x}_1), \ n = \text{int}\left(\frac{2\omega t}{\pi}\right) \tag{14.75}$$

$$f_2 = C_n^{m_2} \left(\frac{1 - \bar{x}_1}{2}\right)^{m_2} \left(\frac{\bar{x}_1 - 1}{2}\right)^{n - m_2}, \ m_2 = \frac{1}{2}(n + \bar{x}_2) \tag{14.76}$$

Each of them represents a non-symmetric random walk before the reflections from the boundaries $|x_1| = 1$, $|x_2| = 1$ take place.

Now the following question can be asked: how to find an underlying joint probability density $\Phi(x_1, x_2)$. It turns out that this is a hard question even from a conceptual viewpoint. Indeed, the relationships between $f_1(x_1 \mid x_2)$, $f_2(x_2 \mid x_1)$ and $\Phi(x_1, x_2)$ are the following:

$$\Phi(x_1, x_{2)} = f_1(x_1 \mid x_2) \int_{-\infty}^{\infty} \Phi(z, x_2) dz = f_2(x_2 \mid x_1) \int_{-\infty}^{\infty} \Phi(x_1, z) dz$$

whence

$$\frac{f_1(x_1 \mid x_2)}{f_2(x_2 \mid x_1)} = \frac{\int_{-\infty}^{\infty} \Phi(x_1, z) dz}{\int_{-\infty}^{\infty} \Phi(z, x_2) dz}$$

i.e.

$$\ln \frac{f_1(x_1 \mid x_2)}{f_2(x_2 \mid x_1)} = \ln \int_{-\infty}^{\infty} \Phi(x_1, z) dz - \ln \int_{-\infty}^{\infty} \Phi(z, x_2) dz$$

and therefore

$$\frac{\partial^2}{\partial x_1 \partial x_2} \ln \frac{f_1(x_1 \mid x_2)}{f_2(x_2 \mid x_1)} \equiv 0 \tag{14.77}$$

Thus the existence of the joint probability density $\Phi(x_1, x_2)$ requires that the conditional probability densities must satisfy the compatibility equation (14.77). But it is easily verifiable that the solutions (14.75–14.76) do not satisfy this equation, i.e., they are incompatible:

$$\text{ink } (\varphi_1, \varphi_2) = \frac{\partial^2}{\partial x, \partial x_2} \ln \frac{f_1}{f_2} \neq 0 \tag{14.78}$$

At the same time, there is nothing wrong with these solutions since they describe two stochastic processes which can be implemented by dynamical simulations. Hence, the only conclusion which can be made is that the joint probability in this particular case does not exist! But how "particular" is this case? Based upon the degree of arbitrariness to which the feedbacks in the system (14.69–14.70) can be set up, it is obvious that the incompatibility of the conditional probabilities is a rule rather than an exception. In other words, there is a class of coupled stochastic processes for which a joint probability does not exist, and therefore, they are inseparable, i.e., there is no such transformation of variables which would break them down into independent components. A similar conclusion (but in a different context) was made by Cavello and George (1992).

Let us modify the feedbacks in (14.69–14.70) as follows:

$$\varepsilon_1 = \varepsilon \left[\tilde{y} - (\alpha_{11} x_1 + \alpha_{12} x_2) \right], \; \varepsilon_1 = \varepsilon_0 \left[\lambda \tilde{y} - (\alpha_{21} x_1 + \alpha_{22} x_2) \right]$$

$$\lambda = \text{Const} > 0, \; \alpha_{11}\alpha_{22} - \alpha_{12}\alpha_{21} \neq 0 \tag{14.79}$$

Then the probability equations (14.71–14.72) are replaced by

$$f_1(x_1, x_2, t+\tau) = p_1(x_1, x_2) f_1(x_1 - h, x_2, t) + [1 - p(x_1, x_2)] f_1(x_1 + h, x_2, t) \tag{14.80}$$

$$f_2(x_1, x_2, t+\tau) = p_2(x_1, x_2) f_2(x_1, x_2 - h, t) + [1 - p(x_1, x_2)] f_2(x_1, x + h, t) \tag{14.81}$$

where

$$p_1 = \begin{cases} 0 \text{ if} & \alpha_{11}\bar{x}_1 + \alpha_{12}\bar{x}_2 > 1 \\ \frac{1}{2}[1 - (\alpha_{11}\bar{x}_1 + \alpha_{12}\bar{x}_2)] \text{ if} & |\alpha_{11}\bar{x}_1 + \alpha_{12}\bar{x}_2| \leq 1 \\ 1 \text{ if} & \alpha_{11}\bar{x}_1 + \alpha_{12}\bar{x}_2 < 1 \end{cases} \tag{14.82}$$

$$p_2 = \begin{cases} 0 \text{ if} & \alpha_{21}\bar{x}_1 + \alpha_{22}\bar{x}_2 > \lambda \\ \frac{1}{2}[\lambda - (\alpha_{21}\bar{x}_1 + \alpha_{22}\bar{x}_2)] \text{ if} & |\alpha_{11}\bar{x}_1 + \alpha_{12}\bar{x}_2| \leq \lambda \\ 1 \text{ if} & \alpha_{21}\bar{x}_1 + \alpha_{22}\bar{x}_2 < \lambda \end{cases} \tag{14.83}$$

Following the same line of argumentation as that applied for the one-dimensional case [see (14.39–14.45)], let us find conditions for the existence of coupled stochastic attractors.

For

$$\bar{x}_1 = \bar{x}_1^0, \bar{x}_2 = \bar{x}_2^0 \tag{14.84}$$

where $\bar{x}_1^0$ and $\bar{x}_2^0$ are solutions to the system:

$$\alpha_{11}\bar{x}_1^0 + \alpha_{12}\bar{x}_2^0 = 1, \ \alpha_{21}\bar{x}_1^0 + \alpha_{22}\bar{x}_2^0 = \lambda \tag{14.85}$$

the conditions similar to (14.41) will be satisfied:

$$f_1(\bar{x}_1^0, \bar{x}_2^0, t+\tau) = f_1(\bar{x}_1^0 + h, \bar{x}_2^0, t) \tag{14.86}$$

$$f_1(\bar{x}_1^0, \bar{x}_2^0, t+\tau) = f_2(\bar{x}_1^0, \bar{x}_2^0 + h, t) \tag{14.87}$$

and therefore, if

$$f_1(\bar{x}_1^0 + h, \bar{x}_2^0, t) = 0 \text{ and } f_2(\bar{x}_1^0, \bar{x}_2^0 + h, t) = 0 \tag{14.88}$$

then

$$f_1(\bar{x}_1^0, \bar{x}_2^0, t+\tau) \equiv 0, \ f_2(x_1^0, x_2^0, t+\tau) \equiv 0 \tag{14.89}$$

This means that with initial conditions (14.26) [applied for both (14.80) and (14.81)],

$$f_1(\bar{x}_1^0, \bar{x}_2^0, t) \equiv 0, \ f_2(x_1^0, x_2^0, t) \equiv 0 \tag{14.90}$$

For the same reason:

$$f_1(\bar{x}_1^{00}, \bar{x}_2^{00}, t) \equiv 0, \ f_2(x_1^{00}, x_2^{00}, t) \equiv 0 \tag{14.91}$$

where $\bar{x}_1^{00}$, and $\bar{x}_2^{00}$ are solutions to the system:

$$\alpha_{11}\bar{x}_1^{00} + \alpha_{12}\bar{x}_2^{00} = -1, \ \alpha_{21}\bar{x}_1^{00} + \alpha_{22}\bar{x}_2^{00} = -\lambda \tag{14.92}$$

Hence, as in the one-dimensional case [see (14.49) and (14.45)], the system may be trapped within the region:

$$\bar{x}_1^{00} \leq \bar{x}_1 \leq \bar{x}_1^0, \ \bar{x}_2^{00} \leq \bar{x}_2 \leq \bar{x}_2^0 \tag{14.93}$$

if the inequalities following from (14.93)

$$\bar{x}_1^{00} \leq \bar{x}_1^0, \ \bar{x}_2^{00} \leq \bar{x}_2^0 \tag{14.94}$$

also follow from the solutions to the systems (14.85) and (14.92), and this imposes certain constraints upon the coefficients α_{11}, α_{12}, α_{21}, α_{22} and λ.

Indeed, these solutions are:

$$\bar{x}_1^0 = \frac{\alpha_{22} - \lambda\alpha_{12}}{\alpha_{11}\alpha_{22} - \alpha_{12}\alpha_{21}}, \ \bar{x}_2^0 = \frac{\alpha_{11} - \lambda\alpha_{21}}{\alpha_{11}\alpha_{22} - \alpha_{12}\alpha_{21}}$$

$$\bar{x}_1^{00} = \frac{\alpha_{22} - \lambda\alpha_{12}}{\alpha_{11}\alpha_{22} - \alpha_{12}\alpha_{21}} = -\bar{x}_1^0, \ \bar{x}_2^0 = \frac{\alpha_{11} - \lambda\alpha_{21}}{\alpha_{11}\alpha_{22} - \alpha_{12}\alpha_{21}} = -\bar{x}_2^{00} \tag{14.95}$$

and therefore, the inequality (14.94) is satisfied if

$$\frac{\alpha_{22} - \lambda\alpha_{12}}{\alpha_{11}\alpha_{22} - \alpha_{12}\alpha_{21}} > 0, \ \frac{\alpha_{11} - \lambda\alpha_{21}}{\alpha_{11}\alpha_{22} - \alpha_{12}\alpha_{21}} > 0 \tag{14.96}$$

Thus, the inequalities (14.96) guarantee that if the initial conditions for the system (14.80–14.81) are within the region (14.93), then the solutions will be trapped there. But if the initial conditions are outside of this region, then following the same line of argumentation as those for one-dimensional case [see (14.46–14.48)], one concludes that eventually the solutions will approach the region (14.93) and will remain there. Hence, the inequalities (14.96) represent the necessary and sufficient conditions that the region (14.93) is a two-dimensional stochastic attractor. It should be noted that in contradistinction to the stochastic attractors introduced by Zak (1996), this one is inseparable, i.e., it cannot be broken down into two independent one-dimensional stochastic attractors by any change of variables.

It can easily be verified that opposite signs in (14.96) convert the region (14.93) into a stochastic repeller, while different signs

$$\frac{\alpha_{22} - \lambda\alpha_{12}}{\alpha_{11}\alpha_{22} - \alpha_{12}\alpha_{21}} > 0, \ \frac{\alpha_{11}\lambda - \alpha_{21}}{\alpha_{11}\alpha_{22} - \alpha_{12}\alpha_{21}} < 0 \tag{14.97}$$

lead to attraction in the x_1 direction, and repulsion in the x_2 direction, i.e., to a mixed type of limit set. If

$$\alpha_{22} = \lambda\alpha_{12}, \ \alpha_{21} = \lambda\alpha_{11} \tag{14.98}$$

then the region (14.93) degenerates into a point, while if the only one equality in (14.98) holds, then this region degenerates into a line $x_2 =$ Const, or $x_1 =$ Const, respectively.

A general form of multi-dimensional simulated conditional probabilities can be derived from (14.69–14.73):

$$\dot{v}_i = \nu v_i^{1/3} \sin \omega t + \varepsilon_0 \left\{ \tilde{y} - \varphi_i \left[x_1, ..., x_n, \Phi\left(f_1, ..., f_n\right) \right] \right\}, \ \dot{x}_i = v_i \tag{14.99}$$

$$f_1(x_1, ..., x_n, t+\tau) = p_i f_i(x_1, ..., x_i - h, ...x_n, t) + (1-p_i) f_i(x_1, ..., x_i + h, ..., x_n, t) \quad (14.100)$$

where

$$p_i = \begin{cases} 0 \text{ if} & \varphi_i \geq y_{\max} \\ \frac{y_{\max} - \varphi_i}{y_{\max} - y_{\min}} \text{ if} & y_{\min} \leq \varphi_i \leq y_{\max} \\ 1 \text{ if} & \varphi_i \leq y_{\min} \end{cases} \quad (14.101)$$

and $\varphi_i(f_1, ..., f_n)$ are functionals of the probabilities, $f_1, ..., f_n$. Non-Markovian effects can be incorporated in this system by including time-delay variables $x_i(t-\tau)$ as additional arguments for the functions φ_i .

It should be emphasized that as in the two-dimensional case, all the stochastic processes here, in general, are inseparable.

14.5 Simulations of Probabilistic Turing Machine

All the simulated stochastic processes discussed above had the following limitation: they could perform a random jump only to the adjacent neighboring states, thereby representing a simple random walk. However, for the purpose of universal computations, such a performance is not sufficient: random jumps from an arbitrary state to any other state with a prescribed probability is required. In this section we will demonstrate how to modify the models introduced above to attain such a property.

Let us turn to (14.21–14.22) and assume that

$$\tau = \frac{\pi}{4\omega} n_1, \ \frac{\pi}{4\omega} n_2, ..., \text{etc.} \ -\frac{\pi}{4\omega} n < t^* < \frac{\pi}{4\omega} n \quad (14.102)$$

Now the solution to (14.20) changes its values at

$$t = \frac{\pi}{4\omega} n_1, \ \frac{\pi}{4\omega} n_2, ..., \text{etc.} \quad (14.103)$$

This means that the solution to (14.18), and (14.2) cannot have random jumps between the intervals (14.103), i.e., the length of a step in the random walk will be

$$h_n = nh \quad (14.104)$$

where h is expressed by (14.9).

Suppose that n is a random variable which is simulated by the dynamics:

$$\dot{v}_n = \nu_n v_n^{1/3} \sin \omega t + \varepsilon_0 \left[\tilde{y} - \varphi(n, x) \right], \ \dot{n} = v_n \quad (14.105)$$

where φ is an arbitrary function of n and x. Then, analogously to (14.9):

$$n_i - n_{i-1} = 64(3\omega)^{-5/2} \nu_n^{3/2} = \ell \quad (14.106)$$

where ℓ is the unit step of change of the variable n.

This step can be set

$$\ell = 1 \tag{14.107}$$

by an appropriate choice of ν_n

$$\nu_n = \frac{(3\omega)^{5/3}}{16} \tag{14.108}$$

Hence,

$$n = \pm 1, \pm 1, \pm 1, \pm \text{ etc.} \tag{14.109}$$

while the positive and negative signs in (14.109) alternate randomly with the probabilities p_n and $(1 - p_n)$, respectively, where

$$p_n = \begin{cases} 0 \text{ if} & \varphi(n,x) > y_{\max} \\ \frac{y_{\max} - \varphi(n,x)}{y_{\max} - y_{\min}} \text{ if} & y_{\min} \leq \varphi(n,x) \leq y_{\max} \\ 1 \text{ if} & \varphi(n,x) < y_{\min} \end{cases} \tag{14.110}$$

The probability density $f_n(n,x,t)$ is governed by the following difference equation :

$$f_n(n,x,t+\tau) = p_n f_n(n-1,x,t) + (1-p_n) f_n(n+1,x,t) \tag{14.111}$$

$$\sum_{n=1}^{n} f_n(n,x,t) = 1 \tag{14.112}$$

where p_n is expressed by (14.110).

Thus, now in all of the equations considered above [see (14.12), (14.39), (14.71–14.72),(14.80–14.81)], the unit step of change in x or x_i is expressed by (14.104) where n is a random variable with the probability distribution $f_n(n,x,t)$ governed by (14.111–14.112). Therefore, all the equations listed above should be modified, and we will illustrate this modification based upon (14.39) which now reads:

$$f(x,t+\tau) = \sum_{n=1}^{N} f_n \left[p f(x - nh, t) + (1-p) f(x+nh,t) \right] \tag{14.113}$$

$$\int_{-\infty}^{\infty} f(x,t) dx = 0 \tag{14.114}$$

where p and f_n are expressed by (14.32) and (14.111–14.112), respectively.

The corresponding dynamical system which simulates the probability equations (14.113) and (14.114) is:

$$\dot{v} = \nu v^{1/3} \sin \omega t + \varepsilon_0 (\tilde{y} - x), \quad \dot{x} = v \tag{14.115}$$

$$\dot{v}_n = \nu_n v_n^{1/3} \sin \omega t + \varepsilon_0 \left[\tilde{y} - \varphi(n,x) \right], \quad \dot{n}_* = v_n, \quad n = N - n_* > 0 \tag{14.116}$$

$$\tilde{y} = y(t) - 0.51 \tag{14.117}$$

$$y(t+\tau) = 4y(t)\,[1 - y(t)]\,, \quad \tau = \frac{\pi}{2\omega}n \tag{14.118}$$

$$y(t^*) = 0.2, \quad \frac{\pi}{4\omega} < t^* < \frac{\pi}{4\omega} \tag{14.119}$$

Thus, the variable x simulated by (14.115) performs a restricted non-symmetric, generalized random walk: after each time period $\tau = 2\pi/\omega$ it changes its value on $h_n = \pm nh$. But n, in turn, is also a random variable simulated by (14.116), and its probability density follows from (14.113–14.114). Hence, at each step, the variable x has a probability p or $(1-p)$ to move right or left, respectively, and in a selected direction it has a probability p_n to move from any fixed point $x = x_0$.

Indeed, let us select a point $x = x_0$. This will uniquely define the probability $p(x_0)$ [see (14.32)] that the next step will be directed to the right. Utilizing (14.111–14.112), one can find the probability $f_n(x_0, n_0)$ that the length of the jump will be

$$h_n = n_0 h \tag{14.120}$$

Hence, the following transition probability matrix can be introduced:

$$p = \begin{pmatrix} p_{11} & p_{12} & \cdots & p_{1N} \\ p_{21} & p_{22} & \cdots & p_{2N} \\ \cdots & \cdots & \cdots & \cdots \\ p_{N1} & p_{N2} & \cdots & p_{NN} \end{pmatrix} \tag{14.121}$$

where

$$p_{ij} = p f_n(x_i, n_j) \tag{14.122}$$

is the probability that the system being in the state x_i would move to the right making the step of the length

$$h_n = n_j h \tag{14.123}$$

Conversely,

$$p_T^* = \begin{pmatrix} p_{11}^* & \cdots & p_{1N}^* \\ \cdots & \cdots & \cdots \\ \cdots & \cdots & \cdots \\ p_{N1}^* & \cdots & p_{NN}^* \end{pmatrix} \tag{14.124}$$

where

$$p_{ij}^* = (1-p) f_n(x_i, n_j) \tag{14.125}$$

is the probability that the system being in the state x_i would move to the left making the step of the length (14.123). Obviously, if the initial probability distribution over the states $x_1, x_2, ..., x_N$ is $\alpha_k^{(1)}$:

$$\alpha_k^{(2)} = \alpha_j^{(2)} p_{jk}, \quad \overset{*}{\alpha}{}_k^{(2)} = \alpha_j^{(1)} \overset{*}{p}_{jk} \tag{14.126}$$

where $\alpha_k^{(2)}$ and $\overset{*}{\alpha}_k^{(2)}$ are the probability distributions over the states to the right and to the left of the original states respectively.

The basic property of the relationships (14.126) is that the probability α_k assigned to the state x_k depends only upon transition probabilities p_{jk} to the same state from all other states, and it does not depend upon the transition probabilities $p_{j\ell}(\ell \neq k)$ to different states. Physically it means that if a certain branch of the probabilistic scenario is already chosen in the course of dynamical simulations, then all other branches become irrelevant: they will never effect the evolution of the dynamical system. In other words, there is no interference between classical probabilities, and therefore, one dynamical "device" can process only one probabilistic branch.

In contradistinction to this, quantum probabilities evolve differently: each next step in the quantum Turing machine is attained by multiplying the vector of probability amplitudes of all possible configurations at the current step by a unitary matrix to obtain the vector representing the probability amplitude of each configuration in the next step, while the quantum probability is the square of its amplitude.

Although there is some similarity between the transformations of classical probabilities and quantum probability amplitudes in classical and quantum Turing machines (QTM), respectively, there are three fundamental differences between them.

Firstly, since the matrix of transformations of probability amplitude is unitary, the performance of the QTM is fully reversible.

Secondly, in contradistinction to the classical probabilities, the probability amplitudes can be negative, and that draws a line between dynamical simulations of classical and quantum probabilities; in particular, there is no way that quantum probabilities can be dynamically imitated by a classical device.

Thirdly, in the classical case, the probability that a particular configuration is reached at a certain step k in the computation is the sum of the probabilities of all the nodes corresponding to that configuration at the level k in the computational tree.

In contradistinction to this, in a QTM, the probability of the same configuration is the square of the sum of the amplitudes of all leaf nodes corresponding to that configuration. As a result of this, the probabilities interfere in such a way that two different probabilistic branches can amplify or cancel each other (the last case occurs if some of the probability amplitudes are negative). This means that in quantum computations one cannot follow a selected probabilistic branch as in the classical case: a QTM processes simultaneously all the probabilistic scenarios in the form of a special type of their superposition, and this particular property has been proven to be the most important in reducing the exponential complexity of computations to the polynomial one for some hard problems like factorization of large numbers (Shor 1994).

However, there is another effect of probability interference which can be associated with so called emergent computations (Forrest 1990). Emergent computation is an alternative to parallel computing which exploits the interactions among simultaneous computations to improve efficiency, increase flexibility, or provide a more natural representation. The basic idea behind emerging computation is that if a physical phenomenon is described in terms of its information processing properties, then the information which is absent at lower levels can exist at the level of collective activities. This is why emerging computations can lead to effects of self-organization, and cooperation between primitive components without global control.

It is interesting to note that the property of performing emerging computations has not been well pronounced in the theory of QTM, probably because of limitations imposed upon interactions between different branches of computations by unitary transformations required by quantum mechanics. As will be shown below, emerging computations can be simulated within the framework of the Turing machine introduced above [see (14.115–14.119)], in a more flexible way; the only price paid for this will be the necessity to exploit "calculated" probabilities. In order to demonstrate this, let us turn to (14.115) and modify it as follows:

$$\dot{v} = \nu v^{1/3} \sin \omega t + \varepsilon_0 \left[y - \varphi(x, H, \sigma, ...) \right], \quad \dot{x} = v \tag{14.127}$$

where H and σ are given by expressions (14.36), and (14.64), respectively.

Then, (14.32) should be replaced by

$$p = \begin{cases} 0 \text{ if} & \geq y_{\max} \\ 1 \text{ if} & \leq y_{\min} \\ \frac{y_{\max} - \varphi(x, H, \sigma, ...)}{y_{\max} - y_{\min}} \text{ if} & y_{\min} \leq \varphi(x, H, \sigma, ...) \leq y_{\max} \end{cases} \tag{14.128}$$

and therefore, the components of the transition probability matrix (14.122) will depend upon H, σ and other functionals of the probability distribution $f(x,t)$. This means that now the evolution of a certain probabilistic branch simulated by (14.115–14.119) will depend upon the evolution of other branches via the functionals H, σ,..., etc., and therefore, it will be coupled with (14.111–14.114) which govern the probabilities $f_n(n, x, t)$ and $f(x,t)$.

Thus, the system (14.111–14.114), (14.116–14.119) and (14.127) represents a classical Turing machine performing emerging computations, while the way in which different branches of computations interfere can be set up by an appropriate choice of functionals in (14.29). However, it should be recognized that there is a difference between the role of (14.111–14.114) in the case of (14.115), i.e., without interference, and in the case of (14.127), i.e., with interference of probabilities. In the first case, the dynamical equations (14.115–14.119) simulate the probabilities f and f_n, and therefore they simulate (14.111–14.114). In the last case, (14.116)–(14.127), and (14.111–14.114) are coupled, i.e., the probabilities cannot be simulated without equations describing "calculated" probabilities.

14.6 Simulation of Intelligent Systems

When dynamical systems describe natural phenomena, their basic properties are defined by energy and its time evolution. However, for those dynamical systems which model intellectual activities such as optimal behavior, decision making processes, and games, the more useful characteristic of performance is information rather than energy.

Let us consider, for instance, a dynamical system (14.115–14.119). The evolution of information stored in it is defined as:

$$H(t) = -\int_{-\infty}^{\infty} f(x,t)\log_2 f(x,t)dx \tag{14.129}$$

while $f(x,t)$ is governed by (14.113–14.114), which are uniquely determined by the parameters of the system (14.115–14.119).

However, the fundamental difference between the systems modeling natural phenomena, i.e., physical systems, and man-made, or intelligent systems is not only in the way in which they are described, [as a matter of fact, physical systems can also be described in terms of information (Haken 1985)]. The basic property of intelligent systems is that their structure, and therefore, the structure of the information (14.129), is not given in advance: it is supposed to be created based upon the purpose to which the system serves.

For illustration, consider the system (14.99) and present the functions φ_i in a parameterized form adopting a neural net formalism (Zak, 1994b):

$$\dot{v}_i = \nu v_i^{1/3} \sin \omega t + \varepsilon_0(\tilde{y} - \varphi_i), \;\; \dot{x}_{i=}v_i, \;\; i = 1, 2, ..., n \tag{14.130}$$

$$\varphi_i = \tanh \sum_{j=1}^{n} T_{ij} x_j \tag{14.131}$$

where T_{ij} do not depend upon x_i (but they may depend upon the functionals of $f(x)$; the hyperbolic tangent, tanh, represent nonlinearities in $\varphi_i(x_i, ..., x_n)$. Thus, the dynamical system (14.110) in terms of its probability evolution (14.100) is uniquely defined by the choice of the constants T_{ij}, and therefore, the problem of synthesis of an intelligent system can be reduced to finding these constants based upon the objective of system performance.

Let us assume that the purpose of the system performance is to minimize some functional of the probability distributions $f_i, ..., f_n$:

$$\Phi(f_1, ..., f_n) \to \min \tag{14.132}$$

where f_i depends upon T_{ij} via (14.100–14.101) and (14.131)

Then T_{ij} are found from the system:

$$\frac{\partial \Phi}{\partial T_{ij}} = 0, \;\; i, j = 1, 2, ..., n \tag{14.133}$$

However, despite the conceptual simplicity of this approach, its practical significance is limited since in real life situations the global objective in the form (14.132) is not available. Instead, each dynamical unit, i.e., each variable x_i, has its own local objective which can be partly compatible, and partly contradictory to local objectives of other units. Surprisingly, very often such systems exhibit very interesting properties associated with the concept of a collective brain (see Chap. 9).

Let us assume that each member (or a dynamical unit) characterized by the variable x_i, has its own version of the global objective of the whole dynamical system which can be expressed in the form similar to (14.132):

$$\Phi_i(f_i, ..., f_n) \to \min \tag{14.134}$$

while, in general,

$$\Phi_i \neq \Phi_j \text{ if } i \neq j$$

Each unit can learn (in its own way) the global objective of the system during previous collective tasks. Based upon this, it may "derive" its own version of the system (14.133):

$$\frac{\partial \Phi_k}{\partial T_{ij}^{(k)}} = 0, \quad k = 1, 2, ..., n \tag{14.135}$$

and therefore, its own version of the whole dynamical system in the form similar to (14.130):

$$\dot{v}_i^{(k)} = \nu (v_i^{(k)})^{1/3} \sin \omega t + \varepsilon_0 \left(\tilde{y} - \varphi_i^{(k)}\right), \ \dot{x}_i^{(k)} = v_i^{(k)}, \ i, k = 1, 2, ..., n \tag{14.136}$$

Here $v_i^{(i)}$, $x_i^{(i)}$ are the actual values of the variables, while $v_i^{(k)}$, $x_i^{(k)}$ $(k \neq i)$ are the values of the same variables predicted by the kth dynamical unit.

Hence, as a result of the collective brain paradigm, the original dynamical system (14.130) of $2n$ equations with respect to $2n$ variables x_i and v_i is replaced by the system (14.136) of 2 n^2 equations with respect to $2n^2$ variables $x_i^{(k)}$, $v_i^{(k)}$, $i, k = 1, 2, ..., n$.

Since the last system has the same dynamical structure as the original dynamical system (14.130), its solution can be described by n^2 equations similar to (14.125):

$$f_{in}^{(k)}(x_1^{(k)}, ..., x_n^{(k)}, t + \tau) = p_i^{(k)} f_i^{(k)}(x_1^{(k)}, ..., x_1^{(k)} - h, ..., x_n^{(k)}) +$$
$$(1 - p_i^{(k)}) f_1^{(k)}(x_1^{(k)}, ..., x_i^{(k)} + h, ..., x_n^{(k)} + h, ..., x_n^{(k)}, t) \tag{14.137}$$

where

$$p_i^{(k)} = \begin{cases} 0 \text{ if} & \varphi_i^{(k)} \geq y_{\max} \\ \frac{y_{\max} - \varphi_i^{(k)}}{y_{\max} - y_{\min}} \text{ if} & y_{\min} \leq \varphi_i^{(k)} \leq y_{\max} \\ 1 \text{ if} & \varphi_i^{(k)} \leq y_{\min} \end{cases} \tag{14.138}$$

and

$$\varphi_i^{(k)} = \tanh \sum_{j=1}^{n} T_{ij}^{(k)} x_j^{(k)} \tag{14.139}$$

As follows from (14.137–14.139), the dynamics of the collective brain is less predictable than the original dynamics. However, in contradistinction to the original dynamics which requires global control for its performance, the last version of dynamics is more flexible: it can perform relatively well based upon the autonomy of the dynamical units which can predict the events if the actual values of the variables are not available.

The autonomy of the dynamics with a collective brain can be increased if each unit can have not only its own version of the global objective of the system, but also its versions of the global objectives of others dynamical units. Clearly such an ability will require deeper correlations between the dynamical units which can be achieved by more intensive learning during the previous collective tasks. From the analytical viewpoint, the complexity of this dynamic system will be significantly higher: the system having the same structure as (14.136–14.137), will contain $2n^3$ equations with respect to $2n^3$ variables. In the same way one can introduce more autonomous (but more complex) dynamical systems with a collective brain of higher dimensionalities.

The intelligent systems of the type (14.136–14.137) or of its more complex versions discussed above, can be linked to game theory. Indeed, here each ith player (represented by the corresponding variable x_j), tries to achieve its local objective by taking into account the knowledge about possible local objectives of other players. However, in contradistinction to classical game theory which is based upon "calculated" probabilities, the intelligent dynamical systems discussed above are based upon simulated probabilities: they are capable of learning probabilistic strategies, and the knowledge is acquired and stored in the deterministic coefficients T_{ij}.

14.7 Quantum Intelligence

In previous sections we analyzed the impact of ideas introduced in connection with a hypothetical quantum computer upon concepts of classical computing. Our attention was focused on simulation of classical probabilities and imitation of the quantum effects of probability interference. Actually, the last effect has ignited the scientific community by its potential for powerful non-quantum applications to computational problems of exponential complexity. This kind of problem became an obstacle to progress in many classical areas such as operations research, artificial intelligence, combinatorial optimization, etc. Because of non-quantum nature of these applications, classical simulations which could imitate probability interference can be considered as an attractive alternative to a hypothetical quantum device. Such simulations were introduced in the form of the dynamical system (14.110–14.119), and they can be associated with quantum intelligence.

Let us take a close look at this system. Actually it simulates the evolution of a classical probabilistic Turing machine. If this machine has N possible states, and it starts from an initial root of the Nth leveled "decision" tree, there are 2^N possible probabilistic scenarios to approach the Nth level. By specifying a set of scenarios consequent to each possible "action," a "decision" tree facilitates evaluation of this action. Therefore, in order to evaluate all actions and select the best (based upon a prescribed objective), one has to perform an exhaustive search and run 2^N scenarios, while this number grows exponentially with the linear growth of the dimensionality N.

Let us assume now that there is an interference of these probabilistic scenarios. In our case such an interference is achieved by incorporating probability functionals [such as $E(f)$, $\sigma(f)$, $H(f)$, etc.] into the dynamical equation (14.116). Then only a simple run of (14.115–14.119) will include simultaneously all the contributions from other probabilistic branches via the probability functionals, and due to these contributions the actual trajectory may jump from one branch to another. By selecting an appropriate structure for the feedback in (14.116), this trajectory can be optimized subject to a prescribed objective of the system performance.

The strategy for the trajectory optimization can be drawn out of the methodology proposed in the previous section. Indeed, consider the system (14.110–14.119) and (14.32), and assume that the objective function is expressed in the form of the maximum of the information H:

$$H \to \max, \ H = -\sum_{i=1}^{N} f(x_i) \log_2 f(x_i) \tag{14.140}$$

subject to the constraints:

$$\sum_{i=1}^{N} f(x_i) = 1, \ \sum_{i=1}^{N} x_i f(x_i) = E(t) \tag{14.141}$$

where the expectation E is a prescribed function of time.

Then the optimal probability distribution $f_0(x_i, t)$ is found from the condition:

$$\frac{\partial}{\partial f}\left[H + \lambda_1 \sum_{i=1}^{N} f(x_i) + \lambda_2 \sum_{i=1}^{N} x_i f(x_i)\right] = 0 \tag{14.142}$$

whence

$$f_0(x_i, t) = e^{\lambda_1 + \lambda_2 x_i - 1}, \ \lambda_1 = \lambda_1(t), \ \lambda_2 = \lambda_2(t) \tag{14.143}$$

Here λ_1 and λ_2 are the Lagrange multipliers found from the constraints (14.141). Now one has to find a solution to (14.113–14.114) which would have a minimal deviation (in the least square sense) from (14.143).

This means that for each time step τ:

$$\sum_{i=1}^{N}\left[\tilde{f}\left(x_i,t+\tau\right)-f_0\left(x_i,t+\tau\right)\right]^2 \to \min \tag{14.144}$$

where

$$\tilde{f}\left(x_i,t+\tau\right)=\sum_{i=1}^{N} f_n^0\left[pf_0\left(x_i-nh,t\right)+(1-p)f_0\left(x_i+nh,t\right)\right] \tag{14.145}$$

The minimum in (14.145) should be sought with respect to f_n^0 subject to the constraint (14.114). This leads to minimization of the following function:

$$\theta_f=\sum_{i=1}^{N}\left[\sum_{i=1}^{N} f_n^0\Phi\left(x_i,n,t\right)-f_0\right]^2+\lambda\sum_{i=1}^{N} f_n^0 \to \min \tag{14.146}$$

where λ is a Lagrange multiplier, and

$$\Phi(x_i,n)=pf_0(x_i-nh,t)+(1-p)f_0(x_i+nh,t) \tag{14.147}$$

Parametrizing φ_n^0 as functions of x:

$$f_n^0=\sum_{k=1}^{m} T_{kn}^f x^m \tag{14.148}$$

one reduces the problem to finding the constraints T_{kn}^f from the system of linear equations:

$$\frac{\partial\theta_f}{\partial T_{kn}^f}=0,\ k=1,2,...,m;\ n=1,2,...,N \tag{14.149}$$

which, together with (14.149) define the optimal [subject to the objectives (14.140–14.141)] probability distributions, and therefore, the optimal transitional probabilities (14.122), for each time step τ.

The Lagrange multiplier λ is found from the constraints (14.114).

The next step is to find the appropriate coefficients p_n^0 in (14.111) which provide the optimal distributions $f_n^0(x)$ determined above. Applying the same strategy as those for (14.113), one obtains:

$$\theta_p=\sum_{i=1}^{N}\left[\tilde{f}_n\left(n,x,t+\tau\right)-f_n^0\left(n,x,t+\tau\right)\right]^2 \to \min \tag{14.150}$$

where

$$\tilde{f}_n\,(n, x, t+\tau) = p_n^0 f_n^0(n-1, x, t) + (1-p_n^0) f_n^0(n+1, x, t) \qquad (14.151)$$

The minimum in (14.150) is sought with respect to p_n^0.

Parametrizing p_n^0 as functions of x:

$$p_n^0 = \sum_{k=1}^{m} T_{kn}^p x^m \qquad (14.152)$$

one again reduces the problem to finding the constraints T_{kn}^p from the system of linear equations:

$$\frac{\partial \theta_p}{\partial T_{kn}^f} = 0 \qquad (14.153)$$

which together with (14.152), determine the optimal functions p_n^0 in (14.110–14.111), and therefore, the optimal structure for feedback in (14.116), for each time step τ :

$$\varphi(n, x) = y_{\max} - p_n^0 (y_{\max} - y_{\min}) \qquad (14.154)$$

Thus, starting with the objective (14.140), subject to the constraints (14.141), for each time step τ one can determine the optimal feedback (14.153) for (14.116).

Now a single run of (14.115–14.119) at each time step, will include "thinking," i.e., determining the structure of the feedback (14.154) for the next time step τ via solving (14.149) and (14.153).

It should be understood that the optimal trajectory discussed above has a certain meaning only in the probabilistic sense. Indeed, each run of (14.115–14.119) will result in different optimal trajectory, but all such trajectories will form an ensemble whose probabilistic properties are optimal subject to the objective (14.140–14.141).

In order to evaluate the commutational complexity of the optimization performed by quantum intelligence, we recall that solutions to (14.149) and (14.153) [which implement "thinking" accompanying the single run of (14.115–14.119)] have polynomial complexity, and therefore, the quantum intelligence paradigm eliminates combinatorial explosion.

Obviously the same approach can be applied to more complex intelligent systems where a global objective is replaced by a set of competing local objectives [see (14.136–14.139)]. In these systems, quantum intelligence is implemented in the form of dependence of feedbacks in (14.136) on the probability functionals via the coefficients $T_{ij}^{(k)}$ [see (14.131) and (14.135)]. As a result of this, the system attains some new self-organizing properties which have not been prescribed in advance. The mechanism of such an emerging phenomenon exploits the contribution of the collective brain paradigm when each dynamical unit not only has its own local objective, but also predicts local objectives of other units.

So far, the interference of probabilities was implemented via the dependence of the feedback in the dynamical equations upon the probability functionals $E(f)$, $\sigma(f)$, $H(f)$, etc., i.e., via the global influence between different probabilistic branches.

However, the interference of probabilities can be local if the feedback depends upon the probability distribution, f, and its derivatives, but not upon their functionals. Turning, for instance, to (14.99), one can modify the feedback φ_i in the following way:

$$\varphi_i = \varphi_i\left(x_1, ..., x_n, f_1, ..., f_n, \frac{\partial f_1}{\partial x}, ..., \frac{\partial f_n}{\partial x}, ..., \text{etc.}\right) \tag{14.155}$$

For a qualitative analysis of such a local interference of probabilities, one should note that as a consequence of (14.155), the equations (14.100) describing the evolutions of the probabilities f_i, become nonlinear, and they can be considered as discretized versions of nonlinear parabolo-hyperbolic equations since the shift operator E is expressed via the differential operator D as:

$$E_h = e^{hD}, \quad E_\tau = e^{\tau D} \tag{14.156}$$

where h and τ are space and time shifts, respectively. This is why local interference of probabilities can lead to such fundamental nonlinear effects as shock waves, Burger's waves, solitons, i.e., concentrations of probabilities which can be interpreted as special emerging effects of self-organization.

There are several advantages of classical imitations of probability interference over "natural" quantum simulations for non-quantum applications. Firstly, the interference is not restricted to linear unitary (reversible) transformations; moreover, it can be selected in an optimal way subject to the required objective of the performance.

Secondly, it can be applied to a much broader class of problems, and in particular, to the field of intelligent systems which are fundamentally irreversible.

Thirdly, the classical imitations of the probability interference are based upon existing technology: both chaotic and terminal dynamics can be implemented in circuits.

So far we discussed non-quantum computational applications. Now one may ask how far can we go in imitating a hypothetical quantum computer by classical devices. The border line for such imitations is drawn by the so called hidden-variable problem (Feynman 1982): It is impossible to represent the results of quantum mechanics with a classical universal device. This means that for applications to quantum mechanics, the superiority of a quantum computer over any classical one is unquestionable.

14.8 Simulation of Schrödinger Equation

The mathematical formalism exploited in all the previous sections was based upon the relationships between the terminal dynamics equations (14.1–14.2) simulating a random walk, and the discretized version of the Fokker-Planck equation (14.12) governing the evolution of the probability corresponding to this random walk.

A continuous version of (14.12), i.e., the Fokker-Planck equation, is obtained if

$$\nu \sim \omega^{4/3}, \text{ and } \omega \to \infty \tag{14.157}$$

Indeed, then:

$$\tau = \frac{2\pi}{\omega} \to 0, \ h \sim \frac{1}{\omega^{1/2}} \to 0, \ \frac{h^2}{\tau} \to 2D = \text{Const} \tag{14.158}$$

and (14.12) reduces to the Fokker-Planck equation:

$$\frac{\partial f}{\partial t} = D\frac{\partial^2 f}{\partial x^2} \tag{14.159}$$

There is a formal mathematical similarity between the Fokker-Planck and the Schrödinger equations: replacing real time t in (14.159) by an imaginary time

$$t_{**} = -i\, t, \ i = \sqrt{-1} \tag{14.160}$$

Indeed, after replacing the probability density f in (14.159) by the probability amplitude ψ one arrives at the Schrödinger equation:

$$-i\frac{\partial \psi}{\partial t_{**}} = D\frac{\partial^2 \psi}{\partial x^2} \tag{14.161}$$

Continuing this analogy, one may ask: does there exist a dynamical system which simulates the Schrödinger equation (14.161) in the same way in which the dynamical system (14.1–14.2) simulates the Fokker-Planck equation?

The formal mathematical answer to this question is very simple: yes, it does. Indeed, turning to(14.1–14.2) and introducing an imaginary time

$$t_* = it \tag{14.162}$$

one obtains

$$\frac{dx}{dt_*} = v_*, \ \frac{dv_*}{dt_*} = \nu v_*^{1/3} \sin \omega t_*, \ \nu \sim \omega^{4/3}, \ \omega \to \infty \tag{14.163}$$

Formally this system is identical to (14.1–14.2), and therefore, it describes a random walk whose probability is governed by the Fokker-Planck equation:

$$\frac{\partial f}{\partial t_*} = D\frac{\partial^2 f}{\partial x^2} \tag{14.164}$$

or, after returning to real time t, and replacing f by ψ by the Schrödinger equation (14.161).

Let us establish formal relationships between the parameters ν and ω of the dynamical system (14.55), and the quantum characteristics of a particle. Identifying ω with the wave frequency of the particle, one obtains:

$$\omega = \frac{E}{\hbar} \tag{14.165}$$

where E is the particle energy, and $\hbar$ is Planck's constant.

The actual transition to the continuous limit from (14.12–14.13) to the Fokker-Planck equation [see (14.157–14.158)] is restricted by the uncertainty principle

$$2D = \lim \frac{(\Delta x)^2}{\Delta t} = \lim \Delta x \, \Delta v = \frac{\hbar}{m} \tag{14.166}$$

i.e.

$$D = \frac{\hbar}{2m} \tag{14.167}$$

where m is the mass of the particle.

Then, as follows from (14.9), (14.165–14.166):

$$\nu \sim \frac{E^{7/3}}{\hbar^{5/3} m^{2/3}} \tag{14.168}$$

Thus, (14.165), (14.167–14.168) express the parameters of the dynamical system (14.163) and the corresponding Schrödinger equation via the physical characteristics of the particle.

Surprisingly, the mechanism of instability of (14.163) is explained much easier here than those in the classical case: it just follows from the uncertainty principle which rejects a possibility that initial conditions for both the position and the velocity of a particle are known exactly.

However, for the purpose of actual simulations of the Schrödinger equation (14.161), the dynamical system does not offer much (since it evolves in imaginary time), unless it can be given a meaningful physical interpretation. A mathematical formalism for such an interpretation can be borrowed from the special theory of relativity in which physical events are mapped into a pseudo-Euclidean space with real space coordinates and imaginary time. But the main problem here is not in mathematics, but rather in physics: does the dynamical system (14.163) exist in a real physical world?

The discovery of chaos in classical mechanics raised many questions among quantum physicists about a possibility that there is a deterministic microstructure behind the Schrödinger equation, and as a result of instability, this microstructure loses its determinism and "collapses" into a probabilistic world in the same way in which deterministic Newtonian dynamics attains stochasticity due to chaos. Such speculations were encouraged by views expressed by Einstein who had never been comfortable with the probabilistic origin of quantum mechanics. From this viewpoint, the dynamical system (14.163) represents an alternative to this probabilistic origin: it is fully deterministic (since it does not include any random parameters); it is driven by instability triggered by uncertainties in initial conditions (in this context, the uncertainty principle in quantum mechanics plays the same role as the finite precision of initial conditions does in classical mechanics); and finally, the evolution of probability resulting from instabilities is described by the Schrödinger equation.

At this stage, we cannot prove (or disprove) existence of a deterministic origin of quantum mechanics. But we can make the following statement: if such a deterministic origin exists, its phenomenological structure is likely to be similar to that of (14.163), and then the quantum device for a quantum computer can be based directly upon special "quantum" simulations of (14.163) as was described in the previous sections for classical simulations.

15

Turbulence and Quantum Fields Computations

I have second thoughts. Maybe God is malicious. *–A. Einstein*

15.1 Introduction

Because complex problems of physics involve so many degrees of freedom, their solution requires not only formulating the physical law correctly, but also computation using modern computers. Unfortunately, traditional discretization or numerical methods introduce artifacts, among them are truncation error, roundoff error, extra tendencies toward stability or instability, and even limitations in the degrees of freedom which can be handled. It follows therefore that application of computational or numerical approaches must anticipate these types of problems and develop a strategy for mitigating such unwanted effects in the computation.

Many numerical methods have been developed for space-time problems for which the orders of accuracy and stability are documented (e.g. Fletcher 1991), and this information is used to apply suitable methods to each problem, though considerable skill is still required in application to such difficult problems as quantum theory and turbulence. The traditional development of improved solutions to physical problems generally lies along implementation of higher order Taylor series, or spectral methods. Unfortunately, these traditional methods which perform well on linear or weakly nonlinear problems, have severe limitations in highly nonlinear high Reynolds number direct numerical simulations where even simple flat plate or channel problems can turn into multi-month computations on the best and most expensive computers. Considering all the important problems, say of geophysics, aerodynamics, and laser propagation through turbulence, which go poorly solved for want of sufficient computational power, it therefore is of considerable urgency to find alternative more efficient solution methods. It is for this reason that we turn to the novel stabilization and terminal dynamics techniques. Of course, as we show in the other chapters of this book, these methods are only novel to the numerical techniques, not to the physics which nature implements. It turns out that the stabilization technique, because of its intrinsic control capability, while coping with the physical instability of the solutions, also presents a tool for controlling numerical instability. Furthermore, the terminal dynamics technique introduces a computational engine which mixes the capabilities of digital and analog computers. If fully realized in a physical chip technology, because of the speed of signal amplitude development and the low energy needed, even in a high noise environment, the terminal dynamics approach may surpass all other known chip computational engines for speed. Beyond this is the potential for quantum or quantum inspired computing (Chap. 14).

In the sections below, we present solutions to turbulent flows and quantum wave fields using stabilization and terminal dynamics techniques discussed in other chapters. Because of the newness of the techniques, these solutions are only initial proofs of concept. Impressive in their own right, there is still much room for improvement in further implementation.

15.2 Representation of Turbulence by Stabilization

15.2.1 The Navier-Stokes Equation

High Reynolds Number Simulations

The intent of our application of the stabilization principle to the Navier-Stokes equation is to simulate atmospheric turbulence dynamics over complex terrain and structures at very high Reynolds numbers. In the following we describe the motivation and processes for modeling turbulence with the stabilization principle.

Our approach to turbulent modeling starts from the Navier-Stokes equations, which is the $F = ma$ of atmospheric fluid flow. The Navier-Stokes equations govern the motion of the atmosphere, and allow us to compute turbulence over complex terrain and structures.

The instantaneous Navier-Stokes equations in the form of the momentum and conservation of mass equations are given by

$$\rho(\frac{\partial \mathbf{u}}{\partial t} + \mathbf{u} \cdot \nabla \mathbf{u}) = \rho \mathbf{g} - \nabla P + \mu \nabla^2 \mathbf{u} \tag{15.1}$$

$$\frac{\partial \rho}{\partial t} + \nabla \cdot (\rho \mathbf{u}) = 0 \tag{15.2}$$

Here $\mathbf{u}$ is the velocity, ρ is the density of air, $\mathbf{g}$ is gravity vector, P is the pressure, and μ is the molecular viscosity.

The atmosphere can often be treated as incompressible, and for this case the continuity equation becomes

$$\nabla \cdot \mathbf{u} = 0 \tag{15.3}$$

The equation for the temperature which depends upon the turbulent velocity is given by

$$\frac{\partial T}{\partial t} + \nabla \cdot (T\mathbf{u}) = \nu_T \nabla^2 T + s_T \tag{15.4}$$

where T is the temperature, ν_T is the molecular coefficient of thermal transfer and s_T are sources and sinks of temperature. The temperature generates buoyancy in the momentum equation, e.g. by the ground heating of the sun which heats nearby air, by varying the density through an equation of state, for example

$$P = \rho RT \tag{15.5}$$

Alternatively, equivalent formulas for polyatomic gases can also be used. When the atmosphere is treated as incompressible, the Boussinesque formulation for buoyancy can be used.

An equation for humidity, q, transport and diffusion can be written as

$$\frac{\partial q}{\partial t} + \nabla \cdot (q\mathbf{u}) = \nu_T \nabla^2 q + s_q \tag{15.6}$$

Humidity changes the density of the fluid and therefore also contributes to buoyancy effects.

Large Eddy Simulation and the Stabilization Principle

The Navier-Stokes equations can be solved using finite difference methods. In principle, these are based upon the Taylor series and use analytic continuation between grid points and time intervals in a Eulerian form of the momentum equation. The formal order of accuracy of the method is generally determined by the highest order of the Taylor series which the finite difference scheme preserves in the discretized variable. Alternate discretization forms of solution include using finite elements and spectral methods, Lagrangian forms, and combinations of any of the above. The discretization methods of solution can be by explicit or implicit means (Fletcher 1991). While either method can formally achieve high accuracy, the explicit method is usually much faster for higher order solutions. However, low order implicit methods tend to remain stable at large time steps. Their low order accuracy for large problems however often disqualifies them from use in detailed studies of turbulence.

The direct numerical simulation of the Navier-Stokes equations is called DNS. This method strives to resolve all of the important spatial and temporal scales. The DNS approaches use the same equations as for solving laminar problems, but with great care and computer resources to allow solution to higher Reynolds number. The DNS solution of the Navier-Stokes equations can be obtained up to modest Reynolds numbers, in simple problems using direct numerical simulation. DNS approaches usually use explicit finite difference or spectral methods.

The property of the Navier-Stokes equations is that at each numerical time step the nonlinear advection terms double the spatial wave numbers. In laminar flow the energy at higher wave numbers is dissipated fast enough that the flow stays laminar. The Reynolds number is given by

$$R = \frac{UL}{v} \tag{15.7}$$

where R is the Reynolds number, L is a characteristic length of the flow, and v is the molecular viscosity. At higher Reynolds numbers the flow is turbulent and involves large numbers of degrees of freedom.

To understand the problem consider the following. If we computed all the scales of turbulence larger than 1 mm in volume 1 km on a side we would end up calculating more than 10^{18} degrees of freedom at each time step – clearly prohibitive. Even then, the 1 mm scale is not small enough to capture the dynamics in the laminar sublayer.

Suppose we are only interested in scales of motion larger than some volume size. To reduce the number of degrees of freedom, which must be computed on a grid, from approximately 10^{18} for a turbulence boundary layer calculation at each step to a manageable level, it is necessary to average the turbulence over each grid volume while still keeping track of the subgrid scale Reynolds stress contributions to turbulence. A computational grid cell of size Δx on side can resolve a wave length of $\lambda = 2\Delta x$ or larger. These wavelengths of turbulence fields represent grid resolved wavelengths of the larger eddies. Hence the grid computation cannot resolve wavelengths smaller than $2\Delta x$, and these wavelengths are called subgrid scale wavelengths..

The usual methods of reducing the number of degrees of freedom by volume averaging fall under the name of Large Eddy Simulation (LES) (Smagorinsky 1963, Deardorff 1970, Deardorff 1971, Deardorff 1973, Germano et al. 1991, Germano 1992, Ghosal et al. 1995, Piomelli and Liu 1995). The empirical coefficients attempt to relate functions of the strain to the Reynolds stress. The classic LES approach uses empirical coefficients which can vary as a function of the experiment to be simulated (Smagorinsky 1963, Deardorff 1970, Deardorff 1971, Deardorff 1973). Various dynamic subgrid approximation processes have been developed (Germano et al. 1991, Germano 1992, Ghosal et al. 1995, Piomelli and Liu 1995) which endeavor to seek consistency with dynamics and some to eliminate the need for empirical coefficients. Estimates of coefficients are made by use of mean square estimation theory and scale projections applied to the closures. The dynamic LES methods at this stage lack rigor and various alternative approaches are pending evaluation. Most troubling, the methods are sometimes found to be unstable, particularly when geometric shapes or complex terrain are used. It is particularly in the case of complex geometry and complex terrain that the closures are most suspect.

We have developed the stabilization principle to provide a direct calculation of the Reynolds stress without the need for empirical coefficients. The LES technique simulates the grid resolved wavelengths of turbulence directly for the Navier-Stokes equations, but must model the effects of the subgrid scale turbulence fields using semiempirical coefficients and formulas, and will be shown below. The stabilization principle is new and powerful, and provides a universal turbulence closure, without the need for empirical coefficients. The stabilization principle approach is under intense research and development and its application has been successfully applied to flow over complex structures at very high Reynolds numbers. The stabilization principle takes advantage of the fact that turbulence feedback adjusts the velocity fluctuations to a state of neutral dynamic instability. This is theoretically well-posed, and experimentally verified. From this condition, the wind profile and the turbulence structure can be computed without semiempirical formulas or coefficients.

AIRFLOS (Air Flow Over Structures Model) uses the LES or stabilization principle to calculate the grid resolved and subgrid scale turbulent temperature effects. Normally we generate turbulence from these equations using the large eddy simulation (LES) or the stabilization principle approaches in our model, (AIRFLOS). We calculate the detailed motion of the atmosphere at scales which the numerical grid can resolve, and keep track of subgrid velocity and temperature energies of finer scales within each grid cell.

Large Eddy Simulation

There are several contributions to the stress. The molecular viscosity creates several important effects, including causing dissipation, creating stress terms which couple the air motion to boundaries, and creating vortex interactions. The molecular stress tensor is given by

$$\tau_{ijm} = -\frac{1}{2}\nu_m(\frac{\partial u_i}{\partial x_j} + \frac{\partial u_j}{\partial x_i}) \tag{15.8}$$

where ν_m is the kinematic viscosity coefficient, and i, j are the spatial coordinate indices.

There is another molecular stress term which has viscosity against compression, but its effect is generally negligible for meteorological conditions in the boundary layer, although it can be included into the models.

Subgrid Scale (SGS) Turbulent Stress

In addition, the subgrid velocity fluctuations create a stress due to the spatial averaging of the large eddy simulation process over each grid cell. Using formulations adopted from Deardorff and Leonard the LES sub-grid stress formula for the large eddy simulation can be given as

$$\tau_{ijsgs} = -\frac{1}{2}\nu_T\left(\frac{\partial u_i}{\partial x_j} + \frac{\partial u_j}{\partial x_i}\right) + \frac{1}{3}\delta_{ij}\tau_{kksgs} \tag{15.9}$$

where

$$\nu_T = C_s l^2 \left|S\right| \tag{15.10}$$

and

$$S^2 = \sum_{i=1}^{3}\sum_{j=1}^{3} S_{ij}S_{ij} \tag{15.11}$$

The term δ_{ij} is the Kronecker delta, and the kk symbols in τ_{kksgs} represent summed indices on the diagonal terms of the subgrid stress.

The subgrid kinetic energy per unit mass is modeled as

$$\frac{1}{2}\tau_{kksgs} = \frac{2}{3}\left(\frac{\nu_T}{C_I l}\right)^2 \tag{15.12}$$

where the empirical coefficients

$$C_I = .9C_S \tag{15.13}$$

are used.

The velocity strain is defined as

$$S_{ij} = \frac{1}{2}\left(\frac{\partial u_i}{\partial x_j} + \frac{\partial u_j}{\partial x_i}\right) \tag{15.14}$$

$$C_S = \text{Smagorinsky constant} \tag{15.15}$$

(Deardorff) and

$$l = \text{les filter length} \tag{15.16}$$

The LES subgrid stress is generally much larger than the molecular stress for typical meteorological computational grids.

Generalized Coordinates

To facilitate discretization and subsequent computer solution of physical problems it is often worthwhile to solve the problem in coordinates which are natural to the geometry of the physical problem. Gains in accuracy and efficiency generally result from solution in an optimal coordinate system. For example, cylindrical or spherical coordinates are important with solutions having cylindrical and spherical symmetries respectively. In the cases with which we are concerned it is important to capture the effects of surfaces, such as the ground, buildings, and vehicles upon the generation of turbulence and the deposition of particulates. Body-fitted coordinates obtained from generalized curvilinear coordinate transformations formations satisfy this requirement, and in addition provide structured data access which is much faster than unstructured data access used in typical finite element formulations.

The coordinate transformations use Jacobians

$$J_{ij} = \frac{\partial \xi_i}{\partial x_j} \tag{15.17}$$

where x_j is the jth component of the position coordinate in real space, and ξ_i is the ith component of the position coordinate in the transformed space.

The Jacobian matrix is defined as

$$J = \begin{bmatrix} \frac{\partial \xi_1}{\partial x_1} & \frac{\partial \xi_1}{\partial x_2} & \frac{\partial \xi_1}{\partial x_3} \\ \frac{\partial \xi_2}{\partial x_1} & \frac{\partial \xi_2}{\partial x_2} & \frac{\partial \xi_2}{\partial x_3} \\ \frac{\partial \xi_3}{\partial x_1} & \frac{\partial \xi_3}{\partial x_2} & \frac{\partial \xi_3}{\partial x_3} \end{bmatrix} \tag{15.18}$$

The determinant of the Jacobian is

$$|J| = \left\| \begin{bmatrix} \frac{\partial \xi_1}{\partial x_1} & \frac{\partial \xi_1}{\partial x_2} & \frac{\partial \xi_1}{\partial x_3} \\ \frac{\partial \xi_2}{\partial x_1} & \frac{\partial \xi_2}{\partial x_2} & \frac{\partial \xi_2}{\partial x_3} \\ \frac{\partial \xi_3}{\partial x_1} & \frac{\partial \xi_3}{\partial x_2} & \frac{\partial \xi_3}{\partial x_3} \end{bmatrix} \right\| \tag{15.19}$$

The derivative operators become

$$\frac{\partial}{\partial x_i} = \sum_{j=1}^{3} J_{ij} \frac{\partial}{\partial \xi_j} \tag{15.20}$$

The inverse relations hold

$$\frac{\partial}{\partial \xi_i} = \sum_{j=1}^{3} J_{ij}^{-1} \frac{\partial}{\partial x_j} \tag{15.21}$$

where

$$J_{ij}^{-1} \tag{15.22}$$

is the matrix inverse. The obvious advantage of discretization in the ξ coordinates is that they can be made uniform and regular, and that accurate finite difference or finite volume formulations are easier to construct. The tensor form of our equations automatically result in finite volume formulations in finite difference form (Thompson et al. 1985).

Navier-Stokes Equations in Generalized Coordinates

The Navier-Stokes equations in ξ generalized coordinates can be put in the following form

$$\frac{\partial u_i^*}{\partial t} + \sum_{j=1}^{3} \frac{\partial}{\partial \xi_j}(u_i^* \sum_{k=1}^{3} u_j J_{jk}) = -\frac{1}{\overline{\rho}} \sum_{j=1}^{3} J_{ij} \frac{\partial}{\partial \xi_j} P - \sum_{j=1}^{3} J_{ij} g_j$$
$$- \text{ viscous stress terms} \qquad (15.23)$$

If we let

$$V_j = \sum_{k=1}^{3} u_j J_{jk} \qquad (15.24)$$

then the equation can be put into more compact form

$$\frac{\partial u_i^*}{\partial t} + \sum_{j=1}^{3} \frac{\partial}{\partial \xi_j}(u_i^* V_j) = -\overline{\rho}^{-1} \nabla P - \mathbf{g} - \text{viscous stress terms} \qquad (15.25)$$

Sub-grid Scale Stress Terms

For the large eddy simulation formulation the sub-grid scale Reynolds stress terms have been calculated as follows. First, the Smagorinsky subgrid closure is used.

$$\frac{\partial \tau_{ijsgs}}{\partial x_j} = -\frac{\partial}{\partial x_j}[\nu_T \frac{1}{2}(\frac{\partial u_i}{\partial x_j} + \frac{\partial u_j}{\partial x_i}) - \delta_{ij}\frac{q^2}{3}] \qquad (15.26)$$

Secondly, the equations are put into generalized coordinates, here split into two steps to show the derivation, giving

$$\frac{\partial \tau_{ijsgs}}{\partial x_{jsgs}} = -\frac{\partial}{\partial x_j}[\nu_T \frac{1}{2} \sum_{l=1}^{3} (\frac{\partial u_i}{\partial \xi_l} J_{jl} + \frac{\partial u_j}{\partial \xi_l} J_{il}) - \delta_{ij}\frac{q^2}{3}] \qquad (15.27)$$

and

$$\frac{\partial \tau_{ijsgs}}{\partial x_j} = -\sum_{m=1}^{3} J_{jm} \frac{\partial}{\partial \xi_m}[\nu_T \frac{1}{2} \sum_{l=1}^{3} (\frac{\partial u_i}{\partial \xi_l} J_{jl} + \frac{\partial u_j}{\partial \xi_l} J_{il}) - \delta_{ij}\frac{q^2}{3}] \qquad (15.28)$$

15.2.2 Computational Techniques

This section reviews the numerical methods and associated mathematical and physical reasoning used to solve the turbulence, transport and diffusion models. We solve the large eddy simulation and the stabilization method implementation of the Navier-Stokes with coupled equations using higher order finite difference methods programmed in Fortran. The problem is set up with consistent initial and boundary conditions.

AIRFLOS Discretization

The Continuity Equation

When AIRFLOS is run as incompressible, then the pseudo-compressibility method is used to establish incompressibility (Chorin 1967). This has the advantage of automatically satisfying velocity boundary conditions while solving for the pressure. The following pseudo-compressible equations are used

$$\frac{\partial P}{\partial t} = -\beta(\nabla \cdot u) \tag{15.29}$$

where β is a constant chosen for optimal convergence properties.

Space Discretization

The advection terms in the Navier-Stokes equations are discretized in generalized coordinates using a variation of the strong conservation form. To achieve conservation the flux form of the equations are used.

We let

$$f_{ij} = u_i^* V_j \tag{15.30}$$

for the discretization of the flux form.

Discretizing the fluxes in ξ space yields the modified strong form of the conservation equations.

$$\frac{\partial u_i}{\partial t} = -\sum_{j=1}^{3} \frac{|J|}{\Delta\xi_j}\left[\frac{f_{ij}(\xi_j + \frac{\Delta\xi_j}{2})}{\left|J(\xi_j + \frac{\Delta\xi_j}{2})\right|} - \frac{f_{ij}(\xi_j - \frac{\Delta\xi_j}{2})}{\left|J(\xi_j - \frac{\Delta\xi_j}{2})\right|}\right] \tag{15.31}$$

The unmodified strong conservation form is

$$\frac{\partial u_i^*}{\partial t} = -\sum_{j=1}^{3} \frac{1}{\Delta\xi_j}\left[\frac{f_{ij}(\xi_j + \frac{\Delta\xi_j}{2})}{\left|J(\xi_j + \frac{\Delta\xi_j}{2})\right|} - \frac{f(\xi_j - \frac{\Delta\xi_j}{2})}{\left|J(\xi_j - \frac{\Delta\xi_j}{2})\right|}\right] \tag{15.32}$$

where

$$u_i^* = u_i\,|J|^{-1} \tag{15.33}$$

Flux Form of the Difference Equations

The flux form of the difference equations is used to improve conservation properties.

Sub-Grid Stress

For the large eddy simulation formulation, sub-grid stress terms to be calculated are

$$\frac{\partial \tau_{ijsgs}}{\partial x_j} = -\sum_{m=1}^{3} J_{jm} \frac{\partial}{\partial \xi_m}[\nu_T \frac{1}{2} \sum_{l=1}^{3} (\frac{\partial u_i}{\partial \xi_l} J_{jl} + \frac{\partial u_j}{\partial \xi_l} J_{il}) - \delta_{ij} \frac{q^2}{3}] \quad (15.34)$$

These sub-grid stress terms have been discretized to second order.

Time Discretization and Stability Analyses

For time discretization an explicit third order Runge-Kutta was performed. This was found to perform well and provides third order accurate time integration of all the coupled equations. The advantage of the Runge-Kutta method is that it can be applied to fully nonlinear equations without linearization. Secondly, coupled to our higher order space discretization schemes, good stability and relatively long time steps could be achieved. Full stability analyses were performed on our coupled time and space discretizations in generalized coordinates using complicated Jacobian fields of complex terrain and building structures. The stability analyses showed that present high space and time accurate scheme has good stability properties in complex geometry. The Runge-Kutta scheme that was used is as follows. To solve the equation

$$\frac{\partial u}{\partial t} = f \quad (15.35)$$

we use the following process

$$u_{n+1} = u_n + \frac{1}{4}k_1 + \frac{3}{4}k_3 \quad (15.36)$$

$$k_2 = \Delta t f(t_n, u_n) \quad (15.37)$$

$$k_2 = \Delta t f(t_n + \frac{1}{3}\Delta t, u_n + \frac{1}{3}k_1) \quad (15.38)$$

$$k_3 = \Delta t f(t_n + \frac{2}{3\Delta t}, u_n + \frac{2}{3}k_2) \quad (15.39)$$

Here, f is the right-hand side of the equation, n is the time index, and Δt is the time increment.

Chemical Plume Dispersion

Plume concentration species, c_k, are calculated in CHAFFSIM from the conservation of mass (continuity) equation which runs fully interactively with the AIRFLOS model, and provides feedback to the momentum equations in the case of dense or buoyant gases.

The equation for the time and space evolution of the kth species is given as

$$\frac{\partial c_k}{\partial t} + \frac{\partial}{\partial x_j}(c_k u_i) = \nu_{c_k} \nabla^2 c_k + \text{Source} \tag{15.40}$$

Performing the large eddy simulation averaging yields the following equation

$$\frac{\partial c_k}{\partial t} + \sum_{j=1}^{3} \frac{\partial}{\partial x_j} \cdot (c_k u_j) = -\sum_{j=1}^{3} \frac{\partial}{\partial x_j}(\tau_{\rho_k sgsj}) + \text{Source} \tag{15.41}$$

where

$$\tau_{\rho_k sgsj} = -\frac{1}{2}\nu_{\rho_k sgs}(\frac{\partial c_k}{\partial x_j}) \tag{15.42}$$

$$\nu_{c_k sgs} = C_s l^2 \left|S\right| \tag{15.43}$$

CHAFFSIM uses higher order discretization in space and time. The code has been designed to be able to vary the order of the discretization. Typically third order accurate Total Variation Diminishing (TVD) explicit Runge-Kutta time discretization is used, as well as fifth order upwind biased space discretization. Jacobians are typically set at second order center differenced, and yield excellent turbulent calculations. Mass is conserved using the flux differencing procedure.

Example

The proposed stabilization method was applied to three dimensional space and time numerical computations of the turbulent velocity field past a cylinder by Meyers and Zak (1996) (Fig. 15.1). The three space dimensional non-stationary Navier-Stokes equations are given as

$$\rho\left(\frac{\partial \overline{\mathbf{v}}}{\partial t} + \overline{\mathbf{v}}\nabla\overline{\mathbf{v}}\right) = \nabla \cdot (\sigma + \widetilde{\sigma}) \tag{15.44}$$

$$\nabla \cdot \overline{v} = 0 \tag{15.45}$$

$$\sigma = -\nabla\overline{p} + \mu\nabla^2\overline{v} \tag{15.46}$$

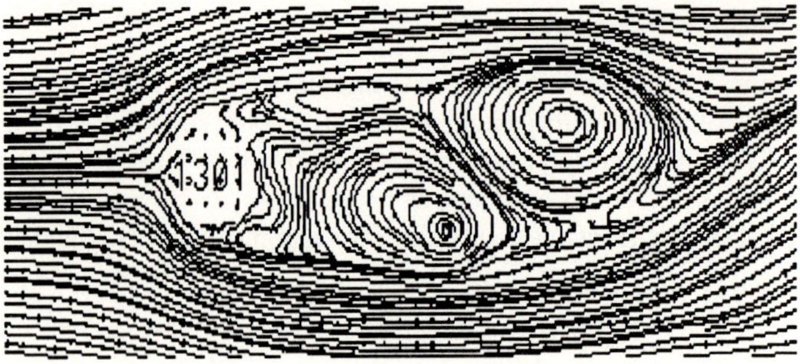

Fig. 15.1. Velocity streamlines of a cylinder in a turbulent flow (The number "130" refers to frame number)

$$\widetilde{\sigma}_{ij} = -\rho \overline{v_i' v_j'} \qquad (15.47)$$

in which $\widetilde{\sigma}$ is the Reynolds stress tensor, where $\overline{\mathbf{v}}$ and $\overline{p}$ are the mean values of velocity and pressure, respectively, v_i' is the velocity fluctuations, μ is the viscosity of the fluid, were solved subject to non-slip boundary conditions at the surface of the cylinder. The stabilization principle algorithm was built in as an expansion of the existing computational codes for laminar flow based upon finite difference techniques. At each time-step, the Reynolds stresses acted as actuators in a control system: they were chosen from the condition that the original instability of the governing equation (15.44) is suppressed down to neutral stability. Computations were performed for two Reynolds numbers: $R = 10,000$ and $R = 500,000$. Calculations demonstrated that the mean velocity field initially formed two symmetric vortices behind a cylinder and asymmetric vortex shedding existed at $R = 10,000$ and $T = 48$ which evolved to the famous Karman street. Similar effects are captured for $R = 500,000$. The drag computed for $R = 10,000$ agreed closely with experiments and those at $R = 500,000$ are only slightly higher that those obtained from the measurements. This deviation may be a result of a small amount of numerical "viscosity."

In the next figure (Fig. 15.2) Navier-Stokes turbulence over a flat plate at $R = 1,000,000$ is computed using the stabilization principle in a version of our AIRFLOS code. The advection numerical scheme is finite difference with 3rd order accuracy in time and 5th order accurate in space. The computer code is implemented in generalized coordinates for use of complex geometries. While the computer code usually uses the large eddy (LES) simulation closures with empirical coefficients, no such closures were used in this computation, and no empirical coefficients were used. This is the main advantage in of the stabilization approach over LES approaches, that no empirical coefficients are needed for solution of different flows. The velocities and pressures were cell averaged as part of the application of the stabilization principle to limit the degrees of freedom. The computations of

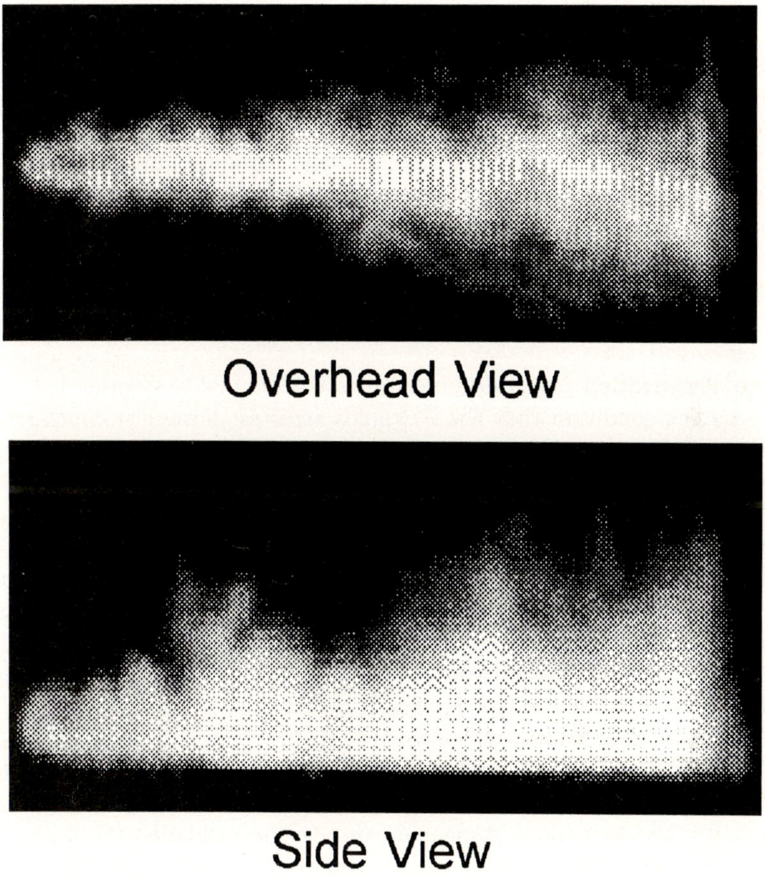

Fig. 15.2. Top and side view of a gaseous plume in a turbulent flow over a flat plate computed by the Stabilization Principle, using AIRFLOS/CHAFFSIM with a plume length of 158 m (R.E. Meyers and K.S. Deacon, U.S. Army Research Laboratory)

the turbulence are visualized by the release of a white neutral tracer chemical plume using our CHAFFSIM advection computer code which solves the continuity equation. The computations on a grid of 500,000 nodes were performed on an SGI Onyx computer with 4 processors.

15.2.3 Discussion

At high Reynolds numbers otherwise accurate numerical schemes endeavoring to solve the Navier-Stokes equation by DNS become unstable or over damp the turbulent solution.

The Large Eddy simulation methods used to approximate volume averaged flow depend upon empirical coefficients of sub-grid scale processes which vary for each flow, or have dynamically suspect closures. The problems are particularly great when complex geometry and terrain are involved. Interestingly enough, when the coefficients or formulation are chosen incorrectly, they damp out the turbulence or become numerically unstable.

The stabilization principle determines the Reynolds stresses in turbulence by the condition that the Reynolds stresses through nonlinear feedback suppress the velocity fluctuations to neutral instability. The stabilization principle is a universal closure, for which no empirical closure is needed. It follows that the stabilization principle, should also be able to predict suitable LES closures.

The stabilization principle is designed to stabilize the physical instability of the velocities. However, in numerical experiment the stabilization principle has also been applied to stabilize the numerical methods, which lets some operate in higher Reynolds number and Courant number ranges, although the accuracy of this approach needs to be determined.

The theory and numerical discretization procedures used in the development of the AIRFLOS and CHAFFSIM models have been described. AIRFLOS is a versatile ultra fine scale meteorological model capable of generating atmospheric turbulence over complex structures and complex terrain. CHAFFSIM uses the winds computed in AIRFLOS and interactively computes the time and space evolution of gases, aerosols, and particulates. These models use either the LES closure or stabilization principle approach to solution of turbulence. Figure 15.3 shows the application of the stabilization principle approach of flow past a circular cylinder at high Reynolds number as depicted in Fig. 15.1. The computed drag coefficient is only slightly higher than the experimental results and displays a reduction in drag at the correct place.

In conclusion, it has been demonstrated that application of the stabilization principle provides a universal closure for the Navier-Stokes equations: it does not depend upon particular boundaries, and it does not require any experimental coefficients. The application of the computation code implementing the stabilization principle to the turbulent flow past a cylinder and

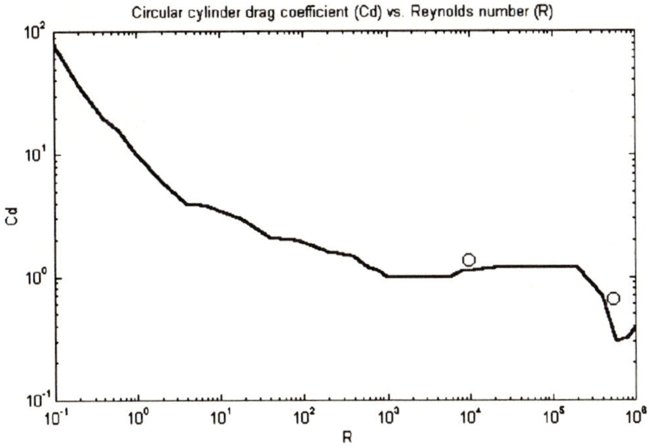

Fig. 15.3. Stabilization drag coefficient calculated solutions (circles o) plotted over experimental results of drag coefficients versus Reynolds number (solid line)

over a flat plate demonstrated good qualitative and quantitative agreement with known experimental results. It should be noticed that stabilization principle can be applied to other turbulent flows such as in aerodynamics and meteorology and any dynamical system which exhibits a chaotic behavior. Examples of the application of the stabilization principle to orbital unstable dynamical systems as well as to Hadamard unstable continua are described in Chap. 5.

15.3 Terminal Dynamics Schrödinger Equation Simulation

The Schrödinger equation is given as

$$\frac{\partial \psi}{\partial t} = i \frac{\hbar}{2m} \frac{\partial^2 \psi}{\partial x^2} \tag{15.48}$$

This equation and its solution are important for several diverse reasons. First, by representing matter waves it describes the non-relativistic behavior of electrons and some other elementary particles where Newton's laws fail. This provides the basis of classical quantum physics dealing with elementary particles as well as modern chemistry concerned with the evolution of large molecules such as proteins in biological systems. While the equation has many noticeable successes it does not have as simple an intuitive basis as does Newtonian physics, and this limits its ultimate success. A clarifying interpretation of the Schrödinger equation in terms of terminal dynamics and probabilities can no doubt greatly expand it usefulness.

Secondly, the Schrödinger equation is a useful model for laser light propagation through turbulence. With a potential mapped to index of refraction changes, the light goes down beam in the t direction and spreads in the transverse directions. This is a useful model for laser remote sensing and communication.

Third, the Schrödinger equation provides a model for quantum computing and quantum inspired computing which are discussed in Chap. 14. Using the wave function properties of the Schrödinger equations Feynman (1982) and later Deutsch (1989) have shown the potential for extraordinarily efficient quantum simulation of physical problems.

15.3.1 Numerical Simulation

As shown in Chap. 14, the Schrödinger equation can be obtained from the solution of the Fokker-Planck equation with time t replaced with imaginary time it . The Fokker-Planck equation is solved as

$$\frac{dx}{dt} = u(x,t) \tag{15.49}$$

and

$$\frac{du}{dt} = Au^{\frac{1}{3}} sin(\omega t) + \varepsilon \tag{15.50}$$

As $\omega \to \infty$ then the multivalued path solutions go to the trajectories of the Fokker-Planck equation. By phase space continuity

$$\frac{\partial f}{\partial t} + \frac{\partial fu}{\partial x} + \frac{\partial f(Au^{\frac{1}{3}} sin(\omega t) + \varepsilon)}{\partial u} = 0 \tag{15.51}$$

it is clear that solutions averaged over all velocities and initial conditions yield

$$\overline{\frac{\partial f}{\partial t}} + \overline{\frac{\partial fu}{\partial x}} = 0 \tag{15.52}$$

Chapter 14 shows that the average of the values of this equation obtained from transition probabilities yields the equation

$$\frac{\partial \overline{f}}{\partial t} = \frac{\hbar}{2m} \frac{\partial^2 \overline{f}}{\partial x^2} \tag{15.53}$$

This Fokker-Planck equation, also known as the diffusion equation, can be solved by numerically integrating the trajectories at a sufficient order of accuracy. We obtained solutions using several different numerical schemes, including the following simple one. The paths could be solved by simply integrating the trajectories as

$$x(t + \Delta t) = x(t) + u(t)\Delta t \tag{15.54}$$

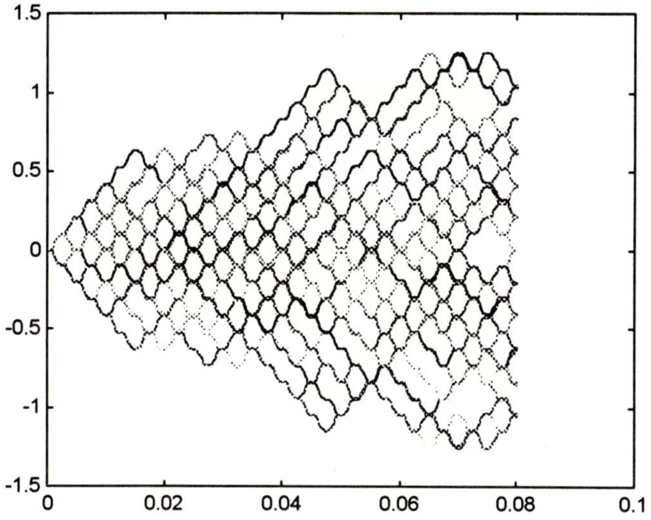

Fig. 15.4. Paths of terminal dynamics trajectories. Vertical axis is x position, horizontal axis is time

$$u(t+\Delta t) = u(t) + \Delta t A u^{\frac{1}{3}}(t) sin(\omega t) + \varepsilon \Delta t \tag{15.55}$$

The term ε is typically a very small noise term which can be the result of numerical truncation, or added from a chaos process to produce this effect. In the following computations we used the logistic equation to generate ε. A sample of the terminal dynamics paths $x(t)$ versus t is given as follows (Fig. 15.4). These solutions correspond to a Gaussian density initial condition.

Note the characteristic hopping patterns. In this case the mirror image was added to obtain a symmetric pattern. Having obtained the Fokker-Planck equation solution to a certain precision, it is now necessary to replace t by it to obtain the equivalent Schrödinger equation solution. Although there are several ways to do this, including the spectral in time method, a Taylor series expansion of the time derivative operator was utilized as follows. The wave function and density relation

$$\psi(t) = \overline{f}(it) \tag{15.56}$$

can be written as a Taylor series expansion and iterated operations as

$$\overline{f}(it) = \overline{f}(t + (i-1)t) \tag{15.57}$$

$$\overline{f}(it) \quad = \quad e^{(i-1)\frac{t\partial}{\partial t}} \overline{f}(t) \tag{15.58}$$

$$= \sum_{n=0}^{\infty} \left[(i-1) \frac{t\partial}{\partial t} \right]^n \overline{f}(t)$$

$$\simeq \sum_{n=0}^{n} \left[(i-1) \frac{t\partial}{\partial t} \right]^n \overline{f}(t).$$

The time derivative operator, because it operates on $\overline{f}(t)$, can be replaced by the equivalent space operator from the Fokker-Planck equation

$$\frac{\partial}{\partial t} = \frac{\hbar}{2m} \frac{\partial^2}{\partial x^2} \tag{15.59}$$

$$\begin{aligned} \overline{f}(it) &= e^{(i-1)t\frac{\hbar}{2m}\frac{\partial^2}{\partial x^2}} \overline{f}(t) \\ &= \sum_{n=0}^{\infty} \left[(i-1)t \frac{\hbar}{2m} \frac{\partial^2}{\partial x^2} \right]^n \overline{f}(t) \\ &\simeq \sum_{n=0}^{n} \left[(i-1)t \frac{\hbar}{2m} \frac{\partial^2}{\partial x^2} \right]^n \overline{f}(t). \end{aligned} \tag{15.60}$$

This last equation was used upon the terminal dynamics Fokker-Planck solution to obtain the real and imaginary parts of the Schrödinger wave equation at a time t after the initial conditions. Fourth order spatial differences were used along with a twelfth order expansion. The analytic and computational solutions of the real and imaginary $\psi(t)$ respectively are displayed in each of the following figures (Fig. 15.5 and Fig. 15.6) for a small sample size of 80 particle trajectories. Narrow Gaussian weight kernels were used as delta function sources. The figures show that the terminal dynamics technique can solve the Schrödinger equation for both the real and imaginary parts of the wave function. Of course the density $\psi\psi^*$also follows from our computation. Increased numbers of particles will provide more accuracy.

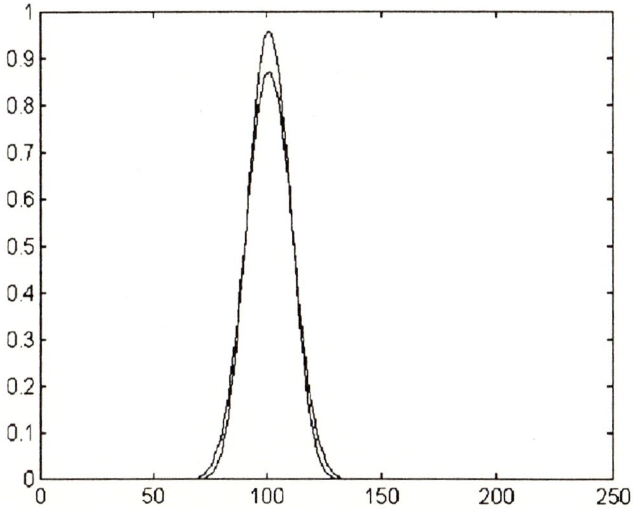

Fig. 15.5. Vertical shows real part of Schrödinger wave function. Horizontal is space coordinate x. Analytic solution is slightly taller and narrower, terminal dynamics solution is shorter and wider

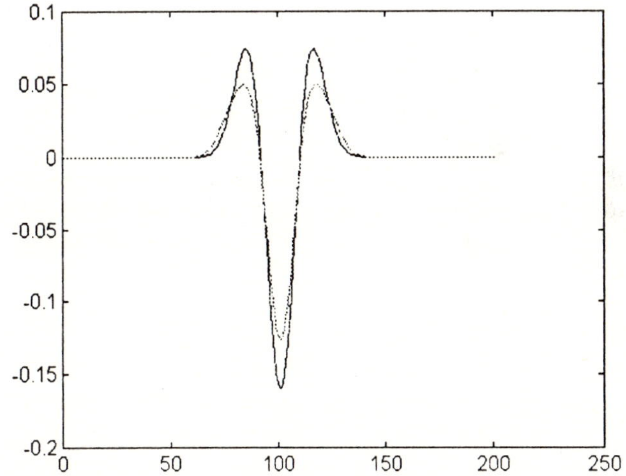

Fig. 15.6. Vertical axis is imaginary part of Schrödinger wave function, horizontal axis is space coordinate x. Analytic and terminal dynamics computations. Analytic solution has greater vertical and horizontal extremes

16
Epilogue

I believe the brain plays a game... –*S. Ulam*

Recently, it has been suggested that quantum effects have a role in the workings of the brain. Quite naturally, coupled with the interest in quantum computing, this has engendered speculation regarding the role of a "quantum computer" and basic neuronal processes. To date however, there has been no demonstration that such quantum effects are necessary to describe neuronal dynamics. Indeed, it has been pointed out by Scott (1995) that the time intervals and energies involved in neuronal firings produce a product of the time and energy of 10^{-15} joule-seconds, which is larger than Planck's constant by a factor of more than 10^{18}. This, of course, would allow for an error uncertainty in a classical description of time and energy of one in 10^9 without violating the quantum mechanical uncertainty principle. Such an observation, however, deals only with the gross neuronal signals. If one considers the scales of ion channels which help generate the impulses, one finds that they have a time-energy product on the order of 10^{-30}. This of course, places these processes on the border of the quantum world, then the possibilities become more interesting (Eisenberg 1996, Chen et al. 1995). Suddenly, thermodynamics, van der Waals forces, and charges acquire a new perspective in light of what has been presented in this monograph. Whether or not these signals in anyway participate in a "computational" paradigm remains to be demonstrated.

Even if future experimentation confirmed the importance of quantum neuronal effects, it must be remembered, that based on the idea of quantum superposition, all that such quantum "computations" could provide would be "choices" – it could not make determinations for decisions – the quantum world is linear and symmetric. Categorization, decision making, etc., on the other hand, are complex properties of the human biological world – classically nonlinear. For the microscopic quantum effects to exert their influence, there needs to be a nexus with the biological world. In this respect, non-Lipschitz dynamics, and stochastic attractors (cf. Chap. 14) could provide the connection, which would describe how the choices are to be processed. Indeed, if quantum computation were to become a reality, it would be a remarkable feat in taming combinatorial problems, and would be part of an overall algorithm involving non-Lipschitz dynamics to complete the scenario.

References

[1] Amari, S. (1983): Field theory of self-organizing neural nets. IEEE Trans. Syst. Man Cybern. **13**, 741–748

[2] Appel, P. (1953): Traite de mecanique rationelle, Part II, Sect. XXV. Gauthier-Villars, Paris

[3] Arnold, V.I. (1988): Mathematical methods of classical mechanics. Springer, Berlin Heidelberg New York, 331

[4] Babloyantz, A., Destexhe, A. (1988): Is the heart a periodic oscillator. Biol. Cybern. **58**, 203–211

[5] Barhen, J., Gulati, S., Zak, M. (1989): Neural learning of constrained nonlinear transformations. Computer **22**, 67–76

[6] Barone, S.R. (1993): Newtonian chaos + Heisenberg uncertainty = macroscopic indeterminacy. Amer. J. Physics **61**, 423–427

[7] Barth, T.J. (1987): Analysis of implicit local linearization techniques for upwind, and TVD algorithms. AIAA Paper 87–0595

[8] Campbell, L., Garnett, W. (1884): The life of James Clerk Maxwell, with selections from his correspondence and occasional writings. MacMillan and Co., London

[9] Cassandras, C.G. (1989): Discrete event systems. Irwin, Homewood Boston

[10] Cavello, G., George, E. (1992): Explaining the Gibbs sampler. The American Statistician **46**, 167–174

[11] Cetin, B., Barhen, J., Burdick, J. (1990): TRUST for fast global optimization, robotics and mechanical system, Report No. RMS–90–03, School of Engineering and Applied Sciences, Caltech

[12] Cetin, B.C., Kerns, D.A., Burdick, J.W., Barhen, J. (1991): Analog circuits for terminal attractors, repellers and gradient decent. Robotics and Mechanical Systems Report No. RMS–92–01, Department of Mechanical Engineering, Caltech

[13] Chen, D., Nonner, W., Eisenberg, R.S. (1995): PNP theory fits current-voltage relations of a neuronal anion channel in 13 solutions. Biophys. J. **68**, A370

[14] Chen, Z.-Y. (1990): Noise-induced instability. Phys. Rev. A **42**, 5837–5843

[15] Chorin, A.J (1967): A numerical method for solving incompressible viscous flow problems. J. Comp. Physics **2**, 12–26

[16] Coddington, E.A., Levinson, N. (1955): Theory of ordinary differential equations. McGraw-Hill, New York

[17] Cohen, M.A., Grossberg, S. (1983): Absolute stability of global pattern formation and parallel memory storage by competitive neural network. IEEE Trans. SMC **13**, 815–826

[18] Crutcfield, J.P., Kaneko, K. (1988): Are attractors relevant to fluid turbulence? Phys. Rev. Let. **60**, 2715 –2718

[19] Crutchfield, J.P. (1993): Observational complexity and the complexity of observation. In: Atmanspacher, H. (ed.) Inside versus outside. Springer, Berlin, pp. 234–272

[20] da Costa, N., Doria, F., (1991): Undecidability and incompleteness in classical mechanics. Int. J. of Theoretical Physics **30**, 1041–1073

[21] Deardorff, J.W. (1970): A numerical study of three dimensional turbulent channel flow at large reynolds number. J. Fluid Mech. **41**, 452–480

[22] Deardorff, J.W. (1971): On the magnitude of the subgrid-scale eddy coefficient. J. Comput. Phys. **7**, 120–133

[23] Deardorff, J.W. (1973): Three dimensional numerical study of turbulence in an entrained mixing layer. AMS workshop in meteorology. AMS, p. 271

[24] DeFelice, L.J., Isaac, A. (1992): Chaotic states in a random world: relationship between the nonlinear differential equations of excitability and the stochastic properties of ion channels. J. Stat. Phys. **70**, 338–354

[25] Deutsch, D. (1989): Quantum computational networks. Proc. Roy. Soc. Lond. A **425**, 73–90

[26] Dixon, D.D., Cummings, F.W., Kaus, P.E. (1993): Continuous "chaotic" dynamics in two dimensions. Physica D **65**, 109–116

[27] Dixon, D.D. (1995): Piecewise deterministic dynamics from the application of noise to singular equations of motion. J. Physics A **28**, 5539–5551

[28] Drazin, P.G., Reid, W.H. (1984): Hydrodynamic stability. Cambridge Univ. Press, London

[29] Einstein, A. (1983): Ather und relativitats – theorie. [Sidelinghts on relativity.] Dover, New York

[30] Eisenberg, R.S. (1996): Atomic biology, electrostatics and ionic channels. In Elber, R. (ed.) New developments and theoretical studies of proteins. World Scientific, Philadelphia, Chap. 5

[31] Feynman, R. (1982): Simulating physics with computers. Int. J. of Theoretical Physics **21**, 467–488

[32] Fletcher, C.A.J. (1991): Computational techniques for fluid dynamics. Vol. 1 and 2. Springer series in Computational Physics. Springer, Berlin Heidelberg New York

[33] Fluge, N., (1962): Handbook on engineering mechanics. McGraw-Hill, N.Y., pp. 33–9

[34] Ford, J. (1988): Quantum chaos, is there any? In: Bai-lin, H. (ed.) Directions in chaos. World Scientific, Singapore, pp. 128–147

[35] Forrest, S. (1990): Emergent computation – self-organizing collective, and cooperative phenomena in natural and artificial computing networks. Physica D **42**, 1–11

[36] Fraser, R.S., Harley, C., Wiley, T. (1967): Electrocardiogram in the normal rat. J. Appl. Physiol. **23**, 401–402

[37] Freeman, W. (1987): Simulation of chaotic EEG patterns with a dynamic model of the olfactory system. Biol. Cybern. **56,** 139–150

[38] Gantmacher, F., (1970): Lectures in analytical mechanics. MIR Publishers, Moscow, pp. 80, 81

[39] Garfinkel, A., Spano, M.L., Ditto, W.L., Weiss, J.N. (1992): Controlling cardiac chaos. Science **257**, 1230–1235

[40] Geissler, H. (1987): The temporal architecture of central information processing. Pshychol. Res. **49**, 99–116

[41] Germano, M. (1992): Turbulence: the filtering approach. J. Fluid Mech. **238**, 325

[42] Germano, M., Piomelli, U., Moin, P., Cabot, W.H. (1991): A dynamic subgrid-scale eddy viscosity model. Phys. Fluids A **3**, 1760–1765

[43] Ghosal, S., Lund, T., Moin, P. (1995): A dynamic localization model for large-eddy simulation of turbulent flows. J. Fluid Mech. **286**, 229–255

[44] Giuliani, A., LoGiudice, P., Mancini, A.M., Quatrini, G., Pacifici, L., Webber, Jr. C.L., Zak, M., Zbilut, J.P. (1996): A Markovian formalization of heart rate dynamics evinces a quantum-like hypothesis. Biol. Cybern. **74**, 181–187

[45] Gödel, K. (1931): Über formal unentscheidbare sätze per principia mathematica und verwandter systeme I. Monatshefte für mathematik und physik **38**, 173–198

[46] Gregson, A.M. (1993): Learning in the context of nonlinear psychophysics: the gamma Zak embedding. British Journal of Mathematical and Statistical Psychology **46**, 31–48

[47] Guevara, M.R., Lewis, T.J. (1995): A minimal single channel model for the regularity of beating in the sinoatrial node. Chaos **5**, 174–183

[48] Haken, H. (1985): Information and self-organization. Springer, Berlin Heidelberg New York

[49] Haken, H. (1991): Synergetics – can it help physiology? In: Haken, H., Koepchen H.-P. (eds.) Rhythms in physiological systems. Springer, Berlin Heidelberg New York, pp. 21–31

[50] Harth, E., Csermely, T.J., Beek, B. Lindsay, R.D. (1970): Brain functions and neural dynamics. J. Theoret. Biol. **26**, 93–120

[51] Harth, E. (1993): Electrical activity of the brain. IEEE Trans. Syst. Man, and Cybern. **13**, 782–789

[52] Hebb, D.O. (1949): The organization of behavior. Wiley, New York

[53] Holstein-Rathlou, N.H. (1993): Oscillations and chaos in renal blood flow control. J. Am. Soc. Nephrol. **4**, 1275–1287

[54] Holstein-Rathlou, N.H. (1994): Renal blood flow regulation and arterial pressure fluctuations. Physiol. Rev. **74**, 637–681

[55] Hopfield, J.J. (1982): Neural networks and physical systems with emergent collective computational abilities. Proc. Natl. Acad. Sci. **79**, 2554–2558

[56] Hopfield, J.J. (1984): Neurons with graded response have collective computational properties like those of two-state neurons. Proc. Natl. Acad. Sci. **81**, 3088–92

[57] Huberman, B. (1989): The collective brain. Int. J. of Neural Systems **1**, 41–45

[58] Hübler, A. (1992): Modeling and control of complex systems. In: Lam, L., Naroditsky, V. (eds.). Modeling complex phenomena. Springer, Berlin Heidelberg New York, pp. 5–65

[59] Kanters, J., Holstein-Rathlou, N., Agner, E. (1994): Lack of evidence for low dimensional chaos in heart rate variability. J. Cardiovasc. Electrophysiol. **5**, 591–601

[60] Kitney, R.I., Rompelman, 0. (eds.) (1980): The study of heart rate variability. Clarendon Press, Oxford

[61] Kuwahara, M., Yayou, K., Ishii, K., Hashimoto, S., Tsubone, H., Sugano, S. (1994): Power spectral analysis of heart rate variability as a new method for assessing autonomic activity in the rat. J. of Electrocardiology **27**, 333–337

[62] Lab, M.J. (1996): Mechanoelectric feedback (transduction) in the heart. Cardiovascular Res. **32**, 3–14

[63] Landau, L.D., Lifshitz, E.M. (1959): Fluid mechanics. Vol. 6. Course of theoretical physics. Perg. Press, London

[64] Lapedes, A., Farber, R. (1986): Programming a massively parallel computation universal system. In: Denker, J.S. (ed.). Neural Networks for computing. American Inst. of Physics, New York, pp. 283–298

[65] Lichtenberg, A.J., Lieberman, M.A. (1983): Regular and chaotic dynamics. Springer, Berlin Heidelberg New York

[66] Ljung, L. (1988): System identification. McGraw Hill, New York

[67] Mallat, S., Hwang, W. (1992): Singularity detection and processing with wavelets. IEEE Trans. PAMI **11**, 674–693

[68] McCulloch, W.S., Pitts, W.H. (1943): A logical calculus of ideas immanent in nervous activity. Bulletin of Math. Biophysics **5**, 115–133

[69] Malik, M., Camm, A.J. (1995): Heart rate variability. Futura, Armouk, NY

[70] McCulloch, W.S., Pitts, W.H. (1943): A logical calculus of the ideas immanent in nervous activity. Bull. Math. Biophys. **5**, 115–133

[71] Mayer-Kress, G., Yates, F.E., Benton, L., Keidl, M., Tisch, W., Pöppl, S.J., Geist K.-H. (1988): Dimensional analysis of nonlinear oscillations in brain heart and muscle. Math. Biosci. **90**, 155–182

[72] Meyers, R.E., Zak, M. (1996): Representation of turbulence and chaos by stabilization. Zeitschrift Angewandte für Mechanic und Mathematic **76**, Suppl 5, 337–336

[73] Osovets, S.D. Ginzburg, D., Gurfinkel, V.S., Zenkov, L.P., Latash, L.P., Malkin, V.B., Mel'nichuk, P.V., Pasternak, E.B. (1983): Order and chaos in neural systems. Usp. Fiz. Nauk **141**, 103–150

[74] Pineda, F.J. (1987): Generalization of back-propagation to recurrent neural networks. Phys. Rev. Lett. **59**, 2229–2232

[75] Piomelli, U., Liu, J. (1995): Large eddy simulation of rotating channel flows using a localized dynamic model. Phys. Fluids **7**, 839–848

[76] Prigogine, I. (1980): From being to becoming. Freeman and Co., San Francisco

[77] Prigogine, I., Nicolis, G., Babloyantz, A. (1972a): Thermodynamics and evolution. Part I. Physics Today **25**, 23–28

[78] Prigogine, I., Nicolis, G., Babloyantz, A. (1972b): Thermodynamics and evolution. Part II. Physics Today **25**, 38–44

[79] Pritchard, W.S., Duke, D.W., Krieble, K.K. (1995): Dimensional analysis of resting EEG II: surrogate-data testing indicates nonlinearity but not low-dimensional chaos. Psychophysiology **32**, 486–491

[80] Reynolds, O. (1895): On the dynamical theory of incompressible viscous fluids and the determination of the criterion. Phil. Trans. R. Soc. Lond., Ser. A **186,** 123–164

[81] Richardson, D. (1968): Some undecidable problems involving elementary functions of a real variable. Journal of Symbolic Logic **33**, 514–520

[82] Rubboli, A., Sobotka, P.A., Euler, D.E. (1994): Effect of acute edema on left ventricular function and coronary vasular resistence in the isolated rat heart. Am. J. Physiol. **267**, H1054–1061

[83] Rubenstein, D.S., Zbilut, J.P., Webber, Jr., C.L., Lipsius, S.L. (1993): Phase-dependent properties of the cardiac sarcoplasmic reticulum oscillator in cat right atrium: a mechanism contributing to dysrhythmias induced by Ca2+ overload. Experimental Physiol. **78**, 79–93

[84] Ruelle, D. (1994): Where can one hope to profitably apply the ideas of chaos? Physics Today **47** (July), 24–30

[85] Rumelhart, D. (1987): Parallel distributed processing. Vol. 1, The MIT Press, p.12

[86] Sakaguchi, Y. (1990): Topographic organization of nerve field with teacher signal. Neural Network **3**, 411–421

[87] Sayers, B.McA. (1973): Analysis of heart rate variability. Ergonomics **16**, 85–97

[88] Scott, A. (1995): Stairway to the mind. Springer, Berlin Heidelberg New York

[89] Seeley, T., Levien, R. (1988): A colony of mind. The Sciences, July, 1988, 39–42

[90] Shen, P., Larter, R. (1995): Chaos in intracellular Ca2+ oscillations in a new model for non-excitable cells. Cell Calcium **17**, 225–232

[91] Shirokova, N., Garcia, J., Pizarro, G., Rios, E. (1996): Ca^{2+} release from the sarcoplasmic reticulum compared in amphibian and mammalian skeletal muscle. J. Gen. Physiol. **107**, 1–18

[92] Shor, P. (1994): Algorithms for quantum computations. In Proceedings of the 35th Annual Symposium on the Foundations of Computer Science. IEEE Computer Society Press, New York, pp. 124–134

[93] Smagorinsky, J. (1963): General circulation experiments with the primitive equations I. The basic experiment. Mon. Weather Rev. **91**, 99–165

[94] Synge, J.L. (1926): On the geometry of dynamics. Phil. Trans. R. Soc. Lond., Ser. A **226**, 31–106

[95] Thompson, J. F., Warsi, Z.U.A., Mastin, C.W. (1985): Numerical grid generation. North-Holland, New York

[96] Thorburn, W.M. (1918): The myth of Occam's razor. Mind **217**, 345–353

[97] Toomarian, N.B., Barhen, J. (1992): Learning a trajectory using adjoint functions and teacher forcing. Neural Net **5**, 473–484

[98] Trulla, L.L., Giuliani, A., Zbilut, J.P., Webber, Jr., C.L. (1996): Recurrence quantification analysis of the logistic equation with transients. Phys. Lett. A **223**, 255–260

[99] Tsuda, A., Henry, F.S., Butler, J.P. (1995): Chaotic mixing of alveolated duct flow in rhythmically expanding pulmonary acinus. J. Appl. Physiol. **79**, 1055–1063

[100] Turchette, Q.A., Hood, C.J., Lange, W. Mabuchi, H., Kimble, H.J. (1995): Measurement of conditional phase shifts for quantum logic. Phys. Rev. Lett. **75**, 4710–4713

[101] Turing, A.M. (1937): On computable numbers, with an application to the Entscheidungsproblem. Proc. Lond. Math. Soc. (ser. 2) **42**, 230–265

[102] Weaver, J.C., Astumian, R.D. (1990): The response to very weak electric fields: the thermal noise limit. Science **247**, 459–462

[103] Webber, Jr. C.L., Zbilut, J.P. (1994a): Neural network estimation of cardiac nondeterminsim. In Dagli, C.H., Fernandez, B., Ghosh, J., Kumara. S.R.T. (eds.) Intelligent engineering systems through artificial neural networks. Vol. 4. ASME Press, New York, pp. 695–700

[104] Webber, Jr. C.L., Zbilut, J.P. (1994b): Dynamical assessment of physiological systems and states using recurrence plot strategies. J. Appl. Physiol. **76**, 965–973

[105] Webber, Jr. C.L., Zbilut, J.P. (1996): Assessing deterministic structures in physiological systems using recurrence plot strategies. In: Khoo, M.C.K. (ed.) Bioengineering approaches to pulmonary physiology and medicine. Plenum Press, New York, pp. 137–148

[106] Whitham, G. (1972). Linear and nonlinear waves. Wiley, New York

[107] Wilders, R., Jongsma, H.J. (1993): Beating irregularity of single pacemaker cells isolated from the rabbit sinoatrial node. Biophys. J. **65**, 2001–2013

[108] Wilders, R. (1996): Private communication

[109] Winfree, A.T. (1987): When time breaks down. Princeton Univ. Press, Princeton

[110] Yockey, H.P. (1992): Information theory and molecular biology. Cambridge University Press, Cambridge

[111] Zak, M. (1968): Propagation of waves in elastic threads of three-dimensional shape. Solid Mechanics Journal (Russian) **3**(3), Moscow, pp. 68–53, Transl. Eng. 32–35

[112] Zak, M. (1970): Uniqueness and stability of the solution of the small perturbation problem of a flexible filament with a free end. PMM **39**, 1048–1052

[113] Zak, M. (1974): Non-classical problems in continuum mechanics. Leningrad Univ. Press, Leningrad

[114] Zak, M. (1979a): Dynamics of films. J. of Elasticity **2**, 171–185

[115] Zak, M. (1979b): Dynamics of liquid films and thin jets. SIAM J. Appl. Math. **37**, 276–289

[116] Zak, M. (1980): A mathematical model of post-instability in continuum mechanics. Mechanics Research Communications **7**, 47–53

[117] Zak, M. (1982a): A mathematical model of post-instability in fluid mechanics. Acta Mechanica **43**, 97–117

[118] Zak, M. (1982b): Post instability in continuous systems. Solid Mechanics Archives **7**, 467–503

[119] Zak, M. (1982c): On the failure of hyperbolicity in elasticity. Journal of Elasticity, **12**(2), 219–229

[120] Zak, M. (1983): Cummulative effect at the soil surface due to shear wave propagation. Journal of Applied Mechanics **50**, 227–228

[121] Zak, M. (1984): Inviscid fluid in high frequency excitation field. Acta Mechanica **53**, 245–258

[122] Zak, M. (1985a): Deterministic representation of chaos in classical dynamics. Phys. Letters A **107**, 125–128

[123] Zak, M. (1985b): Two types of chaos in nonlinear dynamics. International Journal of Nonlinear Mechanics, **20**(4), 297–308

[124] Zak, M. (1986a): Deterministic representation of chaos in turbulence. Physica D **18**, 486–487

[125] Zak, M. (1986b): Closure in turbulence theory using stabilization principle. Phys. Letters A **118**, 139–143

[126] Zak, M. (1987): Deterministic representation of chaos with application to turbulence. International Journal of Math. Modelling **9**, 599–612

[127] Zak, M. (1988): Terminal attractors for associative memory in neural networks. Phys. Letters A **133**, 18–22

[128] Zak, M. (1989a): Analysis of turbulence in shear flows using the stabilization principle. Math. and Computer Modelling **12**, 985–990

[129] Zak, M. (1989b): Non-Lipschitzian dynamics for neural net modelling. Appl. Math. Letters **2**, 69–74

[130] Zak, M. (1989c): Spontaneously activated systems in neurodynamics. Complex Systems **3**, 471–492

[131] Zak, M. (1989d): Terminal attractors in neural networks. Neural Network **2**, 259–274

[132] Zak, M. (1989e): The least constraint principle for learning in neurodynamics. Phys. Letters A **135**, 25–28

[133] Zak, M. (1990a): Weakly connected neural nets. Appl. Math. Letters **3**, 131–135

[134] Zak, M. (1990b): Hydrodynamics stability and frames of reference. Mathematical and Computer Modelling **13**, 33–37

[135] Zak, M. (1990c): Creative dynamics approach to neural intelligence. Biol. Cybern. **64**, 15–23

[136] Zak, M. (1991a): Terminal chaos for information processing in neurodynamics. Biol. Cybern. **64**, 343–351

[137] Zak, M. (1991b): An unpredictable dynamics approach to neural intelligence. IEEE Expert (August), 4–10

[138] Zak, M. (1991c): Neurodynamics with spatial self-organization. Biol. Cybern. **65**, 121–127

[139] Zak, M. (1992a): Irreversibility and creativity in neurodynamics. Int. J. Computers and Electrical Engineering **19**, 401–418

[140] Zak, M. (1992b): The problem of irreversibility in Newtonian dynamics. Int. J. of Theoretical Physics **31**, 333–343

[141] Zak, M. (1993a): Terminal model of Newtonian dynamics. Int. J. of Theoretical Physics **32**, 159–190

[142] Zak, M. (1993b): Introduction to terminal dynamics. Complex Systems **7**, 59–87

[143] Zak, M. (1994a): Postinstability models in dynamics. Int. J. of Theor. Phys. **33**, 2215–2280

[144] Zak, M. (1994b): Physical models of cognition. Int. J. Theor. Phys. **33**, 113–116

[145] Zak, M. (1996): Irreversibility in thermodynamics. Int. J. of Theoretical Physics **35**, 347–382

[146] Zak, M. (1997a): Introduction to quantum inspired computing. Int. J. of Math. Modelling and Scientific Comput, (in press)

[147] Zak, M. (1997b): Dynamical simulation of probabilities. Chaos, Solitons & Fractals, (in press)

[148] Zak, M., Barhen, J. (1989): Neural networks with creative dynamics, 7th Int. Conf. on Math. and Computer Modeling, Aug. 2–5, Chicago, IL

[149] Zak, M., Toomarian, N.B. (1990): Unsupervised learning in neurodynamics using the phase velocity field approach. In: Touretzky, D.S. (ed.) Advances in neural information processing systems 2. M. Kaufman Publishers, San Mateo CA, pp. 583–589

[150] Zak, M., Zak, A. (1994): Unpredictable dynamics and collective brain. Int. J. Computers and Math. Appl. **27**, 185–197

[151] Zak, M., Meyers, R.E. (1995): Prediction of chaos using stabilization principle. In: Computational Mechanics '95. Springer, Berlin Heidelberg New York, pp. 935–940

[152] Zak, M., Zak, A., Meyers, R.E. (1996): Stabilization through fluctuations in chaotic systems. In: Millonis, M. (ed.) Fluctuations and order. Springer, Berlin Heidelberg New York, pp. 91–107

[153] Zbilut, J.P., Mayer-Kress, G., Geist, K.-H. (1988): Dimensional analysis of heart rate variability in heart transplant recipients. Math. Biosci. **90**, 155–182

[154] Zbilut, J.P., Mayer-Kress, G., Sobotka, P.A., O'Toole, M., Thomas, Jr. J.X. (1989): Bifurcations and intrinsic chaotic 1/f dynamics in an isolated perfused rat heart. Biol. Cybern. **61**, 371–378

[155] Zbilut, J.P., Eldridge, F., Webber, Jr. C.L. (1991): Noise-induced metastability in a model of respiratory oscillations. Proceedings IEEE-EMBS **13**, 1863–1864

[156] Zbilut, J.P., Zak, M., Webber, Jr. C.L. (1995): Physiological singularities in respiratory and cardiac dynamics. Chaos, Solitons & Fractals **5**, 1509–1516

[157] Zbilut, J.P., Hübler, A., Webber, Jr. C. L. (1996a): Physiological singularities modeled by nonderministic equations of motion and the effect of noise. In: Millonas, M. (ed.) Fluctuations and order. Springer, Berlin Heidelberg New York, pp. 397–417

[158] Zbilut, J.P., Zak, M., Meyers, R. (1996b): A terminal dynamics model of the heartbeat. Biol. Cybern. **75**, 277–280

[159] Ziegler, H. (1963): Some extremum principles in irreversible thermodynamics with application to contimum mechanics. North-Holland Publ., Amsterdam

Index

Lecture Notes in Physics

For information about Vols. 1–455
please contact your bookseller or Springer-Verlag

Vol. 456: H. B. Geyer (Ed.), Field Theory, Topology and Condensed Matter Physics. Proceedings, 1994. XII, 206 pages. 1995.

Vol. 457: P. Garbaczewski, M. Wolf, A. Weron (Eds.), Chaos – The Interplay Between Stochastic and Deterministic Behaviour. Proceedings, 1995. XII, 573 pages. 1995.

Vol. 458: I. W. Roxburgh, J.-L. Masnou (Eds.), Physical Processes in Astrophysics. Proceedings, 1993. XII, 249 pages. 1995.

Vol. 459: G. Winnewisser, G. C. Pelz (Eds.), The Physics and Chemistry of Interstellar Molecular Clouds. Proceedings, 1993. XV, 393 pages. 1995.

Vol. 460: S. Cotsakis, G. W. Gibbons (Eds.), Global Structure and Evolution in General Relativity. Proceedings, 1994. IX, 173 pages. 1996.

Vol. 461: R. López-Pen˜a, R. Capovilla, R. Garcı´a-Pelayo, H. Waelbroeck, F. Zertuche (Eds.), Complex Systems and Binary Networks. Lectures, México 1995. X, 223 pages. 1995.

Vol. 462: M. Meneguzzi, A. Pouquet, P.-L. Sulem (Eds.), Small-Scale Structures in Three-Dimensional Hydrodynamic and Magnetohydrodynamic Turbulence. Proceedings, 1995. IX, 421 pages. 1995.

Vol. 463: H. Hippelein, K. Meisenheimer, H.-J. Röser (Eds.), Galaxies in the Young Universe. Proceedings, 1994. XV, 314 pages. 1995.

Vol. 464: L. Ratke, H. U. Walter, B. Feuerbach (Eds.), Materials and Fluids Under Low Gravity. Proceedings, 1994. XVIII, 424 pages, 1996.

Vol. 465: S. Beckwith, J. Staude, A. Quetz, A. Natta (Eds.), Disks and Outflows Around Young Stars. Proceedings, 1994. XII, 361 pages, 1996.

Vol. 466: H. Ebert, G. Schütz (Eds.), Spin – Orbit-Influenced Spectroscopies of Magnetic Solids. Proceedings, 1995. VII, 287 pages, 1996.

Vol. 467: A. Steinchen (Ed.), Dynamics of Multiphase Flows Across Interfaces. 1994/1995. XII, 267 pages. 1996.

Vol. 468: C. Chiuderi, G. Einaudi (Eds.), Plasma Astrophysics. 1994. VII, 326 pages. 1996.

Vol. 469: H. Grosse, L. Pittner (Eds.), Low-Dimensional Models in Statistical Physics and Quantum Field Theory. Proceedings, 1995. XVII, 339 pages. 1996.

Vol. 470: E. Martı´nez-González, J. L. Sanz (Eds.), The Universe at High-z, Large-Scale Structure and the Cosmic Microwave Background. Proceedings, 1995. VIII, 254 pages. 1996.

Vol. 471: W. Kundt (Ed.), Jets from Stars and Galactic Nuclei. Proceedings, 1995. X, 290 pages. 1996.

Vol. 472: J. Greiner (Ed.), Supersoft X-Ray Sources. Proceedings, 1996. XIII, 350 pages. 1996.

Vol. 473: P. Weingartner, G. Schurz (Eds.), Law and Prediction in the Light of Chaos Research. X, 291 pages. 1996.

Vol. 474: Aa. Sandqvist, P. O. Lindblad (Eds.), Barred Galaxies and Circumnuclear Activity. Proceedings of the Nobel Symposium 98, 1995. XI, 306 pages. 1996.

Vol. 475: J. Klamut, B. W. Veal, B. M. Dabrowski, P. W. Klamut, M. Kazimierski (Eds.), Recent Developments in High Temperature Superconductivity. Proceedings, 1995. XIII, 362 pages. 1996.

Vol. 476: J. Parisi, S. C. Müller, W. Zimmermann (Eds.), Nonlinear Physics of Complex Systems. Current Status and Future Trends. XIII, 388 pages. 1996.

Vol. 477: Z. Petru, J. Przystawa, K. Rapcewicz (Eds.), From Quantum Mechanics to Technology. Proceedings, 1996. IX, 379 pages. 1996.

Vol. 479: H. Latal, W. Schweiger (Eds.), Perturbative and Nonperturbative Aspects of Quantum Field Theory. Proceedings, 1996. X, 430 pages. 1997.

Vol. 480: H. Flyvbjerg, J. Hertz, M. H. Jensen, O. G. Mouritsen, K. Sneppen (Eds.), Physics of Biological Systems. From Molecules to Species. X, 364 pages. 1997.

Vol. 481: F. Lenz, H. Grießhammer, D. Stoll (Eds.), Lectures on QCD. VII, 276 pages. 1997.

Vol. 482: X.-W. Pan, D. H. Feng, M. Vallières (Eds.), Contemporary Nuclear Shell Models. Proceedings, 1996. XII, 309 pages. 1997.

Vol. 483: G. Trottet (Ed.), Coronal Physics from Radio and Space Observations. XVII, 226 pages. 1997.

Vol. 484: L. Schimansky-Geier, T. Pöschel (Eds.), Stochastic Dynamics. XVIII, 386 pages. 1997.

Vol. 485: H. Friedrich, B. Eckhardt (Eds.), Classical, Semiclassical and Quantum Dynamics in Atoms. VIII, 341 pages. 1997.

Vol. 486: G. Chavent, P. C. Sabatier (Eds.), Inverse Problems of Wave Propagation and Diffraction. XV, 379 pages. 1997.

Vol. 488: B. Apagyi, G. Endrédi, P. Lévay (Eds.), Inverse and Algebraic Quantum Scattering Theory. XV, 385 pages. 1997.

Vol. 489: G. M. Simnett, C. E. Alissandrakis, L. Vlahos (Eds.), Solar and Heliospheric Plasma Physics. VIII, 278 pages, 1997.

Vol. 491: O. Boratav, A. Eden, A. Erzan (Eds.), Turbulence Modeling and Vortex Dynamics. XII, 245 pages, 1997.

New Series m: Monographs

Vol. m 1: H. Hora, Plasmas at High Temperature and Density. VIII, 442 pages. 1991.

Vol. m 2: P. Busch, P. J. Lahti, P. Mittelstaedt, The Quantum Theory of Measurement. XIII, 165 pages. 1991. Second Revised Edition: XIII, 181 pages. 1996.

Vol. m 3: A. Heck, J. M. Perdang (Eds.), Applying Fractals in Astronomy. IX, 210 pages. 1991.

Vol. m 4: R. K. Zeytounian, Mécanique des fluides fondamentale. XV, 615 pages, 1991.

Vol. m 5: R. K. Zeytounian, Meteorological Fluid Dynamics. XI, 346 pages. 1991.

Vol. m 6: N. M. J. Woodhouse, Special Relativity. VIII, 86 pages. 1992.

Vol. m 7: G. Morandi, The Role of Topology in Classical and Quantum Physics. XIII, 239 pages. 1992.

Vol. m 8: D. Funaro, Polynomial Approximation of Differential Equations. X, 305 pages. 1992.

Vol. m 9: M. Namiki, Stochastic Quantization. X, 217 pages. 1992.

Vol. m 10: J. Hoppe, Lectures on Integrable Systems. VII, 111 pages. 1992.

Vol. m 11: A. D. Yaghjian, Relativistic Dynamics of a Charged Sphere. XII, 115 pages. 1992.

Vol. m 12: G. Esposito, Quantum Gravity, Quantum Cosmology and Lorentzian Geometries. Second Corrected and Enlarged Edition. XVIII, 349 pages. 1994.

Vol. m 13: M. Klein, A. Knauf, Classical Planar Scattering by Coulombic Potentials. V, 142 pages. 1992.

Vol. m 14: A. Lerda, Anyons. XI, 138 pages. 1992.

Vol. m 15: N. Peters, B. Rogg (Eds.), Reduced Kinetic Mechanisms for Applications in Combustion Systems. X, 360 pages. 1993.

Vol. m 16: P. Christe, M. Henkel, Introduction to Conformal Invariance and Its Applications to Critical Phenomena. XV, 260 pages. 1993.

Vol. m 17: M. Schoen, Computer Simulation of Condensed Phases in Complex Geometries. X, 136 pages. 1993.

Vol. m 18: H. Carmichael, An Open Systems Approach to Quantum Optics. X, 179 pages. 1993.

Vol. m 19: S. D. Bogan, M. K. Hinders, Interface Effects in Elastic Wave Scattering. XII, 182 pages. 1994.

Vol. m 20: E. Abdalla, M. C. B. Abdalla, D. Dalmazi, A. Zadra, 2D-Gravity in Non-Critical Strings. IX, 319 pages. 1994.

Vol. m 21: G. P. Berman, E. N. Bulgakov, D. D. Holm, Crossover-Time in Quantum Boson and Spin Systems. XI, 268 pages. 1994.

Vol. m 22: M.-O. Hongler, Chaotic and Stochastic Behaviour in Automatic Production Lines. V, 85 pages. 1994.

Vol. m 23: V. S. Viswanath, G. Müller, The Recursion Method. X, 259 pages. 1994.

Vol. m 24: A. Ern, V. Giovangigli, Multicomponent Transport Algorithms. XIV, 427 pages. 1994.

Vol. m 25: A. V. Bogdanov, G. V. Dubrovskiy, M. P. Krutikov, D. V. Kulginov, V. M. Strelchenya, Interaction of Gases with Surfaces. XIV, 132 pages. 1995.

Vol. m 26: M. Dineykhan, G. V. Efimov, G. Ganbold, S. N. Nedelko, Oscillator Representation in Quantum Physics. IX, 279 pages. 1995.

Vol. m 27: J. T. Ottesen, Infinite Dimensional Groups and Algebras in Quantum Physics. IX, 218 pages. 1995.

Vol. m 28: O. Piguet, S. P. Sorella, Algebraic Renormalization. IX, 134 pages. 1995.

Vol. m 29: C. Bendjaballah, Introduction to Photon Communication. VII, 193 pages. 1995.

Vol. m 30: A. J. Greer, W. J. Kossler, Low Magnetic Fields in Anisotropic Superconductors. VII, 161 pages. 1995.

Vol. m 31: P. Busch, M. Grabowski, P. J. Lahti, Operational Quantum Physics. XI, 230 pages. 1995.

Vol. m 32: L. de Broglie, Diverses questions de mécanique et de thermodynamique classiques et relativistes. XII, 198 pages. 1995.

Vol. m 33: R. Alkofer, H. Reinhardt, Chiral Quark Dynamics. VIII, 115 pages. 1995.

Vol. m 34: R. Jost, Das Märchen vom Elfenbeinernen Turm. VIII, 286 pages. 1995.

Vol. m 35: E. Elizalde, Ten Physical Applications of Spectral Zeta Functions. XIV, 228 pages. 1995.

Vol. m 36: G. Dunne, Self-Dual Chern-Simons Theories. X, 217 pages. 1995.

Vol. m 37: S. Childress, A.D. Gilbert, Stretch, Twist, Fold: The Fast Dynamo. XI, 410 pages. 1995.

Vol. m 38: J. González, M. A. Martín-Delgado, G. Sierra, A. H. Vozmediano, Quantum Electron Liquids and High-T_c Superconductivity. X, 299 pages. 1995.

Vol. m 39: L. Pittner, Algebraic Foundations of Non-Commutative Differential Geometry and Quantum Groups. XII, 469 pages. 1996.

Vol. m 40: H.-J. Borchers, Translation Group and Particle Representations in Quantum Field Theory. VII, 131 pages. 1996.

Vol. m 41: B. K. Chakrabarti, A. Dutta, P. Sen, Quantum Ising Phases and Transitions in Transverse Ising Models. X, 204 pages. 1996.

Vol. m 42: P. Bouwknegt, J. McCarthy, K. Pilch, The W_3 Algebra. Modules, Semi-infinite Cohomology and BV Algebras. XI, 204 pages. 1996.

Vol. m 43: M. Schottenloher, A Mathematical Introduction to Conformal Field Theory. VIII, 142 pages. 1997.

Vol. m 44: A. Bach, Indistinguishable Classical Particles. VIII, 157 pages. 1997.

Vol. m 45: M. Ferrari, V. T. Granik, A. Imam, J. C. Nadeau (Eds.), Advances in Doublet Mechanics. XVI, 214 pages. 1997.

Vol. m 46: M. Camenzind, Les noyaux actifs de galaxies. XVIII, 218 pages. 1997.

Vol. m 48: P. Kopietz, Bosonization of Interacting Fermions in Arbitrary Dimensions. XII, 259 pages. 1997.

Vol. m 49: M. Zak, J. B. Zbilut, R. E. Meyers, From Instability to Intelligence. Complexity and Predictability in Nonlinear Dynamics. XIV, 552 pages. 1997.